Modern Systems Engineering

Modern Systems Engineering

Edited by **Brian Maxwell**

New York

Published by NY Research Press,
23 West, 55th Street, Suite 816,
New York, NY 10019, USA
www.nyresearchpress.com

Modern Systems Engineering
Edited by Brian Maxwell

International Standard Book Number: 978-1-63238-502-4 (Hardback)

Printed in the United States of America.

Contents

Preface

Modern systems engineering refers to an interdisciplinary branch of engineering which deals with designing and managing complex engineering systems over their life span. It deals with the issues like reliability, logistics, testing and monitoring which are required for the smooth functioning, development, designing, etc. of a complex system. Its objective is to perform these functions in a cost effective way. This book provides significant information to help develop a good understanding of systems engineering and other modern advances in this area. It explores all the important aspects of this subject in the present day scenario. As this subject is emerging at a rapid pace, the contents of this text will help the readers gain in-depth knowledge of the latest concepts and applications of this field.

The researches compiled throughout the book are authentic and of high quality, combining several disciplines and from very diverse regions from around the world. Drawing on the contributions of many researchers from diverse countries, the book's objective is to provide the readers with the latest achievements in the area of research. This book will surely be a source of knowledge to all interested and researching the field.

In the end, I would like to express my deep sense of gratitude to all the authors for meeting the set deadlines in completing and submitting their research chapters. I would also like to thank the publisher for the support offered to us throughout the course of the book. Finally, I extend my sincere thanks to my family for being a constant source of inspiration and encouragement.

Editor

1

Sensorless control for the brushless DC motor: an unscented Kalman filter algorithm

Haidong Lv, Guoliang Wei*, Zhugang Ding and Xueming Ding

Department of Control Science and Engineering, University of Shanghai for Science and Technology, Shanghai 200093, People's Republic of China

In this paper, a new mathematical model is built according to the characteristics of the brushless DC (BLDC) motor and a new filtering algorithm is proposed for the sensorless BLDC motor based on the unscented Kalman filter (UKF). The proposed UKF algorithm is employed to estimate the speed and rotor position of the BLDC motor only using the measurements of terminal voltages and three-phase currents. In order to observe the drive performance, two simulation examples are given and the feasibility and effectiveness of the UKF algorithm are verified through the simulation results, and the accurate estimate performance is shown in simulation figures.

Keywords: BLDC; sensorless control; UKF algorithm; nonlinear wave function; trigonometricseries

1. Introduction

With the rapid development of modern industries, the brushless DC (BLDC) motor is increasingly being used in computer peripherals, office automation and machine tool industries because of its high efficiency, easy control, compact form, etc. (Pillay and Krishnan, 1991), and the BLDC motor also has the characteristics of trapezoidal electromotive force (emf) and quasi-rectangular current waveforms. In order to obtain the appropriate commutation signals every 60 electrical degrees, rotor position sensors must be used, such as hall sensors and photoelectric encoders. However, speed or position sensors require additional mounting space, they increase the cost and the complexity of the system as well as reduce the reliability of the system. In recent years, the sensorless control technology has received wide attention. The sensorless technology improves the system reliability and it is of great significance to further expand the application fields of the BLDC motor.

At present, there are so many papers that have reported the sensorless control technology. Among them, the most popular and widely used method is the back electromotive force (back-emf) method (Iizuka, Uzuhashi, Kano, Endo, and Mohri, 1985). However, this method generally has a drawback that the back-emf cannot be detected exactly under low-speed conditions. In addition, some other methods have been discussed, such as the estimation flux method (Ji and Li, 2008) and the freewheeling diodes detection method. Nevertheless, these methods cannot provide continual rotor position information when a high accuracy of the rotor speed and position are required. The extended Kalman filter (EKF) is an optimal recursive estimation algorithm for nonlinear systems and has been applied for estimating the state variable of the BLDC motor (Lenine, Reddy, and Kumar, 2007; Terzic and Jadric, 2001). However, there are also some limitations for the EKF algorithm, such as the complex computation of the Jacobian matrices, only first-order accuracy, etc.

In this paper, a new method has been introduced to overcome the above drawbacks by means of the unscented Kalman filter (UKF) algorithm. The UKF algorithm uses a deterministic sampling approach to capture the mean and covariance estimates with a minimal set of sample points. The related applications of the UKF have been reported in many papers, such as permanent magnet synchronous motor drive (Chan and Borsje, 2009) and induction motor drive (Rigatos and Siano, 2012). The simulation results by MATLAB/Simulink indicate that the speed could be tracked and adjusted precisely, and the dynamical property of systems is evidently improved.

2. Mathematical model of the BLDC

In this paper, the three-phase BLDC motor with star connection is considered, and the voltage equation of the phase A of the stator is as follows (Pillay and Krishnan, 1989):

$$u_a = Ri_a + (L - M)\frac{di_a}{dt} + e_a + u_n, \tag{1}$$

*Corresponding author. Email: guoliang.wei1973@gmail.com

where u_a and u_n are the terminal voltage of phase A and neutral point voltage, respectively, R, i_a, L and e_a are the phase resistance, the phase current, the phase inductance and the phase back-emf of the phase A, respectively, and M is the mutual inductance. Especially, we have similar voltage equations for phases B and C.

According to the operation principle of the BLDC motor, it is known that only two phases conduct in three-phase stator winding at each time point. Thus, the neutral point voltage can be deduced as follows:

$$u_n = \tfrac{1}{3}[(u_a + u_b + u_c) - (e_a + e_b + e_c)], \tag{2}$$

where u_b and e_b are the voltage and the back-emf of the phase B, respectively, u_c and e_c are the voltage and the back-emf of the phase C, respectively.

By substituting Equation (2) into the voltage equation of phase A, we can obtain the following equation:

$$\frac{di_a}{dt} = \frac{u_{ab} - u_{ca}}{3(L-M)} - \frac{R}{L-M} i_a + \frac{e_b + e_c - 2e_a}{3(L-M)}, \tag{3}$$

where $u_{ab} = u_a - u_b$ and $u_{ca} = u_c - u_a$.

According to the structure of the BLDC motor, the back-emf of phase A can be written as (Chen, Huang, Wang, and Wu, 2011)

$$e_a = \omega\varphi_m f_a(\theta), \tag{4}$$

where ω is the motor angular velocity, φ_m is the magnet flux linkage of the stator winding, $f_a(\theta)$ changeing along with the rotor position is the wave function of the back-emf of phase A and its maximum value and minimum value are 1 and -1, respectively.

The nonlinear function $f_a(\theta)$ can be described as follows:

$$f_a(\theta) = \begin{cases} \frac{6}{\pi}\cdot\theta, & 2k\pi \le \theta \le \frac{\pi}{6} + 2k\pi, \\ 1, & \frac{\pi}{6} + 2k\pi \le \theta \le \frac{5\pi}{6} + 2k\pi, \\ -\frac{6}{\pi}\cdot(\theta - \pi), & \frac{5\pi}{6} + 2k\pi \le \theta \le \frac{7\pi}{6} + 2k\pi, \\ -1, & \frac{7\pi}{6} + 2k\pi \le \theta \le \frac{11\pi}{6} + 2k\pi, \\ \frac{6}{\pi}\cdot(\theta - 2\pi), & \frac{11\pi}{6} + 2k\pi \le \theta \le 2\pi + 2k\pi \end{cases} \tag{5}$$

and represented by the trigonometric series as follows:

$$f_a(\theta) = 1.21\sin\theta + 0.27\sin 3\theta + 0.05\sin 5\theta - 0.02\sin 7\theta - 0.03\sin 9\theta \ldots. \tag{6}$$

For the symmetry structure of the BLDC motor, we have $f_b(\theta) = f_a(\theta - 2\pi/3)$ and $f_c(\theta) = f_a(\theta + 2\pi/3)$.

From Equations (3) and (4), it can be derived that

$$\frac{di_a}{dt} = \frac{u_{ab} - u_{ca}}{3(L-M)} - \frac{R}{L-M} i_a + \frac{\omega\varphi_m[f_b(\theta) + f_c(\theta) - 2f_a(\theta)]}{3(L-M)}. \tag{7}$$

In the motor control community, because of the extensive application of the digital control, the discrete-time system is widely used to describe the motor motions, and hence we consider the corresponding discrete-time equation of Equation (7) in this paper:

$$i_a(k+1) = \frac{T[u_{ab}(k) - u_{ca}(k)]}{3(L-M)} + \left(1 - \frac{TR}{L-M}\right) i_a(k) + \frac{T\omega(k)\varphi_m[f_b(\theta_k) + f_c(\theta_k) - 2f_a(\theta_k)]}{3(L-M)}, \tag{8}$$

where T is the sampling time.

Similarly, we have

$$i_b(k+1) = \frac{T[u_{bc}(k) - u_{ab}(k)]}{3(L-M)} + \left(1 - \frac{TR}{L-M}\right) i_b(k) + \frac{T\omega(k)\varphi_m[f_a(\theta_k) + f_c(\theta_k) - 2f_b(\theta_k)]}{3(L-M)}, \tag{9}$$

$$i_c(k+1) = \frac{T[u_{ca}(k) - u_{bc}(k)]}{3(L-M)} + \left(1 - \frac{TR}{L-M}\right) i_c(k) + \frac{T\omega(k)\varphi_m[f_a(\theta_k) + f_b(\theta_k) - 2f_c(\theta_k)]}{3(L-M)}, \tag{10}$$

where i_b and i_c are the phase current of phases B and C, respectively.

According to Equation (4), we can obtain the following torque equation of the BLDC motor:

$$\begin{aligned} T_e &= \frac{e_a i_a + e_b i_b + e_c i_c}{\Omega} \\ &= \frac{\omega\varphi_m[f_a(\theta)i_a + f_b(\theta)i_b + f_c(\theta)i_c]}{\omega/p} \\ &= p\varphi_m[f_a(\theta)i_a + f_b(\theta)i_b + f_c(\theta)i_c], \end{aligned} \tag{11}$$

where T_e is the electromagnetic torque, Ω is the mechanical angular velocity and p is the number of pole pairs of the BLDC motor.

Consider the motion equation of the BLDC motor:

$$T_e - T_L = J\frac{d\Omega}{dt} + B_v\Omega, \tag{12}$$

where T_L is the load torque of motor, J is the rotational inertia and B_v is the viscous friction coefficient.

From Equations (11) and (12) and by converting the mechanical angular velocity into the electrical angular velocity, we have

$$\omega(k+1) = \frac{p^2\varphi_m T}{J}[f_a(\theta)i_a + f_b(\theta)i_b + f_c(\theta)i_c] - \frac{pT}{J}T_L + \left(1 - \frac{B_v T}{J}\right)\omega(k). \tag{13}$$

According to Newton's law of motion, we can have

$$\theta(k+1) = T\omega(k) + \theta(k). \tag{14}$$

By combining the relevant equations (7), (9), (10), (13) and (14), the nonlinear state equations can be expressed in the following form:

$$\begin{aligned} x_{k+1} &= F_k(x_k)x_k + G_k u_k, \\ y_k &= Hx_k, \end{aligned} \tag{15}$$

where $x_k = [i_a\ i_b\ i_c\ \omega\ \theta]_k^T$; $y_k = [i_a\ i_b\ i_c]_k^T$; $u_k = [u_{ab} - u_{ca}\ u_{bc} - u_{ab}\ u_{ca} - u_{bc}\ T_L]_k^T$

$$F_k(x_k) = \begin{bmatrix} 1 - \frac{RT}{L-M} & 0 & 0 & F14 & 0 \\ 0 & 1 - \frac{RT}{L-M} & 0 & F24 & 0 \\ 0 & 0 & 1 - \frac{RT}{L-M} & F34 & 0 \\ F41 & F42 & F43 & 1 - \frac{B_v \cdot T}{J} & 0 \\ 0 & 0 & 0 & T & 1 \end{bmatrix},$$

$$F14 = \frac{T\omega(k)\varphi_m[f_b(\theta_k) + f_c(\theta_k) - 2f_a(\theta_k)]}{3(L-M)},$$

$$F24 = \frac{T\omega(k)\varphi_m[f_a(\theta_k) + f_c(\theta_k) - 2f_b(\theta_k)]}{3(L-M)},$$

$$F34 = \frac{T\omega(k)\varphi_m[f_a(\theta_k) + f_b(\theta_k) - 2f_c(\theta_k)]}{3(L-M)},$$

$$F41 = \frac{p^2\varphi_m T}{J}f_a(\theta_k),$$

$$F42 = \frac{p^2\varphi_m T}{J}f_b(\theta_k),$$

$$F43 = \frac{p^2\varphi_m T}{J}f_c(\theta_k),$$

$$G_k = \begin{bmatrix} \frac{T}{3(L-M)} & 0 & 0 & 0 \\ 0 & \frac{T}{3(L-M)} & 0 & 0 \\ 0 & 0 & \frac{T}{3(L-M)} & 0 \\ 0 & 0 & 0 & -\frac{T \cdot p}{J} \\ 0 & 0 & 0 & 0 \end{bmatrix},$$

$$H = \begin{bmatrix} 1 & 0 & 0 & 0 & 0 \\ 0 & 1 & 0 & 0 & 0 \\ 0 & 0 & 1 & 0 & 0 \end{bmatrix}.$$

3. UKF algorithm

For the Kalman filtering problem of a nonlinear system, although the EKF algorithm maintains the efficient recursive update form of the Kalman filter, it suffers a number of serious limitations. For instance, the calculation of the Jacobian matrices may be a difficult and error-prone process. For the purpose of overcoming the limitations of the EKF algorithm, the unscented transformation (UT) was proposed by Julier and Uhlman and a new Kalman filter algorithm (UKF) was presented in Julier and Uhlmann (2004). In this algorithm, a set of sample points are used to parameterize the mean and covariance of the probability distribution of the state variables.

In this section, the basic UT is reviewed first and the UKF algorithm is then introduced in detail.

3.1. *Basic UT*

Consider the following nonlinear function:

$$y = h(x), \tag{16}$$

where x is the n-dimensional variable with the mean $\bar{x}$ and covariance matrix P_1, y is the m-dimensional variable with the mean $\bar{y}$ and covariance matrix P_2.

For the variable x in Equation (16), a set of $2n$ sigma points are selected as follows (Simon, 2006):

$$\begin{aligned} x^{(i)} &= \bar{x} + \tilde{x}^{(i)}, \quad i = 1, \dots, 2n, \\ \tilde{x}^{(i)} &= (\sqrt{nP_1})_i^T, \quad i = 1, \dots, n, \\ \tilde{x}^{(n+i)} &= -(\sqrt{nP_1})_i^T, \quad i = 1, \dots, n, \end{aligned} \tag{17}$$

where $\sqrt{nP_1}$ is the matrix square root of nP_1 with $\sqrt{nP_1}^T \cdot \sqrt{nP_1} = nP_1$, $(\sqrt{nP_1})_i$ is the ith row of $\sqrt{nP_1}$. Since nP_1 is a positive-definite symmetric matrix, the square root can be calculated using the Cholesky factorization which can simplify the calculation procedure.

Each sample point $x^{(i)}$ is propagated through the nonlinear function to yield corresponding transformed sigma points $y^{(i)}$, that is

$$y^{(i)} = h(x^{(i)}) \quad i = 1, \dots, 2n.$$

The mean $\bar{y}$ and covariance P_2 are approximated by the average mean and covariance of the transformed sigma points.

$$\bar{y} = \frac{1}{2n}\sum_{i=1}^{2n} y^{(i)},$$

$$P_2 = \frac{1}{2n}\sum_{i=1}^{2n}(y^{(i)} - y)(y^{(i)} - y)^T,$$

where $1/2n$ is the weight being used to calculate the mean and covariance.

3.2. *UKF algorithm*

Consider the following BLDC motor model:

$$x_{k+1} = F_k(x_k)x_k + G_k u_k + w_k, \\ y_k = Hx_k + v_k, \tag{18}$$

where w_k and v_k are, respectively, the process noise and the measurement noise, which is assumed as Gaussian white noise with covariance matrices Q_k and R_k.

In the actual motor systems, the noises are unavoidable, such as the unmodeled noise, the detecting noise and so on. And these noises are likely to affect the normal operation of the motor. In the state-space model (18), all kinds of noises are described by the random variables w_k and v_k.

Based on Equation (18), the UKF algorithm can be summed up as follows:

(1) Compute the set of sigma points $x_{k-1}^{(i)}$ based on the current optimal state estimation $\hat{x}_{k-1}^{+}$ and the covariance estimation P_{k-1}^{+} according to Equation (17), that is,

$$\hat{x}_{k-1}^{(i)} = \hat{x}_{k-1}^{+} + \tilde{x}^{(i)}, \quad i = 1, \ldots, 2n,$$

$$\tilde{x}^{(i)} = (\sqrt{nP_{k-1}^{+}})_i^{\mathrm{T}}, \quad i = 1, \ldots, n,$$

$$\tilde{x}^{(n+i)} = -(\sqrt{nP_{k-1}^{+}})_i^{\mathrm{T}}, \quad i = 1, \ldots, n.$$

(2) Propagate the sigma points $\hat{x}_{k-1}^{(i)}$ to $\hat{x}_k^{(i)}$ through the nonlinear systems of the BLDC motor

$$\hat{x}_k^{(i)} = F_{k-1}(\hat{x}_{k-1}^{(i)})\hat{x}_{k-1}^{(i)} + G_k u_k, \quad i = 1, \ldots, 2n.$$

(3) Obtain the predicted mean $\hat{x}_k^{-}$

$$\hat{x}_k^{-} = \frac{1}{2n}\sum_{i=1}^{2n} \hat{x}_k^{(i)}.$$

(4) Compute the predicted covariance P_k^{-}

$$P_k^{-} = \frac{1}{2n}\sum_{i=1}^{2n} (\hat{x}_k^{(i)} - \hat{x}_k^{-})(\hat{x}_k^{(i)} - \hat{x}_k^{-})^{\mathrm{T}} + Q_{k-1}.$$

(5) Select the sigma points $\hat{x}_k^{(i)}$ based on the predicted mean and covariance

$$\hat{x}_k^{(i)} = \hat{x}_k^{+} + \tilde{x}^{(i)}, \quad i = 1, \ldots, 2n,$$

$$\tilde{x}^{(i)} = (\sqrt{nP_k^{-}})_i^{\mathrm{T}}, \quad i = 1, \ldots, n,$$

$$\tilde{x}^{(n+i)} = -(\sqrt{nP_k^{-}})_i^{\mathrm{T}}, \quad i = 1, \ldots, n.$$

(6) Transform the new sigma points through the measurement model

$$\hat{y}_k^{(i)} = H\hat{x}_k^{(i)}, \quad i = 1, \ldots, 2n.$$

(7) Calculate the predicted mean of the observation $\hat{y}_k$

$$\hat{y}_k = \frac{1}{2n}\sum_{i=1}^{2n} \hat{y}_k^{(i)}.$$

(8) Compute the predicted covariance matrices of the observation P_y

$$P_y = \frac{1}{2n}\sum_{i=1}^{2n} (\hat{y}_k^{(i)} - \hat{y}_k)(\hat{y}_k^{(i)} - \hat{y}_k)^{\mathrm{T}} + R_k.$$

(9) Compute the cross covariance matrices

$$P_{xy} = \frac{1}{2n}\sum_{i=1}^{2n} (\hat{x}_k^{(i)} - \hat{x}_k^{-})(\hat{y}_k^{(i)} - \hat{y}_k)^{\mathrm{T}}.$$

(10) Update the estimation using the Kalman filter algorithm

$$K_k = P_{xy} \cdot P_y^{-1},$$

$$\hat{x}_k^{+} = \hat{x}_k^{-} + K_k(y_k - \hat{y}_k),$$

$$P_k^{+} = P_k^{-} - K_k P_y K_k^{\mathrm{T}}.$$

4. Simulation results

In this section, under the environment of MATLAB/Simulink, the control system model of sensorless BLDC motor is built in Figure 1 and simulation examples are presented to illustrate the effectiveness of the UKF filter design method developed in this paper.

The parameters of the BLDC motor are given as follows: the stator resistance $R = 0.62\ \Omega$, the equivalent inductance of the stator $L - M = 1.0 \times 10^{-3}$ H, the maximum of each phase winding permanent magnet flux $\varphi_m = 0.066$ Wb, the inertia $J = 0.362 \times 10^{-3}$ kg m^2, the viscous friction coefficient $B_v = 9.444 \times 10^{-5}$ N m s, poles of the permanent magnet $p = 4$ and simulation step length $T = 5 \times 10^{-7}$ s, $x_0 = [0\ 0\ 0\ 0\ 0]^{\mathrm{T}}$.

According to the UKF algorithm, the results will be more accurate by using the appropriate noise covariance matrices Q_k and R_k. After repeated experiments, they are chosen as follows: $Q_k = \mathrm{diag}(0.01\ \ 0.01\ \ 0.01\ \ 0.001\ \ 0)$, $R_k = \mathrm{diag}(0.1\ 0.1\ 0.1)$.

In order to verify the estimate performance of our algorithm, two different experiments are given to check the speed tracking performance by the effect of the constant load and the changing load.

4.1. *Constant load*

In this example, the constant load torque is assumed to be 2 N m, and the reference speed changes from 2000 to 2500 rpm at time $t = 0.2$ s. Then, the experiment results (i.e. the performance curves) are obtained and presented in Figure 2.

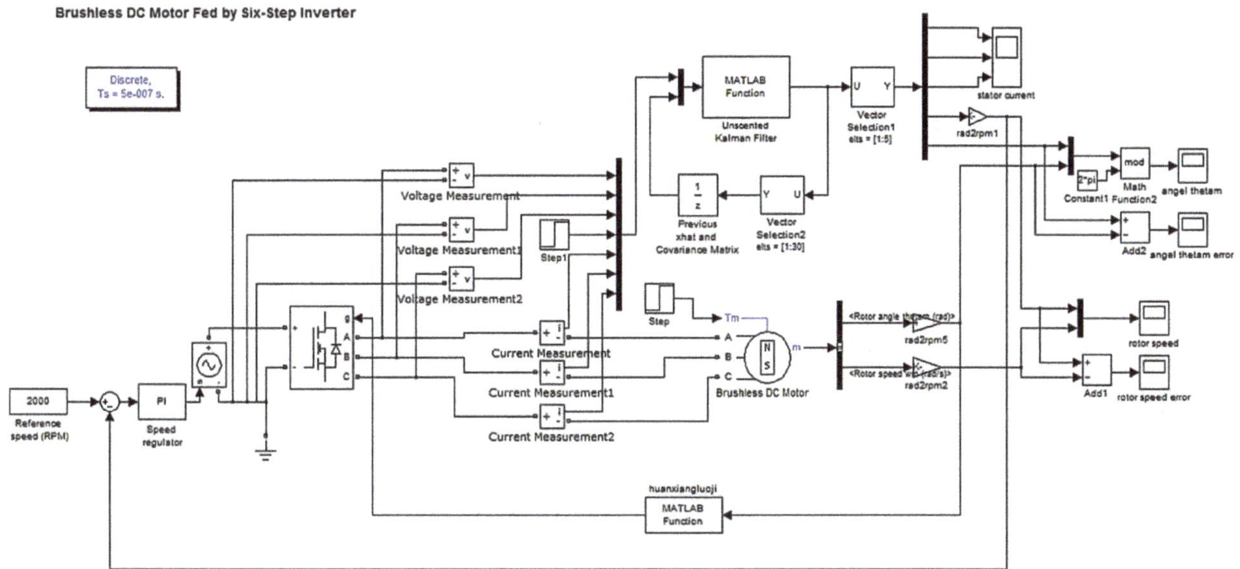

Figure 1. The system model.

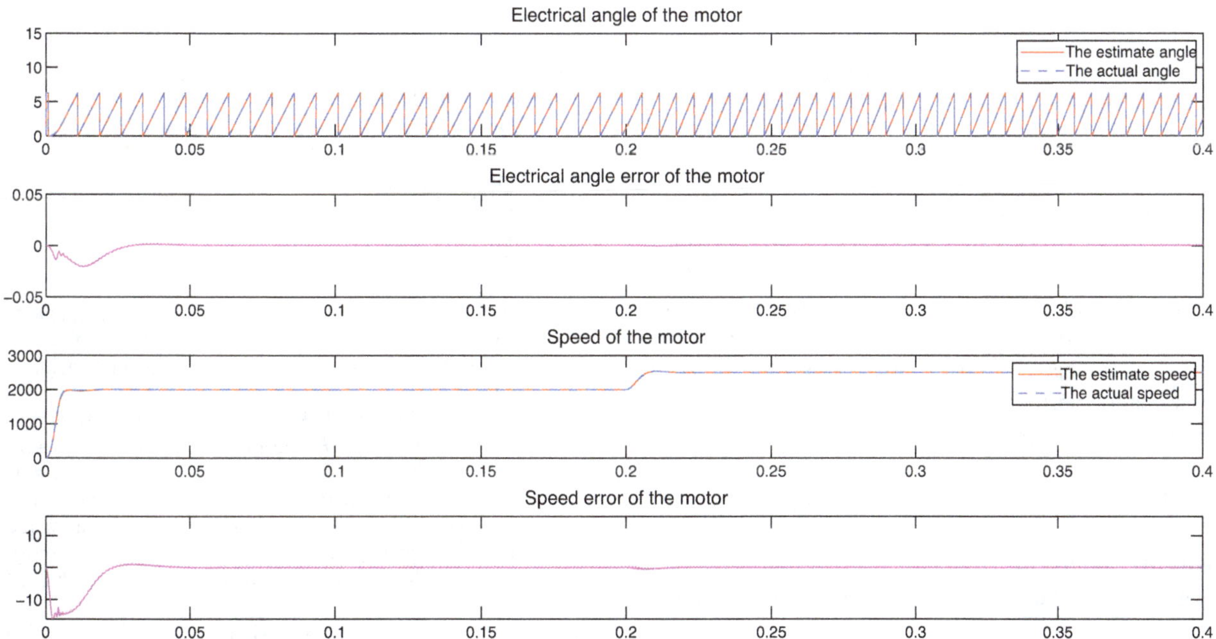

Figure 2. Performance curves of speed change.

From Figure 2, we can see that the estimated speed can track the actual speed accurately when the reference speed changes. The error between the estimated speed and the actual speed is about 0.2 rpm. From the first performance curve, we can see that the estimated position and the actual position of the rotor are almost the same. And the error is about 1×10^{-3} electrical angle, which confirms that the motor can operate normally with little torque ripple. The experiment results illustrate the effectiveness of our filtering algorithm.

4.2. *Changing load*

In this simulation experiment, the speed changes from 0 to reference speed 2000 rpm. Then, the load torque $T_L = 2$ N m is added to this motor at time $t = 0.1$ s and removed at $t = 0.3$ s. The performance curves are presented in Figure 3.

From Figure 3, we can see that the estimated and the actual speed are almost the same when the load torque suddenly changes. Moreover, the error curve changes slightly.

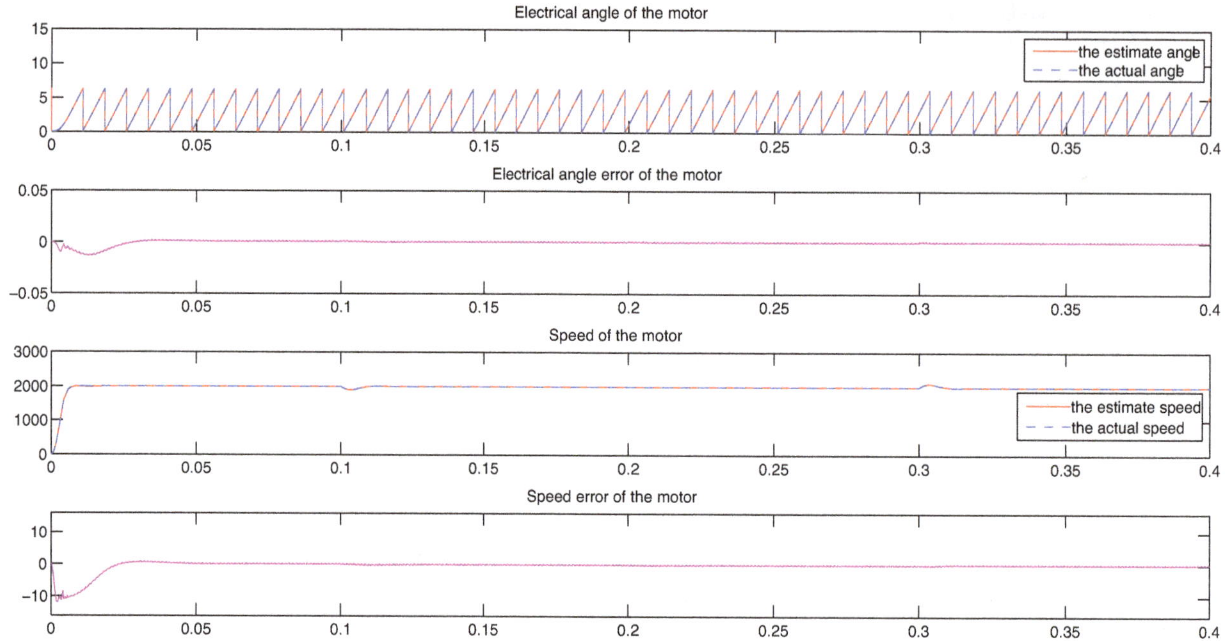

Figure 3. Performance curves of load torque change.

It can be concluded that the designed UKF algorithm is very effective when the torque suddenly changes.

5. Conclusion

In this paper, a mathematical model of a sensorless BLDC motor system has been built and a new filtering problem has been considered for this model based on the UKF algorithm. In order to evaluate the estimate performance, simulation experiments are presented in the paper. It is obvious to see that, from the simulation results, the accurate estimation performance can be obtained and the effectiveness of our designed algorithm can be demonstrated. Moreover, the sensorless BLDC motor can be controlled precisely according to the designed UKF algorithm.

Disclosure statement

No potential conflict of interest was reported by the author(s).

Funding

This work was supported in part by the National Natural Science Foundation of China [grant number 61374039]; the Program for Professor of Special Appointment (Eastern Scholar) at Shanghai Institutions of Higher Learning; the Program for New Century Excellent Talents in University [grant number NCET-11-1051] and Shanghai Pujiang Program under [grant number 13PJ1406300].

References

Chan T., & Borsje P. (2009). *Application of unscented Kalman filter to sensorless permanent-magnet synchronous motor drive*. Paper presented at IEEE international conference on electric machines and drives conference, Miami, Florida, USA (pp. 631–638).

Chen Y., Huang S., Wang S., & Wu F. (2011). *Dynamic equations algorithm of the sensorless BLDC motor drive control with Back-EMF filtering based on dsPIC30F2010*. Paper presented at the 30th Chinese control conference (CCC), Yantai, China (pp. 3626–3630).

Iizuka K., Uzuhashi H., Kano M., Endo T., & Mohri K. (1985). Microcomputer control for sensorless brushless motor. *IEEE Transactions on Industry Applications*, *IA-21*(3), 595–601.

Ji H., & Li Z. (2008). *A new position detecting method for brushless DC motor*. Paper presented at IEEE international conference on automation and logistics, Qingdao, China (pp. 1110–1114).

Julier S. J., & Uhlmann J. K. (2004). Unscented filtering and nonlinear estimation. *Proceedings of the IEEE*, 401–422.

Lenine D., Reddy B.R., & Kumar S.V. (2007). *Estimation of speed and rotor position of BLDC motor using extended Kalman filter*. Paper presented at IET-UK international conference on information and communication technology in electrical sciences (ICTES 2007), Tamil Nadu, India (pp. 433–440).

Pillay P., & Krishnan R. (1989). Modeling, simulation, and analysis of permanent-magnet motor drives. II. The brushless DC motor drive. *IEEE Transactions on Industry Electronics*, *25*(2), 274–279.

Pillay P., & Krishnan R. (1991). Application characteristics of permanent magnet synchronous and brushless DC motors for servo drives. *IEEE Transactions on Industry Applications*, *27*(5), 986–996.

Rigatos G., & Siano P. (2012). *Sensorless nonlinear control of induction motors using unscented Kalman filtering*. Paper presented at the 38th annual conference on IEEE industrial electronics society, Montreal, Canada (pp. 4654–4659).

Simon D. (2006). *Optimal state estimation: Kalman, H_∞, and nonlinear approaches*. Hoboken, NJ: Wiley.

Terzic B., & Jadric M. (2001). Design and implementation of the extended Kalman filter for the speed and rotor position estimation of brushless DC motor. *IEEE Transactions on Industry Electronics*, *48*(6), 1065–1073.

2

Enhanced OCV prediction mechanism for a stand-alone PV-lithium ion renewable energy system

Thomas Stockley*, Kary Thanapalan, Mark Bowkett and Jonathan Williams

Faculty of Computer, Engineering and Sciences, University of South Wales, Pontypridd, Wales

This paper aims to improve the estimation of state of charge (SoC) of the battery component for a small-scale photovoltaic stand-alone system through the use of a simple summing equation, at a set measurement interval. The system uses a pre-defined parameter to accurately predict the open-circuit voltage (OCV) of a cell at a much reduced measurement time of 5 minutes, while maintaining a maximum prediction error of less than 1%SoC. A simulation model has been provided that allows measurement of the cell voltage and current for prediction of the equilibrated OCV. The simulation can be used for single cell, modules and battery packs which use lithium-based technologies. Validation of the model has been performed using experimental data from tests conducted at the Centre for Automotive and Power System Engineering (CAPSE) laboratories, at the University of South Wales. An application has been proposed for this work, which includes a photovoltaic module for energy generation to power an illuminated advertizing sign. The energy is stored in a lithium-based battery model which uses a combination of a battery management system and remote monitoring for real-time data acquisition.

Keywords: PV-lithium; smart monitoring; prediction mechanism; BMS; module simulation

1. Introduction

The abundant sources of power in the form of solar (photovoltaic) and wind energy has been an important tool in the past decade, as efforts are made to reduce global warming while preserving precious dwindling fossil fuel supplies. Popularity of these energy generation tools is still seeing a steady increase, especially in the case of photovoltaic (PV), with reports showing an increase in European capacity by approximately 11 GW in 2013 (European Photovoltaic Industry Association). This is due to advances in the solar technology and government incentives to ensure that growth continues to achieve a target of 20% of European energy being generated by renewable sources by 2020 (European Union. The European Commission, 2007).

A PV array can be operated as a stand-alone or grid-connected system (Salas, Olías, Barrado, & Lázaro, 2006). Mobile stand-alone PV systems are mainly used in the conservation of the environment by using solar energy in locations without access to electricity, and therefore, act as an indispensable electricity source for remote areas. The technology for harvesting the solar energy shows much promise; however, the unreliability of the generation methods mean that they need to be coupled with an energy storage technique for efficient and practical use (Hadjipaschalis, Poullikkas, & Efthimiou, 2008). Due to the emerging advantages of lithium-ion battery technologies, these are being investigated as an alternative energy storage component for renewable energy stations. It is possible to consider lithium batteries for a wider range of applications because of falling manufacturing costs and increasing performance benefits (Nair Nirmal-Kumar & Garimella, 2010).

Although the manufacturing costs of the lithium cell are dropping, the requirement for internal monitoring electronics results in an inflated price when compared to alternative battery technologies. This in turn makes use of lithium technologies in small-scale energy generation systems as a promising next step rather than a common practice, with designers favouring the cheaper more robust lead acid battery.

However, the "expensive" monitoring and management equipment is a necessary addition to the lithium battery pack to stop the battery causing damage to itself, the generation equipment, or a threat of harm to the end user (Daowd, Omar, Bossche, & Mierlo, 2011). This is a wise precaution as was experienced in 2002, where lithium cells were causing damage in high-end electrical appliances (Sima, 2006). The battery management system (BMS) is the equipment charged with keeping each cell within its safe operation limits, and does so by carefully monitoring the cell voltage, temperature and state of charge (SoC). The SoC is the main focal point on a modern BMS, where the knowledge of remaining battery charge is becoming more and more essential.

*Corresponding author. Email: thomas.stockley@southwales.ac.uk

Coulomb counting is the simplest, albeit crudest form of SoC estimation. It allows a very quick to market implementation phase, and requires very little mathematics or background processing (Ng, Moo, Chen, & Hsieh, 2009). Added to these advantages is the fact that the technique can be applied during a charge or discharge state irrelevant of current magnitude and it is easy to see why it is the most common SoC estimation method. The method operates by counting the Amp-hour (Ah) that have been added or removed from a cell during a charge or discharge state and comparing the rolling total to a predefined cell capacity. The cell capacity is usually measured by a low current capacity test. There are, however, disadvantages to using the coulomb counting technique. Unavoidable accumulation errors are common and an unknown initial SoC renders the method useless; however, correction mechanisms are being researched to reduce these errors (Piller, Perrin, & Jossen, 2001). Therefore, the coulomb counting method is seldom used by itself in high-end BMSs where the SoC is requested as a highly accurate estimation.

A second method for the SoC estimation of a lithium cell is the use of an open-circuit voltage (OCV)–SoC lookup table. The OCV–SoC estimation method works by taking a voltage measurement at an equilibrated state (after a 3-hour rest period) and comparing it to a predefined lookup table containing the equivalent SoC of the cell for each voltage value. This method has the disadvantages of being a more complicated technique with a longer implementation time, as the lookup table needs to be generated, and a long open-circuit period is needed to ensure that the cell is at an equilibrated state (Chiang, Sean, & Ke, 2011). These disadvantages mean that the OCV–SoC estimation technique is not one which can be used solely on applications such as PV or wind microstations, where the battery bank is likely to be in a charge or discharge for a majority of the time. The high accuracy recorded by these systems in previous works makes the technique an ideal complement to the more simplistic coulomb counting method, where OCV–SoC can be used to find the initial SoC and then the coulomb counting method can be used to provide an updated running SoC for the battery (Piller et al., 2001; Stockley, Thanapalan, Bowkett, & Williams, 2013).

The need for the OCV measurement to be taken during an equilibrated state is due to hysteresis of the terminal voltage, which produces a false OCV measurement when the cell initially finishes a charge or discharge state. The cell voltage relaxes quickly initially when open-circuit commences, but the rate of the voltage relaxation decays with time as the terminal voltage becomes closer to the OCV. This is an issue that could cause a delay of up to 4 hours as the cell reaches an equilibrated state. This disadvantage is not only common to lithium cells, as alternative chemistries also require the need for a long open-circuit period before an accurate measurement can be made. One solution for predicting the OCV from a 30-minute measurement was proposed by Aylor, Thieme, and Johnso (1992). A simple summing technique was firstly proposed by Aylor et al. (1992), but it was a second technique using asymptotes and a semi-logarithmic scale that was finally implemented. The first method (a simple summing technique) was disregarded due to the fact that lead acid cell properties tend to differ greatly, even between cells in the same battery. However, Stockley et al. (2013) proved that lithium cells are much more uniform because of their chemical construction (the electrodes do not physically break down like a lead acid cell) and a much more automated manufacturing process. This allowed the simple summing method to be used following a charge or discharge rate for both $LiNiMnCoO_2$ and $LiFePO_4$ cells. Stockley et al. (2013) and Stockley, Thanapalan, Bowkett, and Williams (2014a, 2014b) proved that the summing method could be used on lithium cells using a 30-minute measurement, and then reduced this time to 8 minutes to make the prediction mechanism more available to a wider range of applications. The maximum error obtained from the tests conducted in Stockley et al. (2013, 2014a, 2014b) was just 6 mV for the 30-minute rate and 8.3 mV for the 8-minute rate, less than a 1%SoC error.

The second method of OCV estimation using the asymptotes on a logarithmic scale proved a success for Aylor et al. (1992), with a maximum error of just 5%SoC when using a measurement time of 6.6 minutes. Pop, Bergveld, Danilov, Regiten, and Notten (2008) used the asymptote method produced by Aylor to validate a new OCV prediction model and in doing so proved that the asymptote method could be used on lithium-based cells with an error of 0.92%. A combinational model using voltage change over time and temperature was also used to validate the new model, giving a very large error of 20.19%. The model proposed by Houseman (2005) relies on the voltage relaxation curve and achieved an error of just 0.19%. Added to this very low error rate, the model also accounts for the cell temperature, charge/discharge rate and state of health of the cell.

Several other OCV prediction mechanisms are available including: (i) polynomial-based curve estimators as used by Hu, Li, Peng, and Sun (2012), which rely on an extended Kalman filter for prediction. (ii) A sigmoid function approach as used by Weng, Sun, and Peng, (2013), which was compared to several polynomial-based models to produce an error of 2.5 mV, an improvement of 4.8 mV on the most accurate polynomial system. (iii) Linear regression models such as the work by Pei, Wang, Lu, and Zhu (2014), where the diffusion process of a lithium cell was found to have a linear relationship with the OCV, a relationship found through modelling of the cell using equivalent Resistve–Capacitive circuits.

Alternative techniques to estimate the SoC of the lithium cell under load have been proposed (Kutluay, Cadirci, Ozkazanc, & Cadirci, 2005). These methods measure the voltage while the cell is in use and thus the current has a large factor in the cell voltage. For this reason, the loaded voltage technique is more difficult to implement and calls for a much larger voltage to SoC relationship modelling phase. Therefore, the OCV–SoC technique has been chosen as the SoC estimation technique for this scope of work.

Generation and storage of "clean energy" is the focus of reducing the reliance on non-renewable energy sources, but the efficiency of how energy is being stored and consumed also needs to be considered. For this purpose, remote systems or smart metres are now being employed to monitor the power consumption of a specific load (Houseman, 2005). The smart metre was initially prioritized for home installations as part of the Smart Metering Implementation Programme to install metres in every home by 2020 (Great Briton. Department of Energy and Climate Change, 2012), in an aim to reduce the carbon consumption of the UK. Attention is now turning to monitoring energy usage in renewable energy systems both grid connected and stand-alone (Wolfe, 2008; Yu, Zhang, Xiao, & Choudhury, 2011). Wolfe (2008) suggests that by using remote monitoring and management equipment, the renewable energy system could greatly improve the conservation of the generated energy. However, as was discussed by Stefanakos and Thexton (1997) and Zezhong, Hongliang, and Ting (2010), a battery system also benefits from remote monitoring. Operating lithium cells at elevated temperatures, too high charge or discharge rates or in SoC ranges outside of the 80% to 20% boundary results in storage efficiency or cycle life becoming significantly lower (Guena & Leblanc, 2006; Shim, Kostecki, Richardson, Song, & Striebel, 2002). For this reason, the system shown further in this paper will contain a remote energy management and monitoring system, developed at the University of South Wales, to allow instant feedback of incorrect operating parameters.

The remaining sections of this work are organized as follows: Section 2 gives an overview of the OCV–SoC theory, an explanation of previous work and the test setup. Section 3 explains the results of the OCV–SoC prediction tests conducted on the lithium module. Section 4 provides the OCV–SoC model developed in Simulink, while the implemented BMS and design information can be found in Section 5. Lastly, a discussion and conclusion to this paper is provided in Section 6.

2. Research methodology and theory

As mentioned in the Introduction, the OCV–SoC estimation technique usually requires the cell to relax for a period of up to 3 hours after a charge or discharge state. The relaxation time can be reduced by the use of the simple formula provided in the following equation:

$$V_{OC} = V_{tr} \pm K_V, \tag{1}$$

where V_{OC} is the equilibrated OCV, V_{tr} is the voltage at the measurement interval and K_V is a predefined parameter

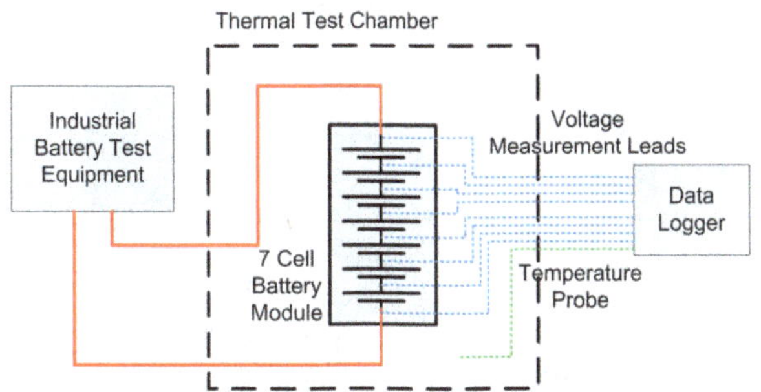

Figure 1. Block diagram of experimental setup for module testing.

found by applying the equation $V_{OC} \pm V_{tr}$ to a set of controlled tests. The voltage at the measurement interval (V_{tr}) was initially set as the voltage after 30 minutes of an open-circuit condition (Stockley et al., 2013, 2014a). However, advances in the research allowed the simple equation to be successfully tested at an improved measurement time of 8 minutes with a maximum SoC error of less than 1% (Stockley et al., 2014b). The work in this paper investigates the further improvement in the rate of prediction with an aim to reduce the time to within 5 minutes.

Testing of the OCV prediction method was conducted at the Centre for Automotive and Power Systems Engineering (CAPSE) labs at the University of South Wales. The tests were carried out using industrial battery test equipment, a thermal chamber to control the boundary conditions and a datalogger to measure the voltage of each of the cells. The setup can be seen in Figure 1. The module under test was made for use in a PV microsystem, and was constructed of seven 40 Ah $LiNiMnCoO_2$ cells as shown in Figure 2. Single cell tests were conducted on a 20-Ah $LiNiMnCoO_2$ cell and although the 40 Ah cells are the focus of this work, the previous work in Stockley et al. (2013, 2014a, 2014b) was conducted on the 20 Ah cells and they are referenced throughout this paper.

The test process included a discharge relaxation test, a charge relaxation test and a mixed state relaxation test. It was expected that by using the OCV prediction mechanism to aid the OCV–SoC technique, a system could be easily implemented to monitor and accurately track the SoC of not only the full battery module, but also each individual cell, ensuring that the module remains correctly balanced.

Each test consisted of a conditioning stage, where the cell was charged to 100%SoC or discharged to 0%SoC depending on the test requirement. The conditioning stage made use of a constant current/constant voltage charge or a constant current discharge. A constant current step was used to alter the cell's SoC by 20%, followed by a four-hour open-circuit measurement period to record the relaxation curve. Initial testing was conducted on the smaller capacity 20 Ah $LiNiMnCoO_2$ cell at 0.33, 1 and 3 C for both the charge and discharge states, resulting in

Figure 2. Seven-cell 40 Ah $LiNiMnCoO_2$ module.

a prediction error of just 6 mV using a 30-minute measurement. This equates to a SoC error of 0.4%SoC, using the fact that previous OCV–SoC tests resulted in a relationship of 9.9 mV is equal to a 1%SoC. Although the prediction error was very low, the application of this mechanism seemed to be very limited with many devices not being able to afford the cells to be out of use for a period of 30 minutes. To resolve this issue, the prediction error was found at several intervals to find the best compromise between measurement time and prediction error. The results can be seen in Figure 3.

As it can be seen in Figure 3, the ideal time for prediction is at 5 minutes where the measurement error is 9.25 mV, slightly less than the 9.9 mV level, where the error would equal 1%SoC. This was used in the new OCV prediction work for the 20 Ah lithium pouch cell, producing a maximum error of just 9.5 mV (0.96%SoC) while reducing the prediction time by 25 minutes from the work conducted in Stockley et al. (2013, 2014a).

3. Module OCV prediction testing

To ensure that the prediction mechanism would transfer from single-cell level to modules and even battery packs, relaxation tests were conducted on the seven-cell $LiNiMnCoO_2$ module. The module voltage was monitored throughout a 0.3 C discharge test, a 0.3 C charge test and a 0.3 C mixed state test, with four-hour open-circuit periods at 80%, 60%, 40% and 20%SoC. Figure 4 provides the relaxation curves for the mixed state test, where each SoC adjust involved both charge and discharge steps. From the shape of the relaxation curves in Figure 4, it can be seen that the open-circuit periods at 80% and 40%SoC followed a discharge state, whereas 60% and 20%SoC relaxed from a charge state. The duty cycle used to achieve these results can be seen in Figure 5.

The value of the parameter K_V for the module was derived by previous testing as explained in Stockley et al. (2014a) as 0.055 and 0.035 following a discharge and charge state, respectively. This allowed a fast yet accurate OCV prediction when using Equation (1). Table 1 contains the mixed state prediction results as an example, where a maximum error of just 45 mV was observed. The module charge and discharge relaxation tests also provided promising results with a maximum error of just 15 and 5 mV for the discharge and charge test, respectively. With an

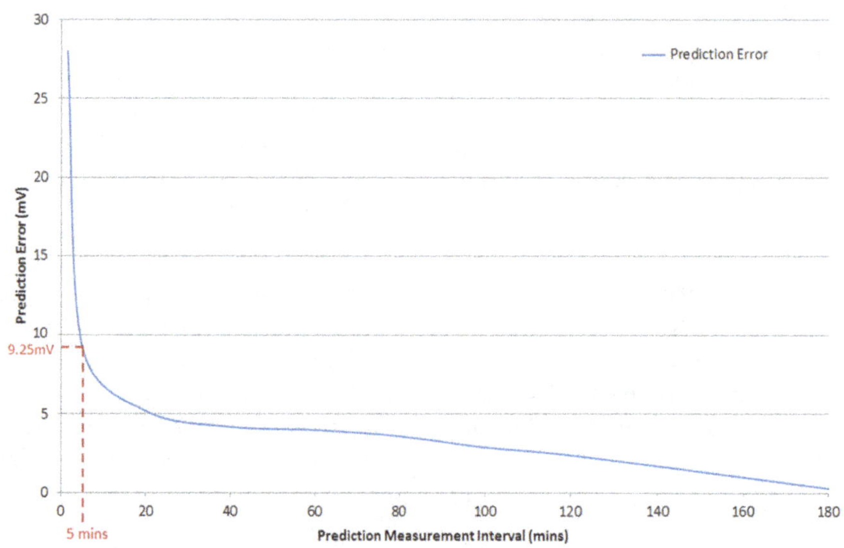

Figure 3. Error vs prediction time for lithium 20 Ah pouch cell.

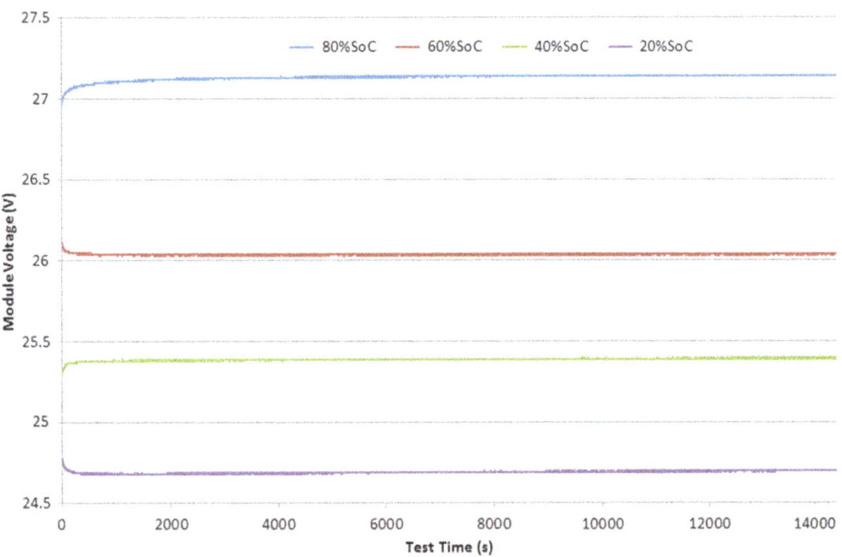

Figure 4. Relaxation curves for module mixed state test.

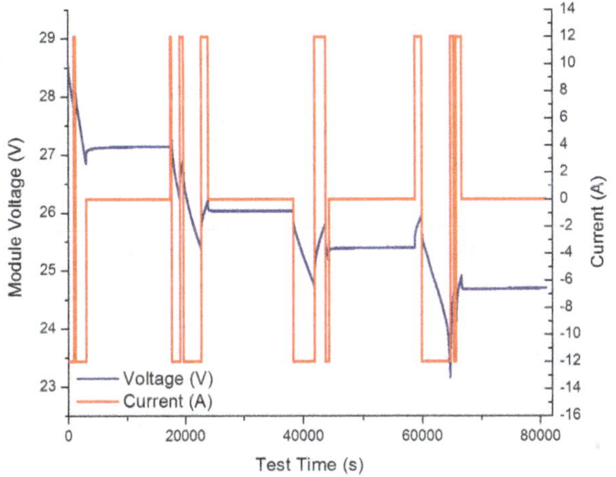

Figure 5. Mixed relaxation test current and voltage duty cycle.

OCV–SoC relationship of approximately 45.5 mV equals 1%SoC, a maximum error of 45 mV is proof that the prediction mechanism can be successfully adapted to lithium modules.

To ensure that the values of parameter K_V for a charge and discharge relaxation curve were chosen to provide the minimum error possible, the OCV prediction model was adapted for design optimization. The functions used for the optimization of parameter K_V can be seen in the following equations:

Table 2. Test parameters for K_V optimization.

Parameter	Value	Unit
Temperature	25	°C
C-rate	0.3	C
State of charge	20–80%	%SoC

$$\min_{(kv)} = f, \tag{2}$$

$$f = \sum_{t=0}^{t=ts} (V_{\text{predict}} - V_{\text{measured}})^2, \tag{3}$$

where f is the objective function, V_{predict} is the predicted OCV using the parameters of 0.55 and 0.35 for a charge and discharge, respectively, and V_{measured} is the measured equilibrated OCV. The boundary model parameters for the optimization can be seen in Table 2.

To optimize the K_V values for a charge and discharge relaxation curve, two optimization techniques were used. The simplex search method algorithm (SSM) was chosen

Table 1. Comparison of recorded data and OCV prediction for 7-cell module mixed state 0.3 C test.

	Open-circuit voltage (OCV) (V)						
Cell SoC (%)	0 minute	5 minute	8 minute	30 minute	180 minute real	180 minute calc.	Error (mV)
80	26.97	27.08	27.09	27.12	27.14	27.135	5
60	26.11	26.04	26.05	26.04	26.04	26.005	35
40	25.3	25.38	25.38	25.38	25.39	24.435	45
20	24.78	24.69	24.68	24.68	24.7	24.655	45

Table 3. Optimized values of K_V from SSM and PSM algorithms.

Cell condition	Derived K_V	SSM-optimized K_V	PSM-optimized K_V
Discharge relaxation	0.55	0.56	0.572
Charge relaxation	0.35	0.36	0.41

due to its high optimization speed which as Thanapalan, Wang, Williams, Liu, and Rees (2008) explain is due to the fact that the SSM optimization is a hill climbing technique. The method basically works by altering the value of the optimization parameter (n) for each iteration and then measuring the response of the model error. If the error produced by the new parameter (n) is less than the previous parameter value ($n-1$), then the new value (n) is adopted. If the error of n is greater than the value resulting from the use of $n-1$, then the new value is rejected.

The second optimization parameter used was a genetic algorithm by use of the pattern search method (PSM). The GAs are computational programmes based on the natural interactions of the genetics seen in nature. As Thanapalan et al. (2008) explain, the method incorporates a survival of the fittest where any parameter values outside of the normal scope is disregarded as the system parameters move closer to the optimized value.

The optimized parameters for K_V can be seen in Table 3 where they are compared to the calculated value of K_V from Table 1.

From the results in Table 3, it can be seen that the optimized results are very close to the initial results produced as in Stockley et al. (2014a, 2014b) and previous in this paper. The use of the optimized values from the SSM algorithm increased the error by 1 mV when compared to the derived K_V parameter values. The PSM-optimized parameter values also resulted in a higher error for the 60%, 40% and 20%SoC tests. The 80%SoC test resulted in a lower error value, however, a reduction in the error of 2.3 mV is negligible when considering that the OCV–SoC relationship of 45 mV is equal to 1%SoC. The results of the optimization parameter values in OCV prediction can be seen in Table 4.

The results in Table 4 show that the derived values of K_V from previous works calculated from the average curves provide the best error value. This means that

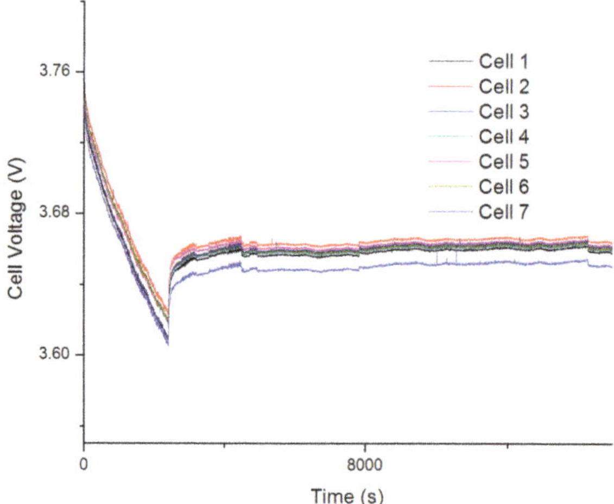

Figure 6. Single cell relaxation curved following 0.3 C discharge to 40%SoC.

optimization does not need to be carried out on further cell testing which makes the prediction mechanism simpler and quicker to transfer to other cell types.

The use of the derived values of the K_V parameter also has the advantage of simpler implementation when moving between single cells and modules. This can be seen by the work in Section 4.

4. OCV prediction testing of module's cells

As the module contains seven $LiNiMnCoO_2$ cells, each cell's relaxation voltage had to be monitored through the use of a datalogger, which allowed results such as the curves in Figure 6 to be created. The relaxation curves in Figure 6 follow a 0.3 C discharge to 80%SoC. It should be noted at this point that there is a slight discrepancy between some of the cells in the module. A capacity test and impedance test on each of the cells suggest that the module was not fully balanced prior to the testing phase, rather than some of the cells being damaged or aged. The slight voltage imbalance in the module can be seen in Figure 6, with the maximum equilibrated voltage of 3.667 V held by Cell 2 and the lowest equilibrated voltage of 3.653 V by Cell 3. However, the difference in cell voltage was negligible for the purpose of this research and had little effect on the further results.

The parameter (K_V) for each individual cell was calculated from the parameter value for the seven-cell module.

Table 4. 5-minute OCV prediction results using derived, SSM and PSM K_V parameter values.

Cell SoC (%)	Measured OCV (V)	Derived K_V calc. (V)	Derived K_V error (mV)	SSM K_V calc. (V)	SSM K_V error (mV)	PSM K_V calc. (V)	PSM K_V error (mV)
80	27.14	27.135	5	27.136	4	27.1373	2.7
60	26.04	26.005	35	26.004	36	25.999	41
40	25.39	25.435	45	25.436	46	25.437	47
20	24.7	24.655	45	24.654	46	24.649	51

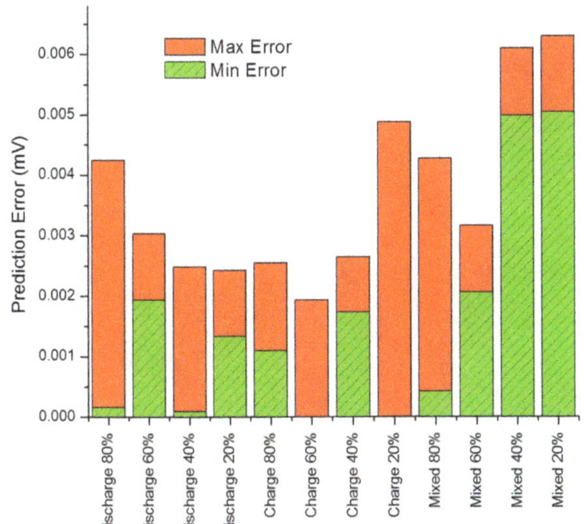

Figure 7. Comparison of minimum and maximum error for each of the single-cell tests.

As the module voltage is a simple summation of the single cells, it stands to reason that the relaxation voltage of a module is equal to the summation of the relaxation voltage of the single cells. Therefore, the module parameter of 0.055 for a discharge and 0.035 for a charge relaxation curve were divided by the number of cells to give a single-cell parameter of 0.0078 and 0.005 for the single-cell discharge and charge curves. As the amount of collected data is large, Figure 7 has been provided as a summary of the tests performed. Figure 7 shows the maximum and minimum errors from each 0.33 C relaxation test noted at the start of Section 3.

The error is greatest for the mixed relaxation tests at 40% and 20%SoC, giving the only prediction errors above 5 mV. From OCV–SoC relationship tests conducted on the 40 Ah $LiNiMnCoO_2$ pouch cell, a relationship of 6.5 mV that equals 1%SoC was found. This results in a maximum SoC error of 0.8% for a measurement time of just 5 minutes.

5. Modelling of OCV prediction mechanism

The OCV prediction mechanism was modelled in Matlab-Simulink which will allow it to be used for simulations and implemented into a real-world application. The developed prediction mechanism consists of several subsystems and is displayed in Figure 8. The top branch of the model is used to control the model, whereas the bottom branch is used to carry out the required action.

The State_Determination subsystem is used to determine whether the battery module is recovering from a charge or discharge state based on the current drawn from or applied to the battery module. This is important as the relaxation curve following a discharge state rises, whereas it falls after a charge state. The KV_Assignment subsystem block is used to assign the correct K_V value to the KvSelect storage variable. A second function of the State_Determination subsystem is to ensure that the prediction mechanism is only used in an open-circuit state. An open-circuit state is identified by a current measurement of less than 0.5 A, a value chosen to account for any measurement errors or system noise. The Timing_System uses the result of the State_Determination subsystem as a trigger to start an internal clock when an open-circuit period is entered. The Timing_System also recognizes when the measurement interval is reached and triggers the system to measure the OCV value. The Process_System monitors the module voltage and predicts the equilibrated 3-hour OCV voltage. The prediction is made based on the OCV value at the measurement interval and the correct K_V value provided by the KV_Assignment subsystem block. The

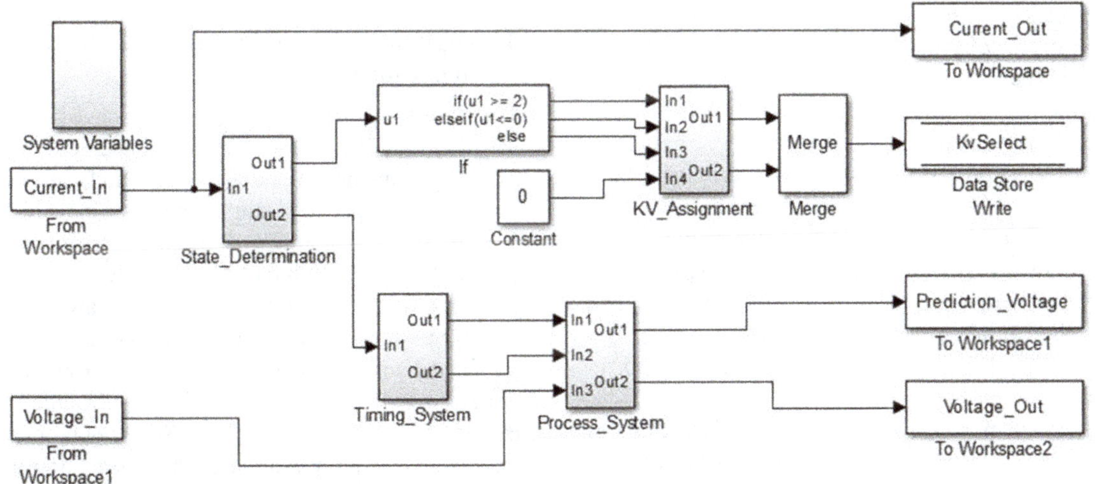

Figure 8. Simulink OCV prediction mechanism model.

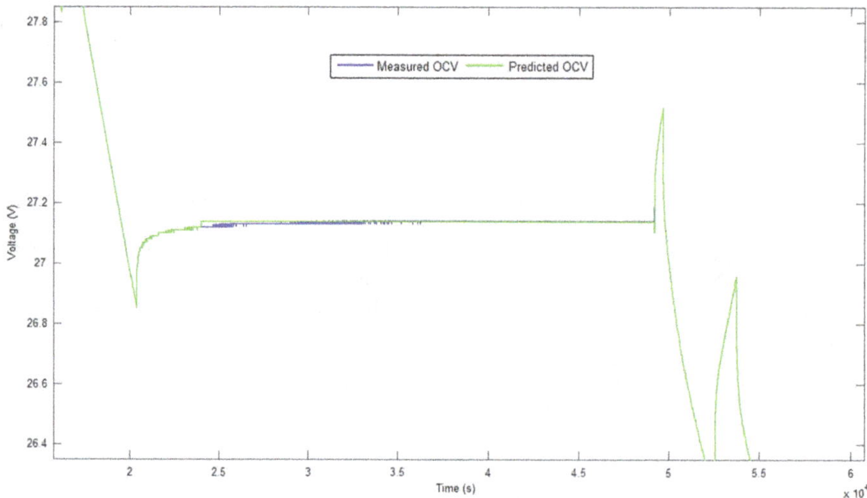

Figure 9. Comparison of measured and simulated relaxation curve at 80%SoC.

results of the model are exported to the Matlab workspace using the components Current_Out, Voltage_Out and Prediction_Voltage.

Figure 9 shows the relaxation curve of the 80%SoC discharge state relaxation test on the seven-cell module. As it can be seen in Figure 9 and noted in Section 3, the relaxation curve follows a discharge and therefore is rising. It is clear to see that the voltage of the module is monitored closely by the prediction model. When the open-circuit period reaches the 5-minute measurement interval, the prediction voltage is calculated and shown as the small jump in the green curve. At this point, the measured OCV can be seen to rise gently towards the predicted OCV measurement value. At the 3-hour mark, the measured and predicted voltages are re-aligned, proving that the model successfully predicted the OCV from the 5-minute measurement.

Although the model in Figure 8 is being used to predict the OCV of the full seven-cell module, it can be easily adapted for single cell or pack prediction. To use the model to predict a single cell, the parameter values of K_V (stored in the KV_Assignment subsystem) are changed from the module parameters in Section 3 to the single-cell K_V values in Section 4. For example, the parameter value would change from 0.055 to 0.0078 for a relaxation curve following a discharge state. As the module K_V value is proven to equal the single-cell K_V value multiplied by the number of cells in the module, it is recognized that a module model could also be accurately represented by summing the output of seven single-cell models. This method allows prediction of each individual cell and the full module.

This mechanism can also be applied to a larger energy storage system if required. As modules can be installed in series to increase the voltage of the battery system, the parameter K_V of the module can be multiplied by the amount of modules to find the K_V value for the battery pack. Likewise with the single-cell-based model, a set of module models could be used to represent a battery pack.

Figure 10. Prototype BMS with embedded OCV prediction mechanism.

6. OCV prediction mechanism in a real-world application

To prove that the prediction mechanism works in a real-world system, the model has been integrated into a BMS currently being developed in the University of South Wales CAPSE labs. The BMS can be seen in Figure 10.

The BMS is a low cost design which monitors the cell and module voltage, the module current and cell temperature. As the voltage and current are already being monitored by the BMS, the prediction mechanism can be monitored without the addition of any hardware. As mentioned in the Introduction, the BMS uses the prediction mechanism coupled to an OCV–SoC lookup table for the initial high accuracy estimation, and the coulomb counting method as a running SoC estimation. Figure 11 shows the flow chart for the prediction mechanism of the BMS. The

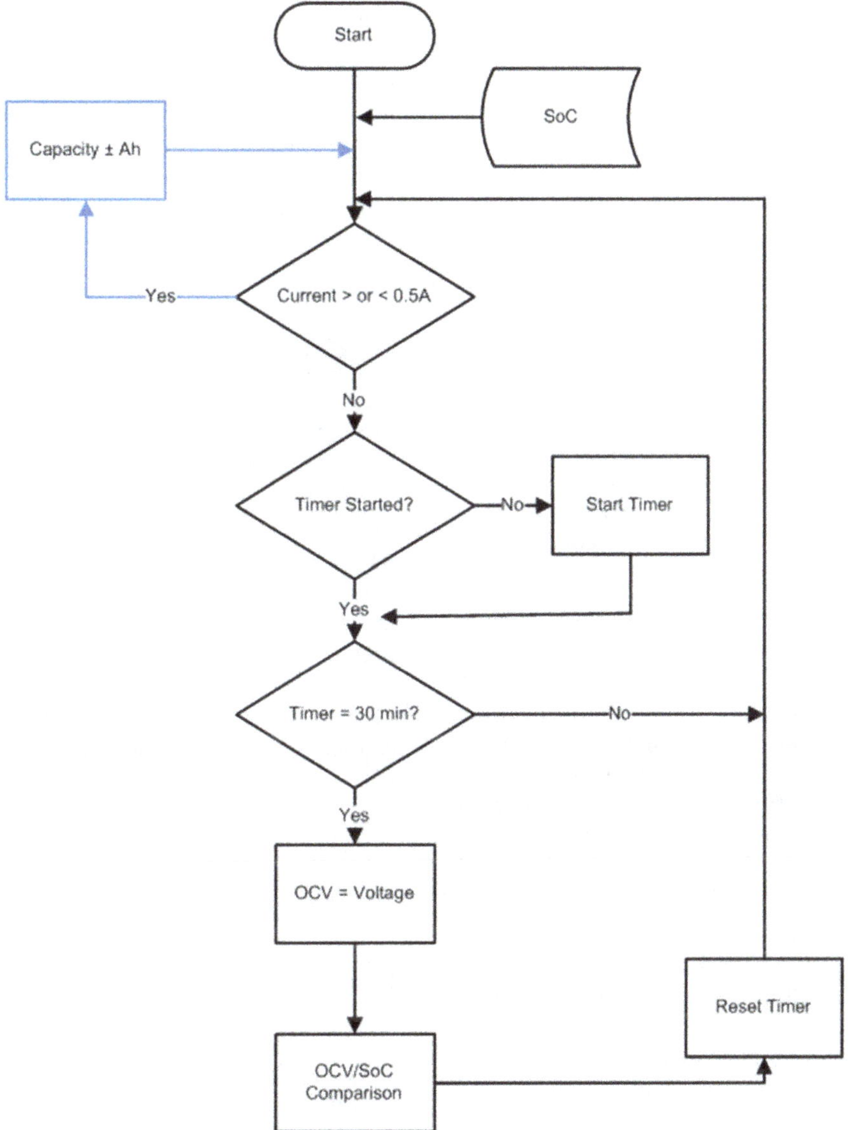

Figure 11. OCV prediction mechanism BMS flow chart.

section of the flow chart that is highlighted blue (dashed) is the coulomb counting section of the code. The SoC block at the top of the flow chart is the SoC value prior to the current iteration of the code.

It should be noted that the BMS is to be placed in a real-world application to monitor performance of a seven-cell lithium module, similar to the module shown in Figure 3. The module can be used as an energy storage system to power a stand-alone system. For example in this case the module is used to power an advertizing sign for a local company, which uses a PV panel as its primary energy source. Figure 12 shows a block diagram of the test rig for this particular case.

As the application for the prediction mechanism and seven-cell module is to be remote, monitoring of the battery condition and performance is difficult. As has been

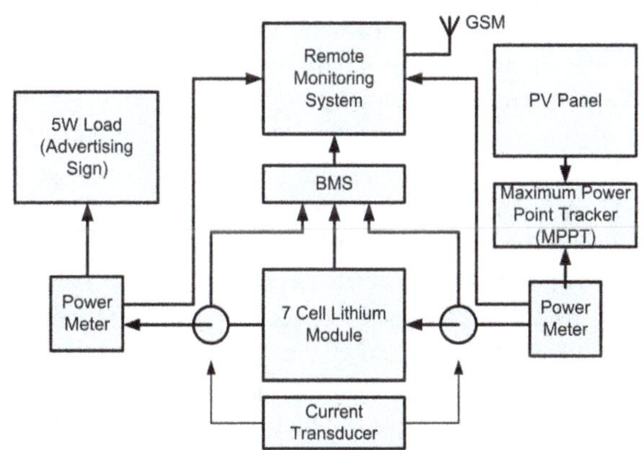

Figure 12. Block diagram of PV powered advertizing sign application for OCV prediction mechanism.

shown in the Introduction, regular monitoring of the battery parameters can lead to improved cycle life and energy storage efficiency. For this purpose, a remote management system is used to provide up-to-date information. The remote management system makes use of a microcontroller as the metre control system, which can read measurements from the BMS using a Modbus connection and sends information to a server using a global system for mobile communications (GSM) connection. The metre is a low cost design and has been implemented at the university for use in this research.

7. Discussion and concluding remarks

The work in this paper shows the advantages of the OCV prediction technique in both a theoretical and practical application. The results achieved from the testing work in Section 3 concur greatly with the tests conducted previously in Stockley et al. (2013, 2014a, 2014b). This is a positive feat as not only were the tests conducted on a seven-cell-connected module, but also on a different cell type to previous cell samples. The cell tested here was a 40-Ah (double the capacity of the previous pouch cell) $LiNiMnCoO_2$ cell. The maximum error from the module OCV prediction tests was well within the acceptable tolerance range of 1%SoC error, by producing a maximum error of just 45, 15 and 5 mV for the mixed, discharge- and charge-based tests as described in Section 3. Added to these low prediction errors is the fact that the OCV has been predicted at a measurement interval of 5 minutes.

While the OCV prediction mechanism was being tested on the module, single-cell measurements were being carried out on the individual cells in the module. This allowed the prediction mechanism to be applied on a module and cell level. Promising results were also seen in this step of the research as a maximum error of 6 mV at a measurement interval of 5 minutes was attained. Excluding the mixed tests would give a further reduced prediction error of 5 mV. The errors seen in this work can be put into perspective by considering the fact that a 1%SoC value has been attributed to a voltage of approximately 6.5 mV. Therefore, the single-cell tests would have equalled a maximum error of approximately 0.8%SoC. A noteworthy point in this work is that the value of the parameter K_V, that is used in Equation (1), has been set to 0.055 for a discharge and 0.035 for a charge relaxation curve. As the voltage for the module is the sum of the cell voltages, the K_V used in the single-cell tests was calculated as 0.0078 and 0.005 for the discharge and charge tests, respectively, which is in fact the module K_V values divided by the number of cells (7).

The prediction mechanism has been modelled in Section 5 which allowed simulations to be run using the results obtained from the full module tests. From Figure 9, it can be seen that the model follows the voltage profile throughout the discharge and until the measurement interval of 5 minutes when the open-circuit period is entered. When the measurement interval is triggered after an open-circuit period of 5 minutes the voltage is predicted. The measured voltage tapers up to the predicted value towards the 3-hour mark, where the cell is judged to be in an equilibrated state. Swapping the module's charge and discharge values of K_V with the cell level K_V allows the single-cell relaxation curve to be modelled, and the OCV to be predicted. Alternatively, using several single-cell models would allow the single cell and module OCV to be predicted, an advantage that is made possible by the simplicity of Equation (2).

The real-world application for the BMS which has the OCV prediction mechanism inbuilt is shown in Figure 10 in Section 6. The system would allow a very simple, low cost BMS to be used to protect the battery system in the stand-alone PV system. An energy management system is included in the system setup which can monitor all battery parameters and then upload the information using a GSM signal. This will allow the company responsible for the battery to be alerted very quickly if the module is being used in an unsafe way or inefficient manor. The flow chart is provided to give a brief understanding of how the software works and how the OCV prediction mechanism and coulomb counting method can be used in conjunction.

Acknowledgements

The authors would also like to thank the staff of CAPSE for their assistance.

Disclosure statement

No potential conflict of interest was reported by the authors.

Funding

The first author would like to acknowledge the financial support from KESS, RUMM Ltd and the University of South Wales.

References

Aylor, J. H., Thieme, A., & Johnso, B. W. (1992). A battery state-of-charge indicator for electric wheelchairs. *IEEE Transactions on Industrial Electronics*, *39*(5), 398–409.

Chiang, Y.-H., Sean, W.-Y., & Ke, J.-C. (2011). Online estimation of internal resistance and open-circuit voltage of lithium-ion batteries in electric vehicles. *Journal of Power Sources*, *196*(8), 3921–3932.

Daowd, M., Omar, N., Bossche, P., & Mierlo, J. (2011). *Passive and active battery balancing comparison based on MATLAB simulation*. The 7th IEEE vehicle power and propulsion conference.

European Photovoltaic Industry Association. Global market outlook for photovoltaics 2014–2018. Retrieved from February 27, 2015. http://files.ctctcdn.com/15d8d5a7001/3f338a6a-eece-4303-b8c4-c007181a59ad.pdf/

European Union. The European Commission. (2007). *The climate and energy package* [Online]. Retrieved from February 27, 2015. http://ec.europa.eu/clima/policies/package/index_en.htm

Great Briton. Department of Energy and Climate Change. (2012). *Smart metering implementation programme* [Online]. Retrieved from February 27, 2015. https://www.gov.uk/government/publications/smart-metering-implementation-programme-first-annual-progress-report-on-the-roll-out-of-smart-meters

Guena, T., & Leblanc, P. (2006). *How depth of discharge affects the cycle life of lithium–metal–polymer batteries*. The 28th annual international telecommunications energy conference 2006 (INTELEC '06) (pp. 1–8).

Hadjipaschalis, I., Poullikkas, A., & Efthimiou, V. (2008). Overview of current and future energy storage technologies for electric power applications. *Renewable and Sustainable Energy Reviews*, *13*, 1513–1522.

Houseman, D. (2005). Smart metering: The holy grail of demand-side energy management? *Refocus Journal*, *6*(5), 50–51.

Hu, X., Li, S., Peng, H., & Sun, F. (2012). Robustness analysis of state-of-charge estimation methods for two types of Li-ion batteries. *Journal of Power Sources*, *217*, 209–219.

Kutluay, K., Cadirci, Y., Ozkazanc, Y. S., & Cadirci, I. (2005). A new online state-of-charge estimation and monitoring system for sealed leadacid batteries in telecommunication power supplies. *IEEE Transactions on Industrial Electronics*, *52*(5), 1315–1327.

Nair Nirmal-Kumar, C., & Garimella, N. (2010). Battery energy storage systems: Assessment for small-scale renewable energy integration. *Energy and Buildings*, *42*(11), 2124–2130.

Ng, K. S., Moo, C.-S., Chen, Y.-P., & Hsieh, Y.-C. (2009). Enhanced coulomb counting method for estimating state-of-charge and state- of-health of lithium-ion batteries. *Applied Energy*, *86*(9), 1506–1511.

Pei, L., Wang, T., Lu, R., & Zhu, C. (2014). Development of a voltage relaxation model for rapid open-circuit voltage prediction in lithium-ion batteries. *Journal of Power Sources*, *253*, 412–418.

Piller, S., Perrin, M., & Jossen, A. (2001). Methods for state-of-charge determination and their applications. *Proceedings of the 22nd International Power Sources Symposium*, *96*(1), 113–120.

Pop, V., Bergveld, H. J., Danilov, D., Regiten, P. P. L., & Notten, P. H. L. (2008). Battery management systems. In V. Pop, H. J. Bergveld, D. Danilov, P. P. L. Regiten, & P. H. L. Notten (Eds.), *Methods for measuring and modelling a battery's electro-motive force* (Vol. *9*, pp. 63–94). Eindhoven: Springer.

Salas, V., Olías, E., Barrado, A., & Lázaro, A. (2006). Review of the maximum power point tracking algorithms for stand-alone photovoltaic systems. *Solar Energy Materials and Solar Cells*, *90*(11), 1555–1578.

Shim, J., Kostecki, R., Richardson, T., Song, X., & Striebel, K. A. (2002). Electrochemical analysis for cucle performance and capacity fading of a lithium-ion battery cycled at elevated temperature. *Journal of Power Sources*, *112*(1), 222–230.

Sima, A. (2006). *Sony exploding batteries – the chronicles*. Softpedia, October 13. Retrieved from February 25, 2015. http://news.softpedia.com/news/Sony-Exploding-Batteries-Chronicles-37848.shtml

Stefanakos, E. K., & Thexton, A. S. (1997). *Remote battery monitoring and management field trial*. The 19th international telecommunications energy conference (INTELEC) (pp. 653–657).

Stockley, T., Thanapalan, K., Bowkett, M., & Williams, J. (2013). *Development of an OCV prediction mechanism for lithium-ion battery system*. The 19th international conference on automation and computing (ICAC) (pp. 48–53).

Stockley, T., Thanapalan, K., Bowkett, M., & Williams, J. (2014a). *Design and implementation of OCV prediction mechanism for PV-lithium ion battery system*. The 20th international conference on automation and computing (ICAC) (pp. 48–54).

Stockley, T., Thanapalan, K., Bowkett, M., & Williams, J. (2014b). Design and implementation of an open circuit voltage prediction mechanism for lithium-ion battery systems. *Systems Science & Control Engineering: An Open Access Journal*, *2*(1), 707–717.

Thanapalan, K., Wang, B., Williams, J., Liu, G., & Rees, D. (2008). *Modelling, parameter estimation and validation of a 300W pem fuel cell system*. Proceedings of the UKACC international conference on control, Manchester (pp. 1–6).

Weng, C., Sun, J., & Peng, H. (2013, October 21–23). *An open-circuit-voltage model of lithium-ion batteries for effective incremental capacity analysis*. Proceedings of the ASME 2013 dynamic systems control conference, CA, USA.

Wolfe, P. (2008). The implications of an increasingly decentralised energy system. *Energy Policy Journal*, *36*(12), 4509–4513.

Yu, F. R., Zhang, P.Xiao, W., & Choudhury, P. (2011). Communication systems for grid integration of renewable energy resources. *IEEE Network*, *25*(5), 22–29.

Zezhong, X., Hongliang, S., & Ting, L. (2010). *Remote monitoring system of lead-acid battery group based on GPRS*. International conference on electrical and control engineering (ICECE) (pp. 4023–4026).

3

Simultaneous state and input estimation with partial information on the inputs

Jinya Su[a], Baibing Li[b]* and Wen-Hua Chen[a]

[a]*Department of Aeronautical and Automotive Engineering, Loughborough University, Loughborough LE11 3TU, UK;*
[b]*School of Business and Economics, Loughborough University, Loughborough LE11 3TU, UK*

This paper investigates the problem of simultaneous state and input estimation (SSIE) for discrete-time linear stochastic systems when the information on the inputs is partially available. To incorporate the partial information on the inputs, matrix manipulation is used to obtain an equivalent system with reduced-order inputs. Then Bayesian inference is drawn to obtain a recursive filter for both state and input variables. The proposed filter is an extension of the recently developed state filter with partially observed inputs to the case where the input filter is also of interest, and an extension of the SSIE to the case where the information on the inputs is partially available. A numerical example is given to illustrate the proposed method. It is shown that, due to the additional information on the inputs being incorporated in the filter design, the performances of both state and input estimation are substantially improved in comparison with the conventional SSIE without partial input information.

Keywords: Bayesian inference; partial information; state filter; unknown input filter

1. Introduction

State estimation for discrete-time linear stochastic systems with unknown inputs has been receiving increasing attention (see, e.g. Cheng, Ye, Wang, & Zhou, 2009; Darouach & Zasadzinski, 1997; Darouach, Zasadzinski, & Boutayeb, 2003; Hsieh, 2000; Kitanidis, 1987, among many others) due to its widespread applications in the fields of weather forecasting (Kitanidis, 1987), fault diagnosis (Mann & Hwang, 2013), etc.

In some applications such as population estimation, traffic management (Li, 2013), and chemical engineering (Mann & Hwang, 2013), however, information on the input variables is not completely unknown; rather, it is available at an aggregate level. Li (2013) has recently proposed a unified filtering approach to incorporate this kind of information. It is shown that this approach includes two extreme scenarios as its special cases, that is, the filter where all the inputs are unknown (i.e. the scenario investigated in Kitanidis, 1987; Gillins & De Moor, 2007, etc.) and the filter where the inputs are completely available (i.e. the classical Kalman filter can be applied). Later, Su, Li, and Chen (2015a) further investigated some properties of the aforementioned unified filter such as existence, optimality and asymptotic stability. However, Li (2013) and Su et al. (2015a) only considered the problem of sole state estimation; the problem of simultaneous state and input estimation (SSIE) with partial information on the inputs has not been investigated.

Gillins and De Moor (2007) developed a SSIE method using the approach of minimum-variance unbiased estimation (MVUE), then Fang and Callafon (2012) further investigated its asymptotic stability. Potentially, the SSIE can be applied to a wide range of problems such as fault diagnosis (see, e.g. Gao & Ding, 2007; Patton, Clark, & Frank, 1989), fault-tolerant control (Jiang & Fahmida, 2005), disturbance rejection control (Profeta, Joseph, William, & Marin, 1990). In the field of fault detection and fault-tolerant control, for example, actuator, sensor and/or structure faults are usually modelled as inputs to the system with unknown dynamics. One can monitor system status by estimating the inputs for fault diagnosis purposes where the estimated inputs can provide valuable information for the fault-tolerant control system; see, for example, Jiang and Fahmida (2005) and Su, Chen, and Li (2014). In the field of disturbance rejection control, the uncertainties in system model are usually modelled as lumped system inputs (which may include system mismatches, parameter uncertainties, external disturbances); see Chen, Ballance, Gawthrop, and O'Reilly (2000), Yang, Li, Su, and Yu (2013), and Yang, Su, Li, and Yu (2014) for a detailed discussion. When inputs are approximately obtained based on disturbance estimation algorithms, one can attenuate their effects on dynamic systems by directly feedthrough of the estimated value.

In this paper, we investigate the problem of SSIE. Unlike Gillins and De Moor (2007) where the inputs are

*Corresponding author. Email: b.li2@lboro.ac.uk

assumed to be completely unknown, we consider the scenario where the information on the inputs is partially available. To incorporate the partial information on the inputs, the original inputs are decoupled into two parts, where the first part is completely known based on the available information on the original inputs, whereas the second part is completely unknown which will serve as the unknown inputs of the new system. On this basis, we draw Bayesian inference (see, e.g. Li, 2009, 2013) and obtain simultaneous estimates of the state and new unknown inputs. According to the Bayesian theory, the obtained estimates are optimal in the sense of minimum mean square estimation under the assumption of Gaussian noise terms (Li, 2013). Finally, the estimates of the original inputs can be worked out by pooling together all the available information on the inputs.

Compared with the filter in Li (2013) where only state estimate is of interest, the proposed method obtains SSIE, and hence the estimated inputs can be used in fault detection and other applications. Compared to the results in Gillins and De Moor (2007), this paper takes into account the additional information on the inputs, and hence it results in a better estimate of state and input vectors. In addition, we show that Bayesian inference can provide an alternative derivation of the filter in Gillins and De Moor (2007) for the SSIE problem. We further investigate the relationships of the proposed filter with some existing approaches. In particular, we show in this paper that (a) when the inputs are completely available, the proposed filter reduces to the classical Kalman filter (Simon, 2006); (b) when no information on the unknown inputs is available, it reduces to the results of Gillins and De Moor (2007) where both state and input estimation are concerned; and (c) if only state estimation is of interest, it is equivalent to the filter for partially available inputs developed in Li (2013).

The rest of paper is structured as follows. Section 2 formulates the considered problem. The main results of the paper are provided in Section 3. In Section 4, a simulation study is carried out to illustrate the proposed filter. Finally, Section 5 concludes the paper.

2. Problem formulation

Consider a discrete-time linear stochastic time-varying system with unknown inputs:

$$\begin{aligned} x_{k+1} &= A_k x_k + G_k d_k + \omega_k, \\ y_k &= C_k x_k + \upsilon_k, \end{aligned} \tag{1}$$

where $x_k \in R^n, d_k \in R^m, y_k \in R^p$ are the state vector, input vector, and measurement vector at each time step k with $p \geq m$ and $n \geq m$. Following Li (2013), the process noise $\omega_k \in R^n$ and the measurement noise $\upsilon_k \in R^p$ are assumed to be mutually independent, and each follows a Gaussian distribution with zero mean and a known covariance matrix, $Q_k = E[\omega_k \omega_k^{\mathrm{T}}] > 0$ and $R_k = E[\upsilon_k \upsilon_k^{\mathrm{T}}] > 0$, respectively. A_k, G_k, C_k are known matrices. Following the existing researches (e.g. Gillins & De Moor, 2007; Kitanidis, 1987; Li, 2013; Su et al., 2015a), G_k is assumed to have a full column-rank; otherwise, the redundant input variables can be removed.

We consider the scenario where the input vector d_k is not fully observed at the level of interest but rather it is available only at an aggregate level. Specifically, let D_k be a $q_k \times m$ known matrix with $0 \leq q_k \leq m$ and F_{0k} an orthogonal complement of D_k^{T} such that $D_k F_{0k} = O_{q_k \times (m-q_k)}$. It is assumed that the input data are available only on some linear combinations:

$$r_k = D_k d_k, \tag{2}$$

where r_k is available at each time step k, whereas no information on $\delta_k = F_{0k}^{\mathrm{T}} d_k$ is available. Hence, δ_k is assumed to have a non-informative probability density function $f(\delta_k)$ such that all possible values of δ_k are equally likely to occur:

$$f(\delta_k) \propto 1. \tag{3}$$

Without loss of generality, we assume that D_k has a full row-rank; otherwise, the redundant rows of D_k can be removed from the analysis (see Su et al., 2015a).

As pointed out in Li (2013), the matrix D_k characterizes the availability of input information at each time step k. It includes two extreme scenarios that are usually considered: (a) $q_k = 0$, that is, no information on the input variables is available; this is the problem investigated in Kitanidis (1987) and Gillins and De Moor (2007); (b) $q_k = m$ and D_k is an identity matrix, that is, the complete input information is available. This is the case that the classical Kalman filter can be applied (Simon, 2006).

The objective of this paper is to simultaneously estimate the state and input vectors based on Equations (1) and (2).

3. Main results

In this section, the main results of the paper will be given. To incorporate the partial information on the inputs d_k, $G_k d_k$ is decoupled into two parts based on a decoupling matrix, that is, the known part given by the prior information (2) and unknown part δ_k. Note that δ_k has a lower dimension than the original input vector d_k and it plays the role of unknown inputs in the new system. Next, Bayesian inference is drawn to obtain recursive estimates of both state variables x_k and unknown inputs δ_k, upon which the estimate of the original input vector d_k can be worked out. Finally, the relationships between the proposed method and the relevant existing filters are discussed. The diagram of the system and the proposed filter structure is shown in Figure 1.

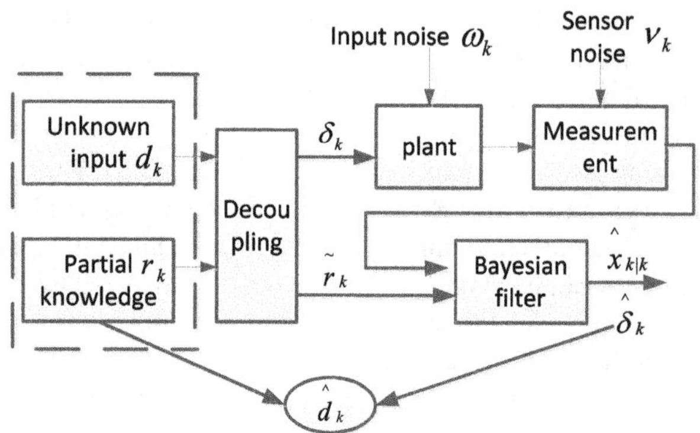

Figure 1. Diagram of the system and filter structure.

3.1. *Transformation*

To incorporate the information $r_k = D_k d_k$, a decoupling method is used here (see Su et al., 2015a). Define a non-singular decoupling matrix M_k of appropriate dimension as follows:

$$M_k = \begin{bmatrix} D_k & O \\ O & I \\ F_{0k}^{\mathrm{T}} & O \end{bmatrix} [G_k, \ G_k^{\perp}]^{-1},$$

where $G_k^{\perp}$ denotes an orthogonal complement of G_k, O and I represent the zero matrix and identity matrix of appropriate dimensions, respectively. F_{0k} is the orthogonal complement of D_k^{T} such that $D_k F_{0k} = O$ and $F_{0k}^{\mathrm{T}} F_{0k} = I$.

Then, $M_k G_k d_k$ can be expressed as follows:

$$\begin{aligned} M_k G_k d_k &= [D_k^{\mathrm{T}}, O, F_{0k}]^{\mathrm{T}} d_k \\ &= [(D_k d_k)^{\mathrm{T}}, O, O]^{\mathrm{T}} + [O, O, I]^{\mathrm{T}} F_{0k}^{\mathrm{T}} d_k \qquad (4) \\ &= \tilde{r}_k + \tilde{G}_k \delta_k, \end{aligned}$$

where $\tilde{r}_k := [r_k^{\mathrm{T}} \ O \ O]^{\mathrm{T}}$ is completely available due to the available information on the inputs, $\tilde{G}_k := [O \ O \ I]^{\mathrm{T}}$ and $\delta_k := F_{0k}^{\mathrm{T}} d_k$.

Multiplying M_k^{-1} (the explicit form of M_k^{-1} is given in the appendix) on both sides of Equation (4), $G_k d_k$ can be decoupled into two parts:

$$G_k d_k = M_k^{-1} \tilde{r}_k + M_k^{-1} \tilde{G}_k \delta_k. \qquad (5)$$

Consequently, the dynamics of x_{k+1} can be written as

$$\begin{aligned} x_{k+1} &= A_k x_k + M_k^{-1} \tilde{r}_k + M_k^{-1} \tilde{G}_k \delta_k + \omega_k \\ &= A_k x_k + M_k^{-1} \tilde{r}_k + F_k \delta_k + \omega_k, \end{aligned}$$

where $F_k := M_k^{-1} \tilde{G}_k = [G_k, \ G_k^{\perp}] \left[\begin{smallmatrix} F_{0k} \\ O \end{smallmatrix} \right] = G_k F_{0k}$.

Hence, the linear system (1) with the additional information on the inputs, $r_k = D_k d_k$, can equivalently be represented by the following system:

$$\begin{aligned} x_{k+1} &= A_k x_k + M_k^{-1} \tilde{r}_k + F_k \delta_k + \omega_k, \\ y_k &= C_k x_k + v_k. \end{aligned} \qquad (6)$$

Remark 1 An alternative approach to incorporating the unknown input information is to use pseudo-inverse theory. From Equation (2), one can obtain the general solution of d_k

$$d_k = D_k^{+} r_k + F_{0k} \bar{\delta}_k, \qquad (7)$$

where $D_k^{+} = D_k^{\mathrm{T}} (D_k D_k^{\mathrm{T}})^{-1}$ and $\bar{\delta}_k$ is completely unknown. If we select $\bar{\delta}_k := \delta_k = F_{0k}^{\mathrm{T}} d_k$, we can show that this approach is equivalent to the decoupling matrix based method.

Remark 2 It should be noted that the partial information on the inputs $r_k = D_k d_k$ has been fully incorporated into the system (6). We also note that the dimension of the inputs has been reduced from m to $m - q_k$.

3.2. *Filter design*

It can be seen from Equation (6) that y_k is a function of x_k, and x_k is related to the inputs δ_{k-1}. Hence, the input estimate of δ_k is delayed by one time unit (Gillins and De Moor, 2007). The objective of filter design is to obtain the estimate of x_k and δ_{k-1} based on the available measurement sequence $Y_k = \{y_1, y_2, \ldots, y_k\}$. For the new system (6), we can either solve the filtering problem based on the approach of MVUE (e.g. Gillins & De Moor, 2007) or Bayesian inference (e.g. Li, 2009, 2013). In the paper, we use the Bayesian method that can be seen as an alternative approach to that of Gillins and De Moor (2007).

In the context of Bayesian inference, the first step is to predict the dynamics of x_k and δ_{k-1} based on the available measurement sequence $Y_{k-1} = \{y_1, y_2, \ldots, y_{k-1}\}$. Since we do not assume that the unknown input vector δ_k satisfies any transition dynamics, prediction is only performed to determine the dynamics of x_k, that is, $p(x_k | Y_{k-1})$. The likelihood function can be determined based on the observation equation of system (6). The second step is to obtain the posterior distribution of the concerned variables after the measurement vector y_k is received based on Bayes' chain rule:

$$p(x_k, \delta_{k-1} | Y_k) \propto p(y_k | x_k) p(x_k, \delta_{k-1} | Y_{k-1}). \qquad (8)$$

The main results on filtering design are summarized in Theorem 1.

THEOREM 1 *For state space model* (6), *suppose the matrix* $C_k F_{k-1}$ *has a full column-rank, then the prior and posterior distributions for* x_k *and* δ_{k-1} *at any time step* k *can be obtained sequentially as follows:*

(i) *Posterior of* x_{k-1} *for given* Y_{k-1} :

$$x_{k-1} \sim N(\hat{x}_{k-1|k-1}, P_{k-1|k-1}^{x}).$$

(ii) *Prediction for* x_k :

$$N(\hat{x}_{k|k-1}, P^x_{k|k-1}),$$

with $\hat{x}_{k|k-1} = A_{k-1}\hat{x}_{k-1|k-1} + M^{-1}_{k-1}\tilde{r}_{k-1}$,

$$P^x_{k|k-1} = A_{k-1}P^x_{k-1|k-1}A^{\mathrm{T}}_{k-1} + Q_{k-1}. \quad (9)$$

(iii) *Posterior of* δ_{k-1} *for given* Y_k :

$$\delta_{k-1} \sim N(\hat{\delta}_{k-1}, P^{\delta}_{k|k}),$$

where the posterior mean is given by

$$\hat{\delta}_{k-1} = P^{\delta}_{k|k}(C_kF_{k-1})^{\mathrm{T}}\tilde{R}^{-1}_k(y_k - C_k\hat{x}_{k|k-1}), \quad (10)$$

and the posterior covariance matrix is given by

$$P^{\delta}_{k|k} = (F^{\mathrm{T}}_{k-1}C^{\mathrm{T}}_k\tilde{R}^{-1}_kC_kF_{k-1})^{-1}, \quad (11)$$

while the posterior of x_k *for given* Y_k *is*

$$x_k \sim N(\hat{x}_{k|k}, P^x_{k|k}),$$

where the posterior mean is given by

$$\begin{aligned}\hat{x}_{k|k} &= \hat{x}_{k|k-1} + P^x_{k|k-1}C^{\mathrm{T}}_k\tilde{R}^{-1}_k(y_k - C_k\hat{x}_{k|k-1})\\ &\quad + (F_k - P^x_{k|k-1}C^{\mathrm{T}}_k\tilde{R}^{-1}_kC_kF_k)\hat{\delta}_{k-1}, \quad (12)\end{aligned}$$

and the posterior covariance matrix is given by

$$\begin{aligned}P^x_{k|k} &= P^x_{k|k-1} - P^x_{k|k-1}C^{\mathrm{T}}_k\tilde{R}^{-1}_kC_kP^x_{k|k-1}\\ &\quad + (F_{k-1} - P^x_{k|k-1}C^{\mathrm{T}}_k\tilde{R}^{-1}_kC_kF_{k-1})(P^{\delta}_{k|k})^{-1}()^{\mathrm{T}}, \quad (13)\end{aligned}$$

where $\tilde{R}_k = C_kP^x_{k|k-1}C^{\mathrm{T}}_k + R_k$, $()^{\mathrm{T}}$ *in* $(*)A()^{\mathrm{T}}$ *stands for the transpose of* $*$.

Proof From Equation (8), the posterior distribution $p(x_k, \delta_{k-1}|Y_k)$ is governed by

$$\begin{aligned}p(x_k, \delta_{k-1}|Y_k) &\propto \exp\{-(y_k - C_kx_k)^{\mathrm{T}}R^{-1}_k()\\ &\quad - (x_k - \hat{x}_{k|k-1} - F_{k-1}\delta_{k-1})^{\mathrm{T}}(P^x_{k|k-1})^{-1}()\}.\end{aligned}$$

By completing the square on $[x^{\mathrm{T}}_k, \delta^{\mathrm{T}}_{k-1}]^{\mathrm{T}}$, the exponent can be rewritten as $-([x^{\mathrm{T}}_k, \delta^{\mathrm{T}}_{k-1}] - [\hat{x}^{\mathrm{T}}_{k|k}, \hat{\delta}^{\mathrm{T}}_{k-1}])P^{-1}_{k|k}()^{\mathrm{T}}$, where

$$\begin{bmatrix}\hat{x}_{k|k}\\ \hat{\delta}_{k-1}\end{bmatrix} = P_{k|k}\begin{bmatrix}C^{\mathrm{T}}_kR^{-1}_ky_k + (P^x_{k|k-1})^{-1}\hat{x}_{k|k-1}\\ -F^{\mathrm{T}}_{k-1}(P^x_{k|k-1})^{-1}x_{k|k-1}\end{bmatrix}$$

and

$$P_{k|k} = \begin{bmatrix}C^{\mathrm{T}}_kR^{-1}_kC_k + (P^x_{k|k-1})^{-1} & -(P^x_{k|k-1})^{-1}F_{k-1}\\ -F^{\mathrm{T}}_{k-1}(P^x_{k|k-1})^{-1} & F^{\mathrm{T}}_{k-1}(P^x_{k|k-1})^{-1}F_{k-1}\end{bmatrix}^{-1}.$$

This indicates that the posterior distribution is a Gaussian distribution with mean $[\hat{x}^{\mathrm{T}}_{k|k}, \hat{\delta}^{\mathrm{T}}_{k-1}]^{\mathrm{T}}$ and covariance matrix $P_{k|k}$. When C_kF_{k-1} is of full row-rank, based on the inverse of partitioned matrix, we can obtain the recursive estimation of both x_k and δ_{k-1} as shown in Equations (9)–(13).

So far, we have obtained the state estimate $\hat{x}_{k|k}$ and estimate $\hat{\delta}_{k-1}$ for the transformed system. When $F^{\mathrm{T}}_{0k-1}d_{k-1} = \hat{\delta}_{k-1}$ is obtained, based on Equation (5), we can further obtain the estimate of the original inputs d_{k-1} as follows:

$$\hat{d}_k = (G^{\mathrm{T}}_kG_k)^{-1}G^{\mathrm{T}}_k(M^{-1}_k\tilde{r}_k + M^{-1}_k\tilde{G}_k\hat{\delta}_k).$$

It can be verified that the obtained unknown input estimate satisfies the unknown input information (Equation (2)), that is,

$$D_k\hat{d}_k = r_k. \quad (14)$$

The proof is given in the appendix. ■

3.3. *Relationships with the existing results*

In this section, we investigate the relationships between the proposed approach and the relevant results in the existing literature. This is summarized in the following theorem.

THEOREM 2 *The set of recursive formulas* (9)–(13) *reduces to*

(1) *the classical Kalman filter when all entries of the input vector* d_k *are available;*
(2) *the filter in* Gillins and De Moor (2007) *when no information on the unknown inputs* d_k *is available;*
(3) *the filter in* Li (2013) *when only state estimation is concerned.*

Proof For the case where all the input variables are available at the level of interest, D_k becomes an $m \times m$ identity matrix, and F^{T}_{0k} becomes an zero-by-zero empty matrix. Consequently, the last term on the right-hand side of Equations (12) and (13) vanishes, and Equations (12) and (13) reduce to

$$P^x_{k|k} = P^x_{k|k-1} - P^x_{k|k-1}C^{\mathrm{T}}_kH^{-1}_kC_kP^x_{k|k-1}.$$

Since $M^{-1}_{k-1}\tilde{r}_{k-1} = G_{k-1}d_{k-1}$, Equation (12) becomes

$$\begin{aligned}\hat{x}_{k|k} &= A_{k-1}\hat{x}_{k-1|k-1} + G_{k-1}d_{k-1}\\ &\quad + P^x_{k|k-1}C^{\mathrm{T}}_kH^{-1}_k(y_k - C_k(A_{k-1}\hat{x}_{k-1|k-1}\\ &\quad + G_{k-1}d_{k-1})).\end{aligned}$$

Clearly, these recursive formulas are identical to the classical Kalman filter equations (see Simon, 2006).

Next, we consider the case where no input information is available. Clearly $\tilde{r}_k$ in Equation (4) is an empty vector, F_k becomes G_k, and $\delta_k = d_k$. Hence, Equation (13) reduces to

$$P^x_{k|k} = P^x_{k|k-1} - P^x_{k|k-1}C_k^T\tilde{R}_k^{-1}C_kP^x_{k|k-1} + [G_k - P^x_{k|k-1}C_k^TH_k^{-1}C_kG_{k-1}]P^\delta_{k|k}[]^T$$

and the unknown input covariance matrix (11) becomes

$$P^\delta_{k|k} = (G^T_{k-1}C_k^T\tilde{R}_k^{-1}C_kG_{k-1})^{-1}.$$

In addition, Equation (12) becomes

$$\hat{x}_{k|k} = \hat{x}_{k|k-1} + P^x_{k|k-1}C_k^T\tilde{R}_k^{-1}(y_k - C_k\hat{x}_{k|k-1}) + (G_k - P^x_{k|k-1}C_k^T\tilde{R}_k^{-1}C_kG_k)\hat{\delta}_{k-1}$$

and the unknown input estimation Equation (10) becomes

$$\hat{\delta}_{k-1} = P^\delta_{k|k}(C_kG_{k-1})^T\tilde{R}_k^{-1}(y_k - C_k\hat{x}_{k|k-1}).$$

These recursive formulas are identical to (a) the results in Kitanidis (1987) when only state filtering is of interest; and (b) the results in Gillins and De Moor (2007) for both unknown input and state estimations obtained using the approach of MVUE.

Finally, if only state estimation is concerned, the proposed method leads to the same results as those in Li (2013). To show this, we note that the state estimation error covariance matrix Equation (13) is the same as the one in Li (2013). In addition, inserting Equations (10) and (11) into Equation (12), Equation (12) can be rewritten in the following form:

$$\hat{x}_{k|k} = \hat{x}_{k|k-1} + K_k(y_k - C_k\hat{x}_{k|k-1}),$$

where the gain matrix K_k is defined as

$$K_k = P^x_{k|k-1}C_k^T\tilde{R}_k^{-1} + [F_{k-1} - P^x_{k|k-1}C_k^T\tilde{R}_k^{-1}C_kF_{k-1}](P^\delta_{k|k})^{-1}F^T_{k-1}C_k^T\tilde{R}_k^{-1}.$$

We can further show that (see the appendix for details)

$$M_{k-1}^{-1}\tilde{r}_{k-1} - K_kC_kM_{k-1}^{-1}\tilde{r}_{k-1} = P_{k|k}\bar{M}^T_{k-1}(\bar{M}_{k-1}P_{k|k-1}\bar{M}^T_{k-1})^{-1}\tilde{r}_{k-1}, \quad (15)$$

where the left-hand side of Equation (15) is the term associated with the prior information of the proposed filter, whereas the right-hand side of Equation (15) is the term associated with the prior information of the filter in Li (2013). This completes the proof. ∎

4. Simulation study

In this section, we use a simple numerical example to illustrate the developed filter. First, we will show that, when only state estimation is of interest, the proposed filter can obtain the same result as that of Li (2013). Next we further demonstrate that incorporating the partially available information on the inputs can effectively improve both state estimation and unknown input estimation in comparison with the one without using the input information (Gillins & De Moor, 2007).

The system for the simulation is chosen the same as that of Su, Li, and Chen (2015b) that has been widely used in many previous studies (see, e.g. Cheng et al., 2009). However, to better assess the performance of the proposed filter under uncertainties, we considered a system subject to larger random variation: the covariance matrices Q_k and R_k of the system and measurement noises were taken 10 times as those of Cheng et al. (2009). The initial values of system model is chosen as $x_0 = [3,\ 1,\ 2,\ 2,\ 1]^T$, the initial state and covariance matrix of filter are chosen as $\hat{x}_0 = 0_{5\times1}$ and $P^x_{0|0} = 0.2 \times I_5$.

We applied the recursive formulas in this paper to estimate the state and unknown input vectors at each time step. To evaluate the quality of the state estimate and unknown input estimate obtained using the developed filter, we calculated the trace of the error covariance matrix $P^x_{k|k}$ and the trace of the error covariance matrix $P^\delta_{k|k}$ at each time step, as displayed in Figure 2(a) and Figure 2(b) (real line), respectively. For comparison, we also considered the state estimation using the filter in Li (2013) (only state estimation is concerned) and Gillins and De Moor (2007) (assuming that the inputs were completely unknown). The traces of $P^x_{k|k}$ are superimposed in Figure 2(a) (dotted line for Li, 2013 and dashed line for Gillins & De Moor, 2007), and the trace of $P^\delta_{k|k}$ is superimposed in Figure 2(b) (dashed line for Gillins & De Moor, 2007).

It can be seen from Figure 2(a) that the trace of state estimation error covariance using the proposed filter is the same as that of Li (2013). Both the method in Li (2013) and the proposed method have a smaller trace of the covariance matrix than that of Gillins and De Moor (2007).

In addition, Figure 2(b) shows that the trace of the error covariance matrix of the unknown input estimate using the proposed filter is smaller in comparison with that of Gillins and De Moor (2007). This is because more information on the unknown inputs was used by the filter developed in this paper. This demonstrates that when the unknown inputs are of practical interest, the proposed method in this paper will have a better performance than Gillins and De Moor (2007) if there is additional information available on the unknown inputs for filtering.

We also compared the state estimates obtained using the three filters, that is, the filter in Li (2013) (Figure 3), the proposed filter (Figure 4) and the filter in Gillins and De Moor (2007) (Figure 5). The upper graphs of Figures 3–5

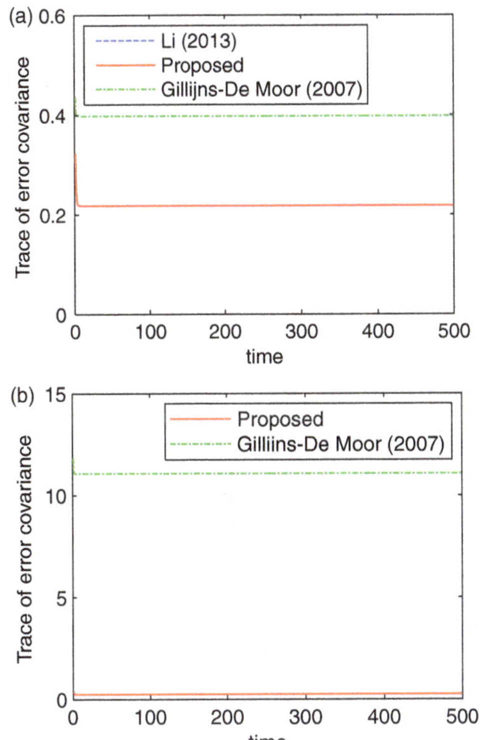

Figure 2. (a) Traces of the covariance matrix $P^x_{k|k}$ for three different filters; (b) traces of the covariance matrix $P^\delta_{k|k}$ for the proposed approach and the filter in Gillins and De Moor (2007).

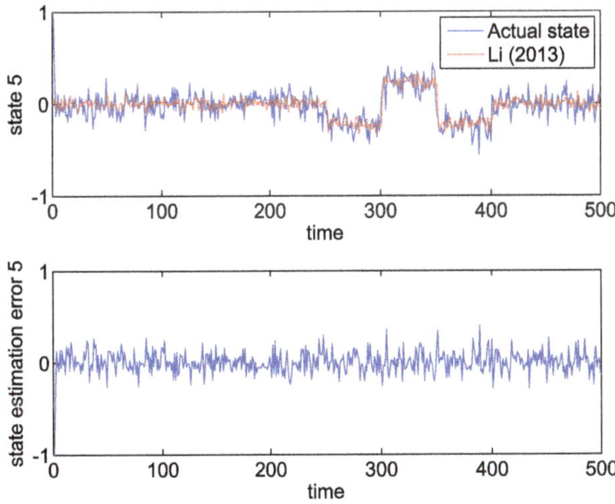

Figure 3. State estimation of the filter in Li (2013) and its estimation error.

display the simulated true values of the fifth state variable (real line) and the estimated state using the filters (dotted line), while the lower graphs plot the corresponding state estimation error for each filter.

It can be seen from Figures 3–5 that the three methods can provide a reasonably good estimate of the state vector. However, overall the state estimation errors using the

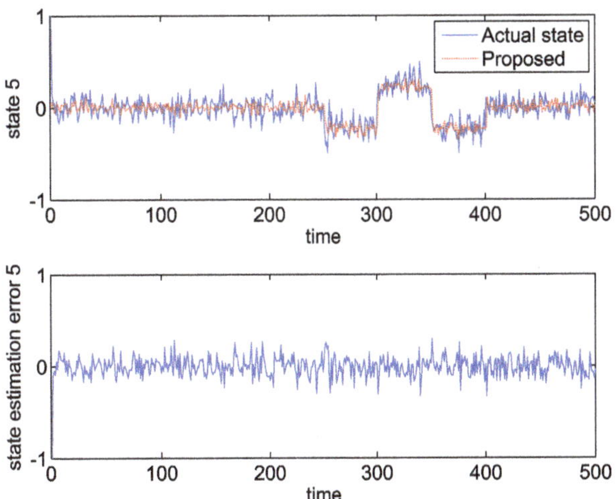

Figure 4. State estimation of the proposed filter and its estimation error.

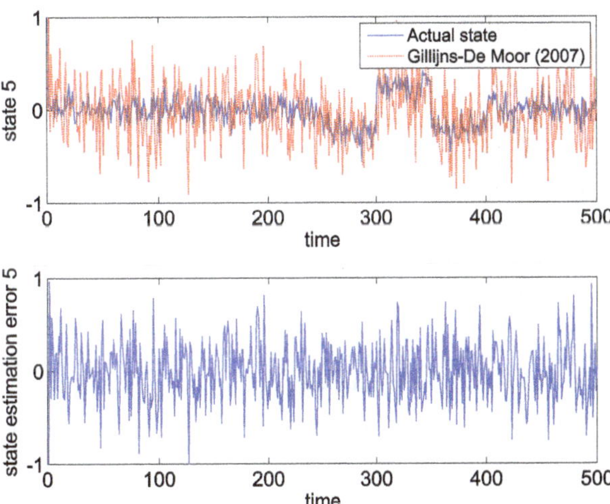

Figure 5. State estimation of the filter in Gillins and De Moor (2007) and its estimation error.

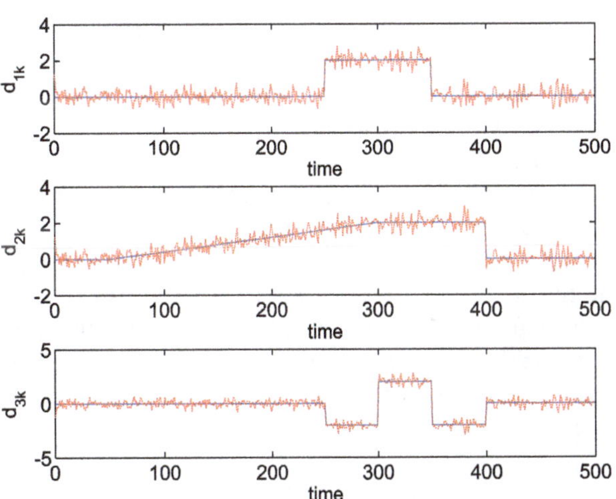

Figure 6. Unknown input estimation based on the proposed filter.

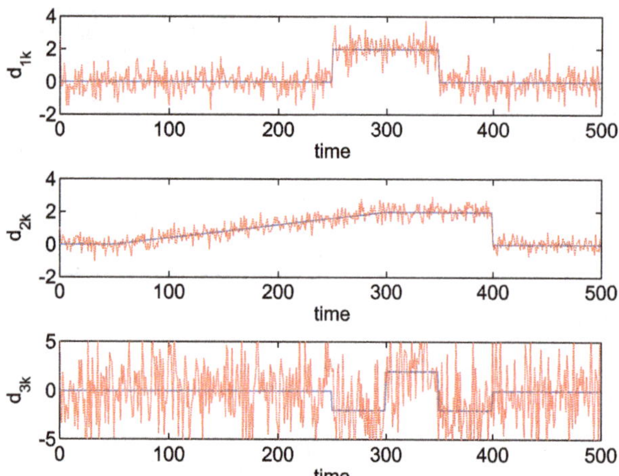

Figure 7. Unknown input estimation based on the filter in Gillins and De Moor (2007).

proposed filter and the filter in Li (2013) are smaller compared with that of Gillins and De Moor (2007) because the additional unknown input information was incorporated into the proposed filter and that of Li (2013).

Finally, we further compared our proposed method with the results in Gillins and De Moor (2007) for the purpose of unknown inputs estimation. The comparison results are shown in Figure 6 (the proposed method) and Figure 7 (the method in Gillins and De Moor (2007)), where real unknown inputs are depicted by real lines, and the unknown input estimations are depicted by the dotted lines.

We can see from Figures 6 and 7 that, by incorporating the information on the unknown inputs, the proposed method can obtain a much better performance for the unknown input estimation.

5. Conclusions

In this paper, the problem of SSIE has been investigated when partial information on the unknown inputs is available at an aggregate level. A decoupling approach is used to incorporate the unknown input information into the system dynamics. Then Bayesian inference is drawn to obtain the recursive state and input filter. The relationships of the proposed approach with the existing results are also discussed. Finally, the numerical example shows that, in comparison with the filter without using any input information, the proposed filter that makes use of the input information available at an aggregate level can substantially improve on the quality of both the state and input estimations. Future research can be done to extend the result to the case where there exists direct feedthrough of the partially observed inputs.

Disclosure statement

No potential conflict of interest was reported by the author(s).

Funding

This work was jointly funded by UK Engineering and Physical Sciences Research Council (EPSRC) and BAE System (EP/H501401/1).

References

Chen, W., Ballance, D., Gawthrop, P., & O'Reilly, J. (2000). A nonlinear disturbance observer for robotic manipulators. *IEEE Transaction on Industrial Electronics*, *47*(4), 932–938.

Cheng, Y., Ye, H., Wang, Y. Q., & Zhou, D. H. (2009). Unbiased minimum-variance state estimation for linear systems with unknown input. *Automatica*, *45*(2), 485–491.

Darouach, M., & Zasadzinski, M. (1997). Unbiased minimum variance estimation for systems with unknown exogenous inputs. *Automatica*, *33*(4), 717–719.

Darouach, M., Zasadzinski, M., & Boutayeb, M. (2003). Extension of minimum variance estimation for systems with unknown inputs. *Automatica*, *39*(6), 867–876.

Fang, H.-Z., & Callafon, R. A. D. (2012). On the asymptotic stability of minimum-variance unbiased input and state estimation. *Automatica*, *48*(12), 3183–3186.

Gao, Z. W., & Ding, S. X. (2007). State and disturbance estimator for time-delay systems with application to fault estimation and signal compensation. *IEEE Transaction on Signal Processing*, *55*(12), 5541–5551.

Gillijns, S., & De Moor, B. (2007). Unbiased minimum-variance input and state estimation for linear discrete-time systems. *Automatica*, *43*(1), 111–116.

Hsieh, C. S. (2000). Robust two-stage Kalman filters for systems with unknown inputs. *IEEE Transaction on Automatic Control*, *45*(2), 2374–2378.

Jiang, B., & Fahmida, N. C. (2005). Fault estimation and accommodation for linear MIMO discrete-time systems. *IEEE Transaction on Control System Technology*, *13*(3), 493–499.

Kitanidis, P. K. (1987). Unbiased minimum-variance linear state estimation. *Automatica*, *23*(6), 775–778.

Li, B. (2009). A non-Gaussian Kalman filter with application to the estimation of vehicular speed. *Technometrics*, *51*(2), 162–172.

Li, B. (2013). State estimation with partially observed inputs: A unified Kalman filtering approach. *Automatica*, *49*(3), 816–820.

Mann, G., & Hwang, I. (2013). State estimation and fault detection and identification for constrained stochastic linear hybrid systems. *IET Control Theory and Application*, *7*(1), 1–15.

Patton, P., Clark, T., & Frank, P. M. (1989). *Fault diagnosis in dynamic systems: Theory and applications*. Upper Saddle River, NJ: Prentice-Hall International Series in Systems and Control Engineering, Prentice-Hall.

Profeta, III., Joseph, A., William, G. Vogt, & Marin, H. Mickle (1990). Disturbance estimation and compensation in linear systems. *IEEE Transaction on Aerospace and Electronics Systems*, *26*(2), 225–231.

Simon, D. (2006). *Optimal state estimation: Kalman, H_∞, and non-linear approaches*. New York: Wiley.

Su, J., Chen, W., & Li, B. (2014). Disturbance observer based fault diagnosis. *Proceedings of the 33rd Chinese control conference*, Nanjing, China, pp. 3024–3029.

Su, J., Li, B., & W.-H. Chen (2015a). On existence, optimality and asymptotic stability of the Kalman filter with partially observed inputs. *Automatica*, *53*, 149–154.

Su, J., Li, B., & W.-H. Chen (2015b). Recursive filter with partial knowledge on inputs and outputs. *International Journal of Automation and Computing*, *12*(1), 35–42.

Yang, J., Li, S., Su, J., & Yu, X. (2013). Continuous nonsingular terminal sliding mode control for systems with mismatched disturbances. *Automatica*, *49*(7), 2287–2291.

Yang, J., Su, J., Li, S., & Yu, X. (2014). High-order mismatched disturbance compensation for motion control systems via a continuous dynamic sliding-mode approach. *IEEE Transactions on Industrial Informatics*, *10*(1), 604–614.

Appendix

A.1. Proof of Equation (14)

First, we can obtain the inverse of M_k as follows:

$$M_k^{-1} = [G_k,\ G_k^{\perp}]\begin{bmatrix}(I - F_{0k}F_{0k}^{\mathrm{T}})D_k^{\mathrm{T}}(D_kD_k^{\mathrm{T}})^{-1} & O & F_{0k}\\ O & I & O\end{bmatrix}.$$

Then, Equation (14) can be obtained as follows:

$$\begin{aligned}D_k\hat{d}_k &= D_k(G_k^{\mathrm{T}}G_k)^{-1}G_k^{\mathrm{T}}M_k^{-1}[r_k\ O\ O]^{\mathrm{T}}\\ &\quad + D_k(G_k^{\mathrm{T}}G_k)^{-1}G_k^{\mathrm{T}}M_k^{-1}[O\ O\ I]^{\mathrm{T}}\hat{\delta}_k\\ &= D_k(G_k^{\mathrm{T}}G_k)^{-1}G_k^{\mathrm{T}}G_k(I - F_{0k}F_{0k}^{\mathrm{T}})D_k^{\mathrm{T}}(D_kD_k^{\mathrm{T}})^{-1}r_k\\ &\quad + D_k(G_k^{\mathrm{T}}G_k)^{-1}G_k^{\mathrm{T}}G_kF_{0k}\hat{\delta}_k\\ &= r_k.\end{aligned}$$

A.2. Proof of Equation (15)

Define $M_{k-1}^P = P_{k|k}\bar{M}_{k-1}^{\mathrm{T}}(\bar{M}_{k-1}P_{k|k-1}\bar{M}_{k-1}^{\mathrm{T}})^{-1}$. Then, we have

$$\begin{aligned}&M_{k-1}^{-1}\tilde{r}_{k-1} - K_kC_kM_{k-1}^{-1}\tilde{r}_{k-1}\\ &= (I - K_kC_k)M_{k-1}^{-1}\tilde{r}_{k-1}\\ &= M_{k-1}^P\bar{M}_{k-1}M_{k-1}^{-1}\tilde{r}_{k-1}\\ &= M_{k-1}^P\bar{M}_{k-1}G_{k-1}D_{k-1}^{\mathrm{T}}(D_{k-1}D_{k-1}^{\mathrm{T}})^{-1}D_{k-1}d_{k-1}\\ &= M_{k-1}^P\begin{bmatrix}D_{k-1} & O\\ O & I\end{bmatrix}[G_{k-1},\ G_{k-1}^{\perp}]^{-1}\\ &\quad\times G_{k-1}D_{k-1}^{\mathrm{T}}(D_{k-1}D_{k-1}^{\mathrm{T}})^{-1}D_{k-1}d_{k-1}\\ &= M_{k-1}^P\begin{bmatrix}r_{k-1}\\ O\end{bmatrix} = M_{k-1}^P\bar{M}_{k-1}G_{k-1}d_{k-1}\\ &= P_{k|k}\bar{M}_{k-1}^{\mathrm{T}}(\bar{M}_{k-1}P_{k|k-1}\bar{M}_{k-1}^{\mathrm{T}})^{-1}\tilde{r}_{k-1},\end{aligned}$$

where in the above derivation, we have used the following identities:

$$M_{k-1}^{-1}\tilde{r}_{k-1} = G_{k-1}D_{k-1}^{\mathrm{T}}(D_{k-1}D_{k-1}^{\mathrm{T}})^{-1}D_{k-1}d_{k-1}, \tag{A1}$$

$$I - K_kC_k = M_{k-1}^P\bar{M}_{k-1}. \tag{A2}$$

Now we show Equation (A2):

$$\begin{aligned}&I - K_kC_k - M_{k-1}^P\bar{M}_{k-1}\\ &= I - K_kC_k - P_{k|k}\bar{M}_{k-1}^{\mathrm{T}}(\bar{M}_{k-1}P_{k|k-1}\bar{M}_{k-1}^{\mathrm{T}})^{-1}\bar{M}_{k-1}\\ &= I - P_{k|k}C_k^{\mathrm{T}}R_k^{-1}C_k - P_{k|k}[\bar{M}_{k-1}^{\mathrm{T}}(\bar{M}_{k-1}P_{k|k-1}\bar{M}_{k-1}^{\mathrm{T}})^{-1}\\ &\quad\times\bar{M}_{k-1} + C_k^{\mathrm{T}}R_k^{-1}C_k - C_k^{\mathrm{T}}R_k^{-1}C_k]\\ &= I - P_{k|k}C_k^{\mathrm{T}}R_k^{-1}C_k - [I - P_{k|k}C_k^{\mathrm{T}}R_k^{-1}C_k]\\ &= O,\end{aligned}$$

where $\bar{M}_k = \begin{bmatrix}D_k & O\\ O & I\end{bmatrix}[G_k,\ G_k^{\perp}]^{-1}$.

4

Neural network-based shoulder instability diagnosis modelling for robot-assisted rehabilitation systems

Esam H. Abdelhameed[a,b*], Noritaka Sato[b] and Yoshifumi Morita[b]

[a]*Faculty of Energy Engineering, Aswan University, Aswan, Egypt;* [b]*Department of Electrical and Computer Engineering, Nagoya Institute of Technology, Nagoya, Japan*

Many researchers are anticipating that robotic systems will contribute to compensating for the shortage of providing therapy for age-related injuries. Therefore, any prospective approach to robot-assisted therapy has to offer systems that can practically and objectively evaluate a patient's physical functions and apply evidence-based rehabilitation protocols. One of the most common disorders, among the general population particularly seniors, 40–60 years of age, is the frozen shoulder. Clinically, frozen shoulder can be diagnosed based on two shoulder functions: instability and cooperativeness of the shoulder joint. The purpose of the present study is to introduce a shoulder instability diagnosis model using artificial neural networks (ANN), which can be applicable using robotic systems. Training, validation, and testing of the neural network were achieved using the force exerted by a subject measured during predesigned clinical examinations. The proposed method has the ability to produce shoulder joint instability evaluations equivalent to those clinically obtained by therapists.

Keywords: robotic-assisted therapy; neural networks; rehabilitation; evaluation models; age-related injuries; shoulder joint instability

Introduction

With age, one is more vulnerable to contracting diseases and suffering from disabilities, as ageing negatively affecting the body basics, i.e., muscles, joints, and bones. According to the World Health Organization, senior citizens, at least 65 years of age, will increase in number by 88% in the coming years (Morales et al., 2010). Consequently, the incidence of age-related diagnoses, such as shoulder injuries, will increase. Rehabilitation services become a pressing necessity with an ageing population. In many cases and due to various causes such as medical service reforms, patients may be forced to leave the hospital before sufficiently recovering. This situation creates both a need and an opportunity to deploy technologies such as robotics and robotic therapy to assist recovery. Generally, rehabilitation programs are suggested based on the observation and evaluation of each patient's case. Consequently, introducing an efficient rehabilitation process requires quantitative and objective assessment of the physical function (Sato et al., 2013). Evaluation of physical functions through conventional methods depends on a therapist's clinical experience and skills (Kikuchi et al. 2007; Mochizumi, 2007). Therefore, there is a necessity to introduce evaluation methods and systems that have the capability to quantitatively evaluate physical functions.

In the last decade, interest in robotic therapy research has exponentially increased. Robot therapy systems have been developed worldwide for the rehabilitation of both upper and lower extremities. For example, the design and characterization of a robotic-assisted rehabilitation system for patients with upper limb disability was introduced by the third author of this study (Maeda et al., 2003). As the shoulder is one of the most complex joints in the human body, the correct diagnosis in shoulder injuries normally requires a thorough history and examination by an experienced professional. Robot-assisted rehabilitation attempted to offer therapeutic systems for the shoulder joint (Noritsugu, Tanaka, & Yamanaka, 1997; Ozawa et al., 2010; Yamaji, Yoshii, Wada, Tanaka, & Tsukamoto, 2001); however, most of the research focused on designing devices that support rehabilitation training, and do not address the quantitative evaluation of human joint functions. Few studies have been conducted on extrapolating quantitative methods for the evaluation of rehabilitative therapies. Sato et al. (2013) developed a quantitative evaluation of shoulder joint functions, which are the flexibility/stability and the cooperativeness. For the flexibility/stability shoulder joint functions, they focused on the ratios between the maximum isometric forces exerted by the patient's left and right arms, and reproduced the clinical test results using

*Corresponding author. Email: ehhameed@yahoo.com

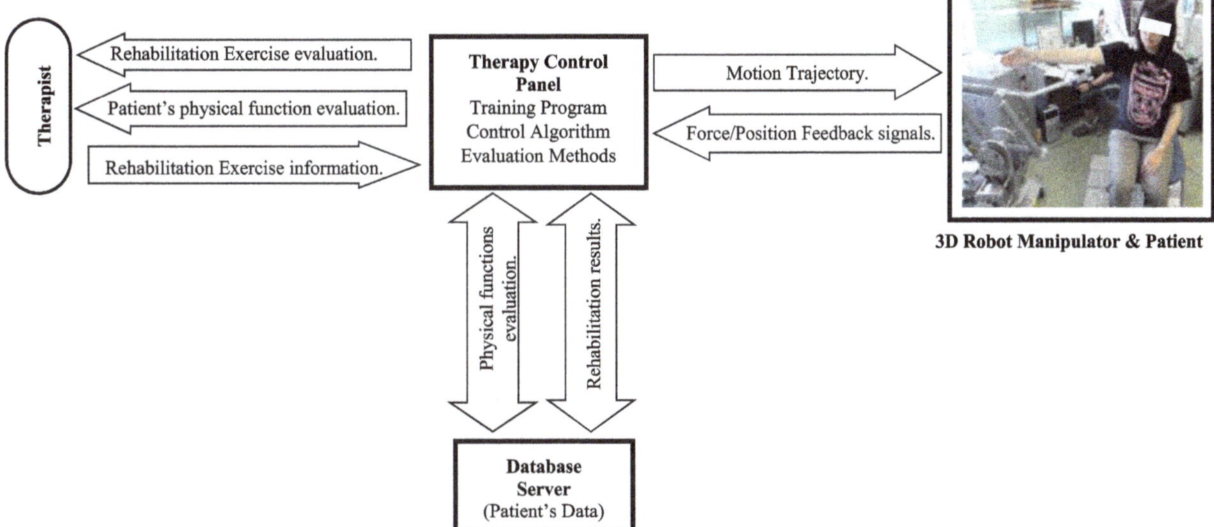

Figure 1. 3D robotic-assisted therapy system.

a regression equation. For the cooperativeness evaluation, they derived threshold values of the upper arm rotation angle and the exerted force. These are optimum values for determining the continuity of the exerted force. Kamada et al. (2014) proposed a screening test robot for the prevention and early detection of the functional decline of the shoulder joint.

This study aims to introduce a robotic-assisted therapy system, which provides a quantitative and objective assessment of the upper extremity physical function, shown in Figure 1, in order to improve upper extremity function and reduce pain in patients with shoulder problems. This robotic-assisted therapy system is a three-dimensional reaction force display robot (3D robot). The design and characterization of this robot was introduced by the third author at Nagoya Institute of Technology. This robot has potential as an evaluation tool for therapeutic effect (Maeda et al., 2003). Morita, Akagawa, Yamamoto, Ukai, and Matsui (2002) proposed a method to enable movement in subjects' upper arms by providing impedance control. Yasukita et al. (2012) introduced a method that allows the robot to replace the therapist to provide load resistance training. Abdelhameed, Kamada, Sato, and Morita (2015) proposed a control algorithm for robotic-assisted isotonic training through circular trajectory.

To the best of our knowledge, quantitative evaluation of shoulder joint functions cannot be performed by any existing device. In this study, the clinical diagnoses obtained by the therapist are defined as correct diagnoses. The proposed evaluation method can be applied to the 3D robot and reproduces diagnoses equivalent to those obtained by therapists. Kamada et al. (2014), in a screening test of flexibility/stability, used the 3D robot to measure the maximum isometric force exerted on the brace of the 3D robot by a subject. In the screening test for functional decline in cooperativeness, the 3D robot was used to measure the maximum isokinetic force exerted by the subject as well as the rotational angle of the upper limb. Sato et al. (2013) proposed a linear regression algorithm, multiple regression analysis (MRA), using Statistical Package for Social Sciences to diagnose shoulder instability. When the MRA algorithm was applied for diagnosing new subjects, the predictive accuracy was 70%. In this study, in order to improve the predictive accuracy of the shoulder instability diagnosis, obtained by Sato et al. (2013), the authors proposed a predictive algorithm with specific attributes. These attributes can be summarized as follows. Firstly, a nonlinear regression algorithm (a neural network-based algorithm) is proposed instead of the MPA algorithm to extract a nonlinear model from the data, as the shoulder mechanism is a complicated model and a nonlinear behaviour is expected. Secondly, the direct measured data of both right and left arms are used as a database for the diagnosis model, whereas the resulting ratio of left to right arm data was used by Sato et al. (2013). Finally, to improve the generalization ability of the proposed model, the database has been divided into three data sets to train, validate, and test the proposed model. This is in contrast to the previous reference wherein the entire database was used to estimate the parameters of their diagnosis model. Consequently, the final objective of this article is to propose a neural network-based evaluation method to reproduce the diagnosis of shoulder joint instability that is obtained clinically by a therapist.

Clinical shoulder joint instability diagnosis

In this section, a representation of shoulder joint anatomy has been introduced, followed by an explanation of the pathogenic mechanism of scapulohumeral periarthritis.

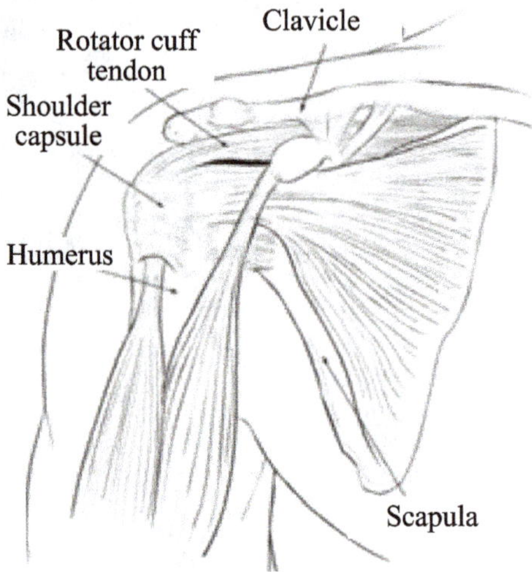

Figure 2. Shoulder joint anatomy.

Finally, the clinical examination method conducted by a therapist for shoulder instability, a scapulohumeral periarthritis-related shoulder joint physical function, is given.

Anatomy of shoulder joint

The shoulder joint plays a major role in most of daily life activities, as it is the most mobile joint in the body. The flexibility of the shoulder joint allows the shoulder's range of motion. Figure 2, which is reproduced and modified by Price and Beaty (2003), illustrates that the shoulder (a ball-and-socket joint) is made up of three bones: the humerus, the scapula, and the clavicle, which are the upper arm bone, the shoulder blade, and the collarbone, respectively. The head of the humerus fits into a shallow socket in the shoulder blade. The joint is surrounded by shoulder capsule which is a strong connective tissue. Synovial fluid lubricates the shoulder capsule and the joint to facilitate shoulder movement. The muscles and tendons of the rotator cuff provide the stability of the shoulder joint and allow the shoulder to rotate. The muscles in the rotator cuff include four muscles: teres minor, infraspinatus, supraspinatus, and subscapularis. Each muscle of the rotator cuff inserts at the scapula, and has a tendon that attaches to the humerus. Together, the tendons and other tissues form a cuff around the humerus. The tendons of these muscles surround and support the humerus while the contraction of the muscles rotates, adducts, or abducts the humerus. Due to its flexibility, the shoulder is not particularly stable which may easily lead to injury.

Frozen shoulder injury

A frozen shoulder, which is also called adhesive capsulitis, is a shoulder joint that has lost a substantial amount of its range of motion due to scarring around the joint. The frozen shoulder range of motion is limited in both active and in passive motion. The active range of motion is the moving of a shoulder joint through its range of motion when the patient attempts motion and exerts forces, while the passive range of motion is conducted without the patient exerting any force. The latter is usually conducted by a therapist who attempts to move the joint while the patient relaxes.

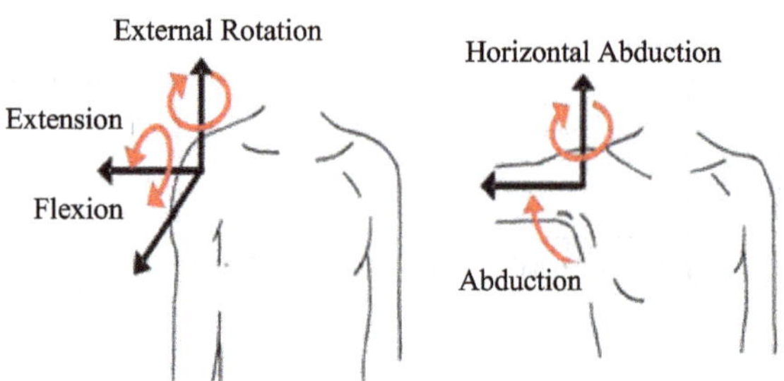

Figure 3. Basic motions of shoulder joint.

The frozen shoulder is a common condition, especially senior citizens. It has an incidence of 3–5% in the general population (Manske & Prohaska, 2008). Manual labourers such as construction workers and house painters are among those with a higher risk. Frozen shoulder occurs more prevalently in women, as mothers are subject to years of stress on their shoulders. In addition, shoulder muscles are affected by ageing, that is, a reduction in muscle force and an abnormality in muscle force balance. This can result in the head of the humerus becoming dislocated and the subacromial bursa and the rotator cuff becoming sandwiched between the acromion and the greater tubercle, resulting in a chance of bones collision. The authors in this study focus on one of the important shoulder functions, that is, the flexibility and stability function. This function is an index indicating whether the head of the humerus is stabilized at its proper position.

The shoulder joint has the following normal ranges of movement (basic motions): external rotation, flexion, extension, horizontal abduction, and abduction, as illustrated in Figure 3. In this research, the motion of external rotation adopted at clinical sites is a horizontal rotation of the upper arm from front to side, with the elbow bent at 90°, in contact with the trunk of the body. As noted, one's dominant extremity is often more susceptible to injury than the non-dominant side. Additionally, the complex series of articulations of the shoulder allows for a wide range of motion. Consequently, to determine the patient's normal range of motion, the affected extremity should be compared with the unaffected one. In this study, ranges of the shoulder joint basic motions have been specified based on a therapist's opinion, as shown in Table 1.

There are various clinical examinations of shoulder stability. The test developed by Sato et al. (2013) has been used to get the clinical evaluations of the shoulder joint instability for this study. The physical examination included inspection and palpation, as well as assessing

Table 1. Specifications of basic motions.

No.	Motion	Range of motion [°]
1	External rotation	0–45
2	Flexion	0–60
3	Extension	5–45
4	Horizontal abduction	0–75
5	Abduction	0–65

the range of motion for possible impingement syndrome and shoulder joint instability. In the instability test, the patient is sitting down and the therapist checks, by touching the patient's shoulder joint, if the head of the humerus is in a proper position with respect to the glenoid cavity of the scapula. More than 90% of patients improve with relatively simple treatments to control pain and improve range of motion (American Academy of Orthopaedic Surgeons, n.d.). Restoring shoulder joint mobility and function can be achieved by continued exercise of the shoulder. Robot-assisted therapy can play a part by providing exercises targeting the improvement of shoulder joint range of motion in addition to evaluating its physical functions.

Shoulder joint instability diagnosis method using neural network

In this research, artificial neural network (ANN) has been used to assist the shoulder instability diagnosis by means of iterative training of data obtained from designed clinical examinations. ANNs are a learning system based on a computational technique that can simulate the neurological processing ability of the human brain, and they can be applied to handle nonlinear problems. They were successfully applied to solve a variety of classification problems, including scientific and medical applications (Kamruzzaman, Hasan, Siddiquee, & Mazumder, 2004; Kamruzzaman & Islam, 2006). Landi, Piaggi, Laurino, and Menicucci (2010) stated that, even though the linear regression algorithms are common for statistically analyzing the data, their capability of extracting data only from linear models limits their applicability to real problems. In order to improve their linear model prediction using the same selected variables, they used ANNs to benefit from their nonlinear modelling capability. Whereas Srivastava and Tripathi (2012) noted that ANNs are being widely used for nonlinear regression and classification applications because of their advantages in data analysis and prediction.

Generally, during the ANN modelling process, the following proceedings are common: collecting ANN training data (database of the studied problem), creating the network structure, training the network, and finally simulating the network response to new inputs.

Database of shoulder joint stability function

The database of the shoulder instability diagnosis using ANN can be classified into two groups of data. The first group is input data for training ANN, which is collected from the clinical examinations conducted by the therapist, as shown in Figure 4(a)–(c), that is, the patient's maximum exerted isometric forces during the basic motions of both right and left shoulder joints. The second group is the ANN targets, which are produced by a therapist through a diagnostic test of the shoulder instability, as shown in Figure 4(d), that is, the clinical test identifies the infected shoulder as a positive result, and the non-infected one as a negative result.

In this study, the maximum exerted isometric force during external rotation, abduction, and horizontal abduction motions is assumed, and used as the explanatory variables. These three motions were adopted as medical findings, which indicate that functions of the inner muscle of the shoulder joint are closely related to scapulohumeral periarthritis and the inner muscle of the shoulder joint contributes to these motions. The clinical examination of external rotation, abduction, and horizontal abduction motions is shown in Figure 4(a)–(c), respectively. During this examination, the subject carries out the basic motions with maximum effort against the therapist's resistance. The therapist holds the patient's shoulder with one of his hands

(a)

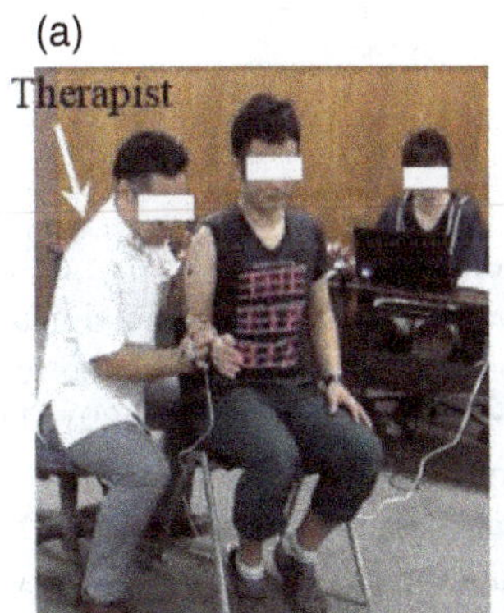

(b)

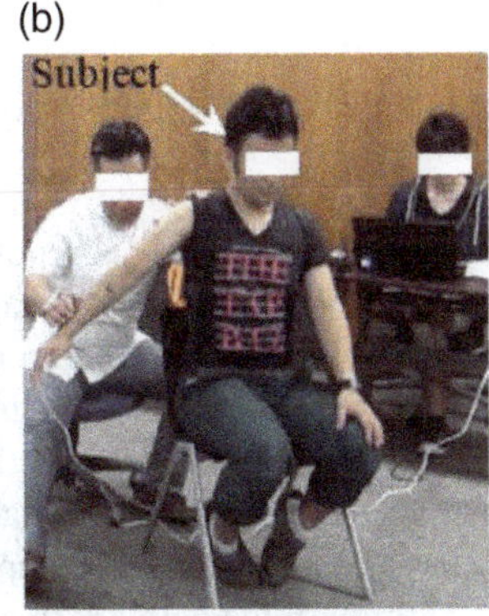

(c)

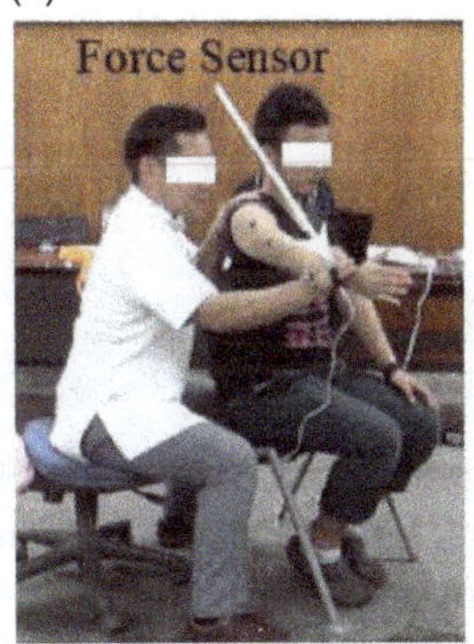

(d)

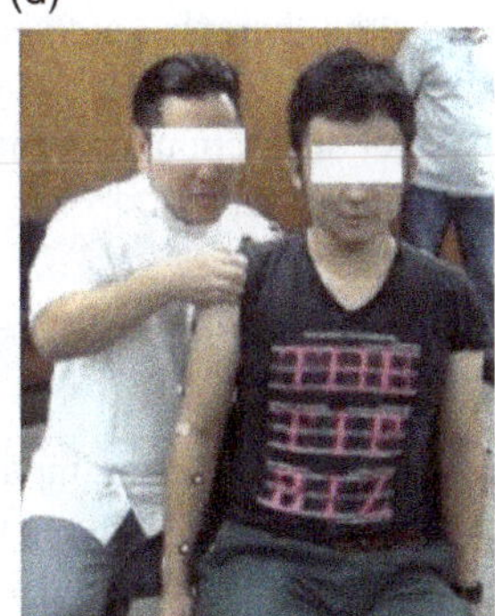

Figure 4. Clinical examination for ANN database collection.

Table 2. Binary equivalents for shoulder instability test and ANN outputs representations.

	Clinical instability examination results		Binary equivalents of the results		Equivalent binary number	ANN output representations
	Left shoulder	Right shoulder	Left shoulder	Right shoulder		
1	Negative	Negative	0	0	00	0
2	Negative	Positive	0	1	01	1
3	Positive	Negative	1	0	10	2
4	Positive	Positive	1	1	11	3

to keep the shoulder from moving. With the therapist's other hand, a resistance is applied to the subject's upper arm in order to keep it from moving at the initial position of basic motion. A force sensor is sandwiched between the therapist's hand and the subject's arm to measure the forces exerted by the subject. The clinical instability examination results by the therapist were described by 1 for the shoulder, which is evaluated as positive, and 0 for that evaluated as negative. Because shoulder instability mainly affects the dominant extremity of a person, the following strategy has been proposed to diagnose it: the maximum exerted isometric forces during the selected three basic motions of both right and left shoulders (six variables per subject) were collected to be used as the input data of the ANN model. The diagnostic test results of both the left and right shoulders are expressed as binary numbers, that is, the equivalent digits of the right and left shoulders are considered as the units and twos digits of the resultant binary number, respectively. The binary equivalents for these descriptions and their representations as ANN outputs (their decimal equivalents) are illustrated in Table 2.

Finally, the previously collected data are applied to train and test a neural network model to produce results equivalent to those produced by the therapist. A total of 16 data sets (from 16 subjects) were used in this research. The collected data sets are divided into 3 groups: 60% of the collected data were used to train the ANN model, 20% were used to cross-validate the relationships established during training process, and the remaining 20% were used to test the ANN model to evaluate its prediction accuracy through final analysis. These three sets of data, the measured isometric forces, and the instability examination results by the therapist, are shown in Figures 5–9. In the figures, 'Abd.', 'Ext. Rot.', and 'Horiz. Abd.' are abbreviations for abduction, external rotation, and horizontal abduction, respectively.

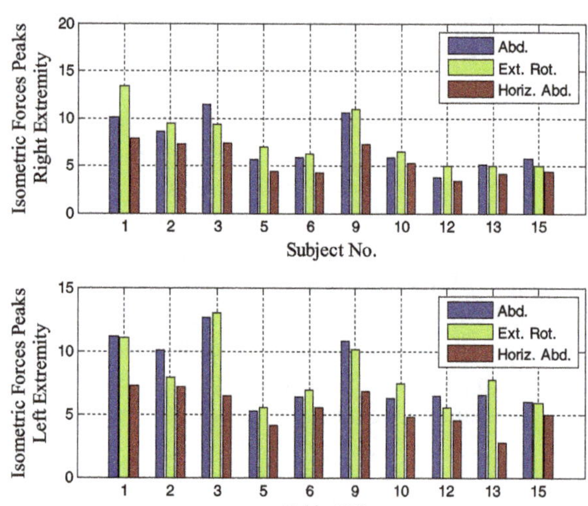

Figure 5. Input data of training set.

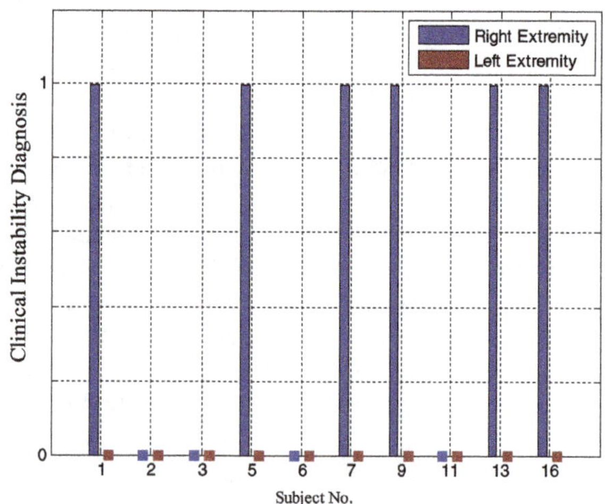

Figure 6. Targeted data of training set.

Neural network structure

For mapping the selecting input and output data, an ANN consisting of a series of layers was selected. One common ANN algorithm is a feed-forward back-propagation neural network (FFBPNN). These networks have the ability to map their inputs and outputs through a training procedure. The FFBPNN is composed of an input layer that delivers ANN inputs, a hidden layer that contains artificial neurons to adopt inputs, and an output layer that offers the results of the mapping procedure (Hecht-Nielsen, 1989). There are no feedback connections on the FFBPNN, but network errors are back-propagated during training to modify its parameters, that is, computations are performed backwards through the network to compute the gradient. The selected FFBPNN structure for the shoulder instability diagnosis

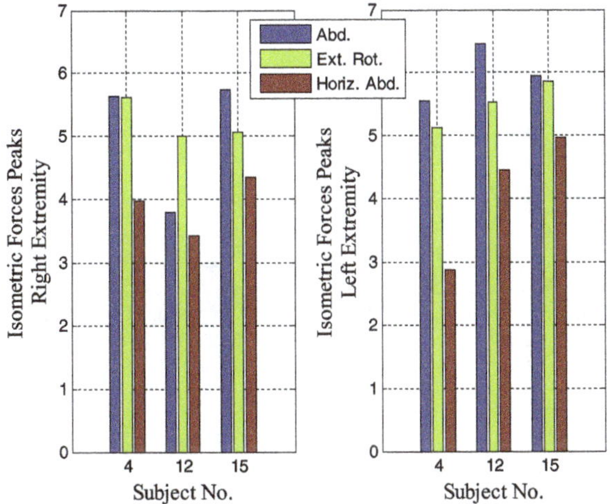

Figure 7. Input data of validation set.

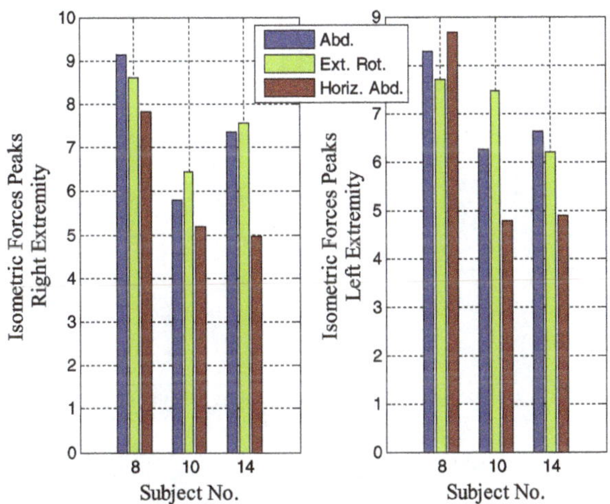

Figure 8. Input data of test set.

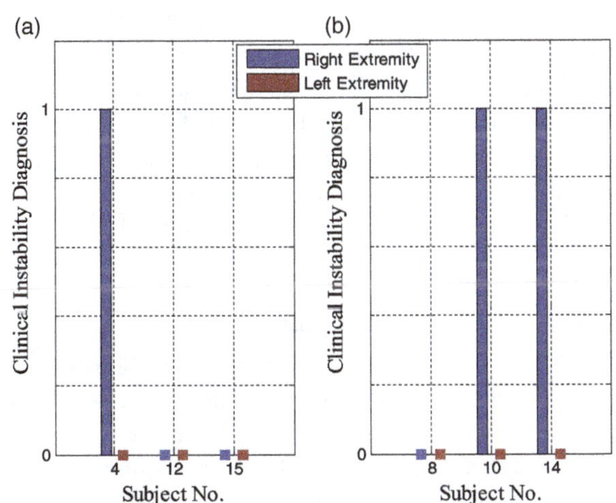

Figure 9. Targeted data of validation and test sets: (a) validation set and (b) test set.

was as follows: an input layer, two hidden layers (chosen by trial and error), and an output layer. The input layer consists of six neurons, which is equal to the number of input data points. As the number of collected data points is limited, and with a view to avoid network over-fitting by minimizing the number of the network's parameters, the number of neurons in the first hidden layer was chosen to be one. The second hidden layer consists of three neurons (chosen by trial and error), and according to the number of output data points, the number of the output layer neurons has been chosen to be only one, as shown in Figure 10. The total number of network's parameters is 17, which can be described as follows: the weights of the input layer, $w_{i1} \sim w_{i6}$; the weights of the first hidden layer, $w_{h11} \sim w_{h13}$; the weights of the second hidden layer, $w_{h21} \sim w_{h23}$; the bias of the first hidden layer, b_1; a three element vector representing the biases of the second hidden layer, b_2; and the bias of the output layer, b_o.

Typically, in order to scale the output of the neural network into proper ranges, the ANN pass the output of their layers through activation or transfer functions. Because the proposed shoulder instability diagnosis model is a nonlinear model, the activation functions of the ANN have to be nonlinear to express its nonlinearity. Hyperbolic tangent and sigmoid functions are common nonlinear activation functions. In order to allow the output layer to receive only positive values, the activation functions of the second hidden layer were chosen as sigmoid functions, whereas the hyperbolic tangent functions were chosen by trial and error for the first hidden layer. Finally, the output layer was selected as a linear transfer function. These activation functions are expressed in Equations (1)–(3), where u is the output of a neuron in the previous layer, which is calculated as the sum of the weighted inputs and the bias of that neuron. The general structure of the proposed ANN is illustrated in Figure 10.

$$\text{Hyperbolic tangent function:} \; f(u) = \frac{e^{2u}-1}{e^{2u}+1}, \qquad (1)$$

$$\text{Sigmoid function:} \; f(u) = \frac{1}{1+e^{-u}}, \qquad (2)$$

$$\text{Linear function:} \; f(u) = u. \qquad (3)$$

Neural network training

The FFBPNN training process involves adjusting network parameter values, weights, and biases, to improve network performance. This adjustment occurs through a performance function. The mean square error (MSE) function, which is the average squared error between the network outputs and the ANN target, is selected in this study. The MSE is defined by the following equation:

$$\text{MSE} = \frac{1}{N_t} \sum_{i=1}^{N_t} (e_i)^2, \qquad (4)$$

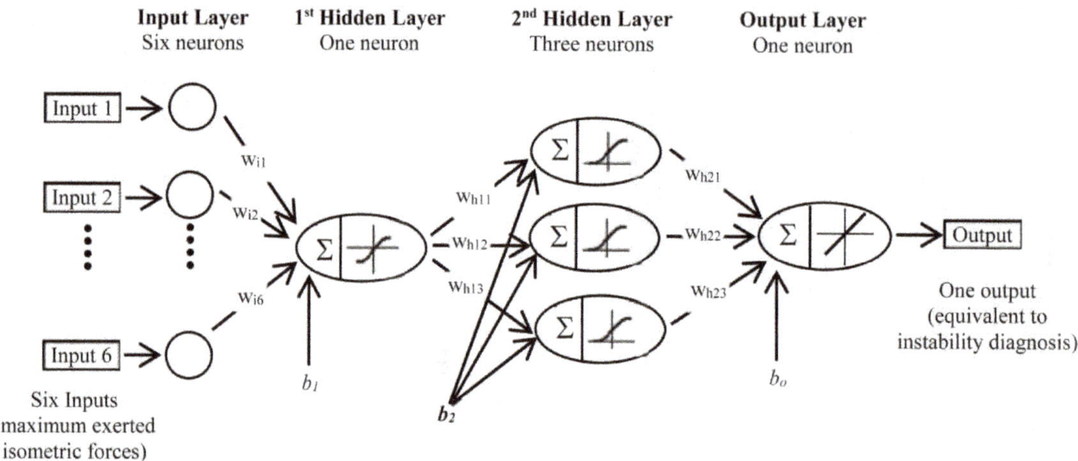

Figure 10. ANN structure.

where e_i is the error between the ith network output and its related target, and N_t is the number of training data points. Arbitrarily, the optimization techniques attempted to minimize this performance function. The back-propagation learning algorithm is based on the minimization of the MSE of the training data, which is known as least mean squares method. The basic idea is to adjust the network parameters in order to minimize the MSE of the input data through an optimization function. The Levenberg–Marquardt numerical optimization algorithm has been used to optimize the performance function during FFBPNN training. This optimization method uses the Jacobian of the network errors with respect to the weights. During the training process, the back-propagation algorithm is used to search for network parameter values that generate neural network outputs that closely correlate to the targeted data. Training with back-propagation is an iterative process.

In order to improve the ANN generalization, a technique called early stopping is used. In this technique, the error in the validation set is calculated during the training process to give an indication for training improvement; that is, if the validation error decreases during the initial phase of training, as the error of the training set does, this indicates normal training. However, when the error on the validation set typically begins to rise, this indicates that the network is beginning to overfit the data (MathWorks Company, 1994–2015). Moreover, a stopping criterion has been evaluated by the optimization algorithm as it calculates the reciprocal of the MSE from the neural network during training. The training process stops occurring for one of the following reasons: performance goal is reached (i.e. the MSE term drops below 10^{-6}), maximum number of epochs to train is reached (i.e. after 1000 iterations), the validation error for 10 iterations is increased (in this case, the parameters at minimum validation error are returned).

However, in order to test the generalization ability of the ANN model, the training process is followed by a testing process, which does not affect the ANN training. The

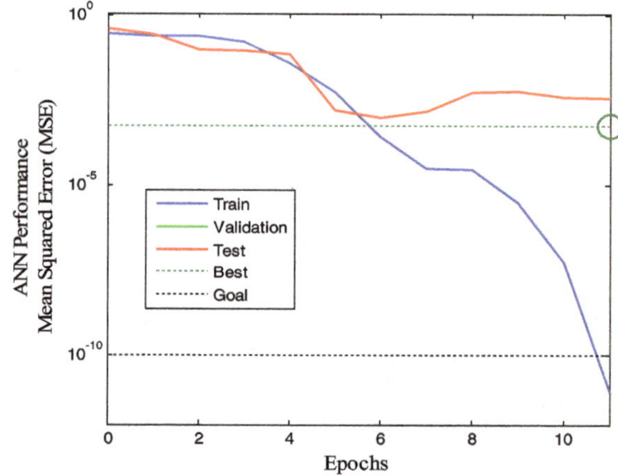

Figure 11. ANN performance.

test set error is calculated during the training process to indicate the quality of the division of the data set; that is, if the error in the test set reaches a minimum at a significantly different iteration number than the validation set error, this might indicate a poor division of the data set (MathWorks Company, 1994–2015). Training, validation, and test errors are indicated in Figure 11 as blue, green, and red lines, respectively. The figure indicates good training of the ANN model. The best validation performance is 5.5821×10^{-4} which was achieved at epoch 11.

Evaluation of the trained neural network

A FFBPNN is implemented for shoulder instability diagnosis. Once ANN training is complete, the weights and biases are specified and the trained ANN model can be used to generate outputs for new inputs. As an additional quality check of the generalization ability of the ANN model, the prediction accuracy is calculated. The ability of the trained ANN for shoulder instability diagnosis is estimated based on the absolute prediction error (APE) of all data sets,

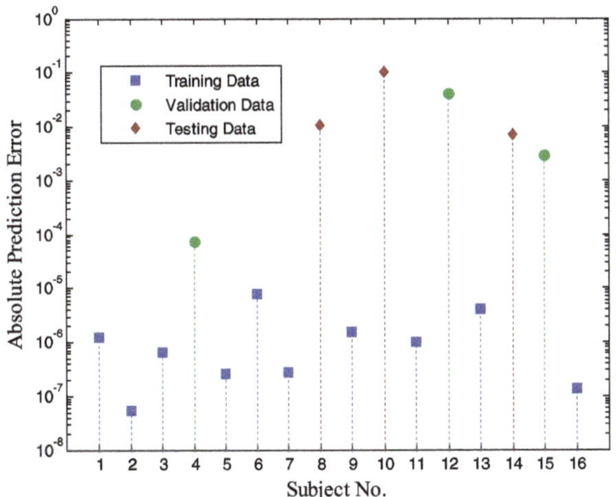

Figure 12. APE during ANN training.

which is defined by the following equation:

$$\mathrm{APE}_i = |y_i - \hat{y}_i|, \tag{5}$$

where APE_i is the APE for the ith network output $\hat{y}_i$, and its related target y_i. Figure 12 shows the APE for the ANN model using the training, validation, and test data sets. The maximum prediction errors are 7.657×10^{-6}, 4.2×10^{-2}, and 10.2×10^{-2} for the training, validation, and test data sets, respectively, which are within the acceptable limit. Furthermore, the mean absolute percentage error (MAPE) for the training, validation, and test data sets (defined by Equation (6)) are 1.677×10^{-6}, 1.4603×10^{-2}, and 3.975×10^{-2}, respectively, where N_s is the number of data points in each set. The MAPE is considered to be a 'robust' measure of predictive accuracy, as it is based on the absolute value of the error. In addition, it is evident that the trained ANN model has a good generalization capability; that is, ANN has good performance on unseen data.

$$\mathrm{MAPE} = \frac{1}{N_s} \sum_{i=1}^{N_s} \mathrm{APE}_i. \tag{6}$$

The high performance of the proposed ANN model is confirmed by a high correlation coefficient R between the clinical shoulder instability diagnosis and those obtained from the ANN model, as indicated by the R value in the regression plots shown in Figure 13. These plots have been created for the training, validation, and test subsets as shown in Figure 13(a)–(c), respectively. In the regression figure, dashed lines in each plot represent the ideal result, whereas the solid lines represent the best fit linear regression line between ANN outputs and its targets. As shown in Figure 13, the training, validation, and test results indicate high correlation, $R > 0.99$.

The proposed ANN model was applied in shoulder instability diagnosis. The targeted data and ANN outputs from the training data set are illustrated in Figure 14, while those of the validation and test data sets are shown in Figure 15(a) and 15(b), respectively. The final diagnosis of shoulder instability can be obtained from the ANN output through two steps: the first step is the approximation of the output to the nearest integer, the second step is converting the approximated value to its binary equivalent. The digit in the units place is related to the right extremity diagnosis, while the digit in the twos place is related to the left side. If the digit value is 1, the shoulder is identified as a positive case, and if it is 0, the shoulder is identified as a negative case.

To conclude, the predictive accuracy for patients' diagnosis is 100%, in contrast to the 70% achieved using the MRA algorithm. As mentioned previously, the hidden layer number, number of neurons in the second hidden layer, and the first hidden layer activation function have been chosen by trial and error. The best results were achieved from the nominated ANN structure, whereas other trials resulted in poor models. When using a sigmoid activation function for all the hidden layers, the predictive accuracy of the test data set was 33.3%, indicating poor generalization of the ANN model. In addition, when using more than three neurons for the second hidden layer, the total number of network parameters was increased dramatically compared to the available

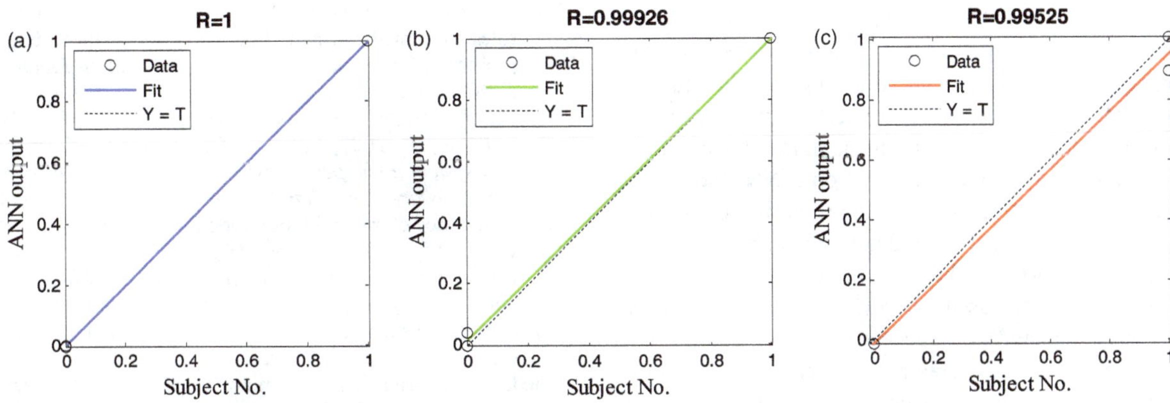

Figure 13. Regression plot of ANN model: (a) training set, (b) validation set, and (c) test set.

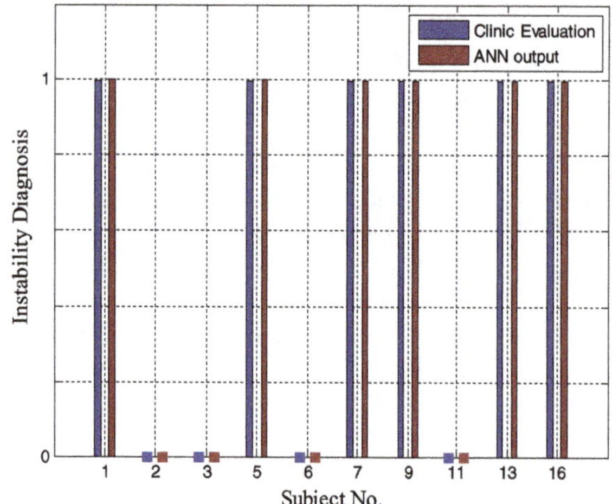

Figure 14. Target data and ANN outputs (training data set).

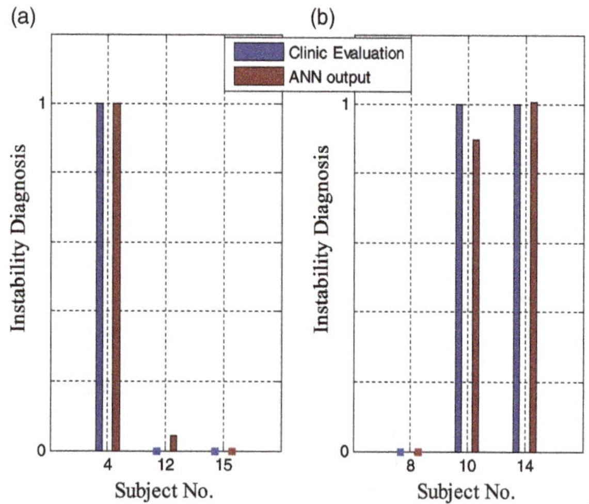

Figure 15. Target data and ANN outputs: (a) validation set and (b) test set.

data base number. Consequently, network over-fitting occurred.

Conclusions

In order to evaluate the shoulder joint instability function, this research has proposed an evaluation method that is applied to a 3D robot in the third author's laboratory. This evaluation method reproduces test results equivalent to those clinically produced by a therapist. For a shoulder joint instability diagnosis based on ANN, the maximum exerted isometric forces of the three basic motions (external rotation, abduction, and horizontal abduction) are used as explanatory variables for the network. The clinical evaluations were used as network targets. An ANN model was simulated and succeeded in generating outputs equivalent to those obtained clinically. As a future work, in order to verify the generality of the proposed ANN diagnosis algorithm, it will be applied for a greater number of subjects. In addition, an evaluation of shoulder cooperativeness based on an artificial technique will be attempted. Consequently, after verifying the reliability of the evaluation techniques of the shoulder stability and its cooperativeness, the proposed methods can be applied to our robot.

Acknowledgement

The author would like to express their sincere appreciation for Mr. Keita Kamada for his assistance in the preparation of this Research database. Mr. Kamada's master graduation from the department of Computer Science and Engineering, Nagoya Institute of Technology, Japan, was on 2015. He is a member of the Institute of Electrical Engineering of Japan (IEEJ). The authors would like to extend their most sincere appreciation to Mr. Kenji Komori for his constructive cooperation during the preparation of the database of this study. Mr. Komori works as the head of the rehabilitation department in Hokuto Hospital, Okazaki, Japan. He is a member of the Japanese Association of Occupational Therapists.

Disclosure statement

No potential conflict of interest was reported by the authors.

References

Abdelhameed, E., Kamada, K., Sato, N., & Morita, Y. (2015, March 6–8). *Post-stroke robotic-assisted therapy: Time-variant damping coefficient based control algorithm for isotonic exercise through circular motion*. IEEE international conference on mechatronics, Nagoya, Japan, pp. 433–437.

American Academy of Orthopaedic Surgeons. (n.d.). Frozen shoulder. Retrieved February, 2015, from orthoinfo.aaos.org/topic.cfm?topic = a00071

Hecht-Nielsen, R. (1989, June 18–22). *Theory of the backpropagation neural network*. International joint conference on neural networks, Washington, DC, Vol. 1, pp. 593–605.

Kamada, K., Sato, N., Morita, Y., Ukai, H., Komori, K., & Taguchi, S. (2014, October 22–25). *Screening test robot for functional decline of shoulder joint*. 14th international conference on control, automation and systems, Korea, pp. 594–598.

Kamruzzaman, S., Hasan, A., Siddiquee, A., & Mazumder, M. (2004, December 28–30). *Medical diagnosis using neural network*. On the 3rd international conference on electrical & computer engineering, BUET, Dhaka, pp. 537–540.

Kamruzzaman, S., & Islam, M. (2006). An algorithm to extract rules from artificial neural networks for medical diagnosis problems. *International Journal of Information Technology, 12*(8), 41–59.

Kikuchi, T., Xinghao, H., Fukushima, K., Oda, K., Furusho, J., & Inoue, A. (2007, June 13–15). *Quasi-3-DOF rehabilitation system for upper limbs: Its force-feedback mechanism and software for rehabilitation*. IEEE 10th international conference on rehabilitation robotics, Netherlands.

Landi, A., Piaggi, P., Laurino, M., & Menicucci, D. (2010, November 29–December 1). *Artificial neural networks for nonlinear regression and classification*. 10th international conference on intelligent systems design and applications, Cairo, Egypt, pp. 115–120.

Maeda, H., Morita, Y., Yamamoto, E., Kakami, H., Ukai, H., & Matsui, N. (2003, July 16–20). *Development of rehabilitation support system for reaching exercise of upper limb.* IEEE international symposium on computational intelligence in robotics and automation, pp. 134–139.

Manske, R., & Prohaska, D. (2008). Diagnosis and management of adhesive capsulitis. *Journal of Current Reviews in Musculoskeletal Medicine*, *1*(3–4), 180–189.

MathWorks Company. (1994–2015). Improve neural network generalization and avoid overfitting. Retrieved February, 2015, from www.mathworks.com

Mochizumi, H. (2007). Impairments and their assessment in stroke patients. *Rigakuryoho Kagaku*, *22*(1), 33–38 (in Japanese).

Morales, R., Badesa, F., Domenech, L., Garcia-Aracil, N., Sabater, J., Mench'on, M., & Fernandez, E. (2010, September 26–29). *Design and control of a rehabilitation robot driven by pneumatic swivel modules.* 3rd IEEE RAS and EMBS international conference on biomedical robotics and bio-mechatronics, Tokyo, Japan, pp. 566–571.

Morita, Y., Akagawa, K., Yamamoto, E., Ukai, H., & Matsui, N. (2002, July 3–5). *Basic study on rehabilitation support system for upper limb motor function.* Proceedings of 7th international workshop on advanced motion control, Maribor, Slovenia, pp. 127–132.

Noritsugu, T., Tanaka, T., & Yamanaka, T. (1997). Application of rubber artificial muscle manipulator as a rehabilitation robot. *IEEE/ASME Transactions on Mechatronics*, *2*(4), 259–267.

Ozawa, T., Furusho, J., Kikuchi, T., Fukushima, K., Tanida, S., & Fujikawa, T. (2010). Development of rehabilitation system for upper limbs; PLEMO-P3 system for hemiplegic subject: Motor function test for assessment and training, and research for development of practical type (mechanical systems). *Transaction of the Japan Society of Mechanical Engineers, C*, *76*(762), 323–330 (in Japanese).

Price, S., & Beaty, J. (2003). *The 2003 body almanac: Your personal guide to bone and joint health at any age.* Rosemont: (c) American Academy of Orthopaedic Surgeons.

Sato, N., Kamada, K., Hiramatsu, Y., Yamazaki, K., Morita, Y., Ukai, H., ... Taguchi, S. (2013). Quantitative evaluation of shoulder joint function to reproduce results of clinical tests by therapist. *Journal of Robotics and Mechatronics*, *25*(6), 983–991.

Srivastava, Sh., & Tripathi, K. (2012). Artificial neural network and non-linear regression: A comparative study. *International Journal of Scientific and Research Publications*, *2*(12), 740–744.

Yamaji, Y., Yoshii, N., Wada, T., Tanaka, S., & Tsukamoto, K. (2001). Realization of rehabilitation by virtual tennis: Effects of tennis on rehabilitation. *Transaction of the Japan Society of Mechanical Engineers*, *6*(1), 19–26 (in Japanese).

Yasukita, M., Iida, Y., Yamazaki, K., Sato, N., Morita, Y., Ukai, H., ... Tanemura, R. (2012, October 17–21). *Evaluation of simplified repeated resistance training system for severe hemiplegic stroke patient.* Proceeding of international conference on control, automation and systems, JeJu Island, pp. 1566–1569.

5

Some relations between left (right) semi-uninorms and coimplications on a complete lattice

Keming Tang[a] and Zhudeng Wang[b]*

[a]*College of Information Science and Technology, Yancheng Teachers University, Jiangsu 224002, People's Republic of China;* [b]*School of Mathematical Sciences, Yancheng Teachers University, Yancheng 224002, People's Republic of China*

Uninorms are important generalizations of triangular norms and conorms, with the neutral elements lying anywhere in the unit interval, left (right) semi-uninorms are non-commutative and non-associative extensions of uninorms, and coimplications are extensions of the Boolean coimplication. In this paper, we study the relationships between left (right) semi-uninorms and coimplications on a complete lattice. We first discuss the residual coimplicators of left and right semi-uninorms and show that the right (left) residual coimplicator of a disjunctive right (left) infinitely $\wedge$-distributive left (right) semi-uninorm is a right infinitely $\vee$-distributive coimplication which satisfies the neutrality principle. Then, we investigate the left and right semi-uninorms induced by a coimplication and demonstrate that the operations induced by right infinitely $\vee$-distributive coimplications, which satisfy the order property or neutrality principle, are left (right) infinitely $\wedge$-distributive left (right) semi-uninorms or right (left) semi-uninorms. Finally, we prove that the meet-semilattice of all disjunctive right (left) infinitely $\wedge$-distributive left (right) semi-uninorms is order-reversing isomorphic to the join-semilattice of all right infinitely $\vee$-distributive coimplications that satisfy the neutrality principle.

Keywords: fuzzy connective; uninorm; semi-uninorm; left (right) semi-uninorm; coimplication

1. Introduction

Uninorms, introduced by Yager & Rybalov (1996) and studied by Fodor, Yager, & Rybalov (1997), are special aggregation operators that have been proven useful in many fields such as fuzzy logic, expert systems, neural networks, aggregation, and fuzzy system modelling (see Gabbay & Metcalfe, 2007; Tsadiras & Margaritis, 1998; Yager, 2001, 2002). This kind of operation is an important generalization of both triangular norms (t-norms for short) and triangular conorms (t-conorms for short) and a special combination of t-norms and t-conorms (see Fodor et al., 1997). But, there are real-life situations when truth functions cannot be associative or commutative (see Flondor, Georgescu, & Iorgulescu, 2001; Fodor & Keresztfalvi, 1995). By throwing away the commutativity from the axioms of uninorms, Mas, Monserrat, & Torrens (2001) introduced the concepts of left and right uninorms on [0, 1] and later on a finite chain (Mas, Monserrat, & Torrens, 2004), Wang and Fang (2009a, 2009b) studied the left and right uninorms on a complete lattice. By removing the associativity and commutativity from the axioms of uninorms, Liu (2012) introduced the concept of semi-uninorms on a complete lattice and Su, Wang, & Tang (2013) discussed the notion of left and right semi-uninorms on a complete lattice. On the other hand, it is well known that a uninorm (semi-uninorm, left and right uninorms) U can be conjunctive or disjunctive whenever $U(0, 1) = 0$ or 1, respectively. This fact allows us to use uninorms (semi-uninorm, left and right uninorms) in defining fuzzy implications and coimplications (see De Baets & Fodor, 1999; Liu, 2012; Mas, Monserrat, & Torrens, 2007; Ruiz & Torrens, 2004; Wang & Fang, 2009a, 2009b).

In this paper, based on De Baets & Fodor (1999), Liu (2012), Ruiz & Torrens (2004), Su & Wang (2013) and Wang & Fang (2009b), we study left (right) semi-uninorms and coimplications on a complete lattice. After recalling some necessary definitions and examples about the left and right semi-uninorms on a complete lattice in Section 2, we discuss the residual coimplicators of left and right semi-uninorms in the third section and show that the right (left) residual coimplicator of a disjunctive right (left) infinitely $\wedge$-distributive left (right) semi-uninorm is a right infinitely $\vee$-distributive coimplication that satisfies the neutrality principle. In Section 4, we investigate the left and right semi-uninorms induced by a coimplication and give some conditions such that the operations induced by a coimplication constitute left or right semi-uninorms. In Section 5, we reveal the relationships between disjunctive right (left) infinitely $\wedge$-distributive left (right)

*Corresponding author. Email: zhudengwang2004@163.com

semi-uninorms and right infinitely $\vee$-distributive coimplications which satisfy the neutrality principle.

The knowledge about lattices required in this paper can be found in Birkhoff (1967).

Throughout this paper, unless otherwise stated, L always represents any given complete lattice with maximal element 1 and minimal element 0; J stands for any index set.

2. Left and right semi-uninorms

Noting that the commutativity and associativity are not desired for aggregation operators in a lot of cases, Liu (2012) introduced the concept of semi-uninorms and Su et al. (2013) studied the notions of left and right semi-uninorms on a complete lattice. Here, we recall some necessary concepts about the left and right semi-uninorms and illustrate these notions by means of two examples.

DEFINITION 2.1 (Su et al., 2013). *A binary operation U on L is called a left (right) semi-uninorm if it satisfies the following two conditions:*

(U1) *there exists a left (right) neutral element, that is, an element $e_{\mathrm{L}} \in L(e_{\mathrm{R}} \in L)$ satisfying $U(e_{\mathrm{L}}, x) = x(U(x, e_{\mathrm{R}}) = x)$ for all $x \in L$,*

(U2) *U is non-decreasing in each variable.*

For any left (right) semi-uninorm U on L, U is said to be left-conjunctive and right-conjunctive if $U(0,1) = 0$ and $U(1,0) = 0$, respectively. U is called conjunctive if both $U(0,1) = 0$ and $U(1,0) = 0$ since it satisfies the classical boundary conditions of AND. U is said to be left-disjunctive and right-disjunctive if $U(1,0) = 1$ and $U(0,1) = 1$, respectively. We call U disjunctive if both $U(1,0) = 1$ and $U(0,1) = 1$ by a similar reason.

If a left (right) semi-uninorm U is associative, then U is the left (right) uninorm on L (see Wang & Fang, 2009a, 2009b).

If a left (right) semi-uninorm U with the left (right) neutral element e_{L} (e_{R}) has a right (left) neutral element e_{R} (e_{L}), then $e_{\mathrm{L}} = U(e_{\mathrm{L}}, e_{\mathrm{R}}) = e_{\mathrm{R}}$. Let $e = e_{\mathrm{L}} = e_{\mathrm{R}}$. Here, U is the semi-uninorm (see Liu, 2012). In particular, if the neutral element $e = 1$, then the semi-uninorm U becomes a t-seminorm (see Suárez García & Gil Álvarez, 1986) or a semi-copula (see Bassan & Spizzichino, 2005; Durante, Klement, Mesiar, & Sempi, 2007); if the neutral element $e = 0$, then the semi-uninorm U becomes a t-semiconorm (see De Cooman & Kerre, 1994).

Clearly, $U(0,0) = 0$ and $U(1,1) = 1$ hold for any left (right) semi-uninorm U on L. Moreover, the left (right) neutral elements need not to be unique. In fact, the projection operator given by $U(x,y) = x$ for all $x, y \in L$ is such that any element in L is a right neutral element. But, left (right) neutral elements are all idempotent (see De Baets, 1999) because $U(e_{\mathrm{L}}, e_{\mathrm{L}}) = e_{\mathrm{L}}$ ($U(e_{\mathrm{R}}, e_{\mathrm{R}}) = e_{\mathrm{R}}$) for any left (right) neutral element e_{L} (e_{R}) of U.

DEFINITION 2.2 (Wang & Fang, 2009a, 2009b). *A binary operation U on L is called left (right) infinitely $\vee$-distributive if*

$$U\left(\bigvee_{j\in J} x_j, y\right) = \bigvee_{j\in J} U(x_j, y)$$

$$\left(U\left(x, \bigvee_{j\in J} y_j\right) = \bigvee_{j\in J} U(x, y_j)\right)$$

$$\forall x, y, x_j, y_j \in L;$$

left (right) infinitely $\wedge$-distributive if

$$U\left(\bigwedge_{j\in J} x_j, y\right) = \bigwedge_{j\in J} U(x_j, y)$$

$$\left(U\left(x, \bigwedge_{j\in J} y_j\right) = \bigwedge_{j\in J} U(x, y_j)\right)$$

$$\forall x, y, x_j, y_j \in L.$$

If a binary operation U is left infinitely $\vee$-distributive ($\wedge$-distributive) and also right infinitely $\vee$-distributive ($\wedge$-distributive), then U is said to be infinitely $\vee$-distributive ($\wedge$-distributive).

Noting that the least upper bound of the empty set is 0 and the greatest lower bound of the empty set is 1 (see Birkhoff, 1967), we have that

$$U(0,y) = U\left(\bigvee_{j\in\emptyset} x_j, y\right) = \bigvee_{j\in\emptyset} U(x_j, y) = 0$$

$$\left(U(x,0) = U\left(x, \bigvee_{j\in\emptyset} y_j\right) = \bigvee_{j\in\emptyset} U(x, y_j) = 0\right)$$

for any $x, y \in L$ when U is left (right) infinitely $\vee$-distributive and

$$U(1,y) = U\left(\bigwedge_{j\in\emptyset} x_j, y\right) = \bigwedge_{j\in\emptyset} U(x_j, y) = 1$$

$$\left(U(x,1) = U\left(x, \bigwedge_{j\in\emptyset} y_j\right) = \bigwedge_{j\in\emptyset} U(x, y_j) = 1\right)$$

for any $x, y \in L$ when U is left (right) infinitely $\wedge$-distributive.

For the sake of convenience, we introduce the following symbols:

$\mathcal{U}_s^{e_L}(L)$: the set of all left semi-uninorms with the left neutral element e_L on L;

$\mathcal{U}_s^{e_R}(L)$: the set of all right semi-uninorms with the right neutral element e_R on L;

$\mathcal{U}_{s\wedge}^{e_L}(L)$: the set of all right infinitely $\wedge$-distributive left semi-uninorms with the left neutral element e_L on L;

$\mathcal{U}_{\wedge s}^{e_L}(L)$: the set of all left infinitely $\wedge$-distributive left semi-uninorms with the left neutral element e_L on L;

$\mathcal{U}_{s\wedge}^{e_R}(L)$: the set of all right infinitely $\wedge$-distributive right semi-uninorms with the right neutral element e_R on L;

$\mathcal{U}_{\wedge s}^{e_R}(L)$: the set of all left infinitely $\wedge$-distributive right semi-uninorms with the right neutral element e_R on L.

Now, we present two examples of left and right semi-uninorms on L.

Example 2.1 (see Su et al., 2013). Let $e_L \in L$,

$$U_{sW}^{e_L}(x,y) = \begin{cases} y & \text{if } x \geq e_L, \\ 0 & \text{otherwise,} \end{cases}$$

$$U_{sM}^{e_L}(x,y) = \begin{cases} y & \text{if } x \leq e_L, \\ 1 & \text{otherwise,} \end{cases}$$

$$U_{sW}^{e_L\,*}(x,y) = \begin{cases} 1 & \text{if } y = 1, \\ y & \text{if } x \geq e_L, \\ 0 & \text{otherwise,} \end{cases}$$

$$U_{dsW}^{e_L}(x,y) = \begin{cases} 1 & \text{if } x = 1 \text{ or } y = 1, \\ y & \text{if } e_L \leq x < 1, \\ 0 & \text{otherwise,} \end{cases}$$

where x and y are elements of L. Then $U_{sW}^{e_L}$ and $U_{sM}^{e_L}$ are, respectively, the smallest and greatest elements of $\mathcal{U}_s^{e_L}(L)$; and $U_{sW}^{e_L\,*}$ and $U_{sM}^{e_L}$ are, respectively, the smallest and greatest elements of $\mathcal{U}_{s\wedge}^{e_L}(L)$. Moreover, it is easy to see that $U_{dsW}^{e_L}$ is the smallest disjunctive right infinitely $\wedge$-distributive left semi-uninorm with the left neutral element e_L.

Example 2.2 (see Su et al., 2013). Let $e_R \in L$,

$$U_{sW}^{e_R}(x,y) = \begin{cases} x & \text{if } y \geq e_R, \\ 0 & \text{otherwise,} \end{cases}$$

$$U_{sM}^{e_R}(x,y) = \begin{cases} x & \text{if } y \leq e_R, \\ 1 & \text{otherwise,} \end{cases}$$

$$U_{sW}^{e_R\,*}(x,y) = \begin{cases} 1 & \text{if } x = 1, \\ x & \text{if } y \geq e_R, \\ 0 & \text{otherwise,} \end{cases}$$

$$U_{dsW}^{e_R}(x,y) = \begin{cases} 1 & \text{if } x = 1 \text{ or } y = 1, \\ x & \text{if } e_R \leq y < 1, \\ 0 & \text{otherwise,} \end{cases}$$

where x and y are elements of L. Then $U_{sW}^{e_R}$ and $U_{sM}^{e_R}$ are, respectively, the smallest and greatest elements of $\mathcal{U}_s^{e_R}(L)$; and $U_{sW}^{e_R\,*}$ and $U_{sM}^{e_R}$ are, respectively, the smallest and greatest elements of $\mathcal{U}_{\wedge s}^{e_R}(L)$. Moreover, it is easy to verify that $U_{dsW}^{e_R}$ is the smallest disjunctive left infinitely $\wedge$-distributive right semi-uninorm with the right neutral element e_R.

3. The residual coimplicators of left and right semi-uninorms

Recently, Mas et al. (2007) and Ruiz & Torrens (2004) studied the implications and coimplications derived from uninorms on [0, 1], Wang & Fang (2009b) discussed the residual coimplicators of left and right uninorms on a complete lattice, Su & Wang (2013) researched pseudo-uninorms and coimplications on a complete lattice. In this section, based on Mas et al. (2007), Ruiz & Torrens (2004), Su & Wang (2013) and Wang & Fang (2009b), we consider the residual coimplications of left and right semi-uninorms on a complete lattice.

First of all, we recall the definitions of implications and coimplications.

DEFINITION 3.1 *(Baczyński & Jayaram, 2008; De Baets, 1997; De Baets & Fodor, 1999). An implication I on L is a hybrid monotonous (with decreasing first and increasing second partial mappings) binary operation that satisfies the corner conditions $I(0,0) = I(1,1) = 1$ and $I(1,0) = 0$.*

A coimplication C on L is a hybrid monotonous binary operation that satisfies the corner conditions $C(0,0) = C(1,1) = 0$ and $C(0,1) = 1$.

Implications are extensions of the Boolean implication $\Rightarrow$ ($P \Rightarrow Q$ meaning that P is sufficient for Q). Coimplications are extensions of the Boolean coimplication $\not\Leftarrow$ ($P \not\Leftarrow Q$ meaning that P is not necessary for Q) (see De Baets, 1997; De Baets, Tsiporkova, & Mesiar, 1999).

Note that for any coimplication C on L, due to the monotonicity, the absorption principle holds, that is, $C(x,0) = C(1,x) = 0$ for any $x \in L$.

We denote the set of all coimplications and the set of all right infinitely $\vee$-distributive coimplications on L by $\mathcal{C}(L)$ and $\mathcal{C}_\vee(L)$, respectively.

Example 3.1 (Su & Wang, 2014). Let

$$C_W(x,y) = \begin{cases} 1 & \text{if } (x,y) = (0,1), \\ 0 & \text{otherwise,} \end{cases}$$

$$C_M(x,y) = \begin{cases} 0 & \text{if } x = 1 \text{ or } y = 0, \\ 1 & \text{otherwise,} \end{cases}$$

where x and y are elements of L. It is easy to see that C_W and C_M are, respectively, the smallest and greatest elements of $\mathcal{C}(L)$ and C_M is also the largest element of $\mathcal{C}_\vee(L)$.

Example 3.2 (*Su & Wang*, 2014). Let $L = \{0, a, b, 1\}$ be a lattice, where $0 < a < 1$, $0 < b < 1$, $a \wedge b = 0$ and $a \vee b = 1$. Define two coimplications C_1 and C_2 as follows:

C_1	0	a	b	1
0	0	a	b	1
a	0	0	0	0
b	0	0	0	0
1	0	0	0	0

C_2	0	a	b	1
0	0	b	a	1
a	0	0	0	0
b	0	0	0	0
1	0	0	0	0

It is easy to see that C_1 and C_2 are two right infinitely $\vee$-distributive coimplications, and $C_1 \wedge C_2 = C_W$. But C_W is not right infinitely $\vee$-distributive. Suppose C is the smallest right infinitely $\vee$-distributive coimplication, then $C_1 \geq C$ and $C_2 \geq C$. Thus, $C_W = C_1 \wedge C_2 \geq C$ and $C_W = C$ since C_W is the smallest coimplication. It leads to the contradiction that C_W is right infinitely $\vee$-distributive. Therefore, there is no the smallest right infinitely $\vee$-distributive coimplication.

This illustrates that $\mathcal{C}_\vee(L)$ is not a $\wedge$-semilattice.

DEFINITION 3.2 (Wang & Fang, 2009b). *Let U be a binary operation on L. Define $C^{\mathrm{L}}_U, C^{\mathrm{R}}_U \in L^{L\times L}$ as follows:*

$$C^{\mathrm{L}}_U(x,y) = \wedge\{z \in L \mid y \leq U(z,x)\} \quad \forall x,y \in L,$$
$$C^{\mathrm{R}}_U(x,y) = \wedge\{z \in L \mid y \leq U(x,z)\} \quad \forall x,y \in L.$$

Here, C^{L}_U and C^{R}_U are, respectively, called the left and right residual coimplicators of U.

For any operation U on L, it is straightforward to verify that (see Wang & Fang, 2009b, Theorems 3.1 and 3.2)

(1) $C^{\mathrm{L}}_U(x,0) = C^{\mathrm{R}}_U(x,0) = 0$ for any $x \in L$.
(2) For any $x,y \in L$, $C^{\mathrm{L}}_U(y, U(x,y)) \leq x$ and $C^{\mathrm{R}}_U(x, U(x,y)) \leq y$.
(3) If U is right-disjunctive, then $C^{\mathrm{L}}_U(1,y) = 0$ and if U is left-disjunctive, then $C^{\mathrm{R}}_U(1,y) = 0$.

By virtue of Definition 3.2, it is easy to see that C^{L}_U and C^{R}_U are all decreasing in the first variable and increasing in the second one when U is a left (right) semi-uninorm; $C^{\mathrm{L}}_U(e,x) = C^{\mathrm{R}}_U(e,x) = x$ for any $x \in L$ when U is a semi-uninorm with the neutral element e on L.

Example 3.3 For those left and right semi-uninorms in Examples 2.1 and 2.2, a simple computation shows that

$$C^{\mathrm{L}}_{U^{e_{\mathrm{R}}}_{sW}}(x,y) = C^{\mathrm{L}}_{U^{e_{\mathrm{R}}*}_{sW}}(x,y) = \begin{cases} 0 & \text{if } y = 0, \\ y & \text{if } x \geq e_{\mathrm{R}}, \\ 1 & \text{otherwise,} \end{cases}$$

$$C^{\mathrm{L}}_{U^{e_{\mathrm{R}}}_{sM}}(x,y) = \begin{cases} y & \text{if } x \leq e_{\mathrm{R}}, \\ 0 & \text{otherwise,} \end{cases}$$

$$C^{\mathrm{L}}_{U^{e_{\mathrm{L}}}_{sW}}(x,y) = \begin{cases} 0 & \text{if } y = 0, \\ e_{\mathrm{L}} & \text{if } 0 < y \leq x, \\ 1 & \text{otherwise,} \end{cases}$$

$$C^{\mathrm{L}}_{U^{e_{\mathrm{L}}*}_{sW}}(x,y) = C^{\mathrm{L}}_{U^{e_{\mathrm{L}}}_{dsW}}(x,y) = \begin{cases} 0 & \text{if } x = 1 \text{ or } y = 0, \\ e_{\mathrm{L}} & \text{if } 0 < y \leq x < 1, \\ 1 & \text{otherwise,} \end{cases}$$

$$C^{\mathrm{L}}_{U^{e_{\mathrm{L}}}_{sM}}(x,y) = \begin{cases} 0 & \text{if } y \leq x, \\ \wedge\{a \mid a \nleq e_{\mathrm{L}}\} & \text{otherwise,} \end{cases}$$

$$C^{\mathrm{R}}_{U^{e_{\mathrm{R}}}_{sM}}(x,y) = \begin{cases} 0 & \text{if } y \leq x, \\ \wedge\{a \mid a \nleq e_{\mathrm{R}}\} & \text{otherwise,} \end{cases}$$

$$C^{\mathrm{R}}_{U^{e_{\mathrm{L}}}_{sW}}(x,y) = C^{\mathrm{R}}_{U^{e_{\mathrm{L}}*}_{sW}}(x,y) = \begin{cases} 0 & \text{if } y = 0, \\ y & \text{if } x \geq e_{\mathrm{L}}, \\ 1 & \text{otherwise,} \end{cases}$$

$$C^{\mathrm{R}}_{U^{e_{\mathrm{L}}}_{sM}}(x,y) = \begin{cases} y & \text{if } x \leq e_{\mathrm{L}}, \\ 0 & \text{otherwise,} \end{cases}$$

$$C^{\mathrm{R}}_{U^{e_{\mathrm{R}}}_{sW}}(x,y) = \begin{cases} 0 & \text{if } y = 0, \\ e_{\mathrm{R}} & \text{if } 0 < y \leq x, \\ 1 & \text{otherwise,} \end{cases}$$

$$C^{\mathrm{R}}_{U^{e_{\mathrm{R}}*}_{sW}}(x,y) = C^{\mathrm{R}}_{U^{e_{\mathrm{R}}}_{dsW}}(x,y) = \begin{cases} 0 & \text{if } x = 1 \text{ or } y = 0, \\ e_{\mathrm{R}} & \text{if } 0 < y \leq x < 1, \\ 1 & \text{otherwise,} \end{cases}$$

$$C^{\mathrm{L}}_{U^{e_{\mathrm{R}}}_{dsW}}(x,y) = \begin{cases} 0 & \text{if } x = 1 \text{ or } y = 0, \\ y & \text{if } e_{\mathrm{R}} \leq x < 1, \\ 1 & \text{otherwise,} \end{cases}$$

$$C^{\mathrm{R}}_{U^{e_{\mathrm{L}}}_{dsW}}(x,y) = \begin{cases} 0 & \text{if } x = 1 \text{ or } y = 0, \\ y & \text{if } e_{\mathrm{L}} \leq x < 1, \\ 1 & \text{otherwise.} \end{cases}$$

When $e_{\mathrm{L}}, e_{\mathrm{R}} \in L \setminus \{0, 1\}$, we see that $C^{\mathrm{L}}_{U^{e_{\mathrm{L}}*}_{sW}}$, $C^{\mathrm{L}}_{U^{e_{\mathrm{L}}}_{dsW}}$, $C^{\mathrm{R}}_{U^{e_{\mathrm{R}}}_{dsW}}$ and $C^{\mathrm{R}}_{U^{e_{\mathrm{R}}*}_{sW}}$ are four coimplications, $C^{\mathrm{L}}_{U^{e_{\mathrm{R}}}_{dsW}}$, $C^{\mathrm{R}}_{U^{e_{\mathrm{L}}}_{dsW}}$, $C^{\mathrm{L}}_{U^{e_{\mathrm{R}}}_{sM}}$ and $C^{\mathrm{R}}_{U^{e_{\mathrm{L}}}_{sM}}$ are four right infinitely $\vee$-distributive coimplications, but $C^{\mathrm{L}}_{U^{e_{\mathrm{R}}}_{sW}}$, $C^{\mathrm{L}}_{U^{e_{\mathrm{R}}*}_{sW}}$, $C^{\mathrm{L}}_{U^{e_{\mathrm{L}}}_{sW}}$, $C^{\mathrm{L}}_{U^{e_{\mathrm{L}}}_{sM}}$, $C^{\mathrm{R}}_{U^{e_{\mathrm{L}}}_{sW}}$, $C^{\mathrm{R}}_{U^{e_{\mathrm{L}}*}_{sW}}$, $C^{\mathrm{R}}_{U^{e_{\mathrm{R}}}_{sW}}$ and $C^{\mathrm{R}}_{U^{e_{\mathrm{R}}}_{sM}}$ are not coimplications.

THEOREM 3.1 *Let* $U \in \mathcal{U}_s^{e_L}(L)$.

(1) *For any* $x, y \in L$, $y \le x \Rightarrow C_U^{\mathrm{L}}(x,y) \le e_{\mathrm{L}}$.
(2) C_U^{R} *satisfies the neutrality principle with respect to* e_{L} *(w.r.t.* e_{L}*, for short), that is,* $C_U^{\mathrm{R}}(e_{\mathrm{L}},y) = y$ *for any* $y \in L$.
(3) *If U is left-disjunctive, then* $C_U^{\mathrm{R}} \in \mathcal{C}(L)$.
(4) *If* $U \in \mathcal{U}_{s\wedge}^{e_L}(L)$ *is left-disjunctive, then* $C_U^{\mathrm{R}} \in \mathcal{C}_{\vee}(L)$ *and*

$$C_U^{\mathrm{R}}(x,y) = \min\{z \in L \mid y \le U(x,z)\}.$$

Here, C_U^{R} *is called the right residual coimplication of the left semi-uninorm U.*

Proof Clearly, statements (1) and (2) hold.

(3) If U is a left-disjunctive left semi-uninorm with the left neutral element e_{L}, then it follows from the statements before Example 3.3 that C_U^{R} is non-increasing in its first and non-decreasing in its second variable and $C_U^{\mathrm{R}}(1,1) = 1$. Moreover,

$$C_U^{\mathrm{R}}(0,0) = \wedge\{z \in L \mid 0 \le U(0,z)\} = 0.$$

By the non-decreasingness of U, we see that

$$C_U^{\mathrm{R}}(0,1) = \wedge\{z \in L \mid U(0,z) = 1\}$$
$$\ge \wedge\{z \in L \mid z = U(e_{\mathrm{L}},z) \ge U(0,z) = 1\} = 1.$$

Thus, C_U^{R} is a coimplication on L.

(4) Refer to the Proofs of Theorem 3.1 in Su & Wang (2013) and Theorem 3.5 in Wang & Fang (2009b). ■

When $e_{\mathrm{L}} < 1$, for the right infinitely $\wedge$-distributive left semi-uninorm $U_{sW}^{e_L *}$, we see that $C^{\mathrm{L}}_{U_{sW}^{e_L *}} \in \mathcal{C}(L)$ by Example 3.3, but $C^{\mathrm{L}}_{U_{sW}^{e_L *}}(e_{\mathrm{L}},y) = e_{\mathrm{L}} \ne y$ when $0 < y < e_{\mathrm{L}}$, that is, $C^{\mathrm{L}}_{U_{sW}^{e_L *}}$ does not satisfy the neutrality principle. This illustrates Theorem 3.1 does not hold for the left residual coimplicator of a left semi-uninorm.

If P and Q are two propositions, then the property $U(x, C_U^{\mathrm{R}}(x,y)) \ge y$ is a generalization of the following tautology $Q \Rightarrow (P \vee (P \nLeftarrow Q))$ in classical logic and is in some sense dual to the modus ponens (see De Baets, 1997). By Theorem 3.2 in Su & Wang (2013), we know that U and C_U^{R} satisfy the generalized dual modus ponens rule and the following right residual principle:

$$y \le U(x,z) \Leftrightarrow C_U^{\mathrm{R}}(x,y) \le z \quad \forall x,y,z \in L,$$

when U is a right infinitely $\wedge$-distributive left semi-uninorm on L. Similarly, U and C_U^{L} satisfy the generalized dual modus ponens rule in the form: $U(C_U^{\mathrm{L}}(x,y),x) \ge y$ and the following left residual principle (see Su & Wang, 2013, Theorem 3.1):

$$y \le U(z,x) \Leftrightarrow C_U^{\mathrm{L}}(x,y) \le z \quad \forall x,y,z \in L,$$

when U is left infinitely $\wedge$-distributive right semi-uninorm on L. Thus, for right semi-uninorms on L, we have a similar result.

THEOREM 3.2 *Let* $U \in \mathcal{U}_s^{e_R}(L)$.

(1) *For any* $x, y \in L$, $y \le x \Rightarrow C_U^{\mathrm{R}}(x,y) \le e_{\mathrm{R}}$.
(2) C_U^{L} *satisfies the neutrality principle with respect to* e_{R} *(w.r.t.* e_{R}*, for short), that is,* $C_U^{\mathrm{L}}(e_{\mathrm{R}},y) = y$ *for any* $y \in L$.
(3) *If U is right-disjunctive, then* $C_U^{\mathrm{L}} \in \mathcal{C}(L)$.
(4) *If* $U \in \mathcal{U}_{\wedge s}^{e_R}(L)$ *is right-disjunctive, then* $C_U^{\mathrm{L}} \in \mathcal{C}_{\vee}(L)$ *and*

$$C_U^{\mathrm{L}}(x,y) = \min\{z \in L \mid y \le U(z,x)\}.$$

Here, C_U^{L} *is called the left residual coimplication of the right semi-uninorm U.*

The neutrality principle w.r.t. e_{L} : $C(e_{\mathrm{L}},y) = y$ and neutrality principle w.r.t. e_{R}: $C(e_{\mathrm{R}},y) = y$ are generalizations of the neutrality principle $C(0,y) = y$ for any $y \in L$ in De Baets (1997).

Combining Theorems 3.1 and 3.2, we know that both C_U^{L} and C_U^{R} are all right infinitely $\vee$-distributive coimplications when U is, respectively, an infinitely $\wedge$-distributive right and left semi-uninorm on L.

THEOREM 3.3 (1) *If* $U \in \mathcal{U}_{\wedge s}^{e_L}(L)$*, then* C_U^{L} *is right infinitely* $\vee$*-distributive and satisfies the left residual principle and order property with respect to* e_{L} *(w.r.t.* e_{L}*, for short)* :

$$y \le x \Leftrightarrow C_U^{\mathrm{L}}(x,y) \le e_{\mathrm{L}} \quad \forall x,y \in L.$$

Moreover, if $U \in \mathcal{U}_{\wedge s}^{e_L}(L)$ *is strict left-disjunctive, that is, it is disjunctive and satisfies the condition:*

$$U(x,0) = 1 \Leftrightarrow x = 1 \quad \forall x \in L,$$

then C_U^{L} *is a right infinitely* $\vee$*-distributive coimplication which satisfies the order property.*

(2) *If* $U \in \mathcal{U}_{s\wedge}^{e_R}(L)$*, then* C_U^{R} *is right infinitely* $\vee$*-distributive and satisfies the right residual principle and order property with respect to* e_{R} *(w.r.t.* e_{R}*, for short)* :

$$y \le x \Leftrightarrow C_U^{\mathrm{R}}(x,y) \le e_{\mathrm{R}} \quad \forall x,y \in L.$$

Moreover, if $U \in \mathcal{U}_{s\wedge}^{e_R}(L)$ *is strict right-disjunctive, that is, it is disjunctive and satisfies the condition:*

$$U(0,x) = 1 \Leftrightarrow x = 1 \quad \forall x \in L,$$

then C_U^{R} *is a right infinitely* $\vee$*-distributive coimplication which satisfies the order property.*

Proof Assume that U is a left infinitely $\wedge$-distributive left semi-uninorm with the left neutral element $e_{\rm L}$. By Theorem 3.2(4), we can see that $C_U^{\rm L}$ is right infinitely $\vee$-distributive and satisfies the left residual principle. If $x, y \in L$ and $y \leq x$, then it follows from Theorem 3.1(1) that $C_U^{\rm L}(x,y) \leq e_{\rm L}$; if $C_U^{\rm L}(x,y) \leq e_{\rm L}$, then

$$x = U(e_{\rm L}, x) \geq U(C_U^{\rm L}(x,y), x) \geq y.$$

Thus, $C_U^{\rm L}$ satisfies the order property. Moreover, if $U \in \mathcal{U}_{\wedge s}^{e_{\rm L}}(L)$ is strict left-disjunctive, then

$$C_U^{\rm L}(0,1) = \wedge\{z \in L \mid U(z,0) = 1\} = 1,$$
$$C_U^{\rm L}(1,1) = \wedge\{z \in L \mid U(z,1) = 1\} = 0.$$

Therefore, $C_U^{\rm L}$ is a right infinitely $\vee$-distributive coimplication which satisfies the order property.

Similarly, we can show that $C_U^{\rm R}$ is a right infinitely $\vee$-distributive coimplication, which satisfies the right residual principle and order property, when U is a strict right-disjunctive right infinitely $\wedge$-distributive right semi-uninorm. ■

In particular, if U is an infinitely $\wedge$-distributive semi-uninorm with the neutral element e (see Liu, 2012), then $C_U^{\rm L}$ and $C_U^{\rm R}$ satisfy the residual principle (RP), neutrality principle (NP) and order property (OP), and are all right infinitely $\vee$-distributive coimplications by Theorems 3.1–3.3.

4. The left and right semi-uninorms induced by coimplications

Liu (2012) discussed the semi-uninorms induced by implications and Su & Wang (2013) studied the pseudo-uninorms induced by coimplications. In this section, based on these works, we investigate the left and right semi-uninorms induced by coimplications on a complete lattice.

DEFINITION 4.1 (Su & Wang, 2013). *Let C be a binary operation on L. Define two induced operators $U_C^{\rm L}$ and $U_C^{\rm R}$ of C as follows:*

$$U_C^{\rm L}(x,y) = \vee\{z \in L \mid C(y,z) \leq x\} \quad \forall x, y \in L,$$
$$U_C^{\rm R}(x,y) = \vee\{z \in L \mid C(x,z) \leq y\} \quad \forall x, y \in L.$$

Clearly, $U_C^{\rm L}(1,x) = U_C^{\rm R}(x,1) = 1; U_C^{\rm L}(0,x) = U_C^{\rm R}(x,0) = \vee\{z \in L \mid C(x,z) = 0\}$ for any $x \in L$; and $U_C^{\rm L} = U_C^{\rm R}$ if I satisfies the condition:

$$C(y,z) \leq x \Leftrightarrow C(x,z) \leq y \quad \forall x, y, z \in L.$$

When C is hybrid monotonous, it is easy to see that $U_C^{\rm L}$ and $U_C^{\rm R}$ are all non-decreasing in its each variable.

Moreover, for any binary operation C, it follows from Definition 4.1 that

$$y \leq U_C^{\rm L}(C(x,y), x), \quad y \leq U_C^{\rm R}(x, C(x,y)) \quad \forall x, y \in L.$$

These explain that $U_C^{\rm L}$ and C, $U_C^{\rm R}$ and C satisfy the generalized dual modus ponens rule.

Example 4.1 For two coimplications C_W and C_M in Example 3.1, we have that

$$U_{C_M}^{\rm L}(x,y) = U_{C_M}^{\rm R}(x,y) = \begin{cases} 1 & \text{if } x = 1 \text{ or } y = 1, \\ 0 & \text{otherwise,} \end{cases}$$

$$U_{C_W}^{\rm L}(x,y) = \begin{cases} \vee_{a \in L\setminus\{1\}} a & \text{if } x < 1 \text{ and } y = 0, \\ 1 & \text{otherwise,} \end{cases}$$

$$U_{C_W}^{\rm R}(x,y) = \begin{cases} \vee_{a \in L\setminus\{1\}} a & \text{if } x = 0 \text{ and } y < 1, \\ 1 & \text{otherwise.} \end{cases}$$

Thus, four operations induced by coimplications C_W and C_M are neither left semi-uninorms nor right semi-uninorms on L.

We know that the right residuum of a disjunctive right infinitely $\wedge$-distributive left semi-uninorm and the left residuum of a disjunctive left infinitely $\wedge$-distributive right semi-uninorm are all right infinitely $\vee$-distributive coimplications. Below, we find some conditions such that these operations induced by coimplications are left or right semi-uninorms.

THEOREM 4.1 *Let $C \in \mathcal{C}(L)$.*

(1) *If C satisfies the order property w.r.t. $e_{\rm L}$, then $U_C^{\rm L} \in \mathcal{U}_s^{e_{\rm L}}(L)$; if C satisfies the neutrality principle w.r.t. $e_{\rm L}$, then $U_C^{\rm R} \in \mathcal{U}_s^{e_{\rm L}}(L)$. Here, $U_C^{\rm L}$ and $U_C^{\rm R}$ are called the left semi-uninorms induced by the coimplication C.*

(2) *If C satisfies the order property w.r.t. $e_{\rm R}$, then $U_C^{\rm R} \in \mathcal{U}_s^{e_{\rm R}}(L)$; if C satisfies the neutrality principle w.r.t. $e_{\rm R}$, then $U_C^{\rm L} \in \mathcal{U}_s^{e_{\rm R}}(L)$. Here, $U_C^{\rm L}$ and $U_C^{\rm R}$ are called the right semi-uninorms induced by the coimplication C.*

(3) *If C satisfies the order property w.r.t. $e_{\rm L}$ and neutrality principle w.r.t. $e_{\rm R}$, then $U_C^{\rm L}$ is a semi-uninorm (see Liu, 2012) on L.*

(4) *If C satisfies the order property w.r.t. $e_{\rm R}$ and neutrality principle w.r.t. $e_{\rm L}$, then $U_C^{\rm R}$ is also a semi-uninorm on L.*

Proof Assume that $C \in \mathcal{C}(L)$. Then $U_C^{\rm L}$ is non-decreasing in each variable. If C satisfies the order property w.r.t. $e_{\rm L}$, then

$$U_C^{\mathrm{L}}(e_{\mathrm{L}},y) = \vee\{z \in L \mid C(y,z) \leq e_{\mathrm{L}}\}$$
$$= \vee\{z \in L \mid z \leq y\} = y \quad \forall y \in L.$$

Thus, $U_C^{\mathrm{L}} \in \mathcal{U}_s^{e_{\mathrm{L}}}(L)$. If C satisfies the neutrality principle w.r.t. e_{L}, then

$$U_C^{\mathrm{R}}(e_{\mathrm{L}},y) = \vee\{z \in L \mid C(e_{\mathrm{L}},z) \leq y\}$$
$$= \vee\{z \in L \mid z \leq y\} = y \quad \forall y \in L.$$

So, $U_C^{\mathrm{R}} \in \mathcal{U}_s^{e_{\mathrm{L}}}(L)$.

Similarly, we can show that $U_C^{\mathrm{R}} \in \mathcal{U}_s^{e_{\mathrm{R}}}(L)$ when the coimplication C satisfies the order property w.r.t. e_{R} and $U_C^{\mathrm{L}} \in \mathcal{U}_s^{e_{\mathrm{R}}}(L)$ when C satisfies the neutrality principle w.r.t. e_{R}.

If C satisfies the order property w.r.t. e_{L} and neutrality principle w.r.t. e_{R}, then $U_C^{\mathrm{L}}(e_{\mathrm{L}},x) = U_C^{\mathrm{L}}(x, e_{\mathrm{R}}) = x$ for any $x \in L$. Thus, $e_{\mathrm{L}} = e_{\mathrm{R}}$. Let $e = e_{\mathrm{L}} = e_{\mathrm{R}}$. Then U_C^{L} is a semi-uninorm with the neutral element e on L.

In a similar way, we can see that U_C^{R} is also a semi-uninorm on L when C satisfies the order property w.r.t. e_{R} and neutrality principle w.r.t. e_{L}. ∎

When $C \in \mathcal{C}(L)$, $C(1,x) = 0$ for any $x \in L$ and hence it follows from Definition 4.1 that $U_C^{\mathrm{L}}(0,1) = U_C^{\mathrm{R}}(1,0) = 1$. Thus, U_C^{L} and U_C^{R} in Theorem 4.1 are all disjunctive left or right semi-uninorms induced by the coimplication C.

THEOREM 4.2 *Let $C \in \mathcal{C}_\vee(L)$.*

(1) *If C satisfies the order property w.r.t. e_{L}, then $U_C^{\mathrm{L}} \in \mathcal{U}_{\wedge s}^{e_{\mathrm{L}}}(L)$; if C satisfies the neutrality principle w.r.t. e_{L}, then $U_C^{\mathrm{R}} \in \mathcal{U}_{s\wedge}^{e_{\mathrm{L}}}(L)$.*

(2) *If C satisfies the order property w.r.t. e_{R}, then $U_C^{\mathrm{R}} \in \mathcal{U}_{s\wedge}^{e_{\mathrm{R}}}(L)$; if C satisfies the neutrality principle w.r.t. e_{R}, then $U_C^{\mathrm{L}} \in \mathcal{U}_{\wedge s}^{e_{\mathrm{R}}}(L)$.*

(3) *If C satisfies the order property w.r.t. e_{L} and neutrality principle w.r.t. e_{R}, then U_C^{L} is a left infinitely $\wedge$-distributive semi-uninorm on L.*

(4) *If C satisfies the order property w.r.t. e_{R} and neutrality principle w.r.t. e_{L}, then U_C^{R} is a right infinitely $\wedge$-distributive semi-uninorm on L.*

Proof Assume that C is a right infinitely $\vee$-distributive coimplication. By the Proof of Theorem 5.1 in Su & Wang (2013), we have that

$$C(y,z) \leq x \Leftrightarrow z \leq U_C^{\mathrm{L}}(x,y) \quad \forall x,y,z \in L.$$

Noting that $x \geq C(y, U_C^{\mathrm{L}}(x,y))$, we know that

$$U_C^{\mathrm{L}}(x,y) = \max\{z \in L \mid C(y,z) \leq x\}.$$

Moreover, when the index set $J \neq \emptyset$, for any $x_j, y \in L$ $(j \in J)$, we have that

$$U_C^{\mathrm{L}}(\wedge_{j\in J}x_j, y) = \vee\{z \in L \mid C(y,z) \leq \wedge_{j\in J}x_j\}$$
$$= \vee\{z \in L \mid C(y,z) \leq x_j \ \forall j \in J\}$$
$$= \vee\{z \in L \mid z \leq U_C^{\mathrm{L}}(x_j,y) \ \forall j \in J\}$$
$$= \vee\{z \in L \mid z \leq \wedge_{j\in J}U_C^{\mathrm{L}}(x_j,y)\}$$
$$= \bigwedge_{j\in J} U_C^{\mathrm{L}}(x_j,y).$$

When $J = \emptyset$, we see that $U_C^{\mathrm{L}}(\wedge_{j\in\emptyset}x_j, y) = U_C^{\mathrm{L}}(1,y) = 1 = \wedge_{j\in\emptyset}U_C^{\mathrm{L}}(x_j,y)$. Thus, U_C^{L} is left infinitely $\wedge$-distributive. Therefore, by virtue of Theorem 4.1, $U_C^{\mathrm{L}} \in \mathcal{U}_{\wedge s}^{e_{\mathrm{L}}}(L)$ when C satisfies the order property w.r.t. e_{L} and $U_C^{\mathrm{L}} \in \mathcal{U}_{\wedge s}^{e_{\mathrm{R}}}(L)$ when C satisfies the neutrality principle w.r.t. e_{R}.

Similarly, we can show that U_C^{R} is a right infinitely $\wedge$-distributive right semi-uninorm and left semi-uninorm when C satisfies the order property w.r.t. e_{R} and the neutrality principle w.r.t. e_{L}, respectively.

Statements (3) and (4) are the direct consequences of statements (1) and (2) and Theorem 4.1. ∎

In Theorem 4.2, if $C \in \mathcal{C}_\vee(L)$ satisfies the order property w.r.t. e_{L} and $U_C^{\mathrm{L}}(x,0) = 1$, then

$$1 = C(0, U_C^{\mathrm{L}}(x,0)) = C(0, \vee\{z \in L \mid C(0,z) \leq x\})$$
$$= \vee\{C(0,z) \mid z \in L, C(0,z) \leq x\} \leq x,$$

that is, $x = 1$. Thus, U_C^{L} is strict left-disjunctive. Similarly, if $C \in \mathcal{C}_\vee(L)$ satisfies the order property w.r.t. e_{R}, then U_C^{R} is strict right-disjunctive.

By virtue of Theorem 4.2, we see that if $C \in \mathcal{C}_\vee(L)$ satisfies the order property (OP) and the neutrality principle (NP), then U_C^{L} and U_C^{R} are, respectively, a left infinitely $\wedge$-distributive semi-uninorm and a right infinitely $\wedge$-distributive semi-uninorm on L.

When C is a right infinitely $\vee$-distributive coimplication on L, by the proof of Theorem 4.2, we know that C and U_C^{L}, C and U_C^{R} satisfy the following adjunction conditions (see Su & Wang, 2013, Theorems 5.1 and 5.2):

$$C(y,z) \leq x \Leftrightarrow z \leq U_C^{\mathrm{L}}(x,y),$$
$$C(x,z) \leq y \Leftrightarrow z \leq U_C^{\mathrm{R}}(x,y) \quad \forall x,y,z \in L.$$

Moreover, we have that $U_C^{\mathrm{R}}(x,y) = \max\{z \in L \mid C(x,z) \leq y\}$.

5. The relationships between left (right) semi-uninorms and coimplications

In the final section, we reveal the relationships between disjunctive right (left) infinitely $\wedge$-distributive left (right) semi-uninorms and right infinitely $\vee$-distributive coimplications which satisfy the neutrality principle on a complete lattice.

THEOREM 5.1 (1) *If* $U \in \mathcal{U}^{e_{\mathrm{R}}}_{\wedge s}(L)$ *is right-disjunctive, then* $C^{\mathrm{L}}_U \in \mathcal{C}_{\vee}(L)$ *satisfies the neutrality principle w.r.t.* e_{R} *and* $U^{\mathrm{L}}_{C^{\mathrm{L}}_U} = U$.

(2) *If* $U \in \mathcal{U}^{e_{\mathrm{L}}}_{s\wedge}(L)$ *is left-disjunctive, then* $C^{\mathrm{R}}_U \in \mathcal{C}_{\vee}(L)$ *satisfies the neutrality principle w.r.t.* e_{L} *and* $U^{\mathrm{R}}_{C^{\mathrm{R}}_U} = U$.

(3) *If* $C \in \mathcal{C}_{\vee}(L)$ *satisfies the neutrality principle w.r.t.* e_{L}, *then* $U^{\mathrm{R}}_C \in \mathcal{U}^{e_{\mathrm{L}}}_{s\wedge}(L)$ *is disjunctive and* $C^{\mathrm{R}}_{U^{\mathrm{R}}_C} = C$.

(4) *If* $C \in \mathcal{C}_{\vee}(L)$ *satisfies the neutrality principle w.r.t.* e_{R}, *then* $U^{\mathrm{L}}_C \in \mathcal{U}^{e_{\mathrm{R}}}_{\wedge s}(L)$ *is disjunctive and* $C^{\mathrm{L}}_{U^{\mathrm{L}}_C} = C$.

Proof We only prove that statements (1) and (3) hold.

(1) If U is a right-disjunctive left infinitely $\wedge$-distributive right semi-uninorm, then $C^{\mathrm{L}}_U \in \mathcal{C}_{\vee}(L)$ and satisfies the neutrality principle w.r.t. e_{R} by Theorem 3.2. Moreover, it follows from the left residual principle that

$$\begin{aligned} U^{\mathrm{L}}_{C^{\mathrm{L}}_U}(x,y) &= \vee\{z \in L \mid C^{\mathrm{L}}_U(y,z) \le x\} \\ &= \vee\{z \in L \mid z \le U(x,y)\} \\ &= U(x,y) \quad \forall x,y \in L. \end{aligned}$$

Thus, $U^{\mathrm{L}}_{C^{\mathrm{L}}_U} = U$.

(3) If $C \in \mathcal{C}_{\vee}(L)$ satisfies the neutrality principle w.r.t. e_{L}, then U^{R}_C is a disjunctive right infinitely $\wedge$-distributive left semi-uninorm by Theorem 4.2. Moreover, it follows from the adjunction conditions that

$$\begin{aligned} C^{\mathrm{R}}_{U^{\mathrm{R}}_C}(x,y) &= \wedge\{z \in L \mid y \le U^{\mathrm{R}}_C(x,z)\} \\ &= \wedge\{z \in L \mid C(x,y) \le z\} \\ &= C(x,y) \quad \forall x,y \in L. \end{aligned}$$

Therefore, $C^{\mathrm{R}}_{U^{\mathrm{R}}_C} = C$. ■

We denote by $\mathcal{U}^{e_{\mathrm{L}}}_{ds\wedge}(L)$ and $\mathcal{U}^{e_{\mathrm{R}}}_{\wedge ds}(L)$, respectively, the set of all disjunctive right infinitely $\wedge$-distributive left semi-uninorms and the set of all disjunctive left infinitely $\wedge$-distributive right semi-uninorms; by $\mathcal{C}^{npe_{\mathrm{L}}}_{\vee}(L)$ and $\mathcal{C}^{npe_{\mathrm{R}}}_{\vee}(L)$, respectively, the set of all right infinitely $\vee$-distributive coimplications which satisfy the neutrality principle w.r.t. e_{L} and the set of all right infinitely $\vee$-distributive coimplications which satisfy the neutrality principle w.r.t. e_{R} on a complete lattice.

For any $U_1, U_2, C_1, C_2 \in L^{L\times L}$, we define $U_1 \wedge U_2$, $C_1 \vee C_2 \in L^{L\times L}$ as follows:

$$\begin{aligned} (U_1 \wedge U_2)(x,y) &= U_1(x,y) \wedge U_2(x,y), \\ (C_1 \vee C_2)(x,y) &= C_1(x,y) \vee C_2(x,y) \quad \forall x,y \in L. \end{aligned}$$

It is straightforward to verify that the following theorem holds.

THEOREM 5.2 (1) $\mathcal{U}^{e_{\mathrm{L}}}_{ds\wedge}(L)$ *is a meet-semilattice with the smallest element* $U^{e_{\mathrm{L}}}_{dsW}$ *and the greatest element* $U^{e_{\mathrm{L}}}_{sM}$.

(2) $\mathcal{U}^{e_{\mathrm{R}}}_{\wedge ds}(L)$ *is a meet-semilattice with the smallest element* $U^{e_{\mathrm{R}}}_{dsW}$ *and the greatest element* $U^{e_{\mathrm{R}}}_{sM}$.

(3) $\mathcal{C}^{npe_{\mathrm{L}}}_{\vee}(L)$ *is a join-semilattice with the smallest element* $C^{\mathrm{R}}_{U^{e_{\mathrm{L}}}_{sM}}$ *and the greatest element* $C^{\mathrm{R}}_{U^{e_{\mathrm{L}}}_{dsW}}$.

(4) $\mathcal{C}^{npe_{\mathrm{R}}}_{\vee}(L)$ *is a join-semilattice with the smallest element* $C^{\mathrm{L}}_{U^{e_{\mathrm{R}}}_{sM}}$ *and the greatest element* $C^{\mathrm{L}}_{U^{e_{\mathrm{R}}}_{dsW}}$.

Define two mappings $\varphi : \mathcal{U}^{e_{\mathrm{L}}}_{ds\wedge}(L) \to \mathcal{C}^{npe_{\mathrm{L}}}_{\vee}(L)$ and $\psi : \mathcal{U}^{e_{\mathrm{R}}}_{\wedge ds}(L) \to \mathcal{C}^{npe_{\mathrm{R}}}_{\vee}(L)$ as follows:

$$\begin{aligned} \varphi(U) &= C^{\mathrm{R}}_U \quad \forall U \in \mathcal{U}^{e_{\mathrm{L}}}_{ds\wedge}(L), \\ \psi(U) &= C^{\mathrm{L}}_U \quad \forall U \in \mathcal{U}^{e_{\mathrm{R}}}_{\wedge ds}(L). \end{aligned}$$

Then it follows from Theorem 5.1 that φ and ψ are all invertible and

$$\begin{aligned} \varphi^{-1}(C) &= U^{\mathrm{R}}_C \quad \forall C \in \mathcal{C}^{npe_{\mathrm{L}}}_{\vee}(L), \\ \psi^{-1}(C) &= U^{\mathrm{L}}_C \quad \forall C \in \mathcal{C}^{npe_{\mathrm{R}}}_{\vee}(L). \end{aligned}$$

Moreover, we have the following theorem.

THEOREM 5.3 (1) *If* $U_1, U_2 \in \mathcal{U}^{e_{\mathrm{L}}}_{ds\wedge}(L)$, *then* $C^{\mathrm{R}}_{(U_1\wedge U_2)} = C^{\mathrm{R}}_{U_1} \vee C^{\mathrm{R}}_{U_2}$.

(2) *If* $U_1, U_2 \in \mathcal{U}^{e_{\mathrm{R}}}_{\wedge ds}(L)$, *then* $C^{\mathrm{L}}_{(U_1\wedge U_2)} = C^{\mathrm{L}}_{U_1} \vee C^{\mathrm{L}}_{U_2}$.

(3) *If* $C_1, C_2 \in \mathcal{C}^{npe_{\mathrm{L}}}_{\vee}(L)$, *then* $U^{\mathrm{R}}_{(C_1\vee C_2)} = U^{\mathrm{R}}_{C_1} \wedge U^{\mathrm{R}}_{C_2}$.

(4) *If* $C_1, C_2 \in \mathcal{C}^{npe_{\mathrm{R}}}_{\vee}(L)$, *then* $U^{\mathrm{L}}_{(C_1\vee C_2)} = U^{\mathrm{L}}_{C_1} \wedge U^{\mathrm{L}}_{C_2}$.

Proof We only prove that statements (2) and (4) hold.

(2) If $U_1, U_2 \in \mathcal{U}^{e_{\mathrm{R}}}_{\wedge ds}(L)$, then it follows from the left residual principle that

$$\begin{aligned} C^{\mathrm{L}}_{(U_1\wedge U_2)}(x,y) &= \wedge\{z \in L \mid y \le (U_1 \wedge U_2)(z,x)\} \\ &= \wedge\{z \in L \mid y \le U_1(z,x) \wedge U_2(z,x)\} \\ &= \wedge\{z \in L \mid y \le U_1(z,x),\ y \le U_2(z,x)\} \\ &= \wedge\{z \in L \mid C^{\mathrm{L}}_{U_1}(x,y) \le z,\ C^{\mathrm{L}}_{U_2}(x,y) \le z\} \\ &= \wedge\{z \in L \mid C^{\mathrm{L}}_{U_1}(x,y) \vee C^{\mathrm{L}}_{U_2}(x,y) \le z\} \\ &= (C^{\mathrm{L}}_{U_1} \vee C^{\mathrm{L}}_{U_2})(x,y) \quad \forall x,y \in L, \end{aligned}$$

that is, $C^{\mathrm{L}}_{(U_1\wedge U_2)} = C^{\mathrm{L}}_{U_1} \vee C^{\mathrm{L}}_{U_2}$.

(4) If $C_1, C_2 \in \mathcal{C}^{npe_{\mathrm{R}}}_{\vee}(L)$, then it follows from the adjunction conditions that

$$\begin{aligned} U^{\mathrm{L}}_{(C_1\vee C_2)}(x,y) &= \vee\{z \in L \mid (C_1 \vee C_2)(y,z) \le x\} \\ &= \vee\{z \in L \mid C_1(y,z) \vee C_2(y,z) \le x\} \\ &= \vee\{z \in L \mid C_1(y,z) \le x,\ C_2(y,z) \le x\} \\ &= \vee\{z \in L \mid z \le U^{\mathrm{L}}_{C_1}(x,y),\ z \le U^{\mathrm{L}}_{C_2}(x,y)\} \\ &= \vee\{z \in L \mid z \le U^{\mathrm{L}}_{C_1}(x,y) \wedge C^{\mathrm{L}}_{U_2}(x,y)\} \\ &= (U^{\mathrm{L}}_{C_1} \wedge U^{\mathrm{L}}_{C_2})(x,y) \quad \forall x,y \in L. \end{aligned}$$

Therefore, $U^{\mathrm{L}}_{(C_1\vee C_2)} = U^{\mathrm{L}}_{C_1} \wedge U^{\mathrm{L}}_{C_2}$. ■

By virtue of Theorem 5.3, we know that

$$\varphi(U_1 \wedge U_2) = \varphi(U_1) \vee \varphi(U_2) \quad \forall U_1, U_2 \in \mathcal{U}^{e_L}_{ds\wedge}(L),$$

$$\psi(U_1 \wedge U_2) = \psi(U_1) \vee \psi(U_2) \quad \forall U_1, U_2 \in \mathcal{U}^{e_R}_{\wedge ds}(L).$$

Thus, φ and ψ are, respectively, order-reversing isomorphisms of the meet-semilattice $\mathcal{U}^{e_L}_{ds\wedge}(L)$ onto the join-semilattice $\mathcal{C}^{npe_L}_{\vee}(L)$ and the meet-semilattice $\mathcal{U}^{e_R}_{\wedge ds}(L)$ onto the join-semilattice $\mathcal{C}^{npe_R}_{\vee}(L)$.

Similarly, we see that

$$\varphi^{-1}(C_1 \vee C_2) = \varphi^{-1}(C_1) \wedge \varphi^{-1}(C_2) \ \forall C_1, C_2 \in \mathcal{C}^{npe_L}_{\vee}(L),$$

$$\psi^{-1}(C_1 \vee C_2) = \psi^{-1}(C_1) \wedge \psi^{-1}(C_2) \ \forall C_1, C_2 \in \mathcal{C}^{npe_R}_{\vee}(L),$$

that is, φ^{-1} and ψ^{-1} are, respectively, order-reversing isomorphisms of the join-semilattice $\mathcal{C}^{npe_L}_{\vee}(L)$ onto the meet-semilattice $\mathcal{U}^{e_L}_{ds\wedge}(L)$ and the join-semilattice $\mathcal{C}^{npe_R}_{\vee}(L)$ onto the meet-semilattice $\mathcal{U}^{e_R}_{\wedge ds}(L)$.

6. Conclusions and future works

Uninorms are important generalizations of triangular norms and conorms, with the neutral elements lying anywhere in the unit interval. Noting that the associative binary operators are often used to generate n-ary aggregation operators and the commutativity is not desired for these aggregation operators in a lot of cases, Mas et al. introduced the concepts of left and right uninorms on [0, 1] in Mas et al. (2001) and later on a finite chain in Mas et al. (2004) by eliminating the commutativity from the axioms of uninorm and Liu (2012) discussed the concept of semi-uninorms on a complete lattice by removing the associativity and commutativity from the axioms of uninorm. On the other hand, Mas et al. (2007) and Ruiz & Torrens (2004) studied the implications and coimplications derived from uninorms on [0, 1], Wang & Fang (2009b) discussed the residual coimplications of left and right uninorms on a complete lattice, and Su & Wang (2013) investigated pseudo-uninorms and coimplications on a complete lattice.

In this paper, motivated by these works, we discuss the residual coimplicators of left and right semi-uninorms and the left and right semi-uninorms induced by coimplications, show that the right (left) residual coimplicator of a disjunctive right (left) infinitely $\wedge$-distributive left (right) semi-uninorm is a right infinitely $\vee$-distributive coimplication which satisfies the neutrality principle, give some conditions such that the operations induced by a coimplication constitute left or right semi-uninorms, demonstrate that the operations induced by a right infinitely $\vee$-distributive coimplication, which satisfies the order property or the neutrality principle, are left (right) infinitely $\wedge$-distributive left (right) semi-uninorms or right (left) semi-uninorms, and prove that the meet-semilattice of all disjunctive right (left) infinitely $\wedge$-distributive left (right) semi-uninorms is order-reversing isomorphic to the join-semilattice of all right infinitely $\vee$-distributive coimplications that satisfy the neutrality principle.

In forthcoming papers, we will investigate the relationships between left (right) semi-uninorms, implications and coimplications on a complete lattice.

Disclosure statement

No potential conflict of interest was reported by the authors.

Funding

This work is supported by the National Natural Science Foundation of China [61379064] and Jiangsu Provincial Natural Science Foundation of China [BK2012672].

References

Baczyński, M., & Jayaram, B. (2008). *Fuzzy implication, studies in fuzziness and soft computing* (Vol. 231). Berlin: Springer.

Bassan, B., & Spizzichino, F. (2005). Relations among univariate aging, bivariate aging and dependence for exchangeable lifetimes. *Journal of Multivariate Analysis, 93*, 313–339.

Birkhoff, G. (1967). *Lattice theory*. Providence: American Mathematical Society Colloquium.

De Baets, B. (1997). Coimplicators, the forgotten connectives. *Tatra Mountains Mathematical Publications, 12*, 229–240.

De Baets, B. (1999). Idempotent uninorms. *European Journal of Operational Research, 118*, 631–642.

De Baets, B., & Fodor, J. (1999). Residual operators of uninorms. *Soft Computing, 3*, 89–100.

De Baets, B., Tsiporkova, E., & Mesiar, R. (1999). Conditioning in possibility theory with strict order norms. *Fuzzy Sets and Systems, 106*, 221–229.

De Cooman, G., & Kerre, E. E. (1994). Order norms on bounded partially ordered sets. *Journal of Fuzzy Mathematics, 2*, 281–310.

Durante, F., Klement, E. P., Mesiar, R., & Sempi, C. (2007). Conjunctors and their residual implicators: Characterizations and construction methods. *Mediterranean Journal of Mathematics, 4*, 343–356.

Flondor, P., Georgescu, G., & Iorgulescu, A. (2001). Pseudo-t-norms and pseudo-BL-algebras. *Soft Computing, 5*, 355–371.

Fodor, J. C., & Keresztfalvi, T. (1995). Nonstandard conjunctions and implications in fuzzy logic. *International Journal of Approximate Reasoning, 12*, 69–84.

Fodor, J. C., Yager, R. R., & Rybalov, A. (1997). Structure of uninorms. *International Journal of Uncertainty, Fuzziness and Knowledge-Based Systems, 5*, 411–427.

Gabbay, D., & Metcalfe, G. (2007). Fuzzy logics based on [0,1)-continuous uninorms. *Archive for Mathematical Logic, 46*, 425–449.

Liu, H.-W. (2012). Semi-uninorms and implications on a complete lattice. *Fuzzy Sets and Systems, 191*, 72–82.

Mas, M., Monserrat, M., & Torrens, J. (2001). On left and right uninorms. *International Journal of Uncertainty, Fuzziness and Knowledge-Based Systems, 9*, 491–507.

Mas, M., Monserrat, M., & Torrens, J. (2004). On left and right uninorms on a finite chain. *Fuzzy Sets and Systems, 146*, 3–17.

Mas, M., Monserrat, M., & Torrens, J. (2007). Two types of implications derived from uninorms. *Fuzzy Sets and Systems*, *158*, 2612–2626.

Ruiz, D., & Torrens, J. (2004). Residual implications and coimplications from idempotent uninorms. *Kybernetika*, *40*, 21–38.

Su, Y., & Wang, Z. D. (2013). Pseudo-uninorms and coimplications on a complete lattice. *Fuzzy Sets and Systems*, *224*, 53–62.

Su, Y., & Wang, Z. D. (2014). Constructing implications and coimplications on a complete lattice. *Fuzzy Sets and Systems*, *247*, 68–80.

Su, Y., Wang, Z. D., & Tang, K. M. (2013). Left and right semi-uninorms on a complete lattice. *Kybernetika*, *49*, 948–961.

Suárez García, F., & Gil Álvarez, P. (1986). Two families of fuzzy integrals. *Fuzzy Sets and Systems*, *18*, 67–81.

Tsadiras, A. K., & Margaritis, K. G. (1998). The MYCIN certainty factor handling function as uninorm operator and its use as a threshold function in artificial neurons. *Fuzzy Sets and Systems*, *93*, 263–274.

Wang, Z. D., & Fang, J.-X. (2009a). Residual operations of left and right uninorms on a complete lattice. *Fuzzy Sets and Systems*, *160*, 22–31.

Wang, Z. D., & Fang, J.-X. (2009b). Residual coimplicators of left and right uninorms on a complete lattice. *Fuzzy Sets and Systems*, *160*, 2086–2096.

Yager, R. R. (2001). Uninorms in fuzzy systems modeling. *Fuzzy Sets and Systems*, *122*, 167–175.

Yager, R. R. (2002). Defending against strategic manipulation in uninorm-based multi-agent decision making. *European Journal of Operational Research*, *141*, 217–232.

Yager, R. R., & Rybalov, A. (1996). Uninorm aggregation operators. *Fuzzy Sets and Systems*, *80*, 111–120.

6

Global stability in Lagrange sense for BAM-type Cohen–Grossberg neural networks with time-varying delays

Jigui Jian* and Zhihua Zhao

College of Science, China Three Gorges University, Yichang, Hubei 443002, People's Republic of China

In this paper, we investigate the positive invariant sets and global exponential attractive sets for a class of bidirectional associative memory (BAM)-type Cohen–Grossberg neural networks with multiple time-varying delays. By applying inequality techniques, some easily verifiable delay-independent criteria for the ultimate boundedness and global exponential attractive sets of BAM-type Cohen–Grossberg neural networks are obtained by constructing appropriate Lyapunov functions. Finally, one example with numerical simulations is given to illustrate the results obtained in this paper.

Keywords: BAM-type Cohen–Grossberg neural networks; Lagrange stability; globally attractive set; inequality

1. Introduction

Recently, the stability of different types of Cohen–Grossberg neural networks and the bidirectional associative memory (BAM) neural networks has been widely studied by many researchers and various interesting results have been reported (Jiang & Cao, 2008; Li, 2009; Li, Fei, Tan, & Zhang, 2009; Li, Zhang, Zhang, & Li, 2010; Liu & Zong, 2009; Wang, Jian, & Guo, 2008; Zhang, Liu, & Zhou, 2012). In many applications, since BAM-type Cohen–Grossberg neural networks consider the interaction between two neural networks, the studies of the stability behavior of BAM-type Cohen–Grossberg neural networks are of greater interest than the studies of the stability of Cohen–Grossberg neural networks and BAM neural networks.

For BAM-type Cohen–Grossberg neural networks, the existence of periodic solution and an equilibrium point and their stability have been investigated in Jiang and Cao (2008), Li (2009), Li et al. (2010) and Zhang et al. (2012). But in many actual applications, these conclusions are no longer appropriate in the multistable dynamics which have multiple equilibrium and so many of them are unstable (Lu, Wang, & Chen, 2011; Wang & Chen, 2012). Such as the Cohen–Grossberg neural network, when applications are taken into account in biology, it is necessary and important to deal with multistable properties. In this context, it is worth mentioning that the Lagrange stability refers to the stability of the total system which does not require the information of equilibrium points, because the Lagrange stability is considered on the basis of the boundedness of solutions and the existence of global attractive sets (Liao, Luo, & Zeng, 2008). Just as verified in (Liao et al., 2008), outside the globally attracting set, there is no equilibrium point, chaos attractor, periodic state or almost periodic state in the neural networks. Therefore, the research on positive invariant sets and globally attractive sets of the neural networks have been done by many scholars (Liao et al., 2008; Luo, Zeng, & Liao, 2011; Song & Zhao, 2005; Tu, Jian, & Wang, 2011; Tu, Wang, Zha, & Jian, 2013; Wang, Jian, & Jiang, 2010). Song and Zhao (2005) investigated the dissipativity of neural networks with both variable and unbounded delays by constructing proper Lyapunov functions and using some analytic techniques. And the global stability in the Lagrange sense for a class of Cohen–Grossberg neural networks with time-varying delays and finite distributed delays was studied in Tu et al. (2011) and Wang et al. (2010). In Tu et al. (2013), the authors study the global dissipativity of a class of BAM neural networks with both time-varying and unbound delays. To our best knowledge, few authors have discussed the global attractive sets for BAM-type Cohen–Grossberg neural networks.

Motived by the above analysis, the aim of this paper is to study Lagrange stability and global exponential attractive sets for BAM-type Cohen–Grossberg neural networks with time-varying delays and some delay-independent criteria for the ultimate boundedness and global exponential attractive sets of BAM-type Cohen–Grossberg neural networks are obtained. And some results here obtained in this paper are more general than that of the existing reference on the globally exponentially attractive (GEA) set as special cases. The remaining paper is organized as follows: Section 2 describes some preliminaries including some necessary notations, definitions, assumptions and some lemmas. The main results are stated in Section 3. Section 4

*Corresponding author. Email: jiguijian@ctgu.du.cn

gives a numerical example to testify the theoretical analysis. Finally, conclusions are drawn in Section 5.

2. Problem statement

Consider the following BAM-type Cohen–Grossberg neural network model

$$\begin{cases} \dot{x}_i(t) = a_i(x_i(t)) \begin{bmatrix} -c_i(x_i(t)) + \sum_{j=1}^{m} a_{ij} f_j(y_j(t)) \\ + \sum_{j=1}^{m} b_{ij} f_j(y_j(t-\tau_j(t))) + I_i \end{bmatrix}, \\ \quad i = 1, 2, \ldots, n, \\ \dot{y}_j(t) = b_j(y_j(t)) \begin{bmatrix} -d_j(y_j(t)) + \sum_{i=1}^{n} m_{ji} g_i(x_i(t)) \\ + \sum_{i=1}^{n} n_{ji} g_i(x_i(t-\sigma_i(t))) + J_j \end{bmatrix}, \\ \quad j = 1, 2, \ldots, m, \end{cases} \tag{1}$$

where $x(t) = (x_1(t), \ldots, x_n(t))^{\mathrm{T}}, y(t) = (y_1(t), \ldots, y_m(t))^{\mathrm{T}}$ are the neuron state vectors of the neural network (1), $a_i(x_i(t)) > 0$ for $i \in \Lambda = \{1, 2, \ldots, n\}$ and $b_j(y_j(t)) > 0$ for $j \in \Gamma = \{1, 2, \ldots, m\}$ represent amplification functions of the ith neurons from the neural field F_x and the jth neurons from the neural field F_y, respectively; $c_i(x_i)$, $d_j(y_j)$ are appropriately behaved functions of the ith neurons from the neural field F_x and the jth neurons from the neural field F_y, respectively; f_j and g_i are the activation functions, I_i and J_j are the exogenous inputs. $A = (a_{ij})_{n\times m}$, $M = (m_{ji})_{m\times n}$; $B = (b_{ij})_{n\times m}$, $N = (n_{ji})_{m\times n}$ are the connection weight matrices and the delayed weight matrices, respectively, $I = (I_1, I_2, \ldots, I_n)^{\mathrm{T}}, J = (J_1, J_2, \ldots, J_m)^{\mathrm{T}}$ are external input vectors. The time-varying delays $\tau_j(t)$, $\sigma_i(t)$ are non-negative and bounded, i.e. $0 \le \tau_j(t) \le \tau_j$, $0 \le \sigma_i(t) \le \sigma_i$. We define the vector functions f, g by

$$g(x(\cdot)) = (g_1(x_1(\cdot)), g_2(x_2(\cdot)), \ldots, g_n(x_n(\cdot)))^{\mathrm{T}} \in R^n,$$
$$f(y(\cdot)) = (f_1(y_1(\cdot), f_2(y_2(\cdot), \ldots, f_m(y_m(\cdot)))^{\mathrm{T}} \in R^m.$$

The initial conditions associated with Equation (1) are given by

$$\begin{cases} x_i(\theta) = \phi_i(\theta), & \theta \in [-\sigma, 0], i \in \Lambda, \\ y_j(\theta) = \psi_j(\theta), & \theta \in [-\tau, 0], j \in \Gamma, \end{cases} \tag{2}$$

where $\phi_i(\theta)$ and $\psi_j(\theta)$ are continuous real valued functions defined on their respective domains, $\sigma = \max_{1\le i\le n}\{\sigma_i\}$, $\tau = \max_{1\le j\le m}\{\tau_j\}$.

Remark 1 It is obvious that system (1) includes neural systems considered in Tu et al. (2013) as its special case. For example, $a_i(x_i(t)) = 1$, $b_j(y_j(t)) = 1$ for $i \in \Lambda$ and $j \in \Gamma$, $c_i(x_i(t)) = \tilde{a}_i x_i(t)$ and $d_j(y_j(t)) = \tilde{c}_j y_j(t)$ with the constants $\tilde{a}_i > 0$ and $\tilde{c}_j > 0$, system (1) reduces to the BAM neural network in Tu et al. (2013).

In order to establish the conditions of main results for the neural networks (1), we have the following assumptions:

(H1) $a_i(u), b_j(u) \in C(R, R^+)$. Furthermore, there exist positive constants $\underline{a}_i$, $\bar{a}_i$, $\underline{b}_j$ and $\bar{b}_j (i \in \Lambda, j \in \Gamma)$ such that $0 < \underline{a}_i \le a_i(u) \le \bar{a}_i$, $0 < \underline{b}_j \le b_j(u) \le \bar{b}_j$, $u \in R$.

(H2) $c_i(u), d_j(u) \in C(R, R^+)$. Moreover, there exist positive constants $\underline{c}_i$, $\bar{c}_i$, $\underline{d}_j$ and $\bar{d}_j (i \in \Lambda, j \in \Gamma)$ such that $\underline{c}_i u^2 \le u c_i(u) \le \bar{c}_i u^2$, $\underline{d}_j u^2 \le u d_j(u) \le \bar{d}_j u^2$, $u \in R$.

The set of bounded activation functions is defined as

$$B = \{p(x) | p_i(x_i) \in C(R, R), \exists k_i > 0, |p_i(x_i)| \le k_i, \forall x_i \in R\}$$

The sigmoid function is defined as

$$S = \left\{ p(x) \;\middle|\; \begin{array}{l} p_i(0) = 0, p_i(x_i) \in C(R, R), \\ D^+ p_i(x_i) \ge 0, |p_i(x_i)| \le k_i, \forall x_i \in R \end{array} \right\}.$$

Remark 2 In this paper, $f(\cdot), g(\cdot) \in B$ represent $|g_i(\cdot)| \le s_i$, $|f_j(\cdot)| \le r_j$ for $i \in \Lambda$ and $j \in \Gamma$, respectively, where s_i, r_j are all positive constants.

Let

$$\tilde{M} = \sum_{i=1}^{n} \bar{a}_i M_i s_i,$$

$$M_i = \bar{a}_i \left(\sum_{j=1}^{m} (|a_{ij}| + |b_{ij}|) r_j + |I_i| \right), \quad i \in \Lambda;$$

$$\tilde{N} = \sum_{j=1}^{m} \bar{b}_j N_j r_j,$$

$$N_j = \bar{b}_j \left(\sum_{i=1}^{n} (|m_{ji}| + |n_{ji}|) s_i + |J_j| \right), \quad j \in \Gamma.$$

Let $\Omega \subset R^{n+m}$ be a compact set in R^{n+m}. Denote the complement of Ω by $R^{n+m} \backslash \Omega$. For any

$$\begin{pmatrix} x(t) \\ y(t) \end{pmatrix} \in R^{n+m}, \ \rho\left(\begin{pmatrix} x \\ y \end{pmatrix}, \Omega \right) = \inf_{(x_1^T y_1^T) \in \Omega} \left\| \begin{pmatrix} x \\ y \end{pmatrix} - \begin{pmatrix} x_1 \\ y_1 \end{pmatrix} \right\|$$

is the distance between $\binom{x}{y}$ and Ω. We call a compact set Ω as a global attractive set of networks (1), if for every solution

$$\begin{pmatrix} x(t) \\ y(t) \end{pmatrix} \in R^{n+m} \backslash \Omega$$

with initial condition (2), we have

$$\lim_{t\to+\infty}\rho\left(\begin{pmatrix}x(t)\\ y(t)\end{pmatrix},\Omega\right)=0.$$

Obviously, if the network (1) has global attractive sets, then the solutions are ultimately bounded.

DEFINITION 1 *A compact set $\Omega\in R^{n+m}$ is said to be a global exponential stable (GES) set of system (1), if there exists a constant α and a non-negative bounded continuous functional K such that for every solution $\begin{pmatrix}x(t)\\ y(t)\end{pmatrix}\in R^{n+m}\backslash\Omega$ with an initial condition (2), we have*

$$\rho\left(\begin{pmatrix}x(t)\\ y(t)\end{pmatrix},\Omega\right)\le K(\phi,\psi)\,e^{-\alpha(t-t_0)}.$$

DEFINITION 2 *If there exists a radially unbounded and positive definite Lyapunov function $V(t)=V(x(t),y(t))$ and positive constants l and α such that for any solution $\begin{pmatrix}x(t)\\ y(t)\end{pmatrix}\in R^{n+m}\backslash\Omega$ of (1), $V(t)>l$ for $t\ge t_0$ implies $V(t)-l\le(V(t_0)-l)\,e^{-\alpha(t-t_0)}$, system (1) is said to be GEA. The compact set*

$$\Omega=\left\{\begin{pmatrix}x\\ y\end{pmatrix}\in R^{n+m}|V(t)\le l.\right\}$$

is called a GEA set of Equation (1).

DEFINITION 3 *Network (1) is called GES in the Lagrange sense, if it is both uniformly bounded and GEA.*

LEMMA 1 *For $\forall x,y\in R$, $a>0$, the inequality $-ax^2+xy\le-\frac{1}{2}ax^2+y^2/2a$ holds.*

LEMMA 2 (Wang et al., 2010) *Let $a\ge0, b\ge0, p>1, q>1$ with $1/p+1/q=1$. Then the inequality $ab\le(1/p)a^p+(1/q)b^q$ holds, and the equality holds if and only if $a^p=b^q$.*

LEMMA 3 (Luo et al., 2011) *Let $V(t)\in C[R^n,R^+]$ be a positive definite and radially unbounded function, and suppose there exist two constants $\alpha>0$, $\beta>0$ such that $D^+V(t)\le-\alpha V(t)+\beta$ for $t\ge t_0$, then $V(t)\ge\beta/\alpha$ implies*

$$V(t)-\frac{\beta}{\alpha}\le\left(V(t_0)-\frac{\beta}{\alpha}\right)e^{-\alpha(t-t_0)}.$$

3. Main results

THEOREM 1 *If the activation functions $f(\cdot),g(\cdot)\in B$ and (H1), (H2) are also satisfied, then system (1) is globally exponentially stable in Lagrange sense and Ω_i for $i=1,2,3,4,5$ are all GES set and the set $\Omega=\bigcap_{i=1}^{5}\Omega_i$ is a better GES set of Equation (1), where*

$$\Omega_1=\left\{\begin{pmatrix}x\\ y\end{pmatrix}\in R^{n+m}\left|\begin{array}{l}\sum_{i=1}^{n}x_i^2(t)+\sum_{j=1}^{m}y_j^2(t)\\ \le\frac{\sum_{i=1}^{n}M_i^2/\underline{a}_i\underline{c}_i+\sum_{j=1}^{m}N_j^2/\underline{b}_j\underline{d}_j}{\min_{\substack{1\le i\le n\\1\le j\le m}}\{\underline{a}_i\underline{c}_i,\underline{b}_j\underline{d}_j\}}\end{array}\right.\right\}.$$

$$\Omega_2=\left\{\begin{pmatrix}x\\ y\end{pmatrix}\in R^{n+m}\left|\begin{array}{l}\sum_{i=1}^{n}|x_i(t)|+\sum_{j=1}^{m}|y_j(t)|\\ \le\frac{\sum_{i=1}^{n}M_i+\sum_{j=1}^{m}N_j}{\min_{\substack{1\le i\le n\\1\le j\le m}}\{\underline{a}_i\underline{c}_i,\underline{b}_j\underline{d}_j\}}\end{array}\right.\right\}.$$

$$\Omega_3=\left\{\begin{pmatrix}x\\ y\end{pmatrix}\in R^{n+m}\left|\begin{array}{l}|x_i(t)|\le\frac{M_i}{\underline{a}_i\underline{c}_i},i\in\Lambda;\\ |y_j(t)|\le\frac{N_j}{\underline{b}_j\underline{d}_j},j\in\Gamma\end{array}\right.\right\}.$$

$$\Omega_4=\left\{\begin{pmatrix}x\\ y\end{pmatrix}\in R^{n+m}\left|\begin{array}{l}\sum_{i=1}^{n}x_i^2(t)\le\frac{\sum_{i=1}^{n}M_i^2/\underline{a}_i\underline{c}_i}{\min_{1\le i\le n}\{\underline{a}_i\underline{c}_i\}},\\ \sum_{j=1}^{m}y_j^2(t)\le\frac{\sum_{j=1}^{m}N_j^2/\underline{b}_j\underline{d}_j}{\min_{1\le i\le n}\{\underline{b}_j\underline{d}_j\}}\end{array}\right.\right\}.$$

$$\Omega_5=\left\{\begin{pmatrix}x\\ y\end{pmatrix}\in R^{n+m}\left|\begin{array}{l}\sum_{i=1}^{n}|x_i(t)|\le\frac{\sum_{i=1}^{n}M_i}{\min_{1\le i\le n}\{\underline{a}_i\underline{c}_i\}},\\ \sum_{j=1}^{m}|y_j(t)|\le\frac{\sum_{j=1}^{m}N_j}{\min_{1\le j\le m}\{\underline{b}_j\underline{d}_j\}}\end{array}\right.\right\}.$$

Proof (1) Employ a radially unbounded and positive definite Lyapunov function as

$$V(t)=\frac{1}{2}\sum_{i=1}^{n}x_i^2(t)+\frac{1}{2}\sum_{j=1}^{m}y_j^2(t).$$

Calculating the Dini derivative of $V(t)$ along the positive semi-trajectory of Equation (1), and by virtue of Lemma 1, we obtain

$$\left.\frac{dV(t)}{dt}\right|_{(1)}\le-\sum_{i=1}^{n}\underline{a}_i\underline{c}_ix_i^2(t)+\sum_{i=1}^{n}\bar{a}_i\left(\sum_{j=1}^{m}(|a_{ij}|+|b_{ij}|)r_j+|I_i|\right)|x_i(t)|-\sum_{j=1}^{m}\underline{b}_j\underline{d}_jy_j^2(t)$$

$$+\sum_{j=1}^{m}\bar{b}_j\left(\sum_{i=1}^{n}(|m_{ji}|+|n_{ji}|)s_i+|J_j|\right)|y_j(t)|$$

$$=-\sum_{i=1}^{n}\underline{a}_i\underline{c}_ix_i^2(t)+\sum_{i=1}^{n}M_i|x_i(t)|-\sum_{j=1}^{m}\underline{b}_j\underline{d}_jy_j^2(t)$$

$$+\sum_{j=1}^{m}N_j|y_j(t)|$$

$$\le -\frac{1}{2}\sum_{i=1}^{n}\underline{a}_i\underline{c}_ix_i^2(t)+\frac{1}{2}\sum_{i=1}^{n}\frac{M_i^2}{\underline{a}_i\underline{c}_i}$$

$$-\frac{1}{2}\sum_{j=1}^{m}\underline{b}_j\underline{d}_jy_j^2(t)+\frac{1}{2}\sum_{j=1}^{m}\frac{N_j^2}{\underline{b}_j\underline{d}_j}$$

$$\le -\min_{\substack{1\le i\le n\\1\le j\le m}}\{\underline{a}_i\underline{c}_i,\underline{b}_j\underline{d}_j\}V(t)+\frac{1}{2}\sum_{i=1}^{n}\frac{M_i^2}{\underline{a}_i\underline{c}_i}+\frac{1}{2}\sum_{j=1}^{m}\frac{N_j^2}{\underline{b}_j\underline{d}_j}$$

$$=-\alpha V(t)+\beta,$$

where

$$\alpha=\min_{\substack{1\le i\le n\\1\le j\le m}}\left\{\underline{a}_i\underline{c}_i,\underline{b}_j\underline{d}_j\right\},$$

$$\beta=\frac{1}{2}\sum_{i=1}^{n}\frac{M_i^2}{\underline{a}_i\underline{c}_i}+\frac{1}{2}\sum_{j=1}^{m}\frac{N_j^2}{\underline{b}_j\underline{d}_j}.$$

According to Lemma 3, for $V(t)\ge\beta/\alpha$, $V(t_0)\ge\beta/\alpha$, we have

$$V(t)-\frac{\beta}{\alpha}\le(V(t_0)-\frac{\beta}{\alpha})\mathrm{e}^{-\alpha(t-t_0)}.$$

Hence Ω_1 is a GES set of Equation (1).

(2) Construct another positive definite and radially unbounded Lyapunov function as

$$V(t)=\sum_{i=1}^{n}|x_i(t)|+\sum_{j=1}^{m}|y_j(t)|.$$

So we can get

$$D^+V(t)|_{(1)}$$

$$\le-\sum_{i=1}^{n}\underline{a}_i\underline{c}_i|x_i(t)|$$

$$+\sum_{i=1}^{n}\left(\bar{a}_i\sum_{j=1}^{m}(|a_{ij}t|+|b_{ij}|)r_j+|I_i|\right)$$

$$-\sum_{j=1}^{m}\underline{b}_j\underline{d}_j|y_j(t)|$$

$$+\sum_{j=1}^{m}\bar{b}_j\left(\sum_{i=1}^{n}(|m_{ji}|+|n_{ji}|)s_i+|J_j|\right)$$

$$\le-\alpha V(t)+\beta,$$

where

$$\alpha=\min_{\substack{1\le i\le n\\1\le j\le m}}\{\underline{a}_i\underline{c}_i,\underline{b}_j\underline{d}_j\},\quad \beta=\sum_{i=1}^{n}M_i+\sum_{j=1}^{m}N_j.$$

So we get

$$V(t)-\frac{\beta}{\alpha}\le\left(V(t_0)-\frac{\beta}{\alpha}\right)\mathrm{e}^{-\alpha(t-t_0)}.$$

And the set Ω_2 is a GES set of Equation (1).

(3) Choose another two positive definite and radially unbounded Lyapunov functions as

$$V_i(t)=|x_i(t)|,\ i\in\Lambda;\quad V_j(t)=|y_j(t)|,\ j\in\Gamma.$$

And we have

$$D^+V_i(t)|_{(1)}\le-\underline{a}_i\underline{c}_iV_i(t)+M_i,\quad i\in\Lambda,$$
$$D^+V_j(t)|_{(1)}\le-\underline{b}_j\underline{d}_jV_j(t)+N_j,\quad j\in\Gamma.$$

So we have

$$V_i(t)-\frac{M_i}{\underline{a}_i\underline{c}_i}\le\left(V_i(t_0)-\frac{M_i}{\underline{a}_i\underline{c}_i}\right)\mathrm{e}^{-\underline{a}_i\underline{c}_i(t-t_0)},\ i\in\Lambda,$$

$$V_j(t)-\frac{N_j}{\underline{b}_j\underline{d}_j}\le\left(V_j(t_0)-\frac{N_j}{\underline{b}_j\underline{d}_j}\right)\mathrm{e}^{-\underline{b}_j\underline{d}_j(t-t_0)},\ j\in\Gamma.$$

So the Ω_3 is a GES set of Equation (1).

(4) Employ only the following two radially unbounded and positive definite Lyapunov functions as

$$V_x(t)=\frac{1}{2}\sum_{i=1}^{n}x_i^2(t),\quad V_y(t)=\frac{1}{2}\sum_{j=1}^{m}y_j^2(t).$$

The remaining proof is similar to the proof in the previous part (1). Meanwhile, consider only the following other two Lyapunov functions

$$V_x(t)=\sum_{i=1}^{n}|x_i(t)|,\quad V_y(t)=\sum_{j=1}^{m}|y_j(t)|.$$

The remaining proof is similar to that in the previous part (2). So the sets Ω_4 and Ω_5 are also GES sets of (1). According to the definition of intersection set, we know that the set $\Omega=\bigcap_{i=1}^{5}\Omega_i$ is a better GES set of NN (1). The proof of Theorem 1 is completed. ∎

Remark 3 When $a_i(x_i(t))=1, b_j(y_j(t))=1$ for $i\in\Lambda$ and $j\in\Gamma$, $c_i(x_i(t))=\tilde{a}_ix_i(t)$ and $d_j(y_j(t))=\tilde{c}_jy_j(t)$ with the constants $\tilde{a}_i > 0$ and $\tilde{c}_j > 0$, the set Ω_5 in Theorem 1 here is just the main result (I) of Theorem 3.2 in Tu et al. (2013).

THEOREM 2 *Let $p > 1$, $q > 1$ and $1/p + 1/q = 1$. Choose $\varepsilon_i > 0$, $\bar{\varepsilon}_j > 0 (i \in \Lambda, j \in \Gamma)$ such that $\mu_i = p\underline{a}_i\underline{c}_i - (p-1)\varepsilon_i > 0$, $\eta_j = q\underline{b}_j\underline{d}_j - (q-1)\bar{\varepsilon}_j > 0$. If the activation functions $f(\cdot), g(\cdot) \in B$ and (H1), (H2) are also satisfied, then NN (1) is globally exponentially stable in Lagrange sense and Ω_6 is a GES set, where*

$$\Omega_6 = \left\{ \begin{pmatrix} x \\ y \end{pmatrix} \in R^{n+m} \left| \begin{array}{l} \frac{1}{p}\sum_{i=1}^{n} |x_i(t)|^p + \frac{1}{q}\sum_{j=1}^{m} |y_j(t)|^q \\ \le \dfrac{\sum_{i=1}^{n} M_i^p / p\varepsilon_i^{p-1} + \sum_{j=1}^{m} N_j^q / q\bar{\varepsilon}_j^{q-1}}{\min_{\substack{1\le i\le n \\ 1\le j\le m}} \{\mu_i, \eta_j\}} \end{array} \right. \right\}.$$

Proof We introduce the following Lyapunov function

$$V(t) = \frac{1}{p}\sum_{i=1}^{n} |x_i(t)|^p + \frac{1}{q}\sum_{j=1}^{m} |y_j(t)|^q.$$

Calculating the Dini derivative of $V(t)$ along (1), and by virtue of Lemma 2, we can obtain

$$\begin{aligned} D^+V(t)|_{(1)} &\le -\sum_{i=1}^{n} \underline{a}_i\underline{c}_i |x_i(t)|^p + \sum_{i=1}^{n} M_i |x_i(t)|^{p-1} \\ &\quad - \sum_{j=1}^{m} \underline{b}_j\underline{d}_j |y_j(t)|^q + \sum_{j=1}^{m} N_j |y_j(t)|^{q-1} \\ &\le -\sum_{i=1}^{n} \underline{a}_i\underline{c}_i |x_i(t)|^p + \sum_{i=1}^{n} \left(\frac{p-1}{p}\varepsilon_i |x_i(t)|^p + \frac{1}{p\varepsilon_i^{p-1}} M_i^p \right) \\ &\quad - \sum_{j=1}^{m} \underline{b}_j\underline{d}_j |y_j(t)|^q + \sum_{j=1}^{m} \left(\frac{q-1}{q}\bar{\varepsilon}_j |y_j(t)|^q + \frac{1}{q\bar{\varepsilon}_j^{q-1}} N_j^q \right) \\ &\le -\sum_{i=1}^{n} \left(\underline{a}_i\underline{c}_i - \frac{p-1}{p}\varepsilon_i \right) |x_i(t)|^p + \sum_{i=1}^{n} \frac{1}{p\varepsilon_i^{p-1}} M_i^p \\ &\quad - \sum_{j=1}^{m} \left(\underline{b}_j\underline{d}_j - \frac{q-1}{q}\bar{\varepsilon}_j \right) |y_j(t)|^q + \sum_{j=1}^{m} \frac{1}{q\bar{\varepsilon}_j^{q-1}} N_j^q \\ &\le -\alpha V(t) + \beta, \end{aligned}$$

where

$$\alpha = \min_{\substack{1\le i\le n \\ 1\le j\le m}} \{\mu_i, \eta_j\}, \quad \beta = \sum_{i=1}^{n} \frac{M_i^p}{p\varepsilon_i^{p-1}} + \sum_{j=1}^{m} \frac{N_j^q}{q\bar{\varepsilon}_j^{q-1}}.$$

And by Lemma 3, we get

$$V(t) - \frac{\beta}{\alpha} \le \left(V(t_0) - \frac{\beta}{\alpha} \right) e^{-\alpha(t-t_0)}.$$

So the set Ω_6 is a GES set of Equation (1). ∎

Remark 4 When $a_i(x_i(t)) = 1, b_j(y_j(t)) = 1$ for $i \in \Lambda$ and $j \in \Gamma$, $c_i(x_i(t)) = \tilde{a}_i x_i(t)$ and $d_j(y_j(t)) = \tilde{c}_j y_j(t)$ with the constants $\tilde{a}_i > 0$ and $\tilde{c}_j > 0$ the sets Ω_6 in Theorem 2 here are just the main result of Theorem 3.1 in Tu et al. (2013).

THEOREM 3 *If the activation functions $f(\cdot), g(\cdot) \in S$ and (H1), (H2) are also satisfied, then NN (1) has positive invariant and globally exponential attractive sets*

$$\Omega_7 = \left\{ \begin{pmatrix} x \\ y \end{pmatrix} \in R^{n+m} \left| \begin{array}{l} \sum_{i=1}^{n} \int_0^{x_i(t)} g_i(s)\,ds \\ + \sum_{j=1}^{m} \int_0^{y_j(t)} f_j(\eta)\,d\eta \\ \le \dfrac{\tilde{M} + \tilde{N}}{\min_{\substack{1\le i\le n \\ 1\le j\le m}} \{\underline{a}_i\underline{c}_i, \underline{b}_j\underline{d}_j\}} \end{array} \right. \right\},$$

$$\Omega_8 = \left\{ \begin{pmatrix} x \\ y \end{pmatrix} \in R^{n+m} \left| \begin{array}{l} \sum_{i=1}^{n} \int_0^{x_i(t)} g_i(s)\,ds \\ \le \dfrac{\tilde{M}}{\min_{1\le i\le n}\{\underline{a}_i\underline{c}_i\}}, \\ \sum_{j=1}^{m} \int_0^{y_j(t)} f_j(\eta)\,d\eta \le \\ \dfrac{\tilde{N}}{\min_{1\le j\le m}\{\underline{b}_j\underline{d}_j\}} \end{array} \right. \right\}.$$

And $\Omega = \Omega_7 \cap \Omega_8$ is a better GES set of Equation (1).

Proof Firstly, employ the following Lyapunov function

$$V(t) = \sum_{i=1}^{n} \int_0^{x_i(t)} g_i(s)\,ds + \sum_{j=1}^{m} \int_0^{y_j(t)} f_j(\eta)\,d\eta.$$

Calculating the derivative of $V(t)$, we have

$$\begin{aligned} \left.\frac{dV(t)}{dt}\right|_{(1)} &= \sum_{i=1}^{n} g_i(x_i(t))\dot{x}_i(t) + \sum_{j=1}^{m} f_j(y_j(t))\dot{y}_j(t) \\ &\le -\sum_{i=1}^{n} \underline{a}_i\underline{c}_i x_i(t) g_i(x_i(t)) + \tilde{M} \\ &\quad - \sum_{j=1}^{m} \underline{b}_j\underline{d}_j y_j(t) f_j(y_j(t)) + \tilde{N} \\ &\le - \min_{\substack{1\le i\le n \\ 1\le j\le m}} \{\underline{a}_i\underline{c}_i, \underline{b}_j\underline{d}_j\} V(t) + \tilde{M} + \tilde{N} = -\alpha V(t) + \beta, \end{aligned}$$

where $\alpha = \min_{\substack{1\le i\le n\\1\le j\le m}} \{\underline{a}_i\underline{c}_i, \underline{b}_j\underline{d}_j\}$, $\beta = \tilde{M} + \tilde{N}$. In the light of Lemma 3, we get

$$V(t) - \frac{\beta}{\alpha} \le \left(V(t_0) - \frac{\beta}{\alpha}\right) \mathrm{e}^{-\alpha(t-t_0)}.$$

So the set Ω_7 is a GES set of Equation (1).

Secondly, consider the following Lyapunov functions

$$V_1(t) = \sum_{i=1}^{n} \int_0^{x_i(t)} g_i(s)\,\mathrm{d}s, \quad V_2(t) = \sum_{j=1}^{m} \int_0^{y_j(t)} f_j(\eta)\,\mathrm{d}\eta.$$

Similar to the proof in the previous part, we can obtain that the set Ω_8 is a GES set of Equation (1). Hence, $\Omega = \Omega_7 \cap \Omega_8$ is a better GES set of neural network (1). ■

4. Illustrative examples

In this section, we will give an example to verify our theoretical results.

Example 4.1 Consider the following example:

$$\begin{cases} \dot{x}_i(t) = a_i(x_i(t)) \begin{bmatrix} -c_i(x_i(t)) + \sum_{j=1}^{2} a_{ij} f_j(y_j(t)) \\ + \sum_{j=1}^{2} b_{ij} f_j(y_j(t - \tau_j(t))) + I_i \end{bmatrix}, \\ i = 1, 2, \\ \dot{y}_j(t) = b_j(y_j(t)) \begin{bmatrix} -d_j(y_j(t)) + \sum_{i=1}^{2} m_{ji} g_i(x_i(t)) \\ + \sum_{i=1}^{2} n_{ji} g_i(x_i(t - \sigma_i(t))) + J_j \end{bmatrix}, \\ j = 1, 2, \end{cases} \tag{3}$$

where $a_i(x_i(t)) = 2 + \cos x_i(t)$, $c_i(x_i(t)) = 3x_i(t)$, $g_i(x_i) = 2x_i/(1 + x_i^2)$; $b_j(y_j(t)) = 2 + \sin y_j(t)$, $d_j(y_j(t)) = 3y_j(t)$, $f(y_j) = \frac{1}{2}(|y_j + 1| - |y_j - 1|)$. Let $A = \left(\begin{smallmatrix}1&2\\2&1\end{smallmatrix}\right)$, $B = \left(\begin{smallmatrix}1&0\\0&1\end{smallmatrix}\right)$, $M = \left(\begin{smallmatrix}2&1\\1&0\end{smallmatrix}\right)$, $N = \left(\begin{smallmatrix}2&0\\0&1\end{smallmatrix}\right)$, $I = (1\ 2)^{\mathrm{T}}$, $J = (2\ 1)^{\mathrm{T}}$. So $\underline{a}_i = \underline{b}_j = 1$, $\bar{a}_i = \bar{b}_j = 3$, $\underline{c}_i = \bar{c}_i = \underline{d}_j = \bar{d}_j = 3$, $s_i = r_j = 1$, $M_1 = 15, M_2 = 18, N_1 = 21, N_2 = 9$. Since $f(\cdot), g(\cdot) \in B$, according to Theorem 1, the neural network model (3) has positive invariant and globally exponential attractive sets as follows:

$$\Omega_1 = \left\{ \begin{pmatrix} x \\ y \end{pmatrix} \in R^4 \left| x_1^2(t) + x_2^2(t) + y_1^2(t) + y_2^2(t) \le 119 \right. \right\},$$

$$\Omega_2 = \left\{ \begin{pmatrix} x \\ y \end{pmatrix} \in R^4 \,||x_1(t)| + |x_2(t)| + |y_1(t)| + |y_2(t)| \le 21 \right\}.$$

Meanwhile, let the initial conditions $x_1(t) = 0.7 + y_2(t), x_2(t) = 1 + y_2(t), y_1(t) = 1.2 + y_2(t), y_2(t) = 0.9 + 0.5\sin 2t$, and the delays $\tau_1 = \tau_2 = 100 - \sin t$,

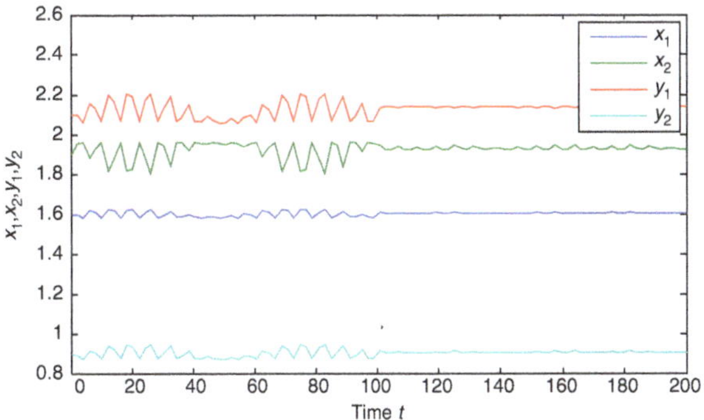

Figure 1. Time response of states $x_1(t), x_2(t), y_1(t)$ and $y_2(t)$ of Equation (3).

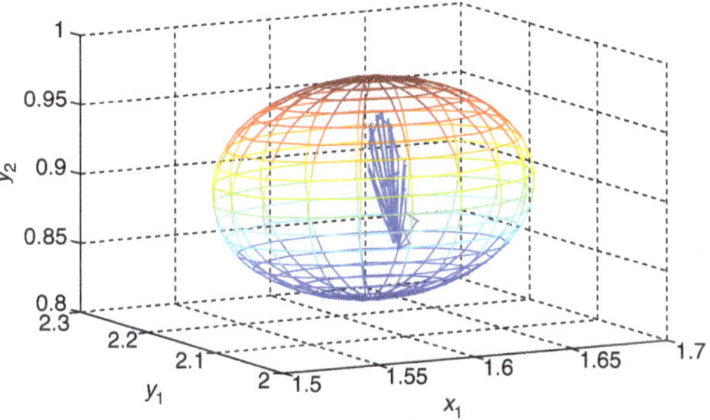

Figure 2. The ultimate bound of Equation (3) in coordinate system (x_1, y_1, y_2).

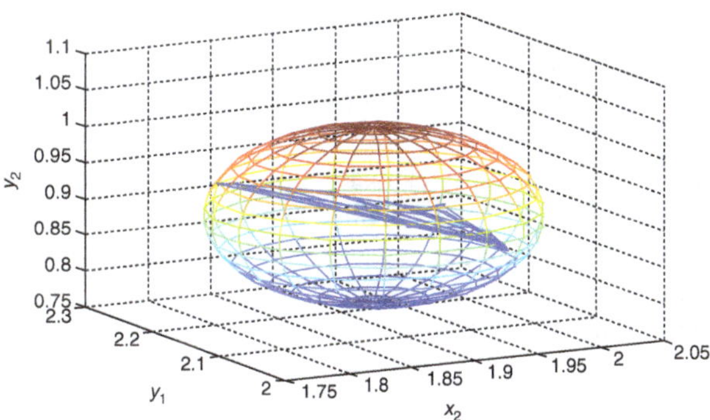

Figure 3. The ultimate bound of Equation (3) in coordinate system (x_2, y_1, y_2).

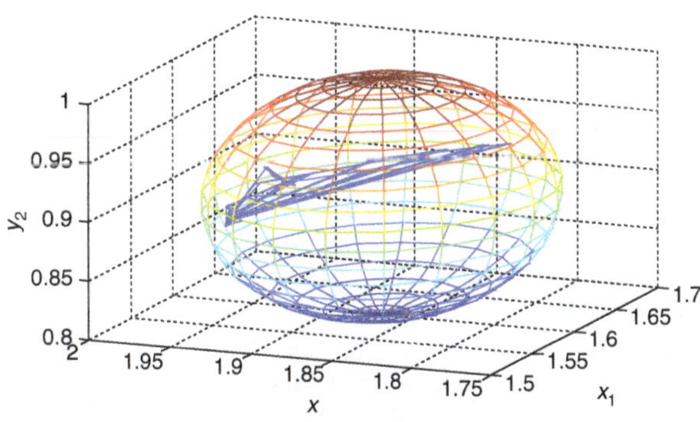

Figure 4. The ultimate bound of Equation (3) in coordinate system (x_1, x_2, y_2).

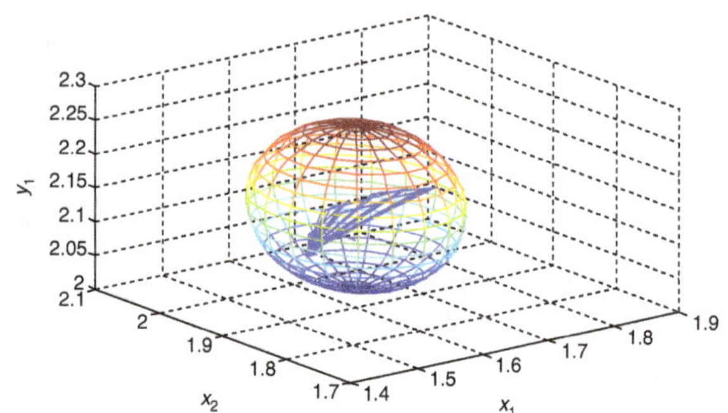

Figure 5. The ultimate bound of Equation (3) in coordinate system (x_1, x_2, y_1).

$\sigma_1 = \sigma_2 = 100 - \sin t$. Figure 1 shows time response of states $x_1(t)$, $x_2(t)$, $y_1(t)$ and $y_2(t)$. Figures 2–5 show the estimations of the ultimate bound of system (3) in the three-dimensional phase space, respectively.

5. Conclusions

Based on the Lyapunov stability theory and some inequalities, this paper has derived some sufficient delay-independent conditions of positive invariant set and globally exponential attractive set for the BAM-type Cohen–Grossberg neural networks with time-varying delays. According to the parameters, the detailed estimations for the positive invariant and globally attractive set of the BAM-type Cohen–Grossberg neural networks have been established without any hypothesis on the existence. Meanwhile, the results obtained in this paper are more general than that of the existing references (Tu et al., 2013) on the GEA set as special cases. Moreover, the proposed methods here can be also applied to nonlinear discrete-time systems with time-varying delays such as that in Dong, Wang, and Gao (2013) and Hu, Wang, Niu, and Stergioulas (2012). Finally, an illustrative example is shown to verify our results.

Acknowledgements

The authors are grateful for the support of the National Natural Science Foundation of China (61174216) and (61273183), the Scientific Innovation Team Project of Hubei Provincial Department of Education (T200809).

Disclosure statement

No potential conflict of interest was reported by the author(s).

References

Dong, H. L., Wang, Z. D., & Gao, H. J. (2013, October). Distributed H_∞ filtering for a class of Markovian jump nonlinear time-delay systems over lossy sensor networks. *IEEE Transactions on Industrial Electronics*, *60*, 4665–4672.

Hu, J., Wang, Z. D., Niu, Y. G., & Stergioulas, L. K. (2012, November). H_∞ sliding mode observer design for a class of nonlinear discrete time-delay systems: A delay-fractioning approach. *International Journal of Robust and Nonliner Control*, *22*, 1806–1826.

Jiang, H. J., & Cao, J. D. (2008, January). BAM-type Cohen–Grossberg neural networks with time delays. *Mathematical and Computer Modelling*, *47*, 92–103.

Li, T., Fei, S. M., Tan, M. C., & Zhang, Y. N. (2009, August). New sufficient conditions for global asymptotic stability of Cohen–Grossberg neural networks with time-varying delays. *Nonlinear Analysis: Real World Applications*, *10*, 2139–2145.

Li, K. L., Zhang, L. P., Zhang, X. H., & Li, Z. A. (2010, February). Stability in impulsive Cohen–Grossberg-type BAM neural networks with distributed delays. *Applied Mathematics and Computation*, *215*, 3970–3984.

Li, X. D. (2009, September). Existence and global exponential stability of periodic solution for impulsive Cohen–Grossberg-type BAM neural networks with continuously distributed delays. *Applied Mathematics and Computation*, *215*, 292–307.

Liao, X. X., Luo, Q., & Zeng, Z. G. (2008, January). Positive invariant and global exponential attractive sets of neural-networks with time-varying delays. *Neurocomputing*, *71*, 513–518.

Liu, J., & Zong, G. D. (2009, June). New delay-dependent asymptotic stability conditions concerning BAM neural networks of neutral type. *Neurocomputing*, *72*, 2549–2555.

Lu, W. L., Wang, L. L., & Chen, T. P. (2011, March). On attracting basins of multiple equilibria of a class of cellular neural networks. *IEEE Transactions on Neural Networks*, *22*, 381–394.

Luo, Q., Zeng, Z. G., & Liao, X. X. (2011, January). Global exponential stability in Lagrange sense for neutral type recurrent neural networks. *Neurocomputing*, *74*, 638–645.

Song, Q. K., & Zhao, Z. J. (2005, July). Global dissipativity of neural networks with both variable and unbounded delays. *Chaos, Solitons and Fractals*, *25*, 393–401.

Tu, Z. W., Jian, J. G., & Wang, W. W. (2011, March). *Global dissipativity for Cohen–Grossberg neural networks with both time-varying delays and infinite distributed delays*. International Conference on Information Science and Technology, Nanjing, 982–985.

Tu, Z. W., Wang, L. W., Zha, Z. W., & Jian, J. G. (2013, September). Global dissipativity of a class of BAM neural networks with time-varying and unbound delays. *Communications in Nonlinear Science and Numerical Simulation*, *18*, 2562–2570.

Wang, B. X., Jian, J. G., & Guo, C. D. (2008, January). Global exponential stability of a class of BAM networks with time-varying delays and continuously distributed delays. *Neurocomputing*, *71*, 495–501.

Wang, B. X., Jian, J. G., & Jiang, M. H. (2010, February). Stability in Lagrange sense for Cohen–Grossberg neural networks with time-varying delays and finite distributed delays. *Nonlinear Analysis: Hybrid Systems*, *4*, 65–78.

Wang, L. L., & Chen, T. P. (2012, December). Complete stability of cellular neural networks with unbounded time-varying delays. *Neural Networks*, *36*, 11–17.

Zhang, Z. Q., Liu, W. B., & Zhou, D. M. (2012, January). T Global asymptotic stability to a generalized Cohen–Grossberg BAM neuralnetworks of neutral type delays. *Journal of Neural Networks*, *25*, 94–105.

7

Effective fault-tolerant control paradigm for path tracking in autonomous vehicles

Afef Fekih* and Shankar Seelem

Department of Electrical and Computer Engineering, University of Louisiana at Lafayette, Lafayette, LA, USA

A novel fault-tolerant control paradigm that integrates fault detection (FD) with optimal control for path tracking is designed to ensure accurate path tracking in the presence of faults. The proposed approach is designed to maintain vehicle stability, dynamics, and maneuverability in the event of a faulty steering system. A sensor fusion-based fault detection and identification approach is proposed to accurately detect and identify sensor faults when they occur. A weight adjustment algorithm is considered to ensure accurate detection while providing robustness against parameter variations and uncertainties. Following FD and using the estimated fault vector, a fault-tolerant controller is designed to guarantee the stability of the closed loop system. The proposed controller incorporates a linear quadratic regulator (LQR)-based algorithm with a feed-forward gain. The LQR-based controller is designed to maintain system stability under faulty conditions while operating the dynamic system at minimum cost. The proposed approach is validated using a ground vehicle required to track various paths while being subject to multiple fault scenarios. For accurate performance analysis, vehicle handling and dynamics were implemented using CarSim, a high-fidelity vehicle simulator. Effective path tracking capabilities, vehicle handling, and stabilization under both fault-free and faulty conditions are the main positive features of the proposed approach.

Keywords: fault detection and identification (FDI); fault-tolerant control (FTC); linear quadratic regulator (LQR); sensor fusion; observer

1. Introduction

As modern day autonomous vehicles become more and more complex and highly integrated, the vulnerability of their components to faults/failures increases (Cheng, 2011; Fekih, 2014). Defects in sensors, actuators, or the system itself can degrade overall system performance. Undetected, faults can develop into failures which probably increase with the increased complexity of the system. Moreover, mitigating unsatisfactory performances or even instability caused by the unpredictable faults in actuators, sensors, or other components is of foremost priority, especially in safety-critical systems such as ground vehicles.

According to the International Federation of Automatic Control SAFEPROCESS technical committee (Gustaffson, 2000), a fault is defined as any unpermitted deviation of at least one characteristic property or parameter of the system from the acceptable/usual/standard condition, while a failure is a permanent interruption of a system's ability to perform its function under specific operating conditions. In order to maintain high levels of performance and guarantee proper system behavior, it is important that faults be promptly detected and identified and appropriate remedies be applied to prevent system malfunctions. Diagnosis is the primary stage of active fault-tolerant control (FTC) systems. Its goal is to perform two main decision tasks: fault detection (FD), consisting of deciding whether or not a fault has occurred, and fault isolation, consisting of deciding which element of the system has failed (Fekih, 2014).

A FTC system is a control system specifically designed to automatically accommodate faults among system components while maintaining system stability along with a desired level of overall performance (Blanke & Staroswiecki, 2006; Noura, Theilliol, Ponsart, & Noura, 2009). The key issue of a FTC system is to prevent local faults from developing into system failures that can end the mission of the system and cause safety hazards for man and environment. Existing efforts in FTC design can be classified into two main approaches: the passive and active approaches (Jiang & Yu, 2012). In the passive approach, robust control techniques are used to ensure that the control loop system remains insensitive to certain faults. The effectiveness of this strategy, which usually assumes very restrictive repertory of faults, depends upon the robustness of the nominal closed-loop system. In the active approach, a new control system is redesigned according to the estimation of the fault performed by the fault detection and identification (FDI) filter and according to the specifications to be met by the faulty system. In contrast to passive approaches that are mostly conservative, active approaches

*Corresponding author. Email: afef.fekih@louisiana.edu

are able to deal with a large number of fault scenarios and can handle a certain number of unforeseen faults that were not considered at the design stage.

A growing body of research in this area has resulted in a number of FD and FTC schemes for ground vehicles (Arogeti, Wang, Low, & Yu, 2008; Chen, Song, & Li, 2011; Dong, Verhaegen, & Holweg, 2008; Fekih & Seelem, 2012; Herpin, Fekih, Golconda, & Lakhotia, 2007; Laureiro, Benmoussa, Touati, Merzouki, & OuldBouamama, 2014; Morteza & Fekih, 2014a, 2014b; Tabbache, Benbouzid, Kheloui, & Bourgeot, 2011; Wang & Wang, 2013; Yang, Cocquempot, & Jiang, 2008). A passive actuator FTC was proposed for Four-Wheel Independently Actuated electric ground vehicle in Wang and Wang (2013). The approach exploits the redundancy of the system and groups actuators with similar faults in one subsystem and applies control allocation to distribute the control effort. Actuator grouping was attempted to reduce the significant computational cost typically associated with control allocation. In Laureiro et al. (2014) a bond graph model-based FD approach and an FTC were designed for an over-actuated heavy size autonomous vehicle. The approach relied heavily on analytical redundancy relations derived from the bond graph model. A robust and adaptive FTC tracking approach was proposed in Chen et al. (2011). An FTC strategy which considers a maximum-likelihood voting algorithm was proposed for sensor faults in Tabbache et al. (2011). Note that most of these works exploited system redundancy and required high computational costs, drawbacks that might prevent their real-time implementation.

In this paper, an FTC framework that implicitly integrates FDI with FTC is designed for the automatic steering of an autonomous ground vehicle subject to sensor faults. The proposed controller is based on a linear quadratic regulator (LQR) augmented with a feed-forward gain. The LQR-based controller is designed to place the system's eigenvalues in the stable region while operating the dynamic system at minimum cost function. An observer-based FDI approach is proposed to detect and identify sensor faults when they occur. Using the estimated fault vector, the fault-tolerant controller is designed to maintain system stability when faults occur. The proposed framework is implemented on a ground vehicle required to follow a given path, while being subject to sensor faults. The steering controller is designed to maintain vehicle stability, dynamics, and maneuvrability in the event of a faulty steering system.

Compared with the existing work already reported in the literature (Arogeti et al., 2008; Chen et al., 2011; Laureiro et al., 2014; Morteza & Fekih, 2014b; Tabbache et al., 2011; Wang & Wang, 2013), the contributions of this paper are in the following aspects:

It presents a complete FTC design with the FDI algorithm as integral part of the framework and applies it to the automatic steering of an autonomous vehicle.

The FDI algorithm incorporates a weight adjustment algorithm to ensure accurate detection, while providing robustness against parameter variations and uncertainties.

It integrates the optimal properties of the LQR framework with an observer-based fault detection scheme to achieve effective fault tolerance.

It provides an easy to implement algorithm which achieves fault-tolerance with optimum computational costs. This is crucial in autonomous vehicles, which often work under tight real-time deadlines and cannot tolerate prolonged delays in control reconfiguration.

The rest of this paper is organized as follows. Section 2 presents the dynamic model of the vehicle and discusses the design specifications. The proposed control paradigm is detailed in Section 3. Section 4 is dedicated to the performance analysis of the proposed algorithm. Finally, some concluding remarks end this paper in Section 5.

2. Vehicle dynamic model and problem formulation

A dynamic model of the vehicle, with the front and rear wheels lumped together into a pair of single wheels at the center of gravity (CG) (Fekih & Deveriste, 2013), is considered as shown in Figure 1.

Assuming constant longitudinal velocity and combining the lateral forces with the available slip angles, the vehicle's dynamic model is defined as follows (Fekih & Deveriste, 2013):

$$\begin{bmatrix} \dot{v}_y \\ \dot{r} \end{bmatrix} = \begin{bmatrix} \dfrac{-(c_f + c_r)}{m v_x} & \dfrac{l_r c_r - l_f c_f}{m v_x} - v_x \\ \dfrac{l_r c_r - l_f c_f}{I_z v_x} & \dfrac{-(l_f^2 c_f + l_r^2)}{I_z v_x} \end{bmatrix} \begin{bmatrix} v_y \\ r \end{bmatrix} + \begin{bmatrix} \dfrac{c_f}{m} \\ \dfrac{l_f c_f}{m} \end{bmatrix} \delta, \tag{1}$$

where $\dot{v}_y$ is the rate of change of lateral velocity, $\dot{r}$ is the yaw rate of the vehicle, δ is the steering angle, θ is the yaw

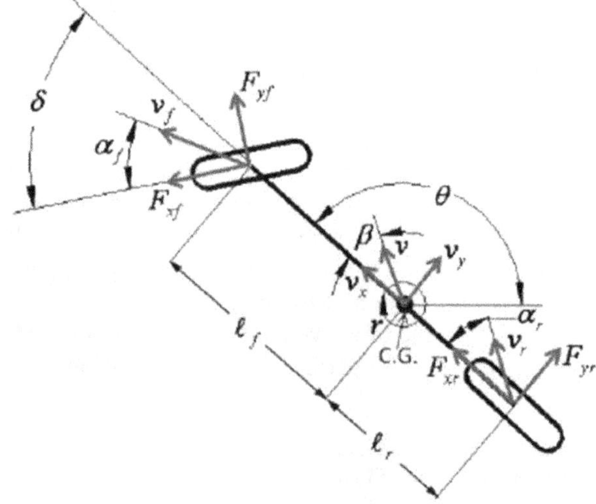

Figure 1. Dynamic bicycle model.

angle (orientation angle of the vehicle with respect to the X axis), and v_y, v_x are the lateral and longitudinal velocity, respectively. c_f and c_r are the cornering stiffness of the front and rear tires, respectively. ℓ_f is the distance from the CG to the front axle and ℓ_r is the distance from the CG to the rear axle. I_z is the vehicle yaw moment of inertia. The remaining variables and parameters are defined in the Appendix (Table A1). Since our objective is to develop a steering control system for automatic lane keeping, the state variables are being expressed in terms of position and orientation error. If we consider a vehicle traveling at a constant velocity on a road of a large radius with curvature k and assume a constant longitudinal velocity, the rate of change of the desired orientation of the vehicle is given by

$$\dot{r}_{\text{des}} = v_x k, \tag{2}$$

where $\dot{r}_{\text{des}}$ is the desired yaw rate and k is the curvature of the road. The desired path lateral acceleration of the vehicle can be written as

$$\dot{v}_y = k v_x^2. \tag{3}$$

Define e as the distance of the CG of the vehicle from the center line of the path and e_1 as the yaw angle error of the vehicle with respect to the path; then, we have

$$\begin{aligned} \ddot{e} &= \ddot{v}_y + v_x(\dot{r} - \dot{r}_{\text{des}}), \\ \dot{e} &= \dot{v}_y + v_x(r - r_{\text{des}}), \\ e_1 &= r - r_{\text{des}}. \end{aligned} \tag{4}$$

The state-space model in tracking error variables is therefore given by

$$\dot{x} = Ax + B_1\delta + B_2\dot{r}_{\text{des}}, \tag{5}$$

with $x = [e \quad \dot{e} \quad e_1 \quad \dot{e}_1]^{\text{T}}$.

$$A = \begin{bmatrix} 0 & 1 & 0 & 0 \\ 0 & -\dfrac{(c_f + c_r)}{mv_x} & \dfrac{c_f + c_r}{m} & \dfrac{l_f c_f - l_r c_r}{mv_x} \\ 0 & 0 & 0 & 1 \\ 1 & \dfrac{l_r c_r - l_f c_f}{I_z v_x} & \dfrac{l_f c_f - l_r c_r}{I_z} & -\dfrac{(l_f^2 c_f + l_r^2 c_r)}{I_z v_x} \end{bmatrix}, \ldots,$$

$$B_1 = \begin{bmatrix} 0 \\ \dfrac{c_f}{m} \\ 0 \\ \dfrac{l_f c_f}{m} \end{bmatrix}, \ldots B_2 = \begin{bmatrix} 0 \\ \dfrac{l_r c_r - l_f c_f}{mv_x} - v_x \\ 0 \\ \dfrac{-(l_f^2 c_f + l_r^2 c_r)}{I_z v_x} \end{bmatrix}.$$

The output vector of the system consists of measurements from the two sets of sensors. Sensor failures are modeled as additive signals to the sensor output as follows:

$$y = [y_1 \quad y_2]^{\text{T}} = Cx + Ff, \tag{6}$$

where y_1 and y_2 are the measurements from the two sensors which measure the lateral deviation of the vehicle and f is the fault signal, which is a function of time and state x, and C is the output matrix, defined as $C = \left[\begin{smallmatrix} C_1 \\ C_2 \end{smallmatrix}\right] = \left[\begin{smallmatrix} 1 & 0 & d_1 & 0 \\ 1 & 0 & -d_2 & 0 \end{smallmatrix}\right]$, where d_1 and d_2 are, respectively, the distances from the CG of the vehicle to the front and rear bumpers where the sensors are located. $F = [1 \quad 0]^{\text{T}}$ in the event of failure of sensor 1 and $F = [0 \quad 1]^{\text{T}}$ in the event of failure of sensor 2, respectively. Here, we consider that the lateral sensing system consists of two sets of sensors which provide the information of the lateral deviation.

3. FTC paradigm

The faults under investigation are sensor faults with varying severities and types. The FTC objectives are to maintain vehicle stability, dynamics, and maneuverability in the event of faulty sensors.

3.1. Observer-based FDI algorithm

The following are some important features of the vehicle's dynamic model that can aid in designing an easy-to-implement FDI algorithm: (1) it has two zero eigenvalues and (2) (A, C_1) and (A, C_2) are observable. This implies that we can estimate the state through either y_1 or y_2. This makes FDI easy to implement with minimum computational cost.

The observability properties of the vehicle imply that we can build two observers, each of which is being driven by a single sensor output. Furthermore, in order to ensure that no erroneous estimates of the state are obtained under sensor failures, we fuse the sensor output and the estimated output from one observer prior to their use by the other observer. Fusion blocks play the role of switches, which select the healthy signal. Then post-filters are designed such that the transfer functions from fault signals to residuals have consistent behavior in order to facilitate fault identification.

Output fusion is a convex combination of the sensor output and the estimated output from the observer. The fused output y_{fi} is given by

$$y_{fi} = (1 - \lambda_i)y_i + \lambda_i \hat{y}_i^j, \tag{7}$$

where $\hat{y}_i^j$ is the estimate of the ith output from the jth observer, and $i, j = 1, 2$, and λ_i is a weight function. The weights $\lambda_i \in [0, 1]$ are adjusted using a weight adjustment algorithm. When there are no faults, that is, $\lambda_i = 0$, then the fused output y_{fi} is identical to the sensor output y_i. When faults occur, the corresponding values of λ_i will increase toward one. When $\lambda_i = 1$, the sensor output is incorrect and therefore is not being taken into account at all. The observers will switch between two configurations according to the relative size of the weights as follows:

If $\lambda_1 < \lambda_2$, then the observers are defined by

$$\begin{aligned}\dot{\hat{x}}_1 &= A\hat{x}_1 + B_1\delta + L_1(y_{f1} - \hat{y}_1^1) + \lambda_1 L_1 C_1(\hat{x}_1 - \hat{x}_2),\\ \dot{\hat{x}}_2 &= A\hat{x}_2 + B_1\delta + L_2(y_{f2} - \hat{y}_2^2).\end{aligned} \tag{8}$$

If $\lambda_1 > \lambda_2$, then

$$\begin{aligned}\dot{\hat{x}}_1 &= A\hat{x}_1 + B_1\delta + L_1(y_{f1} - \hat{y}_1^1)\\ \dot{\hat{x}}_2 &= A\hat{x}_2 + B_1\delta + L_2(y_{f2} - \hat{y}_2^2) + \lambda_2 L_2 C_2(\hat{x}_2 - \hat{x}_1),\end{aligned} \tag{9}$$

where L_1 and L_2 are the solutions of the characteristic equation defined by det(sI-$A + L_1C_1$) and det(sI-$A + L_2C_2$), respectively. Observers (8)–(9) are variations of the Luenberger observer where the fused outputs replace the sensor outputs.

3.2. Threshold generation logic

For accurate FD, define the following fault indicator function or threshold logic:

$$\begin{aligned}\|r_i(t)\| &< T_{\text{h}} \Rightarrow \text{fault-free conditions},\\ \|r_i(t)\| &> T_{\text{h}} \Rightarrow \text{faulty conditions},\end{aligned} \tag{10}$$

where T_{h} is a predefined threshold typically chosen based on the application at hand. Note that setting low thresholds results in high false-positive rates (alarms are issued under no fault conditions), and setting high thresholds increases the false-negative rates (alarms are missed when faults occur). Clearly, the selection of the thresholds is closely related to robustness and sensitivity of the residual generator. Different analysis procedures are used depending on the techniques employed to generate the residual signal. The most widely used approaches to analyze the residual signal generated by observers are threshold logic and limit monitoring. Threshold level selection methods are generally problem specific and are not useful for a general case (Hsiao & Tomizuka, 2005). To avoid improper FD, threshold level selection is often done on the basis of the designer's experience and in response to problem requirements.

3.3. Post-filters

Consider the error vector:

$$e_{yi} = \begin{bmatrix} y_1 - \hat{y}_1^1 & y_1 - \hat{y}_1^2 & y_2 - \hat{y}_2^1 & y_2 - \hat{y}_2^2 \end{bmatrix}^{\text{T}}. \tag{11}$$

Residuals are generated by filtering e_{y} through post-filters $M_i(s)$, that is,

$$r_i = M_i e_{yi} \quad i = 1, 2, 3, 4. \tag{12}$$

$M_i(s)$ define the transfer functions from the faulty signals to the residuals such that the residuals from the two observers are comparable in magnitude. Note that r_1 and r_2 are related to sensor 1 and r_3 and r_4 are related to sensor 2, respectively. Notice that observers (8)–(9) are coupled, that is, faults in either of the two sensors affect all residuals. The problem of identifying the exact fault sensor can be solved by using properly designed post-filters (Zhang, Ding, Lam, & Wang, 2003).

Define the state-space transfer function from fault f to e_{y} by

$$V_i(s) = C(sI - A)^{-1}B + F, \tag{13}$$

where F is a matrix which represents the sensor failure as follows:

$$\begin{aligned}F &= \begin{bmatrix}1 & 0\end{bmatrix}^{\text{T}} \quad \text{If sensor 1 fails},\\ F &= \begin{bmatrix}0 & 1\end{bmatrix}^{\text{T}} \quad \text{If sensor 2 fails}.\end{aligned}$$

Consider the scenarios when sensor 1 has failed and $\lambda_1 < \lambda_2$, then from Equation (12) we have

$$\begin{aligned}V_1(s) &= -(1-\lambda_1)C_1(sI - A + (1-\lambda_1)L_1C_1)^{-1}L_1 + 1\\ &= \frac{n_1(s)}{d(s)},\end{aligned} \tag{14}$$

$$\begin{aligned}V_3(s) &= -(1-\lambda_1)C_2(sI - A + (1-\lambda_1)L_1C_1)^{-1}L_1\\ &= \frac{(1-\lambda_1)n_3(s)}{d(s)},\end{aligned} \tag{15}$$

where $((n_1(s), d(s))$ and $(n_3(s), d(s))$ are the co-prime pairs of the polynomials defined as follows:

$$n_1(s) = \det\left(\begin{bmatrix} sI - A & L_1\\ 0 & 1\end{bmatrix}\begin{bmatrix} I & 0\\ (1-\lambda_1)C_1 & 1\end{bmatrix}\right), \tag{16}$$

$$n_3(s) = \det\left(\begin{bmatrix} sI - A & L_1\\ 0 & 1\end{bmatrix}\begin{bmatrix} I & 0\\ (1-\lambda_1C_2 & 1\end{bmatrix}\right), \tag{17}$$

where $n_1(s)$ and $n_3(s)$ are also independent of λ_1.

Factorizing $n_1(s) = n_1^+(s)n_1^-(s)$ and $n_3(s) = n_3^+(s)n_3^-(s)$, where $n_i^+(s)$ and $n_i^-(s)$, i = 1,3, are the factors of $n_i(s)$ which have their roots in the left half plane.

Choosing:

$$M_1(s) = \frac{n_1{}^+(s)}{n_3^-(s)k(s)}, \tag{18}$$

$$M_3(s) = \frac{n_3^+(s)}{n_1^-(s)k(s)}, \tag{19}$$

where $k(s)$ is a Hurwitz polynomial such that $M_1(s)$ and $M_3(s)$ are proper and stable, yields

$$(1-\lambda_1)M_1(s)V_1(s) = M_3(s)V_3(s). \tag{20}$$

This implies that if we choose post-filters M_i's such that $a_1M_1V_1 = M_3V_3$ and $a_2M_4V_4 = M_2V_2$ for some real numbers $a_1 > 0$ and $a_2 < 1$, we can define the following identification rules:

If $\lambda_1 < \lambda_2$, and a fault was detected, $|r_1| > |r_3|$ indicates that sensor 1 has failed, while $|r_1| < |r_3|$ implies that sensor 2 has failed.

Similarly, if $\lambda_1 > \lambda_2$, and a fault has been detected, $|r_2| > |r_4|$ suggests that sensor 1 has failed, while $|r_2| < |r_4|$ implies that sensor 2 has failed. In order to accommodate faults, when these latter happen, the controller considers the fused outputs, that is, y_{f1} and y_{f2} instead of the sensor outputs y_1 and y_2.

If the fault occurs, the faulty sensor output is replaced by the observer output.

The next step after residual generation is the analysis of the residual signal for FD. The residual generator takes the sensor measurements as inputs and generates residuals. The latter are small, ideally zero, when there are no fault; but when a fault occurs, the residuals are significantly large. Due to the effect of disturbances, model uncertainties, and measurement noise, the residuals are different from zero even when there are no faults. A robust residual generator is proposed next to alleviate these effects while remaining sensitive to faults.

3.4. *Weight adjustment algorithm*

If any fault has been detected and identified, weights λ_i, $i = 1, 2$ in the fusion blocks will be adjusted. Suppose sensor 1 fails, then we adjust the weights according to the following 1st order differential equations:

$$\dot{\lambda}_1 = -\alpha(\lambda_1 - g(|[r_1 \quad r_2]^{\mathrm{T}}|) \tag{21}$$

$$\dot{\lambda}_2 = -\alpha\lambda_2. \tag{22}$$

If sensor 2 fails, then the adaption rule becomes

$$\dot{\lambda}_1 = -\alpha\lambda_1 \tag{23}$$

$$\dot{\lambda}_2 = -\alpha(\lambda_2 - g(|[r_3 \quad r_4]^{\mathrm{T}}|), \tag{24}$$

where g is a sigmoid function defined as follows: g: R $\rightarrow[0,\ 1]$ $g(x) = 1/(1+e^{-ax})$ $a > 0$ with $a \in [-10, 10]$. The value of a is determined based of the fault magnitude. For instance, complete failure of sensor 1 results in $a = -10$, hence the use of information from residuals r_1 and r_2. If no failure is reported, then $a = +10$, resulting in $g = 0$, hence residual signals not being considered in the adjustment algorithm. Values between -10 and 10 represent faults with various magnitudes. The sufficient conditions for convergence of the estimated state are $|\dot{\lambda}_1| < \alpha$ and $\lambda_1 + \lambda_2 \leq 1$. The parameter $\alpha > 0$ and is a trade-off between stability and FDI performance. Large α increases the response of the FDI unit to faults, while small α results in a slowly varying condition which guarantees system stability.

3.5. *FTC strategy*

When a fault is detected, the state variables are reconstructed accordingly and the feedback controller is redesigned as follows:

$$\delta(t) = -K_f\hat{x}(t), \tag{25}$$

where K_f is the feedback controller gain when the lateral control system enters into a degraded mode, that is, when the information is lost in one of the lateral deviation sensors. A fault in any of the sensors results in a change in the output measurements and the state variables are defined as follows:

$$\dot{x}(t) = Ax(t) + B_1\delta(t) + B_2\dot{r}_{\mathrm{des}}, \tag{26}$$

$$y(t) = \bar{C}x(t) + Ff, \tag{27}$$

where A is the state matrix, B_1 is the control matrix, and $\bar{C}$ is the output matrix of the faulty system; $\dot{r}_{\mathrm{des}}$ is the desired yaw rate, f is the additive fault signal, and F represents sensor faults. Now an optimal estimator can easily be designed by solving the relevant Riccati equation associated with the system given by Equations (25) and (26). Assuming the system is observable, the state estimates $\hat{x}$ are defined as follows:

$$\dot{\hat{x}}(t) = (A - L\bar{C})\hat{x}(t) + B_1\delta(t) + Ly(t), \tag{28}$$

where L is the observer gain which is defined by $L = Y\bar{C}B_1^{-1}$ and Y is the positive semi-definite stabilizing solution of the following algebraic Riccati equation:

$$(A - B_1\bar{C})Y + Y(A - \bar{C})^{\mathrm{T}} - Y\bar{C}Y + B_1B_1{}^{\mathrm{T}} = 0. \tag{29}$$

In order to guarantee the stability of the proposed observers, define the estimation error $\varepsilon(t)$ as follows:

$$\varepsilon(t) = x(t) - \hat{x}(t). \tag{30}$$

The error dynamics are stable if and only if the matrix $U = \left[\begin{smallmatrix} A-LC & K \\ -C & 1 \end{smallmatrix}\right]$ is Hurwitz stable.

Note that the matrix U can be written as follows:

$$U = \begin{bmatrix} A & B \\ 0 & 1 \end{bmatrix} - \begin{bmatrix} L \\ 1 \end{bmatrix} [C \quad 0]. \tag{31}$$

Therefore, the poles of the matrix U can be arbitrarily assigned, provided that $\left(\left[\begin{smallmatrix} A & B \\ 0 & 1 \end{smallmatrix}\right], [C \quad 0]\right)$ is observable. Hence, stability of the proposed observer is guaranteed by the proper choice of observer gain L, which is selected in a way such that the matrix U is Hurwitz stable.

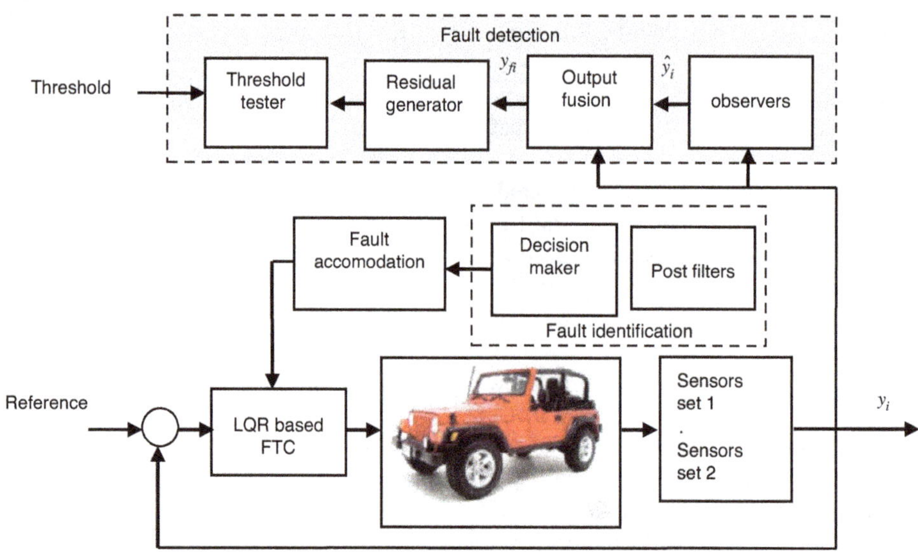

Figure 2. Schematic diagram of the proposed FTC framework.

Define the objective cost function to be minimized by the controller as follows:

$$J = \int_0^\infty \hat{x}^{\mathrm{T}}(t)Q\hat{x}(t) + \dot{\delta}^{\mathrm{T}}(t)R\delta(t), \tag{32}$$

where P satisfies the following Riccati equation:

$$P = A^{\mathrm{T}}PA - A^{\mathrm{T}}PB_1(R + B_1^{\mathrm{T}}PB_1)^{-1}B_1^{\mathrm{T}}PA + Q. \tag{33}$$

Here, Q is a diagonal weighting matrix with an entry for each state corresponding to the performance aspects contributing to the cost function and R is the weighting value corresponding to the control effort contributing to the cost function.

The proposed FTC framework is illustrated in Figure 2.

The approach uses two observers driven by sensor outputs. Following FD, faulty sensors are identified. The state variables are then constructed from the output of the healthy sensor and the controller is updated accordingly to ensure proper steering and maintain system stability when faults occur.

4. Application of the FTC framework to an automotive steering system

To validate the performances of the proposed control paradigm, we provide a series of computer experiments using various paths and fault scenarios. For accurate evaluation, the proposed controller is implemented using CarSim (Mechanical Corporation). The latter is used in the automotive industry as the standard by which vehicle handling and dynamics are tested. It provides a high-fidelity and complete model of the vehicle and its environment. The performance of the proposed control paradigm is compared to that of the CarSim steering controller, which details are illustrated in the Appendix. Two driving maneuvers are chosen to perform the various experiments as detailed in the following.

4.1. Lane change maneuver

Lane change maneuver is a common test for vehicle handling as it represents an essential collision avoidance maneuver. A lane change path is chosen to demonstrate the tracking capability on a straight path as well as the response to a quick, yet continuous transient section (position and curvature). Experiments on this path are performed at a constant longitudinal speed of 30 m/s (108 km/h) and considering a road adhesion factor of one.

A double lane change path is selected to illustrate the tracking capability as well as the steering control of the vehicle on a straight path. Figures 3 and 4 show the vehicle following a lane change maneuver. Computer experiments were first carried out when 90% of the information from the sensor is lost and the front sensor has failed at $t = 4$ sec. For comparison purposes, simulations were carried out with and without FTC. The fault considered here is an abrupt change in the front sensor.

Figures 5 and 6 depict the responses of the vehicle at a longitudinal speed of 40 m/s (144 km/h) and considering a road adhesion factor of 0.75 when 90% of the sensor information is lost ($a = 3$). The observer gains in this case were $L_1 = 10$ and $L_2 = 20$, while the controller gain was $K = 80$. Note a reduction in the friction between the road and the tire of the vehicle in this case. We can observe from the figures that when the FTC approach is considered, the vehicle is able to recover from the fault fairly quickly and follow the prescribed path without much delay. This is important since lane change control operation is high bandwidth in nature and cannot tolerate significant delays in the

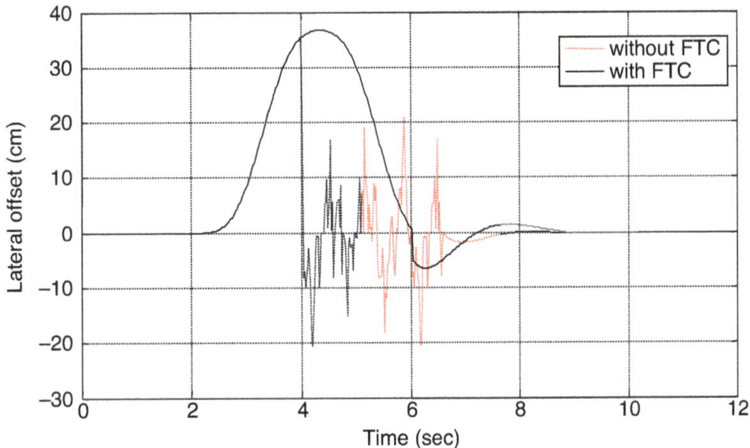

Figure 3. Lateral offset with 90% loss at $v = 30$ m/s and $u = 1$.

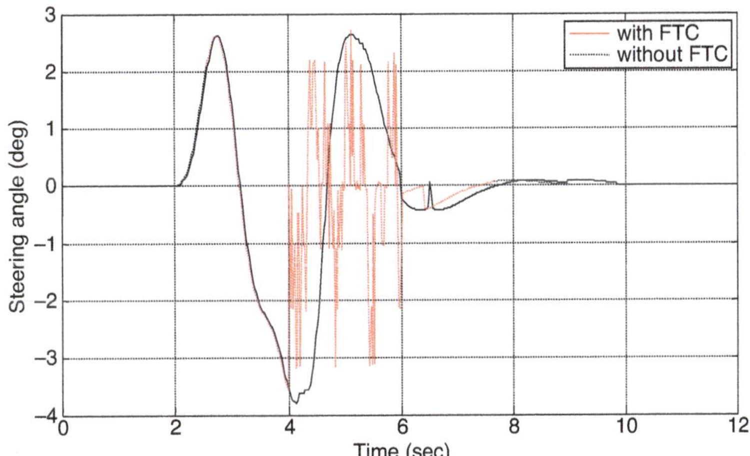

Figure 4. Steering angle with 90% loss at $v = 30$ m/s and $u = 1$.

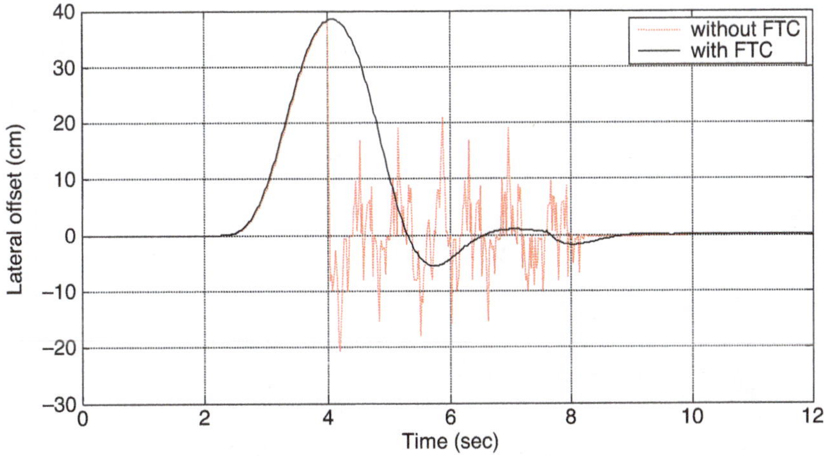

Figure 5. Steering angle with 90% loss at $v = 40$ m/s and $u = 0.75$.

control loop. Therefore, quick accommodation of faults in the lane-keeping control system is an important issue. Note though that due to the increase in speed and decrease in the road adhesion factor, the vehicle takes more time to follow the path. In contrast, the vehicle is not able to track the set path after the fault occurrence in the case without FTC.

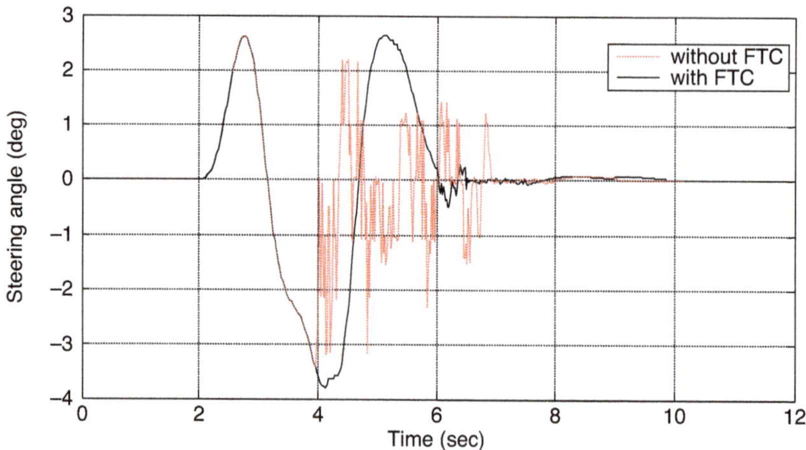

Figure 6. Steering angle with 90% loss at $v = 40$ m/s and $u = 0.75$.

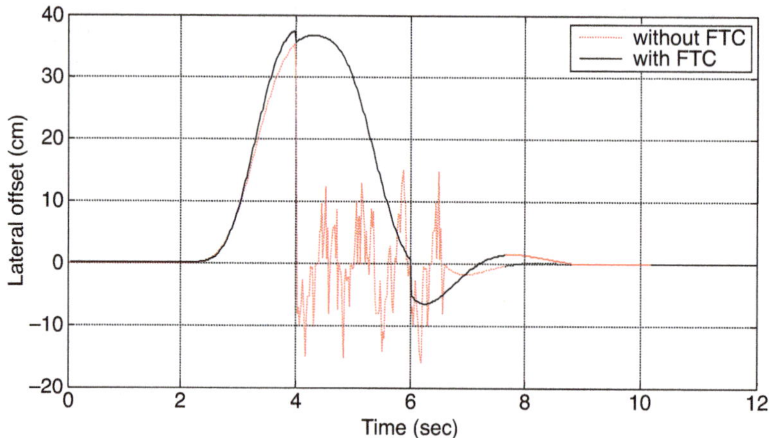

Figure 7. Lateral offset with 70% sensor information loss at $v = 40$ m/s and $u = 0.75$.

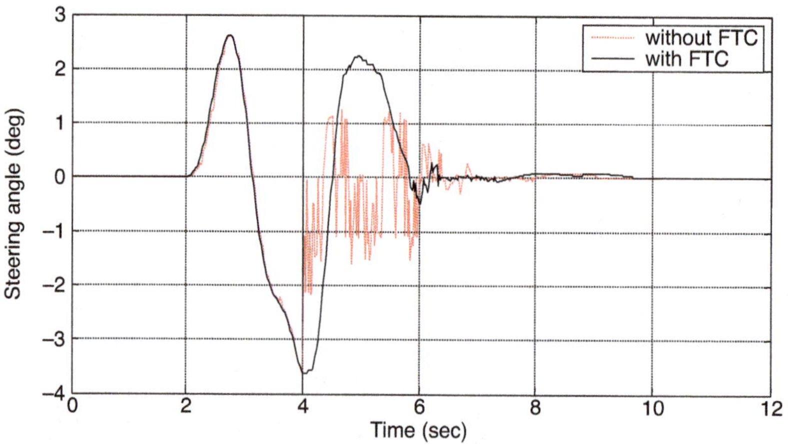

Figure 8. Steering angle at 70% sensor information loss at $v = 40$ m/s and $u = 0.75$.

Next, a 70% sensor information loss is considered ($a = 1$). The observer gains in this case were $L_1 = 8$ and $L_2 = 10$, while the controller gain was $K = 50$. Figures 7 and 8 illustrate the response of the vehicle with a longitudinal speed of 40 m/s (108 km/h) and considering a road adhesion factor of one.

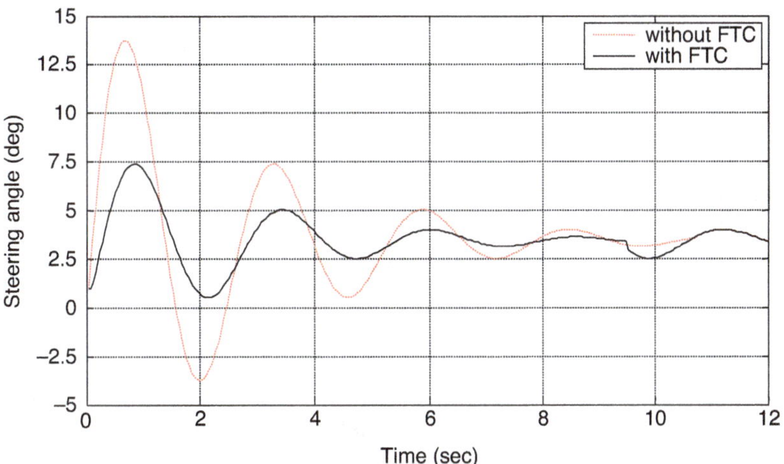

Figure 9. Lateral offset on a circular track with $v = 30$ m/s and $u = 1$.

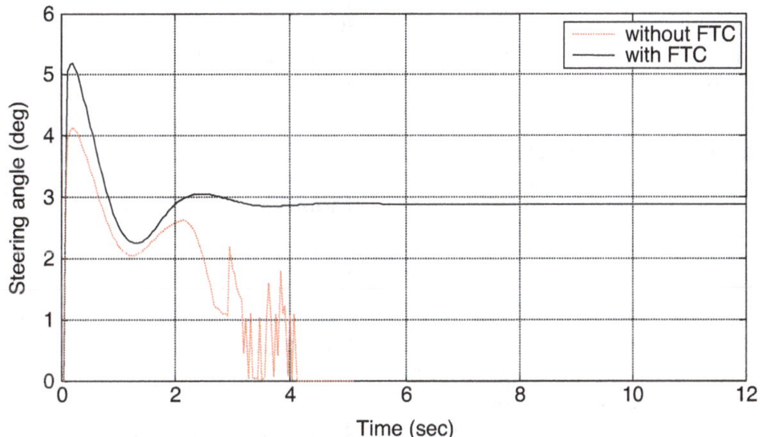

Figure 10. Steering angle on a circular track with $v = 30$ m/s and $u = 1$.

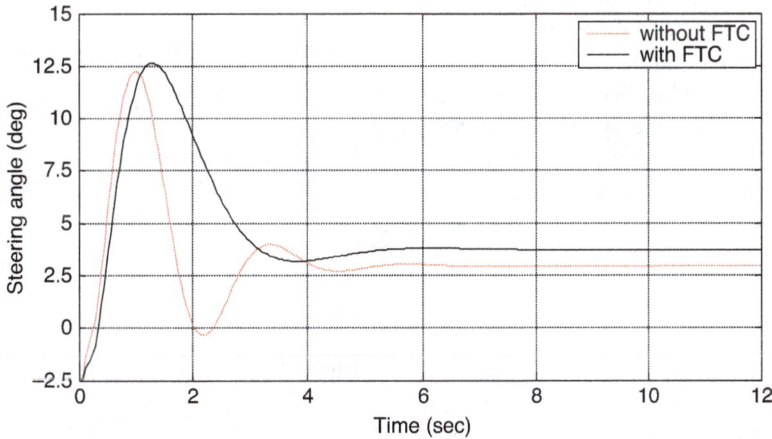

Figure 11. Lateral offset at 70% sensor information loss with $v = 30$ m/s and $u = 1$.

In this case, as expected fewer disturbances can be observed in the presence of faults when compared to the case with 90% information loss.

4.2. Circular track

In this section, we consider a circular track of 500 ft in order to show the performance of the observer-based FTC

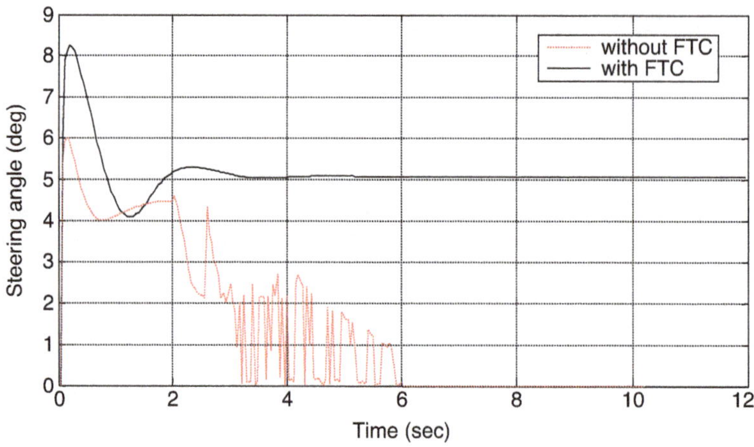

Figure 12. Steering angle at 70% sensor information loss with $v = 30$ m/s and $u = 1$.

algorithm. This path can provide valuable insight into the handling of a vehicle as well as some important characteristics of the proposed controller. Hence, this path was chosen to illustrate the steady-state characteristics of the controller while executing a constant nonzero curvature path. Experiments on this path are performed at a constant longitudinal speed of 30 m/s (108 km/h) and considering a road adhesion factor of one. Figures 9 and 10 illustrate the response of the vehicle when considering 90% of sensor information loss.

Note that information from the sensor is lost after 2 s and the vehicle requires more time to follow the path, as shown in Figure 9. The steering controller is stabilized after 4 s and system stability is maintained as shown in Figure 10. Next, the simulation was carried out with 70% sensor information loss. Figures 11 and 12 illustrate the lateral offset and steering angle in such case.

The simulation results show the effectiveness of the proposed FTC algorithm in improving the vehicle response under different paths and using various driving scenarios under several faulty conditions. Note that compared to the case without an FTC algorithm, the system is not able to recover from the sensor fault and the controller loses its steering capabilities shortly after the occurrence of the sensor fault. This is more prominent when the vehicle is following a circular path.

5. Conclusion

This paper presented an effective FTC paradigm that integrates the optimal properties of the LQR framework with an observer-based FD scheme to maintain vehicle stability and ensure handling in the presence of faults. A weight adjustment algorithm is incorporated in the FDI unit to ensure robust performance in the presence of parameter variations and disturbances. For accurate validation, the control routines were implemented in MATLAB/Simulink environment and tested using CarSim, a high-fidelity vehicle simulator. The results confirmed the ability of the proposed FTC framework to effectively monitor the system and ensure correct tracking performance under faulty conditions. Future work will focus on the integration of the proposed methodology with environment information devices such as radars and vision systems.

Funding

This work is partially supported by the Louisiana Board of Regents Support Fund contract numbers LEQSF (2012-15)-RD-A-26 and LEQSF-EPS (2015)-PFUND-421 and by LaSPACE/NASA [grant number NNX10AI40H] sub awards No. 84415 and No. 89632.

Disclosure statement

No potential conflict of interest was reported by the authors.

References

Arogeti, S., Wang, D., Low, C., & Yu, M. (2008, January). Fault detection isolation and estimation in a vehicle steering system. *IEEE Transaction in Industrial Electronics*, *59*(12), 1–2.

Blanke, M., & Staroswiecki, M. (2006). *Diagnosis and fault-tolerant control* (2nd ed.). Berlin: Springer-Verlag.

Chen, H., Song, Y., & Li, D. (2011). *Fault-tolerant tracking control of FW-steering autonomous vehicles*. Proceedings of the Chinese Control and Decision Conference CCDC, Mianyang, June 25–27, 2008, pp. 92–97.

Cheng, H. (2011). *Autonomous intelligent vehicles. Theory, algorithms and implementation*. London: Springer-Verlag.

Dong, J., Verhaegen, M., & Holweg, E. (2008, July 6–11). *Closed-loop subspace predictive control for fault tolerant MPS design*. Proceedings of the 17th IFAC World Congress, Seoul, pp. 3216–3221.

Fekih, A. (2014, June). *Fault diagnosis and fault tolerant control design for aerospace systems: A bibliographical review*. Proceedings of the 2014 American Control Conference, Portland, OR, pp. 1286–1291.

Fekih, A., & Deveriste, D. (2013, June 17–19). *A Fault-Tolerant Steering Control Design for Automatic Path Tracking in Autonomous Vehicles*. Proceedings of the American Control Conference, Washington, DC.

Fekih, A., & Seelem, S. (2012, October 3–5). *A fault tolerant control design for automatic steering control of ground vehicles*. Proceedings of IEEE Multi-Conference on Systems and Controls, Croatia, Dubrovnic, pp. 1491–1496.
Gustaffson, F. (2000). *Adaptive filtering and change detection*. Chichester: John Wiley & Sons.
Herpin, J., Fekih, A., Golconda, S., & Lakhotia, A. (2007, December). Steering control of the autonomous vehicle: CajunBot. *AIAA Journal of Aerospace Computing, Information, and Communication, 4*, 1134–1142.
Hsiao, T., & Tomizuka, M. (2005, June 8–10). *Threshold selection for timely fault detection in feedback control systems*. Proc. of the American Control Conference, Vol. 5, Portland, pp. 3303–3308.
Jiang, J., & Yu, X. (2012, April). Fault tolerant control systems: A comparative study between active and passive approaches. *Annual Reviews in Control, 36*(1), 60–72.
Laureiro, R., Benmoussa, S., Touati, Y., Merzouki, R., & Ould-Bouamama, B. (2014, January). Integration of fault diagnosis and fault-tolerant control for health monitoring of a class of MIMO intelligent autonomous vehicles. *IEEE Transaction on Vehicular Technology, 63*(1), 30–39.
Mechanical Corporation. Retrieved from www.carsim.com
Morteza, M., & Fekih, A. (2014a). A stability guaranteed robust fault tolerant control design for vehicle suspension systems subject to actuator faults and disturbances. *IEEE Transactions on Control Systems Technology, PP*(99), 1–10.
Morteza, M., & Fekih, A. (2014b, March). Adaptive PID-sliding mode fault tolerant control approach for vehicle suspension systems subject to actuator faults. *IEEE Transactions on Vehicular Technology, 63*(3), 1041–1054.
Noura, H., Theilliol, D., Ponsart, J.-C., & Noura, A. (2009). *Fault-tolerant control systems. Design and practical applications*. London: Springer-Verlag.
Tabbache, B., Benbouzid, M., Kheloui, A., & Bourgeot, J. (2011, June 27–30). *DSP-based sensor fault-tolerant control of electric vehicle power trains*. Proceedings of IEEE International Symposium on Industrial Electronics, Gdansk, pp. 2085–2090.
Wang, R., & Wang, J. (2013, March). Passive actuator fault-tolerant control for a class of over-actuated nonlinear systems and applications to electric vehicles. *IEEE Transaction on Vehicular Technology, 62*(3), 972–985.
Yang, H., Cocquempot, V., & Jiang, B. (2008, June 25–27). *Hybrid fault tolerant tracking control design for electric vehicles*. Proceedings of the Mediterranean Conference on Control and Automation, Ajaccio, pp. 1210–1215.
Zhang, M., Ding, S., Lam, J., & Wang, H. (2003, March). An LMI approach to design robust fault detection filter for uncertain LTI systems. *Automatica, 39*(3), 543–550.

Appendix

A.1. CarSim steering controller

The theory and the algorithm used to steer a road vehicle in CarSim environment are described in this section. The algorithm is intended to provide optimal control for a continuous linear system:

$$\dot{x} = Ax + Bu + Hv \tag{A1}$$

$$y = Cx + Du + Ev. \tag{A2}$$

where x is an array of n state variable, u is control input, y is an output, and A, B, C, D, E, and H are matrices with constant coefficients. The control objective is to determine the value of u to predict output to $y(t)$ to match a target $y_{\text{target}}(t)$ over some preview time t.

The system has initial conditions x_0 at time $t = 0$, a constant input u, and a constant disturbance v, then the time response is defined by

$$x(t) = e^{At}x_0 + \int_0^t e^{A\eta}Bud\eta + \int_0^t e^{A\eta}Hvd\eta. \tag{A3}$$

The term $e^{A\eta}$ is an $(n \times n)$ matrix called the state transition matrix. Each coefficient in the matrix is the portion of state variable i at time t that is linearly related to the state variable j at time zero. The two integrals in Equation (A3) define the forced responses to each state variable due to constant control u and disturbance v over the time interval.

Combining Equations (A2) and (A3), we obtain the following output response:

$$y(t) = Cx = Ce^{At}x_0 + C\left[\int_0^t e^{A\eta}\mathrm{d}\eta\right][Bu + Hv]. \tag{A4}$$

A control response scalar $g(t)$ and a disturbance response scalar $h(t)$ are defined to relate the responses over the time t. A free response array F is defined and relates the state variables at time 0 to the resulting output variable y at time t.

$$F(t) = \mathrm{Ce}^{\mathrm{At}} \tag{A5}$$

The response equation is defined by

$$y(t) = F(t)x_0 + g(t)u + h(t)v \tag{A6}$$

To determine the optimal control, a quadratic performance index J is defined by

$$J = \frac{1}{T}\int_0^T \{y_{\text{target}}(t) - y(t)\}^2 W(t)\mathrm{d}t. \tag{A7}$$

Here, $W(t)$ is an arbitrary weighting function. A optimal control law is designed by minimizing the cost function J, representing the squared deviation of response variable $y(t)$ relative to the target function $y_{\text{target}}(t)$. The control function u minimizing J can be found by substituting Equation (A6) into (A7) and taking a partial derivative of J with respect to u.

$$J = \frac{1}{T}\int_0^T (F(t)x_0 + g(t)u + h(t)v - y_{\text{target}}(t))^2 W(t)\mathrm{d}t. \tag{A8}$$

Solving for u results in

$$u = \frac{\int_0^T \{y_{\text{target}}(t) - F(t)x_0 - h(t)v\}g(t)W(t)\mathrm{d}t}{\int_0^T g(t)^2 W(t)\mathrm{d}t}. \tag{A9}$$

In practice, the integrals over T can be replaced with finite summations

$$u = \frac{\sum_{i=1}^{m}(y_{\text{target}}(t) - F_i x_0 - h_i v)g_i W_i}{\sum_{i=1}^{m} g_i^2 W_i}. \tag{A10}$$

The algorithm is programmed to generate a steering wheel angle in the vehicle solver program for a given target path. The algorithm synthesizes the target path over the preview time and calculates the optimal front steering effort u to minimize deviations from the path. It also delays the driver steering control by a constant time. The geometry of the road is given as a sequence of X and Y coordinates that define a reference line. Station S is defined as the distance along the reference line, typically a road centerline. For each pair of X–Y coordinates, a corresponding increment of S is computed by

$$S_i = S_{i-1} + \sqrt{(X_i - X_{i-1})^2 + (Y_i - Y_{i-1})^2}. \quad \text{(A11)}$$

To calculate the optimal steering control using Equation (A10), the target position is needed at each point considered in the summation. The station target location is

$$S_{\text{targ},i} = S + \frac{iV_x T}{m}, \quad \text{(A12)}$$

where V_x is the vehicle forward speed.

The controller calculations are performed using an axis system where the vehicle is located such that the center of the vehicle front axle is at $X = 0$ and $Y = 0$ and the X and Y axes are aligned with the longitudinal and lateral axes of the vehicle. The target lateral translation is calculated by first getting the inertial X and Y coordinates of the path as a function of the station at the target location (S_{targ}):

$$Y_{\text{target}} = [Y(S_{\text{targ}}) - Y_V]\cos(\psi) - [X(S_{\text{targ}}) - X_V]\sin(\psi). \quad \text{(A13)}$$

A.2. Vehicle parameters

Table A1. Vehicle parameters.

M	Vehicle mass, 1573 kg
c_f, c_r	Cornering stiffness of front/rear wheels 2*60,000 N/rad
l_f	Distance between the front wheels and the center of gravity, 1.137 m
l_r	Distance between the rear wheels and the center of gravity, 1.530 m
I_z	Yaw moment of inertia, 2753 kg m^2

8

Gradient-based step response identification of overdamped processes with time delay

Rui Yan, Tao Liu*, Fengwei Chen and Shijian Dong

Institute of Advanced Control Technology, Dalian University of Technology, Dalian 116024, People's Republic of China

In this paper, a step response identification method is proposed for overdamped industrial processes with time delay from sampled data by developing a gradient searching approach to minimize the output prediction error. Based on establishing a least-squares fitting of the time domain expression of a low-order process model, i.e. first-order-plus-dead-time (FOPDT) or second-order-plus-dead-time (SOPDT), with respect to a step change, the rational model parameters together with the delay parameter can be simultaneously estimated, while the computation effort can be significantly reduced compared to the existing step identification methods based on using the time integral approach for model fitting. Both cases of repetitive poles and distinct poles are considered for the identification of an SOPDT model, along with a guideline for a suitable choice of the model structure for practical applications. The convergence and accuracy of the proposed algorithms are analysed with a strict proof. Four illustrative examples from recent references are used to show the effectiveness of the proposed method.

Keywords: step response identification; overdamped process; time delay; second-order-plus-dead-time model; gradient searching

1. Introduction

For control-oriented model identification in industrial engineering applications, step response tests have been widely used for identifying a transfer function model between the process input and output, owing to its implemental simplicity and economy. In the past decades, the identification of process models with time delay has received increasing attention in view of the fact that time delay is usually involved with industrial processes and system operations (Liu, Wang, & Huang, 2013). Although linear model identification methods have been extensively explored by Rake (1980) and Richard (2003), the extension to identifying a transfer function model with time delay is not straightforward and could become very difficult due to the identification nonlinearity introduced by time delay (Ljung, 1999).

Early references (Gawthrop & Nihtilä, 1985) proposed the approximation of time delay by using Padé approximation and Laguerre expansion, resulting in a higher order rational transfer function model that contains more parameters to be estimated and may cause unacceptable fitting error when the process response has a long time delay. By fitting a few representative points in the transient output response to a step change, Rangaiah and Krishnaswamy (1996) and Huang, Lee, and Chen (2001) reported alternative identification methods for obtaining a first-order-plus-dead-time (FOPDT) or second-order-plus-dead-time (SOPDT) model. Furthermore, based on numerical integral to the time domain expression of step response, Bi et al. (1999) proposed a time integral method for the identification of an FOPDT model, and it has been extended to identify a nth order process model by introducing n-fold multiple integrals in the lecture, e.g. Wang, Guo, and Zhang (2001) and Wang and Zhang (2001). To cope with load disturbance or nonzero initial conditions, modified step response tests were proposed by Wang et al. (2008), Hwang and Lai (2004) and Wang, Liu, and Hang (2005) to develop robust identification algorithms. Liu, Wang, Huang, and Hang (2007) suggested the use of multiple piecewise step tests to establish LS fitting conditions. In contrast, another robust identification algorithm, proposed by Liu and Gao (2008), was developed for obtaining a low-order transfer function model with time delay by using transient system response data from both adding and removing the step change. The idea was further extended to identify the dynamics of inherent-type load disturbance by Liu, Zhou, Yang, and Gao (2010). Note that these multiple-integral-based identification methods can give good accuracy and robustness in comparison with previous methods, but the computation load is relatively high and the use of multiple time integrals for parameter estimation may be sensitive to the test data length. To avoid the use of multiple time integrals for model identification, Ahmed, Huang, and Shah (2006, 2007) developed an iterative identification method based on using a linear filter. By comparison, a frequency domain step identification method was proposed

*Corresponding author. Email: liurouter@ieee.org

by Liu and Gao (2010) to relieve the computation effort by introducing a damping factor to the step response for the computation of the Laplace transform.

For the identification of time delay systems, adaptive identification methods for iteratively estimating the linear model parameters and the delay parameter were proposed in the references, e.g. Orlov, Belkoura, Richard, and Dambrine (2003), Ren, Rad, Chen, and Lo (2005), and Na, Ren, and Xia (2014), obtaining good convergence against measurement noise. A generalized expectation–maximization algorithm was proposed for robust global identification of linear parameter-varying systems with fixed input delay (Yang, Lu, & Yan, 2015). For identifying a canonical state space model with state delay, a recursive least-squares parameter identification algorithm was presented by Gu, Ding, and Li (2014) based on using a state filter. To tackle the identification nonlinearity arising from the time delay, a few numerical optimization methods were developed for estimating all the linear model parameters and the delay parameter, e.g. hierarchical identification strategies (Chen, Garnier, & Gilson, 2015; Previdi & Lovera, 2004; Yang, Iemura, Kanae, & Wada, 2007), Gradient searching algorithm (Ding, Liu, & Chu, 2013), Newton iteration approach (Ji, Xu, Xiong, & Chen, 2015), and the Levenberg–Marquardt optimization method (Sung & Lee, 2001; Baysse, Carrillo, & Habbadi, 2011). Note that these numerical optimization methods were developed based on using persistent excitation tests such as the pseudo-random binary sequence (PRBS), and thus are not suitable for step response tests as preferred in many industrial applications.

In this paper, model identification for overdamped industrial processes with time delay is studied to facilitate model-based control design for such a process in industrial applications. A gradient-based identification method is proposed which can simultaneously identify the linear model parameters together with the delay parameter from the step response data. Based on the time domain expression of the process transfer function model with respect to a step change of the input, a modified Gauss–Newton iteration algorithm for parameter estimation is established with computation efficiency for minimizing the output prediction error. Owing to the use of a step test, it is verified that the cost function of prediction error becomes convex with respect to all the model parameters, such that global convergence can be guaranteed. Moreover, no time integral is required for computation, significantly reducing the computation effort compared with recently developed step response identification methods based on using the time integral approach. For clarity, the paper is organized as follows. Section 2 presents the overdamped process models to be identified. The corresponding identification methods are detailed in Section 3. The convergence of the proposed method is analysed in Section 4, together with some guidelines for the choice of a suitable model structure. Four numerical examples are shown in Section 5 to demonstrate the effectiveness of the proposed identification method. Finally, some conclusions are drawn in Section 6.

2. Overdamped process model

For industrial overdamped processes, e.g. multi-component blending reactors and fermentation tanks, a low-order model of FOPDT or SOPDT is widely used for control system design and tuning (Seborg, Mellichamp, Edgar, & Doyle, 2010). Without loss of generality, two model forms are studied in this paper, one has a single or repetitive poles,

$$G_{\mathrm{m1}}(s) = \frac{k_{\mathrm{p}}}{(\tau_{\mathrm{p}}s+1)^m}\mathrm{e}^{-Ls} \tag{1}$$

and the other has two distinct poles,

$$G_{\mathrm{m2}}(s) = \frac{k_{\mathrm{p}}}{(\tau_1 s+1)(\tau_2 s+1)}\mathrm{e}^{-Ls}, \tag{2}$$

where k_{p} denotes the process static gain, L is the process time delay, τ_{p}, τ_1 and τ_2 are time constants, and m is the number of repetitive poles or the model order.

It should be noted that higher-order overdamped processes can be effectively described by the above models (Seborg et al., 2010; Liu & Gao, 2012).

Under a step test, the input excitation with a magnitude of h can be described by

$$u(t) = \begin{cases} 0 & \text{if} \quad t < 0 \\ h & \text{if} \quad t \geq 0. \end{cases} \tag{3}$$

Correspondingly, the output response subject to measurement noise may be written as

$$y(t) = y_{\mathrm{r}}(t) + v(t), \tag{4}$$

where $y_{\mathrm{r}}(t)$ denotes the true output, and $v(t)$ is usually assumed to be a white noise with zero mean.

Generally, the process static gain, k_{p}, can be directly computed from

$$k_{\mathrm{p}} = \frac{\bar{y}(\infty)}{h}, \tag{5}$$

where $\bar{y}(\infty)$ is obtained by averaging 20–30 measured output values after the process response moves into a steady state in a step test.

Define the model prediction error and cost function, respectively, as

$$\delta(t, \hat{\theta}) = y(t) - \hat{y}(t), \tag{6}$$

$$J(\hat{\theta}) = \frac{1}{2}\varepsilon^{\mathrm{T}}\varepsilon, \tag{7}$$

where $\varepsilon = [\delta(t_1, \hat{\theta}), \ldots, \delta(t_N, \hat{\theta})]^{\mathrm{T}}$, $\hat{\theta}$ denotes the estimation of unknown parameters, $\hat{y}(t)$ is the output prediction, N denotes the collected data length, and t_i ($i = 1, 2, \ldots, N.$) are the sampled instant which should be chosen such as $L \le t_1 < t_2 < \cdots < t_N$. Denote by a column vector θ all the unknown parameters to be identified.

Hence, the identification objective is to estimate the unknown parameters in Equation (1) or Equation (2) from the input–output observations $\{u(t_i),\ y(t_i)\}$ $(i = 1, 2, \ldots, N)$, satisfying

$$\hat{\theta} = \arg\min_{\theta} J(\hat{\theta}). \tag{8}$$

3. Proposed identification algorithms

Two gradient-based identification algorithms are developed to identify overdamped process models in Equations (1) and (2), as detailed in the following two subsections.

3.1. *Time delay model with single or repetitive poles*

Letting $m = 1$ in Equation (1) leads to an FOPDT model. Correspondingly, the time domain step response under zero initial conditions can be expressed by

$$y(t) = \begin{cases} v(t) & \text{if} \quad t < L, \\ k_p h(1 - e^{-\alpha(t-L)}) + v(t) & \text{if} \quad t \ge L. \end{cases} \tag{9}$$

where $\alpha = 1/\tau_p$, and the static gain k_p is computed by Equation (5).

Similarly, for the case of $m = 2$ in Equation (1), the time domain step response can be derived as

$$y(t) = \begin{cases} v(t) & \text{if} \quad t < L, \\ k_p h(1 - e^{-\alpha(t-L)} - \alpha(t-L)e^{-\alpha(t-L)}) + v(t) & \text{if} \quad t \ge L. \end{cases} \tag{10}$$

For brevity, the identification procedure is detailed for the case of $m = 1$, which can be simply extended to a case of $m \ge 2$.

To estimate the unknown model parameters, we let

$$\theta = [\alpha,\ L]^{\mathrm{T}}. \tag{11}$$

The prediction error can be computed from Equation (9) as

$$\delta(t,\ \theta) = y(t) - \bar{y}(\infty)(1 - e^{-\alpha(t-L)}). \tag{12}$$

To minimize the cost function in Equation (7), we take the first-order derivative for Equation (12) with respect to α and L, respectively, obtaining

$$\frac{\partial \delta(t_i,\ \theta)}{\partial \alpha} = -\bar{y}(\infty)(t_i - L)e^{-\alpha(t_i - L)}, \tag{13}$$

$$\frac{\partial \delta(t_i,\ \theta)}{\partial L} = \bar{y}(\infty)\alpha e^{-\alpha(t_i - L)}. \tag{14}$$

The Jacobian matrix of Equation (12) is defined by

$$\Xi(t_i,\ \theta) = \frac{\partial \delta(t_i,\ \theta)}{\partial \theta} = \left[\frac{\partial \delta(t_i,\ \theta)}{\partial \alpha},\ \frac{\partial \delta(t_i,\ \theta)}{\partial L}\right]^{\mathrm{T}}. \tag{15}$$

Denote

$$\Psi(\theta) = [\Xi(t_1,\ \theta),\ \Xi(t_2,\ \theta),\ \ldots,\ \Xi(t_N,\ \theta)]^{\mathrm{T}}. \tag{16}$$

The gradient matrix of $J(\theta)$ with respect to θ is therefore defined by

$$g[J(\theta)] = \Psi^{\mathrm{T}}(\theta)\varepsilon. \tag{17}$$

The corresponding Hessian matrix is formulated by

$$H(\theta) = \Psi^{\mathrm{T}}(\theta)\Psi(\theta). \tag{18}$$

Hence, the parameter vector θ can be estimated using a Gauss–Newton iteration approach, i.e.

$$\hat{\theta}_k = \hat{\theta}_{k-1} - H_k^{-1} g_k, \tag{19}$$

where H_k and g_k can be computed from Equations (17) and (18) in terms of $\hat{\theta}_{k-1}$ estimated from the previous iteration step. The initial estimation of $\hat{\theta}$ may be taken as a vector composed of small positive real numbers while the delay parameter may be taken roughly about the minimal output response delay as can be observed from a step test.

It is well known that the standard Gauss–Newton iteration method may give a very slow convergence rate in the initial stage of the iterative process for parameter estimation, causing a considerable computation effort. To overcome the deficiency, we propose the use of an adaptive step length for iteration based on the Armijo line searching strategy (see, e.g. Wright & Nocedal, 1999), i.e.

$$J(\hat{\theta}_{k-1} + \rho^m d_k) \le J(\hat{\theta}_{k-1}) + \sigma \rho^m g_k^{\mathrm{T}} d_k, \tag{20}$$

where $d_k = -{H_k}^{-1} g_k$ denotes the searching direction, $\rho \in (0,\ 1)$ and $\sigma \in (0,\ 1)$ are generally required for computation, and m is the minimal nonnegative integer satisfying (20).

To accomplish the identification objective in Equation (8) using less computation effort, it is suggested to take the searching step,

$$\rho = 1 - \frac{1}{J_k/J_{k-1} + J_{k-1}/J_k}, \tag{21}$$

$$\rho = \min\{\rho, 0.8\} \tag{22}$$

and $\sigma = \rho$ for the implementation of (20). Note that ρ is step-wise varying with respect to the prediction error. Initially, there is likely to be a larger difference between J_k/J_{k-1} and J_{k-1}/J_k, which may turn out to be a larger value of ρ from Equation (21) that results in a larger step for searching and thus expedites the convergence rate (i.e. reducing the number of external loop iteration relating to the searching step). The lower limit in Equation (22) is set to avoid an over large value of ρ that may cause severe increment of the iteration number of the internal loop iteration for determining m. As the iteration goes on,

J_k/J_{k-1} and J_{k-1}/J_k gradually approach the identity, resulting in $\rho \to 0.5$ that guarantees the searching step not very small to maintain the convergence rate. Owing to the fact that $J_i > 0\ (i = 1, 2, \ldots, N)$, there follows $\rho \in [0.5, 0.8]$ which satisfies the general requirement of the Armijo line searching method.

Another issue involved with the Gauss–Newton iteration method is that a suitable initial value is required for iteration, since a local minimum of solution might be turned out for an arbitrary choice of the initial value. To address the issue, numerical simulations are explored to study the relationship between $J(\theta)$ and the model parameters based on using different input excitations for identification. Consider a second-order process with time delay described by $G(s) = \mathrm{e}^{-Ls}/(s^2 + \alpha_1 s + 1)$, where $\alpha_1 = 2$ and $L = 5$. The input excitation is taken as a unity step change or a PRBS sequence with the variance of $\sigma^2 = 1$, respectively, for comparison. The number of measured output data is $N = 3000$ and the sampling period is $T_s = 0.01(\mathrm{s})$. Assuming that the searching ranges of the model parameters are $\alpha_1 \in [0.1, 4]$ and $L \in [0.1, 30]$, the corresponding $J(\theta)$ between the process response and model response are plotted in Figures 1 and 2 to the input excitations of the step change and PRBS sequence, respectively. It is seen from Figure 1 that there is a unique global minimum, owing to the fact that a step change mainly includes low-frequency components that provoke the output response basically in the low-frequency range. This indicates that using a step test for model identification is not sensitive to the initial choice of model parameters for iteration. In contrast, it is seen from Figure 2 that there are multiple local minimums, which implies that different choices of initial model parameters for iteration will turn out different identification results converging to any local minimums of $J(\theta)$.

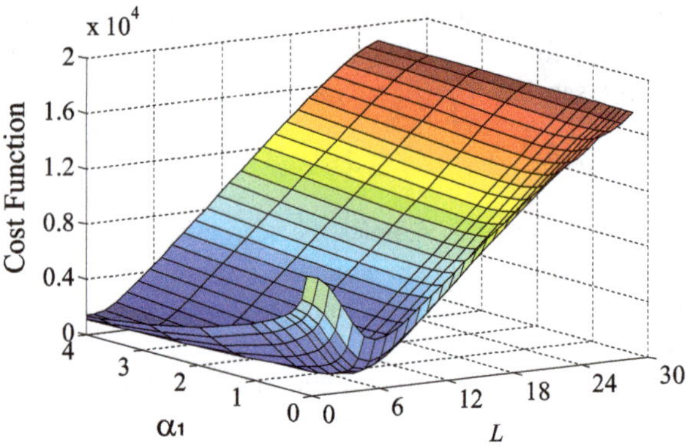

Figure 1. Plot of the identification cost function under a step test.

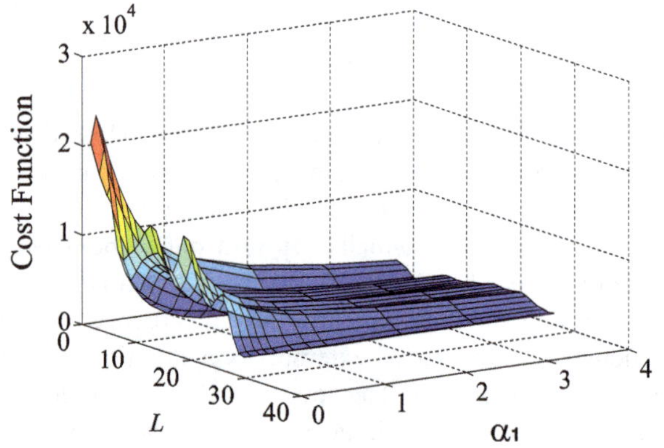

Figure 2. Plot of the identification cost function under a PRBS excitation.

Based on the above analysis, it is suggested to take the initial values of all the model parameters in Equation (1) or Equation (2) as a small positive real number for iteration. To improve the convergence rate, the delay parameter may be taken roughly about the minimum of the possible range as can be observed from a step test.

Hence, the above algorithm named Algorithm I for obtaining a time delay model with a single or repetitive poles can be summarized as

(i) Collect the input–output observations $\{u(t_i), y(t_i)\}$ $(i = 1, 2, \ldots, N)$ from a step test, and compute the process static gain in terms of Equation (5).

(ii) Take the initial estimation of $\hat{\theta}_0$ as a vector composed of small positive real numbers while the delay parameter may be taken roughly about the minimal output response delay as can be observed from a step test.

(iii) Compute the prediction error $\delta(t, \hat{\theta}_{k-1})$ and $J(\hat{\theta}_{k-1})$ using Equations (12) and (7).

(iv) Determine the searching direction $d_k = -H_k{}^{-1}g_k$ by computing the gradient matrix g_k in Equation (17) and the Hessian matrix H_k in Equation (18).

(v) Take the searching step in terms of Equation (22).

(vi) Update $\hat{\theta}_k = \hat{\theta}_{k-1} + \mu_k d_k$, where $\mu_k = \rho^{m_k}$, and m is the minimal nonnegative integer that satisfies (20) with the choice of Equation (22).

(vii) End the algorithm if the fitting condition, $1/N \sum_{i=1}^{N} [y(t_i) - \hat{y}(t_i)]^2 < \mathrm{err}$, is satisfied, where err is a user specified fitting threshold, or the maximum iteration number is attained. Otherwise, let $k = k + 1$ and go to step (iii).

3.2. *Second-order time delay model with distinct poles*

The time domain step response of the process model in Equation (2) is in the form of

$$y(t) = k_{\mathrm{p}}h\left[1 + \frac{\alpha_2}{\alpha_1 - \alpha_2}\mathrm{e}^{-\alpha_1(t-L)} + \frac{\alpha_1}{\alpha_2 - \alpha_1}\mathrm{e}^{-\alpha_2(t-L)}\right] + v(t), t > L, \tag{23}$$

where $\alpha_1 = 1/\tau_1$, $\alpha_2 = 1/\tau_2$, and $k_{\mathrm{p}} = \bar{y}(\infty)/h$.

Denote the parameter vector for estimation,

$$\theta = [\alpha_1,\ \alpha_2,\ L]^{\mathrm{T}}. \tag{24}$$

The prediction error can be computed based on the static gain estimated by Equation (5) as

$$\delta(t,\theta) = y(t) - \bar{y}(\infty)\left[1 + \frac{\alpha_2}{\alpha_1-\alpha_2}e^{-\alpha_1(t-L)} + \frac{\alpha_1}{\alpha_2-\alpha_1}e^{-\alpha_2(t-L)}\right]. \quad (25)$$

To minimize the cost function in Equation (7), we take the first-order derivative for Equation (25) with respect to α_1, α_2 and L, respectively, obtaining

$$\frac{\partial\delta(t_i,\theta)}{\partial\alpha_1} = \bar{y}(\infty)\frac{\alpha_2}{(\alpha_1-\alpha_2)^2}\left[e^{-\alpha_1(t-L)} + (\alpha_1-\alpha_2)(t-L)e^{-\alpha_1(t-L)} - e^{-\alpha_2(t-L)}\right],$$

$$\frac{\partial\delta(t_i,\theta)}{\partial\alpha_2} = \bar{y}(\infty)\frac{\alpha_1}{(\alpha_1-\alpha_2)^2}\left[e^{-\alpha_2(t-L)} + (\alpha_2-\alpha_1)(t-L)e^{-\alpha_2(t-L)} - e^{-\alpha_1(t-L)}\right],$$

$$\frac{\partial\delta(t_i,\theta)}{\partial L} = \bar{y}(\infty)\frac{\alpha_1\alpha_2}{\alpha_1-\alpha_2}\left[e^{-\alpha_2(t-L)} - e^{-\alpha_1(t-L)}\right]. \quad (26)$$

Correspondingly, the gradient matrix of $J(\theta)$ and the Hessian matrix can be computed in terms of Equations (17) and (18), respectively.

Hence, an identification algorithm named Algorithm II for obtaining an SOPDT model with distinct poles can be summarized as

(i) Collect the input–output observations $\{u(t_i), y(t_i)\}$ $(i = 1, 2, \ldots, N)$ from a step test, and compute the process static gain in terms of Equation (5).
(ii) Take the initial estimation of $\hat{\theta}_0$ as a vector composed of small positive real numbers while the delay parameter may be taken roughly about the minimal output response delay as can be observed from a step test.
(iii) Compute the prediction error $\delta(t, \hat{\theta}_{k-1})$ and cost function $J(\hat{\theta}_{k-1})$ using Equations (25) and (7).
(iv) Determine the searching direction $d_k = -H_k{}^{-1}g_k$ by computing the gradient matrix g_k in Equation (17) and the Hessian matrix H_k in Equation (18).
(v) Take the searching step in terms of Equation (22).
(vi) Update $\hat{\theta}_k = \hat{\theta}_{k-1} + \mu_k d_k$, where $\mu_k = \rho^{m_k}$, and m is the minimal nonnegative integer that satisfies (20) with the choice of Equation (22).
(vii) End the algorithm if the fitting condition, $1/N\sum_{i=1}^{N}[y(t_i) - \hat{y}(t_i)]^2 < \text{err}$, is satisfied, where err is a user specified fitting threshold, or the maximum iteration number is attained. Otherwise, let $k = k + 1$ and go to step (iii).

Remark 1 It can be seen from the above algorithms that the identification procedure is relatively independent of the time length of the step response and use no time integral for computation, significantly reducing the computation effort compared with the recently developed step identification methods based on using single or multiple time integrals to the step response data (Ahmed, Huang, & Shah, 2008; Liu et al., 2007, 2010).

4. Analysis on consistency estimation

For practical application in the presence of measurement noise, it should be clarified if consistent estimation could be guaranteed by the proposed method. To address this issue, the following proposition is given accordingly.

PROPOSITION 1 *Assuming that $\delta(t, \theta)$ is differentiable with respect to θ in a neighbourhood of the bounded level set $S = \{\theta | J(\theta) \le J(\theta_0)\}$, and the Jacobian matrix $\Psi(\theta)$ in Equation (16) is full row rank together with the positive-definite Hessian matrix $H(\theta)$ in (18), both Algorithms I and II guarantee uniform convergence satisfying*

$$\lim_{k\to\infty} g_k(\theta) = \lim_{k\to\infty} \Psi^{\mathrm{T}}(\theta_k)\varepsilon(\theta_k) = 0, \quad (27)$$

where $\varepsilon(\theta_k) = [\delta(t_1, \theta_k), \ldots, \delta(t_N, \theta_k)]^{\mathrm{T}}$.

Proof By assuming $\delta(t, \theta)$ is differentiable with respect to θ in the bounded set S, there follows for some positive constants β_1 and β_2 that

$$\|\delta(t,\theta)\| \le \beta_1, \quad \forall\theta \in S \text{ and } \forall t > L, \quad (28)$$

$$\|\Xi(t,\theta)\| \le \beta_2, \quad \forall\theta \in S \text{ and } \forall t > L, \quad (29)$$

where $\|\cdot\|$ denotes the matrix 2-norm. ■

It can be verified that there exists a constant $q > 0$ such that

$$\|\Psi(\theta)\| = \|\Psi^{\mathrm{T}}(\theta)\| \le q. \quad (30)$$

Owing to that $\Psi(\theta)$ is full column rank, there stands for a sufficient large constant, $p > 0$, such that

$$\|\Psi(\theta)z\| \ge p\|z\|. \quad (31)$$

Denote by ω_k the intersection angle between the negative gradient direction, $-g_k$, and the Gauss–Newton searching direction, d_k^{GN}. It can be derived using Equations (30) and (31) that

$$\cos(\omega_k) = -\frac{g_k^{\mathrm{T}}d_k^{\mathrm{GN}}}{\|g_k\|\|d_k^{\mathrm{GN}}\|} = -\frac{(d_k^{\mathrm{GN}})^{\mathrm{T}}\Psi^{\mathrm{T}}(\theta_k)\Psi(\theta_k)d_k^{\mathrm{GN}}}{\|\Psi^{\mathrm{T}}(\theta_k)\Psi(\theta_k)d_k^{\mathrm{GN}}\|\|d_k^{\mathrm{GN}}\|} = \frac{\|\Psi(\theta_k)d_k^{\mathrm{GN}}\|^2}{\|\Psi^{\mathrm{T}}(\theta_k)\Psi(\theta_k)d_k^{\mathrm{GN}}\|\|d_k^{\mathrm{GN}}\|} \ge \frac{p^2\|d_k^{\mathrm{GN}}\|^2}{q^2\|d_k^{\mathrm{GN}}\|^2} = \frac{p^2}{q^2} > 0. \quad (32)$$

From the Armijo condition in (20), we obtain

$$\begin{aligned} J(\hat{\theta}_k) &\le J(\hat{\theta}_{k-1}) + \sigma\rho^{m_k} g_k^{\mathrm{T}} d_k \\ &= J(\hat{\theta}_{k-1}) - c_k \|g_k\| \|d_k^{GN}\| \cos(\omega_k), \end{aligned} \tag{33}$$

where $c_k = \sigma\rho^{m_k}$.

Summing (33) over all indices of k yields

$$J(\hat{\theta}_k) \le J(\hat{\theta}_0) - \sum_{j=0}^{k-1} c_j \|g_j\| \|d_j^{\mathrm{GN}}\| \cos(\omega_j). \tag{34}$$

Considering that $J(\theta)$ is a bounded value, we ensure that $J(\hat{\theta}_0) - J(\hat{\theta}_k)$ is finite for all k, and therefore take the limit to both sides of (34), obtaining

$$\sum_{k=0}^{\infty} c_k \|g_k\| \|d_k^{\mathrm{GN}}\| \cos(\omega_k) < \infty, \tag{35}$$

which implies that

$$\lim_{k\to\infty} c_k \|g_k\| \|d_k^{\mathrm{GN}}\| \cos(\omega_k) = 0 \tag{36}$$

With $d_k^{\mathrm{GN}} = -H_k^{-1} g_k$, H_k is positive definite, $c_k = \sigma\rho^{m_k} > 0$ and $\cos(\omega_k) > 0$, it follows that

$$\lim_{k\to\infty} g_k(\theta) = \lim_{k\to\infty} \Psi^T(\theta_k)\varepsilon(\theta_k) = 0.$$

This completes the proof.

Note that the convergence result shown in Equation (27) indicates that the cost function $J(\theta)$ converges to a steady value when $k \to \infty$, that is to say, the estimated model parameters converge to the true values for perfect model match with the plant, leading to $\lim_{k\to\infty} \tilde{\theta}_k = 0$.

According to Proposition 1, the proposed algorithm guarantees convergence in the presence of measurement noise, owing to the fact that

$$\begin{aligned} \lim_{N\to\infty} g(\theta) &= \lim_{N\to\infty} \Psi^{\mathrm{T}}(\theta)\varepsilon \\ &= \lim_{N\to\infty} \sum_{i=1}^{N} \frac{\partial[y(t_i) - \hat{y}(t_i)]}{\partial\theta}[y(t_i) - \hat{y}(t_i)] \\ &= -\lim_{N\to\infty} \sum_{i=1}^{N} \frac{\partial\hat{y}(t_i)}{\partial\theta}[x(t_i) + v(t_i) - \hat{y}(t_i)] \\ &= -\lim_{N\to\infty} \sum_{i=1}^{N} \frac{\partial\hat{y}(t_i)}{\partial\theta}[x(t_i) - \hat{y}(t_i)] \\ &\quad - \lim_{N\to\infty} \sum_{i=1}^{N} v(t_i)\frac{\partial\hat{y}(t_i)}{\partial\theta} \\ &= -\lim_{N\to\infty} \sum_{i=1}^{N} \frac{\partial\hat{y}(t_i)}{\partial\theta}[x(t_i) - \hat{y}(t_i)] \end{aligned} \tag{37}$$

and

$$H(\theta) = \Psi^{\mathrm{T}}(\theta)\Psi(\theta) = \sum_{i=1}^{N} \frac{\partial\hat{y}(t_i)}{\partial\theta}\frac{\partial\hat{y}(t_i)}{\partial\theta^T}. \tag{38}$$

Therefore, the iterative process of the proposed algorithm will not be affected by the measurement noise, guaranteeing good robustness for convergence.

Note that it is often encountered in practical applications that the true model structure of the process to be identified cannot be known exactly. It is desirable to determine an optimal model structure for representing the process dynamic response characteristics from a step test. The following hypothesis testing condition, which was introduced in Liu and Gao (2010), can be taken for the choice of optimal model order,

$$\frac{\sum_{i=1}^{N} [\hat{y}(t_i) - y(t_i)]^2|_{m_1}}{\sum_{i=1}^{N} [\hat{y}(t_i) - y(t_i)]^2|_{m_2}} \le 0.1, \tag{39}$$

where $\hat{y}(t)$ and $y(t)$ denote, respectively, the measured process output and the model output to the step change, m_1 denotes the current model order and m_2 a higher order to be verified.

It can be seen from Equation (39) that a higher model order m_2 should be accepted only if the output prediction error is larger than one-tenth of that of the current model with m_1. To determine if an SOPDT model with repetitive poles or distinct poles should be chosen, we define the curvature (see, e.g. Rynne & Youngson, 2000) of such a model response to a step change by

$$c = \frac{|y''(t)|}{[1 + y'^2(t)]^{3/2}}, \tag{40}$$

where $y'(t)$ and $y''(t)$ denote the first-order and second-order derivatives, respectively. The maximal curvature for the normalized ranges of $\tau_{\mathrm{p}} \in (0, 1]$, $\tau_1 \in (0, 0.5)$ and $\tau_2 \in (0.5, 1]$ in these two models is plotted in Figures 3 and 4, respectively. It is seen from Figure 3 that an evidently larger curvature occurs for an SOPDT model with repetitive poles compared with Figure 4 for an SOPDT model with distinct poles. Therefore, it is suggested to

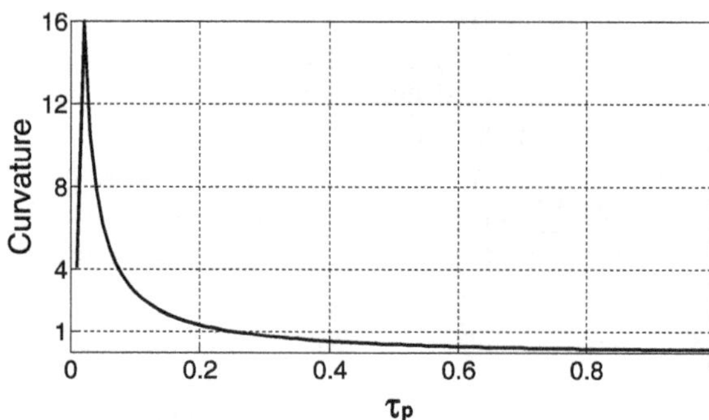

Figure 3. The maximal curvature of an SOPDT model with repetitive poles.

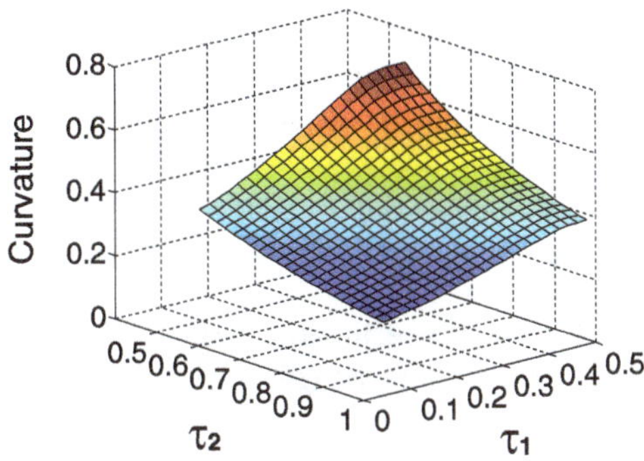

Figure 4. The maximal curvature of an SOPDT model with distinct poles.

take an SOPDT model with repetitive poles if the process response to a step change has an obviously larger curvature.

5. Illustration

Four examples from the recent literatures are studied here to illustrate the performance of the proposed method. Examples 1–3 are given to show good accuracy of the proposed algorithms for identifying low-order processes in terms of the exact model structures, together with measurement noise tests to demonstrate identification robustness. Example 4 is used to show the effectiveness of the proposed algorithm for identification of higher order processes. In the following examples, the measurement noise level is evaluated in terms of the noise-to-signal ratio,

$$\mathrm{NSR} = \frac{\mathrm{mean(abs(noise))}}{\mathrm{mean(abs(signal))}} \times 100\%. \tag{41}$$

The following time domain fitting criterion is used to assess identification accuracy,

$$\mathrm{err} = \frac{1}{N}\sum_{i=1}^{N}[y(t_i) - \hat{y}(t_i)]^2, \tag{42}$$

where $y(t_i)$ is the actual process output and $\hat{y}(t_i)$ is the response of the estimated model.

The parameter estimation error is assessed by the following criterion:

$$\mathrm{ERR} = \sqrt{\frac{\|\hat{\theta}_k - \theta\|_2}{\|\theta\|_2}} \times 100\%, \tag{43}$$

where $\hat{\theta}_k$ is an estimation of the true parameter vector θ.

Example 1 Consider the first-order inertial system with time delay studied by Bi et al. (1999),

$$G(s) = \frac{1}{s+1}\mathrm{e}^{-s}.$$

Based on a unity step response test, Bi et al. (1999) gave a FOPDT model, $G(s) = 1.00\mathrm{e}^{-1.00s}/(0.997s + 1)$, by using a time domain integral of the output response for LS fitting. For illustration, the same step test is performed with the sampling interval $T_s = 0.01(\mathrm{s})$ and the sampling time $T = 50(\mathrm{s})$. According to the proposed guideline for choosing the initial values for iteration, $\theta_0 = [0.01, 0.01]^{\mathrm{T}}$ is taken to apply the proposed Algorithm I, resulting in the exact model parameters based on any data length longer than the transient step response time.

Then assume that the process output measurement is corrupted by a Gaussian white noise with NSR = 10% and NSR = 20%, respectively. The estimation results are listed in Table 1 together with the iteration results. It can be seen from Table 1 that the proposed algorithm converges quickly regardless of poor initial estimation for iteration. Moreover, 100 Monte Carlo tests are conducted under a variety of random measurement noise level (NSR = 10%, 20%) for using the proposed algorithm. Table 2 lists the simulation results where the internal loop indicates the iteration number for determining m in (20), and the external loop indicates the iteration number of searching step for converging to the optimal estimation. It is seen from Table 2 that the proposed algorithm guarantees a fast convergent speed and maintain good robustness against measurement noise. The computation load of the proposed method is listed in Table 3 in comparison with that of the time integral method given by Bi et al. (1999), where N denotes the number of sampled data for identification and K is the number of iterative steps. Owing to the fact that N is much larger than K, the computation load is significantly reduced by the proposed method.

Table 1. Identification results under different noise levels for Example 1.

NSR	Iterative step k	$\hat{\tau}_p$	$\hat{L}$	ERR (%)
10%	1	77.3769	− 90.85	8446.80
	5	30.8218	− 49.83	4167.30
	10	10.5482	− 13.45	1224.50
	20	1.2032	0.91	15.73
	30	0.9816	0.99	1.35
20%	1	77.8236	− 91.11	8481.00
	5	32.1390	− 52.4817	4376.00
	10	12.1918	− 17.06	1502.49
	20	1.5376	0.72	42.97
	30	0.9638	0.99	2.67
	35	0.9633	0.99	2.70
True value		1	1	0

Table 2. Averaged iteration numbers under 100 Monte Carlo tests for Example 1.

NSR	External loop	Internal loop	Success ratio
10%	48	81	100
20%	49	84	99

Table 3. Comparison of the computation load for Example 1.

Methods	Addition	Multiplication
Proposed	$K(11N+18)$	$K(17N+19)$
Bi et al. (1999)	N^2+8N+8	$N^2+13N+48$

Example 2 Consider a second-order process with repetitive poles studied by Liu et al. (2007),

$$G(s) = \frac{e^{-5s}}{s^2+2s+1}. \quad (44)$$

A square wave with an amplitude 1 and a period $T_p = 15(s)$ was adopted as the input excitation in lecture Liu et al. (2007), obtaining an estimated model,

$$G(s) = \frac{0.9958(\pm 0.083)e^{-5.09(\pm 0.066)s}}{0.9958(\pm 0.083)s^2 + 1.9935(\pm 0.118)s + 1}.$$

For illustration, a unity step response test is used in the proposed method where the sampling period is taken as $T_s = 0.01(s)$ and the overall sampling time is $T = 80$ (s).

To test identification robustness against measurement noise, assume that the noise level is NSR = 10% in the step test. By performing 100 Monte Carlo tests in terms of randomly varying the 'seed' of the noise generator, the proposed algorithm gives

$$G(s) = \frac{1.0000(\pm 0.0016)e^{-5.00(\pm 0.049)s}}{1.0058(\pm 0.068)s^2 + 2.0047(\pm 0.067)s + 1},$$

which indicates improved identification accuracy.

To demonstrate consistent estimation against measurement noise, the identification results for different noise levels (NSR = 5%, 10%, 15%, 20%, 25%, 30%) are shown in Figure 5, where the result for each model parameter to a given noise level is shown as a vertical linear segment along with the sample standard deviation in parentheses for 100 Monte Carlo tests. The square solid point in each linear segment denotes the mean of 100 identification results, and the upper and lower bars correspond to the maximum and the minimum of parameter estimation, respectively. It is seen that good identification accuracy and consistent estimation are obtained against different noise levels. Moreover, it can be easily verified that, given a measurement noise level, the sample standard deviation of the proposed estimation error is much smaller than that of the measurement noise.

Figure 6 shows the estimation results of the proposed method for different lengths (N) of data in a range from 165 to 8000 with NSR = 10%. For data collection with respect to different data lengths, the sampling interval is taken as $T_s = T/N$. It is seen that good identification accuracy can be obtained for a wide range of data lengths and the estimation results become better as the number of data points increases or the sampling interval decreases. Moreover, the computation load is listed in Table 4 in comparison with that of Liu et al (2007) based on using the time integral approach, indicating that the computation effort is significantly reduced by the proposed method.

Example 3 Consider a second-order process with distinct poles studied by Wang, Liu, Hang, and Tang (2006),

$$G(s) = \frac{e^{-4.5s}}{(5s+1)(s+1)}. \quad (45)$$

By performing a step test with a sampling period $T_s = 0.01(s)$ and the sampling time $T = 250(s)$, the proposed Algorithm II gives an exact estimation of the process model based on any data length longer than the transient step response time.

Then assume that the process output measurement is corrupted by a white noise causing NSR = 10%. The iteration process of the proposed Algorithm II is listed in Table 5 with an initial choice of $\theta_0 = [0.1,\ 0.3,\ 0.1]^T$, indicating good convergence rate and identification accuracy.

Example 4 Consider the fifth-order system with time delay studied by Wang et al. (2001)

$$G(s) = \frac{1.08}{(s+1)^2(2s+1)^3}e^{-10s}.$$

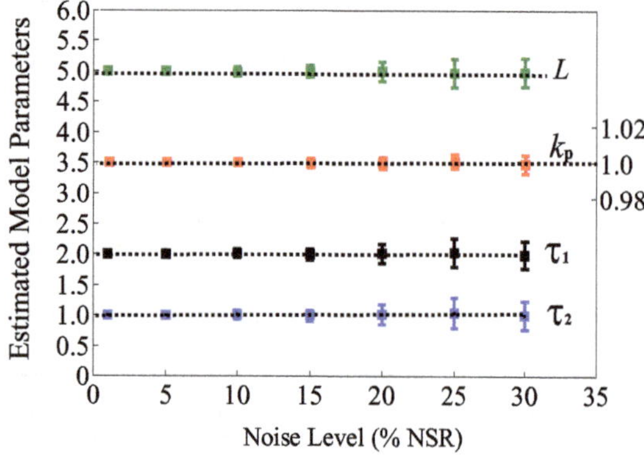

Figure 5. The results of 100 Monte Carlo Tests for Example 2 against measurement noise.

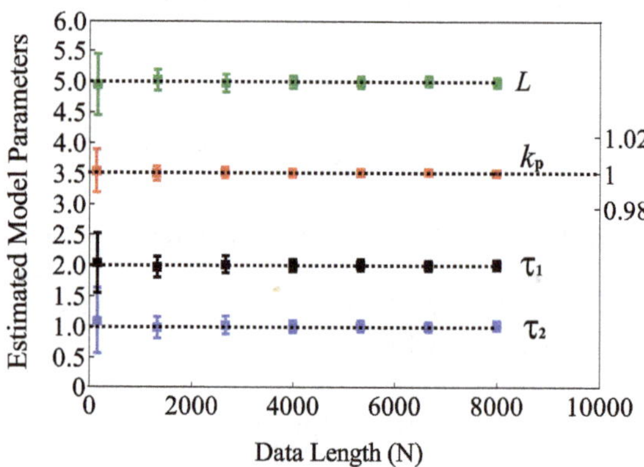

Figure 6. Parameter estimation with respect to data length for Example 2.

Table 4. Comparison of the computation load for Example 2.

Methods	Addition	Multiplication
Proposed	$K(11N+18)$	$K(21N+19)$
Liu et al. (2007)	$\frac{1}{6}N(N+1)(2N+1)+N^2+55N+639$	$\frac{1}{6}N(N+1)(2N+1)+\frac{1}{2}N(N+1)+N^2+66N+1735$

Table 5. Parameter estimation results for Example 3.

Iterative step k	$\hat{\tau}_1$	$\hat{\tau}_2$	$\hat{L}$	ERR (%)
1	10.0000	3.3333	0.10	103.78
10	7.0847	0.6124	10.83	98.21
20	3.6830	0.6836	8.36	60.10
30	5.0930	0.5387	5.68	18.68
40	4.9999	1.0274	4.57	1.20
True value	5	1	4.5	0

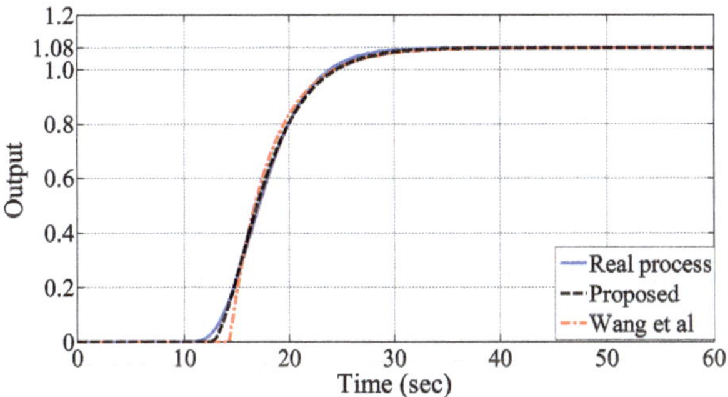

Figure 7. Step response fitting for Example 4.

Based on a step test, Wang et al. (2001) derived an SOPDT model, $G(s) = 1.0800e^{-12.71s}/7.8910s^2 + 5.3231s + 1$, corresponding to the mean-squared output fitting error, err $= 5.71 \times 10^{-4}$. For comparison, the same step test is performed for the sampling time $T = 150$(s) with the sampling interval $T_s = 0.01$(s). The proposed Algorithm II gives an SOPDT model, $G(s) = 1.0800e^{-12.59s}/7.7510s^2 + 5.5679s + 1$, corresponding to err $= 4.88 \times 10^{-4}$.

Then assume that the output measurement noise level is NSR $= 10\%$, the proposed method gives $G(s) = 1.0791e^{-12.49s}/7.8539s^2 + 5.6051s + 1$, corresponding to err $= 3.17 \times 10^{-5}$, compared with that proposed by Wang et al. (2001) $G(s) = 1.0789e^{-14.20s}/0.1111s^2 + 3.8731s + 1$, corresponding to err $= 2.34 \times 10^{-4}$. The step responses of these models are shown in Figure 7, well demonstrating good fitting of the proposed method.

6. Conclusions

A step response identification method has been proposed for overdamped industrial processes with time delay, along with a proof on the convergence and consistency in the presence of measurement noise. By developing a gradient-based searching approach to minimizing the output prediction error, the linear model parameters together with the delay parameter can be simultaneously identified from the time domain expression of a low-order model response to a step change, specifically for the identification of an FOPDT or SOPDT model. A guideline of model structure selection has been given for identifying an SOPDT model with repetitive poles or distinct poles. The computation effort can be significantly reduced compared to recently developed step response identification methods (e.g. Ahmed et al., 2008; Liu et al., 2007, 2010) based on using the time integral to step response data and, moreover, the proposed algorithms are not sensitive to the data length for model identification, therefore facilitating practical applications. Four illustration examples from the literature have well demonstrated that the effectiveness and accuracy of the proposed identification algorithms. It should be noted that the proposed method cannot be applied to an underdamped process with complex poles that hinder the use of a gradient searching strategy for minimizing the output prediction error, which will be studied in our future work.

Funding

This work is supported in part by the National Thousand Talents Program of China, NSF China Grants 61473054, and the Fundamental Research Funds for the Central Universities of China.

References

Ahmed, S., Huang, B., & Shah, S. L. (2006). Parameter and delay estimation of continuous-time models using a linear filter. *Journal of Process Control*, *16*(4), 323–331.

Ahmed, S., Huang, B., & Shah, S. L. (2007). Novel identification method from step response. *Control Engineering Practice*, *15*(5), 545–556.

Ahmed, S., Huang, B., & Shah, S. L. (2008). Identification from step responses with transient initial conditions. *Journal of Process Control*, *18*(2), 121–130.

Baysse, A., Carrillo, F. J., & Habbadi, A. (2011). Time domain identification of continuous-time systems with time delay using output error method from sampled data. The 18th IFAC World Congress, Milano, Italy.

Bi, Q., Cai, W. J., Lee, E. L., Wang, Q. G., Hang, C. C., & Zhang, Y. (1999). Robust identification of first-order plus dead-time model from step response. *Control Engineering Practice*, *7*(1), 71–77.

Chen, F., Garnier, H., & Gilson, M. (2015). Robust identification of continuous-time models with arbitrary time-delay from

irregularly sampled data. *Journal of Process Control, 25*, 19–27.
Ding, F., Liu, X., & Chu, J. (2013). Gradient-based and least-squares-based iterative algorithms for Hammerstein systems using the hierarchical identification principle. *IET Control Theory & Applications*, *7*(2), 176–184.
Gawthrop, P. J., & Nihtilä, M. T. (1985). Identification of time delays using a polynomial identification method. *Systems & Control Letters*, *5*(4), 267–271.
Gu, Y., Ding, F., & Li, J. (2014). State filtering and parameter estimation for linear systems with d-step state-delay. *IET Signal Processing*, *8*(6), 639–646.
Huang, H. P., Lee, M. W., & Chen, C. L. (2001). A system of procedures for identification of simple models using transient step response. *Industrial & Engineering Chemistry Research*, *40*(8), 1903–1915.
Hwang, S. H., & Lai, S. T. (2004). Use of two-stage least-squares algorithms for identification of continuous systems with time delay based on pulse responses. *Automatica*, *40*(9), 1561–1568.
Ji, K., Xu, L., Xiong, W., & Chen, L. (2015). Newton iterative algorithm based modeling and proportional derivative controller design for second-order systems. *Journal of Applied Mathematics and Computing*, *49*(1), 557–572.
Liu, M., Wang, Q. G., Huang, B., & Hang, C. C. (2007). Improved identification of continuous-time delay processes from piecewise step tests. *Journal of Process Control*, *17*(1), 51–57.
Liu, T., & Gao, F. (2008). Robust step-like identification of low-order process model under nonzero initial conditions and disturbance. *IEEE Transactions on Automatic Control*, *53*(11), 2690–2695.
Liu, T., & Gao, F. (2010). A frequency domain step response identification method for continuous-time processes with time delay. *Journal of Process Control*, *20*(7), 800–809.
Liu, T., & Gao, F. (2012). *Industrial process identification and control design: Step-test and Relay-experiment-based Methods*. London: Springer.
Liu, T., Wang, Q. G., & Huang, H. P. (2013). A tutorial review on process identification from step or relay feedback test. *Journal of Process Control*, *23*(10), 1597–1623.
Liu, T., Zhou, F., Yang, Y., & Gao, F. (2010). Step response identification under inherent-type load disturbance with application to injection molding. *Industrial & Engineering Chemistry Research*, *49*(22), 11572–11581.
Ljung, L. (1999). *System identification: Theory for the user*. 2nd ed. Englewood Cliff, NJ: Prentice-Hall.
Na, J., Ren, X. M., & Xia, Y. (2014). Adaptive parameter identification of linear SISO systems with unknown time-delay. *Systems & Control Letters*, *66*, 43–50.
Orlov, Y., Belkoura, L., Richard, J. P., & Dambrine, M. (2003). Adaptive identification of linear time- delay systems. *International Journal of Robust and Nonlinear Control*, *13*(9), 857–872.
Previdi, F., & Lovera, M. (2004). Identification of non-linear parametrically varying models using separable least squares. *International Journal of Control*, *77*(16), 1382–1392.
Rake, H. (1980). Step response and frequency response methods. *Automatica*, *16*, 519–526.
Rangaiah, G. P., & Krishnaswamy, P. R. (1996). Estimating second-order dead time parameters from underdamped process transients. *Chemical Engineering Science*, *51*(7), 1149–1155.
Ren, X. M., Rad, A. B., Chen, P. T., & Lo, W. L. (2005). Online identification of continuous-time systems with unknown time delay. *IEEE Transactions on Automatic Control*, *50*(9), 1418–1422.
Richard, J. P. (2003). Time-delay systems: an overview of some recent advances and open problems. *Automatica*, *39*(10), 1667–1694.
Rynne, B. P., & Youngson, M. A. (2000). *Linear functional analysis*. London: Springer.
Seborg, D. E., Mellichamp, D. A., Edgar, T. F., & Doyle, III F. J. (2010). *Process dynamics and control*. 3rd ed. Hoboken, NJ: John Wiley & Sons.
Sung, S. W., & Lee, I. B. (2001). Prediction error identification method for continuous-time processes with time delay. *Industrial & Engineering Chemistry Research*, *40*(24), 5743–5751.
Wang, Q. G., Guo, X., & Zhang, Y. (2001). Direct identification of continuous time delay systems from step responses. *Journal of Process Control*, *11*(5), 531–542.
Wang, Q. G., Liu, M., & Hang, C. C. (2005). Simplified identification of time-delay systems with nonzero initial conditions from pulse tests. *Industrial & Engineering Chemistry Research*, *44*(19), 7591–7595.
Wang, Q. G., Liu, M., Hang, C. C., & Tang, W. (2006). Robust process identification from relay tests in the presence of nonzero initial conditions and disturbance. *Industrial & Engineering Chemistry Research*, *45*(12), 4063–4070.
Wang, Q. G., Liu, M., Hang, C. C., Zhang, Y., Zhang, Y., & Zheng, W. X. (2008). Integral identification of continuous-time delay systems in the presence of unknown initial conditions and disturbances from step tests. *Industrial & Engineering Chemistry Research*, *47*(14), 4929–4936.
Wang, Q. G., & Zhang, Y. (2001). Robust identification of continuous systems with dead-time from step responses. *Automatica*, *37*(3), 377–390.
Wright, S. J., & Nocedal, J. (1999). *Numerical optimization*. New York, NY: Springer.
Yang, X., Lu, Y., & Yan, Z. (2015). Robust global identification of linear parameter varying systems with generalised expectation–maximisation algorithm. *IET Control Theory & Applications*, *9*(7), 1103–1110.
Yang, Z. J., Iemura, H., Kanae, S., & Wada, K. (2007). Identification of continuous-time systems with multiple unknown time delays by global nonlinear least-squares and instrumental variable methods. *Automatica*, *43*(7), 1257–1264.

9

$\mathcal{L}_1$ adaptive control of a shape memory alloy actuated flexible beam

Bongani Malinga and Gregory D. Buckner*

Department of Mechanical and Aerospace Engineering, North Carolina State University, Campus Box 7910, Raleigh, NC 27695, USA

This paper details the synthesis of an $\mathcal{L}_1$ adaptive controller for a shape memory alloy (SMA) actuated flexible beam. The controller manipulates applied voltage, which alters SMA tendon temperature to track reference bending angles. Simulated and experimental results show that the $\mathcal{L}_1$ adaptive controller provides precise tracking of the reference trajectories and effectively compensates for the nonlinear hysteretic relationship between SMA Joule heating and bending angle without explicitly modelling these characteristics. A simulation model whose results closely resemble the experimental performance results is presented. As a first step towards the development of $\mathcal{L}_1$ adaptive control implementation guidelines, a complete description of the $\mathcal{L}_1$ control parameters and their correlation to tracking performance is presented.

Keywords: control systems; adaptive control; artificial neural networks; shape memory alloys; nonlinear systems; optimization; grid search

1. Introduction

In recent years, there has been an increased focus on the use of shape memory alloys (SMAs) in medical devices, including coronary stents (Kleinstreuer, Li, Basciano, Seelecke, & Farber, 2008) catheter guide wires (Haga, Mineta, Makishi, Matsunaga, & Esashi, 2010), miniature forceps (Nakamura, Matsui, Saito, & Yoshimoto, 1995), eyeglass frames (Wu and Schetky, 2000), energy absorption (Jia, Lalande, & Rogers, 1997), and sensing applications (Stoeckel, 1991). These materials possess the unique ability to 'memorize' shapes through thermally induced phase transitions (Kohl, 2004), making them potentially attractive for micro-actuation applications. Their material composition (e.g. Cu–Zn, Cu–Zn–Al, Cu–Al–Ni, Ni–Ti, Ni–Ti–Fe, Fe–Pt) largely determines critical mechanical properties including ductility, corrosion, and 'memory' properties of the alloy. Three SMA characteristics are associated with crystal reorientation during stress and temperature-induced phase changes: the pseudoelastic effect, the one-way effect, and the two-way effect (Kohl, 2004; Van Humbeeck, Chandrasekaran, & Delaey, 1991). The pseudoelastic effect exhibits a reversible response to deformation by temperature and applied stress. The one-way effect is characterized by shape change upon loading, but no reorientation upon cooling. Once deformed, the 'memorized' shape is recovered when heated above the transition temperature. The two-way effect refers to shape change upon cooling and heating without external loading.

Real-time control of SMA actuated devices is complicated by the highly nonlinear, hysteretic constitutive relationships between material stress, strain, and temperature. Numerous models have been developed to describe the hysteretic behaviour of SMAs; all can be broadly classified as being either physical or empirical in nature. Empirical models date back to Tanaka and Nagaki (1982), with continued development by Liang and Rogers (1990), Brinson (1993) and others. While these models are computationally efficient and adaptable to experimental data, they rely on conditional statements to account for hysteresis in the forward and reverse direction, and tend to be difficult to implement in real-time control algorithms. Physical models have been developed by Achenbach and Müller (1985), Seelecke and Müller (2003) and others. These models are based on statistical thermodynamics and phase transformation probabilities. While single crystal models are computationally efficient, their accuracy is frequently inadequate for control applications. These model-based control design approaches are prone to model errors, which could result from parametric uncertainty and unmodelled dynamics.

Traditional proportional-integral (PI) controllers have been applied to SMA tendon control, but their dependence on integral action to compensate for system hysteresis presents a major drawback. A reduced proportional gain is required to reduce overshoot caused by rapid changes in setpoint. The result is limited bandwidth and limited disturbance rejection characteristics.

This paper introduces an $\mathcal{L}_1$ adaptive controller (Cao and Hovakimyan, 2006) for an SMA-actuated plant: a flexible beam deflected by an offset SMA tendon. The controller manipulates applied voltage, which causes heating in the SMA tendon and bending in the flexible

*Corresponding author. Email: greg_buckner@ncsu.edu

beam; the SMA's two-way effect is used to track reference bending angles. Unlike traditional PI and model-based adaptive control schemes, $\mathcal{L}_1$ adaptive control effectively compensates for hysteresis by considering it a general system uncertainty; it is robust and computationally efficient, enabling real-time implementation. $\mathcal{L}_1$ adaptive control offers higher bandwidth, enhanced robustness, and improved disturbance rejection without the need for exhaustive tuning. Combining $\mathcal{L}_1$ adaptive control with the hysteretic recurrent neural network (HRNN) model and the grid search algorithm provides a systematic framework for adaptive controller design, applicable to a wide range of dynamic systems. The remainder of the paper is organized as follows: Section 2 introduces the plant, model, and controller synthesis, along with computational and experimental methods. Simulated and experimental results are presented in Section 3, with concluding remarks presented in Sections 4 and 5.

2. Methods

2.1. *Experimental methods: plant*

The plant consists of a flexible beam actuated by a single SMA tendon, Figure 1. The SMA tendon is offset from the neutral axis of the beam by a fixed distance a. As the tendon contracts, the actuation force creates a moment about the beam, causing it to bend to an angle θ. While fully maneuverable actuators (like those employed in robotic catheters (Moallem, 2003) require antagonistic actuation, the single-tendon actuator considered here effectively addresses the nonlinearities encountered in SMA actuation. The methods used here can be extended to other single-tendon applications or designs that require antagonistic actuation for increased maneuverability and bandwidth. A complete description of the SMA-actuated flexible beam is presented in (Veeramani, Buckner, Owen, Cook, & Bolotin, 2008).

An experimental realization of this SMA-actuated plant is shown in Figure 2. The central flexible beam (0.5 mm diameter super-elastic Nitinol) features nine equally-spaced plastic collets that maintain a 1 mm offset between the neutral axis and the SMA tendon (0.127 mm diameter Flexinol, Dynalloy, Inc. Tustin, CA). Actuation commands are sent via USB communication from the host PC to a microcontroller, which implements pulse-width modulation to regulate Joule heating and temperature-induced strain in the SMA tendon. Real-time bending is calculated from 3D base and tip measurements (trakSTAR 3D Magnetic Tracking System, Ascension Technology Corporation, Burlington, VT) using constant-curvature deflection relationships (Veeramani et al., 2008).

2.2. *Simulation methods: plant model*

The plant dynamics are simulated using a linear, first-order temperature model cascaded with a single-input, single-output HRNN, as shown in Figure 3. The SMA wire thermal response to the applied control voltage is modelled as an empirical first-order system:

$$G(s) = \frac{1}{\tau s + 1}, \tag{1}$$

where τ is the experimentally determined Joule heating time constant.

A two-phase HRNN is used to map the highly nonlinear and hysteretic relationship between tendon temperature and bending angle. This unique neural network, illustrated in Figure 4, utilizes weighted recurrent neurons, each composed of conjoined sigmoid activation functions, to capture the directional dependencies typical of hysteretic

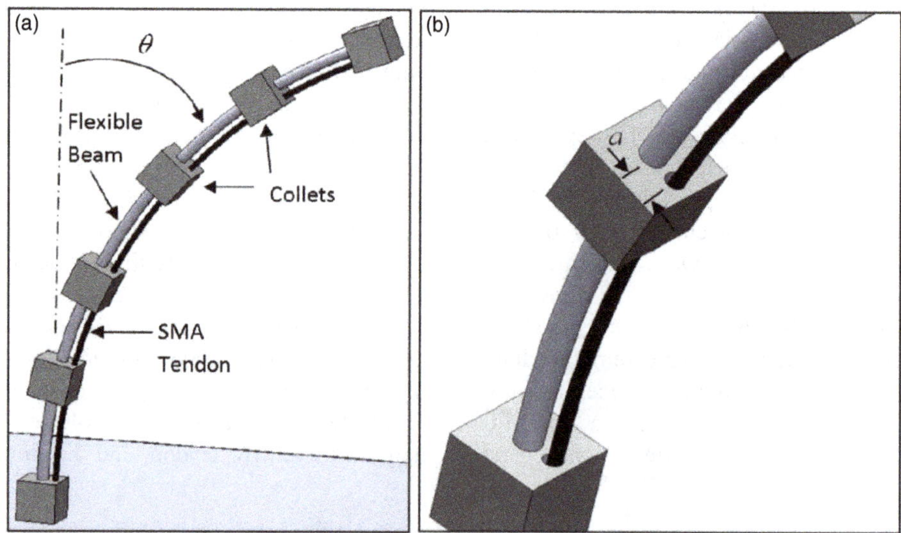

Figure 1. SMA-actuated flexible beam: (a) illustration of the controlled deflection angle; (b) close-up showing collets used to maintain the SMA tendon at a fixed offset a.

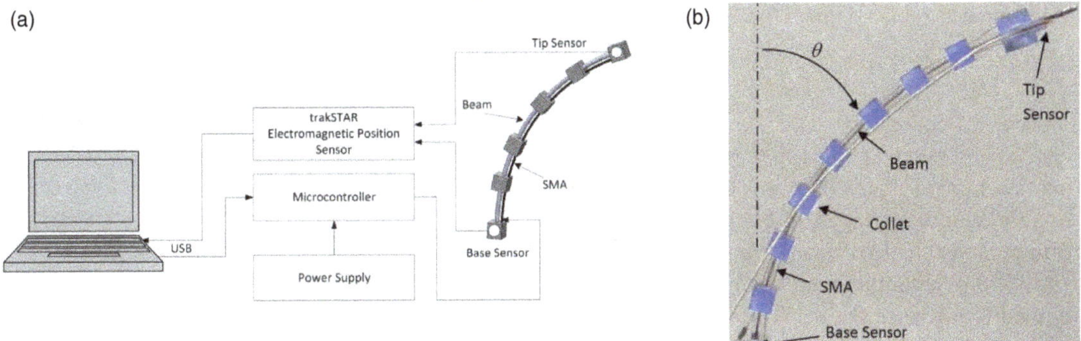

Figure 2. Experimental setup for controller validation: (a) schematic of the test setup; (b) photograph of the flexible beam system with position sensors.

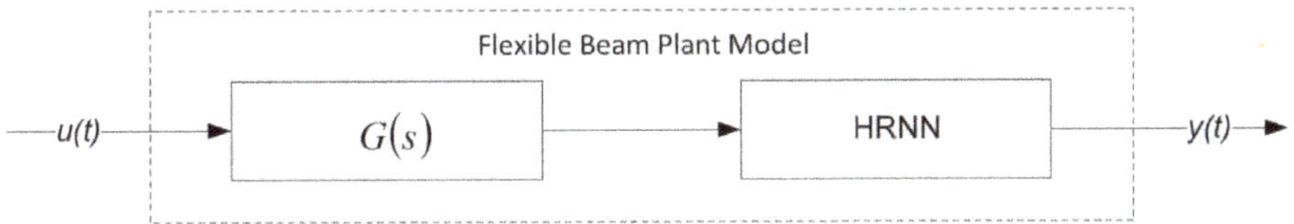

Figure 3. Block diagram of the flexible beam plant model: a first-order empirical model in series with a HRNN.

smart materials. For a complete description of the HRNN and its application to SMA modelling, see (Veeramani, Crews, & Buckner, 2009). At time step q, the output of each activation function is

$$f_i(\hat{T}(q)) = \frac{1 - f_i(\hat{T}(q-1))}{1 + \exp((T_{F,i} - \hat{T}(q))\chi_i)} + \frac{f_i(\hat{T}(q-1))}{1 + \exp((T_{R,i} - \hat{T}(q))\chi_i)}, \quad (2)$$

where $\hat{T}(q)$ refers to the estimated SMA tendon temperature at the qth time step. The activation functions range between 0 and 1, where 0 refers to an inactive neuron and 1 refers to an active neuron. The parameter χ_i controls how quickly the output switches between the two values, with a high value ($\chi_i \gg 1$) leading to step changes. A previously inactive neuron becomes active at the forward transition temperature $T_{F,i}$. A previously active neuron becomes inactive at the reverse transition temperature $T_{R,i}$. The network output

$$\hat{\theta}(q) = \sum_{i=1}^{N} w_i^2 f_i(\hat{T}(q)) \quad (3)$$

is a weighted combination of these activation functions. The network weights w_i are subject to the constraint:

$$\sum_{i=1}^{N} w_i^2 = 1 \quad (4)$$

and are optimized through network training.

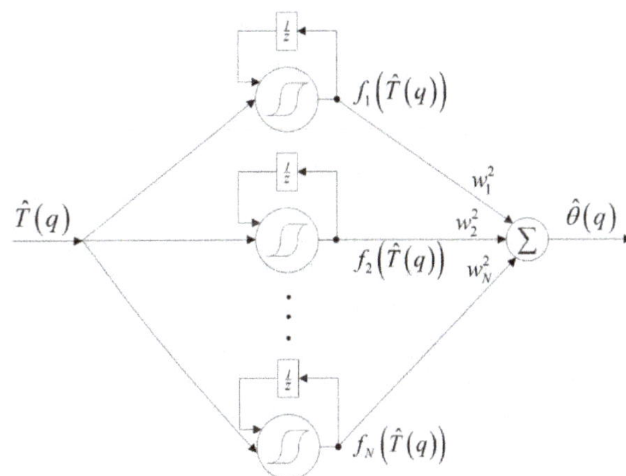

Figure 4. Two-phase HRNN architecture relating SMA tendon temperature to beam bending angle.

The HRNN was trained using open loop data from the physical plant. The Levenberg –Marquardt algorithm (Marquardt, 1963) was used to solve the nonlinear optimization problem governed by the cost function:

$$P = \sum_{q=1}^{Q} e(q)^2, \quad (5)$$

where:

$$e(q) = \theta_m(q) - \hat{\theta}(q) \quad (6)$$

represents the error between an experimentally measured bending angle $\theta_m(q)$ and the HRNN-predicted angle $\hat{\theta}(q)$.

The optimization algorithm was implemented using MATLAB's *lsqnonlin* command (The MathWorks, Inc.,

Natick, MA) using a Jacobian defined by

$$J_{q,i} = \frac{\partial e(q)}{\partial w_i} = -2w_i f_i. \tag{7}$$

Input voltage and bending angle data from a previous investigation (Veeramani et al., 2009) was used for HRNN training using the same plant inputs of Figure 3. SMA tendon temperature was calculated using Equation (1), with the $\mathcal{L}_1$ controller providing the input voltage. The HRNN was then trained to optimize the network weights as outlined in Equations (4)–(7); weights were then imported into the simulation environment for the predicting the bending angle based on the calculated SMA tendon temperature.

2.3. $\mathcal{L}_1$ adaptive control

The $\mathcal{L}_1$ adaptive control objective is to manipulate the input voltage $u(t)$ so that the measured bending angle $y(t)$ tracks the reference bending angle $r(t)$ as shown in Figure 5.

One fundamental difference between $\mathcal{L}_1$ control and traditional model reference adaptive control (MRAC) is the inclusion of a low-pass filter that allows the decoupling of control and adaptation. The filter allows for higher adaptation rates while keeping the control signal within the bandwidth of the system actuator. Key features of $\mathcal{L}_1$ adaptive control include enhanced robustness and transient performance coupled with rapid adaptation, without introducing or enforcing persistence of excitation, without any gain scheduling of controller parameters, and without resorting to high-gain feedback. These features can be achieved by explicitly building the robustness specification into the problem formulation, effectively decoupling adaptation from robustness and increasing the speed of adaptation, subject only to hardware limitations.

2.3.1. $\mathcal{L}_1$ control problem formulation

Consider an SISO system that can be represented by the linear model:

$$y(s) = A(s)(u(s) + d(s)), \tag{8}$$

where $u(s)$ is the system input, $y(s)$ is the system output, and $A(s)$ is a strictly proper unknown transfer function. $d(s)$ represents the time-varying nonlinear uncertainties and disturbances, denoted by $d(t) \triangleq f(t, y(t))$, and f : $\mathbb{R} \times \mathbb{R} \rightarrow \mathbb{R}$ is an unknown map, subject to the following assumptions.

Assumption 1 *Lipschitz continuity: There exist constants $L > 0$ and $L_0 > 0$, possibly arbitrarily large, such that the following inequalities hold uniformly in t:*

$$|f(t, y_1) - f(t, y_2)| \leq L|y_1 - y_2|, \forall t \geq 0,$$
$$|f(t, y)| \leq L|y| + L_0.$$

Assumption 2 *Uniform boundedness of the rate of variation of uncertainties: There exist constants $L_1 > 0$, $L_2 > 0$ and $L_3 > 0$, possibly arbitrarily large, such that for all $t \geq 0$:*

$$|\dot{d}(t)| \leq L_1|\dot{y}(t)| + L_2|y(t)| + L_3.$$

The control objective is to design an adaptive output feedback controller $u(t)$ such that the system output $y(t)$ tracks the given bounded piecewise-continuous reference input $r(t)$ following a desired reference model $M(s)$. For implementation and demonstration simplicity and to streamline parameter selection, in this paper we consider a first-order reference model:

$$M(s) = \frac{m}{s+m}, \quad m > 0. \tag{9}$$

2.3.2. $\mathcal{L}_1$ control architecture

The system output can be expressed in terms of $M(s)$:

$$y(s) = M(s)(u(s) + \sigma(s)), \tag{10}$$

$$u(s) = C(s)(r(s) - \sigma(s)), \tag{11}$$

where uncertainties associated with $A(s)$ and $d(s)$ are lumped into $\sigma(s)$, which is given by

$$\sigma(s) = \frac{(A(s) - M(s))u(s) + A(s)d(s)}{M(s)}. \tag{12}$$

The design of the $\mathcal{L}_1$ adaptive controller proceeds by considering a strictly proper low-pass filter $C(s)$, with $C(0) = 1$, such that

$$H(s) = \frac{A(s)M(s)}{C(s)A(s) + (1 - C(s))M(s)} \tag{13}$$

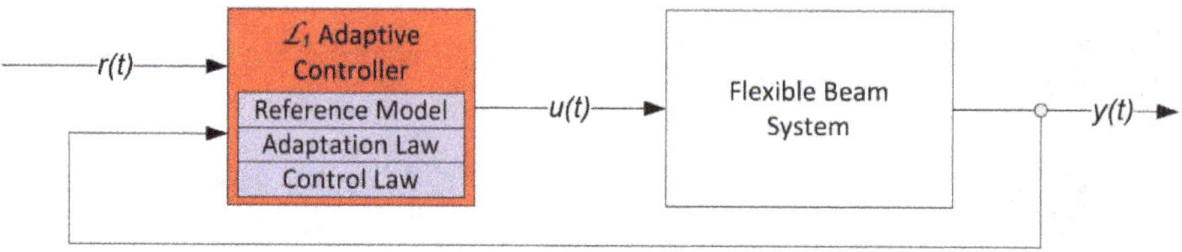

Figure 5. Block diagram of the $\mathcal{L}_1$ adaptive controller.

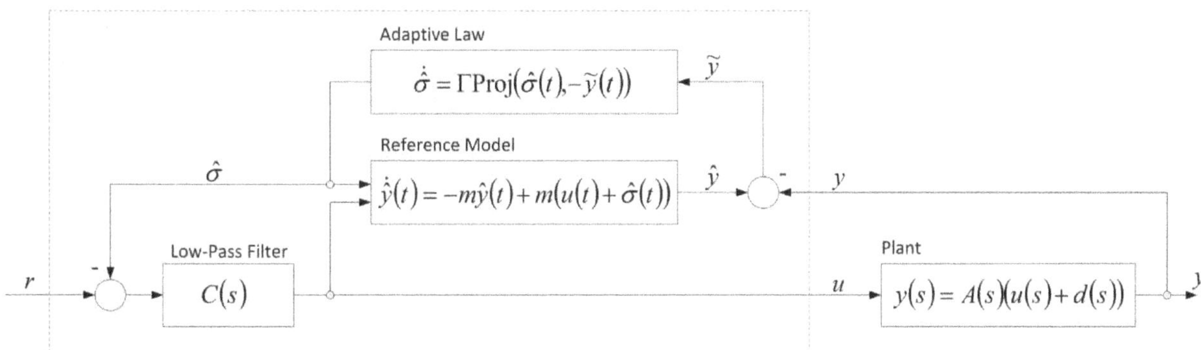

Figure 6. $\mathcal{L}_1$ adaptive control block diagram showing the output predictor, the adaptation law and the low-pass filter applied to the plant $A(s)$.

is stable, and the following $\mathcal{L}_1$ norm condition holds

$$G(s)_{\mathcal{L}_1} L < 1, \tag{14}$$

where $G(s) \triangleq H(s)(1 - C(s))$, L is the maximum of a compact set that represents time-varying unknown parameters, bounded by $\max\{\sigma(s)\}$. The $\mathcal{L}_1$ norm is defined as

$$G(s)_{\mathcal{L}_1} = \int_0^{\infty} |g(t)|\mathrm{d}t, \tag{15}$$

where $g(t)$ is the inverse Laplace transform of $G(s)$ and represents the impulse response of the system.

Reference model/output predictor:

We consider the following output predictor

$$\dot{\hat{y}}(t) = -m\hat{y}(t) + m(u(t) + \hat{\sigma}(t)), \quad \hat{y}(0) = 0, \tag{16}$$

where $\hat{\sigma}(t)$ is the adaptive estimate, governed by the adaptation laws described in the next section.

Adaptation law:

The adaptation of $\hat{\sigma}(t)$ is defined to be:

$$\dot{\hat{\sigma}} = \Gamma\,\mathrm{Proj}(\hat{\sigma}(t), -\tilde{y}(t)), \quad \hat{\sigma}(0) = 0, \tag{17}$$

where Proj is the projection operator, $\Gamma \in \mathbb{R}^+$ is the adaptation rate, $\tilde{y}(t) \triangleq \hat{y}(t) - y(t)$ and the projection is performed with the following bound

$$|\hat{\sigma}(t)| \leq \Delta, \quad \forall t \geq 0. \tag{18}$$

Control law:

The control signal is generated according to the following law

$$u(s) = C(s)(r(s) - \hat{\sigma}(s)), \tag{19}$$

where $C(s)$ is the low-pass filter. In the interest of simplicity, the low-pass filter is parameterized as a first-order transfer function:

$$C(s) = \frac{\omega}{s + \omega}, \quad \omega > 0 \tag{20}$$

assuming zero initialization.

Figure 6 illustrates a closed-loop system with $\mathcal{L}_1$ adaptive control consisting of Equations (16)–(20) subject to the $\mathcal{L}_1$-norm condition in Equation (14).

The ideal control law provides the desired system response by effectively accounting for uncertainties. Thus, the reference system in Equation (16) has a different response as compared to the ideal one. It compensates only for the uncertainties within the bandwidth of $C(s)$, which can be selected to be compatible with the control channel specifications.

2.3.3. *Closed-loop system stability*

Considering the closed-loop system described by Equations (10)–(12), the choice of $M(s)$ and $C(s)$ can be restricted such that $H(s)$ is stable and Equation (14) holds. The condition in Equation (14) restricts the class of systems $A(s)$ in Equation (8) that can be stabilized by the $\mathcal{L}_1$ controller architecture.

Letting

$$A(s) = \frac{A_n(s)}{A_d(s)}, \quad C(s) = \frac{C_n(s)}{C_d(s)}, \quad M(s) = \frac{M_n(s)}{M_d(s)}. \tag{21}$$

It follows from Equation (13) that

$$H(s) = \frac{C_d(s)M_n(s)A_n(s)}{C_n(s)M_d(s)A_n(s) + M_n(s)A_d(s)(C_d(s) - C_n(s))}. \tag{22}$$

A strictly proper $C(s)$ implies that the order of $C_d(s) - C_n(s)$ and $C_d(s)$ is the same. Since the order of $A_d(s)$ is higher than that of $A_n(s)$, the transfer function $H(s)$ is strictly proper.

Using Equations (10)–(12) and (20) it can be shown that

$$y(s) = H(s)(C(s)r(s) + (1 - C(s))d(s)) = H(s)C(s)r(s) + G(s)d(s). \tag{23}$$

Since $H(s)$ is strictly proper and stable, $G(s)$ is also strictly proper and stable and therefore

$$y(s)_{\mathcal{L}_\infty} \leq H(s)C(s)_{\mathcal{L}_1} r(s)_{\mathcal{L}_\infty} + G(s)_{\mathcal{L}_1}(Ly(s)_{\mathcal{L}_\infty} + L_0). \tag{24}$$

Thus,

$$y(s)_{\mathcal{L}_\infty} \leq \frac{H(s)C(s)_{\mathcal{L}_1} r(s)_{\mathcal{L}_\infty} + G(s)_{\mathcal{L}_1} L_0}{(1 - G(s)_{\mathcal{L}_1} L)}. \quad (25)$$

Using the condition in Equation (14), one can write,

$$y(s)_{\mathcal{L}_\infty} L \leq \rho, \quad \rho = \frac{H(s)C(s)_{\mathcal{L}_1} r(s)_{\mathcal{L}_\infty} + G(s)_{\mathcal{L}_1} L_0}{(1 - G(s)_{\mathcal{L}_1} L)} < \infty. \quad (26)$$

Hence, $y(s)_{\mathcal{L}_\infty}$ is bounded. It follows that if $C(s)$ and $M(s)$ verify the condition in Equation (14), the closed-loop reference system in Equations (10)–(12) is stable.

To determine the classes of systems that can satisfy Equation (14) via the choice of $M(s)$ and $C(s)$, we consider the first-order $M(s)$ and $C(s)$ as specified in Equations (9) and (20). It follows from Equations (9) and (20) that

$$H(s) = \frac{m(s+\omega)A_n(s)}{\omega(s+m)A_n(s) + msA_d(s)}. \quad (27)$$

Next, consider stabilization of $A(s)$ by a PI controller, say, of the following structure

$$\mathrm{PI}(s) = \left(\frac{\omega}{m}\right)\frac{(s+m)}{s}, \quad (28)$$

where m and ω are the same as in (9) and (20). The open loop transfer function of the cascaded $A(s)$ with the PI controller will be,

$$H_{\mathrm{PI}}(s) = \left(\frac{\omega}{m}\right)\frac{(s+m)}{s}A(s) \quad (29)$$

resulting in the following closed-loop system:

$$H_{\mathrm{PI}}(s) = \frac{\omega(s+m)A_n(s)}{\omega(s+m)A_n(s) + msA_d(s)}. \quad (30)$$

Hence, the stability of $H(s)$ (27) is equivalent to that of Equation (30), and the problem can be reduced to identifying the class of $A(s)$ that can be stabilized by a PI controller. It also permits the use of root locus methods for checking the stability of $H(s)$ via the open loop transfer function $H_{\mathrm{PI}}(s)$.

The stability equivalence of the $\mathcal{L}_1$ controller with appropriate $M(s)$ and $C(s)$ to that of a PI controller means that any system that can be stabilized by a PI controller belongs to the class of systems that can satisfy Equation (14) via the choice of $M(s)$ and $C(s)$. Noting that SMA systems have been successfully controlled and stabilized using PI controllers (Hannen, Crews, & Buckner, 2012), the choice of $M(s)$ and $C(s)$ in this paper therefore ensures the stability of the $\mathcal{L}_1$ controller-based closed loop system.

2.4. $\mathcal{L}_1$ *control parameter optimization: grid search*

The user-specified parameters of the $\mathcal{L}_1$ controller include the low-pass filter $C(s)$, the reference model $M(s)$, and the adaptation rate Γ. Intuitively, $C(s)$ should be chosen such that its bandwidth does not exceed that of the actuator. In practice however, maximum controller bandwidth does not necessarily translate to optimal tracking performance. The adaptation rate Γ is essentially the gain of the adaptive estimator, and since the control signal is low-pass filtered, relatively large gain values can be used. For real-time controller implementation on a computer, Γ is limited by the stability of the numerical integration method, which is determined largely by available computational capabilities. As discussed in (Michini and How, 2009), the choice of $M(s)$ is not so straightforward for achieving the desired specifications. Given these characteristic limitations, no intuitive method has been established to find combinations of $C(s)$, $M(s)$ and Γ that yield optimal tracking performance for specific application. This brings about the need to employ optimization techniques to determine the best $\mathcal{L}_1$ control parameter set.

Generally, optimization combines the desired performance metrics into a cost function that can be minimized in an effort to find the best combination of $C(s)$, $M(s)$ and Γ. The optimization process calculates a cost function and searches for the parameter set that minimizes the cost function subject to the relevant constraints (like system stability, available bandwidth, etc.). Using the parameterization of $C(s)$ and $M(s)$ given in Equations (9) and (20), a multi-dimensional grid search can be used to find ω, m and Γ that yield the least mean square error (MSE) for a given reference trajectory. The MSE is defined as

$$\mathrm{MSE} = \frac{1}{n}\sum_{i=1}^{n}(\theta_{\mathrm{r}i} - \theta_{\mathrm{m}i})^2, \quad (31)$$

where θ_{r} is the reference bending angle and θ_{m} is the measured or simulated bending angle. Figure 7 shows the general diagram of the optimization process.

In addition to minimizing the cost function, two constraints must be considered. First, filter parameter ω is constrained between $\omega = 0$ (where the filter acts as a pure integrator) and $\omega = 50$ (where the actuator bandwidth becomes limiting). The second constraint addresses stability of the closed-loop system. One simple way to implement this constraint is to augment the cost function with a term that is arbitrarily large if the system is unstable. Instability, in this case, refers to a closed-loop system response that either results in a simulation singularity or diverges from the desired profile as the simulation time increases.

Optimization was conducted using a grid search technique (Ensor and Glynn, 1997), which finds the minimum of a multivariable function over $\Lambda \subseteq \mathbb{R}^d$, the parameter space over which we wish to optimize. Let $\alpha : \Lambda \rightarrow \mathbb{R}$ be a real-valued function that, for each $\lambda \in \Lambda$, measures the performance of the system. The goal, then, is to minimize α over Λ. To numerically optimize α over Λ, we approximate Λ by some finite set of m points $\Lambda_m =$

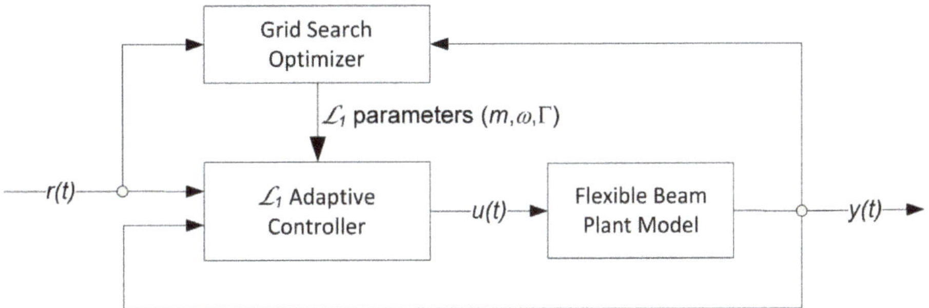

Figure 7. Optimization process block diagram.

$\{\lambda_1, \ldots, \lambda_m\} \subseteq \Lambda$ and then compute α over Λ_m. The minimum of α over Λ_m is then taken as the minimum of α over Λ. Since Λ_m is frequently taken to be a discrete grid (when Λ is a hyper-rectangle), we refer to this approach as a 'grid search' for the minimum. In the bending actuator control application, α is the *MSE* (31) between the reference angle and the simulated angle, evaluated for a single run at each Λ_m.

Each grid dimension has a range of values which are divided into a set of equal-value intervals. The multidimensional grid has a centroid, which locates the optimum point. The search involves multiple passes and in each pass, the method finds the node with the lowest MSE. This node becomes the new centroid and builds a smaller grid around it. Successive passes result in the multidimensional grid shrinking as the centroid keeps moving towards the optimum point. The cost as a function of the three parameters can be visualized using a contour plot, making the process more intuitive to the designer.

2.5. *Implementation environment*

The $\mathcal{L}_1$ adaptive controller and plant model were simulated using Simulink (The MathWorks, Inc., Natick, MA). For real-time implementation, the $\mathcal{L}_1$ controller was programmed in Visual Studio C++ (Microsoft Corporation, Redmond, WA) and set to run at 60 Hz, a sample rate that was governed by the capability of the TrakSTAR position sensor.

3. Results

3.1. *Simulation results*

Transient step response data obtained from the experimental setup, shown in Figure 2, were used to identify the SMA Joule heating time constant: $\tau = 0.25\,\mathrm{s}$. The reader is referred to (Crews, Smith, Pender, Hannen, & Buckner, 2012; Hannen et al., 2012) for more details on the experimental determination of related SMA temperature response model parameters.

The HRNN weights were optimized using 1071 equally spaced neurons. Figure 8 shows the training and testing data, originally published in Hannen et al. (2012), along with the trained HRNN prediction. The optimal solution resulted in a training cost of 2.8×10^{-4} and a testing cost of 2.7×10^{-4}, reduced from initial values of 0.47.

The optimized HRNN weights and transition temperatures were implemented in the $\mathcal{L}_1$ simulation environment to represent the constitutive relationship between SMA temperature and bending angle. The grid search algorithm identified the $\mathcal{L}_1$ control parameter set (m, ω, Γ) that minimized the error cost function for a sinusoidal reference (amplitude 20 degrees, frequency 0.1 Hz).

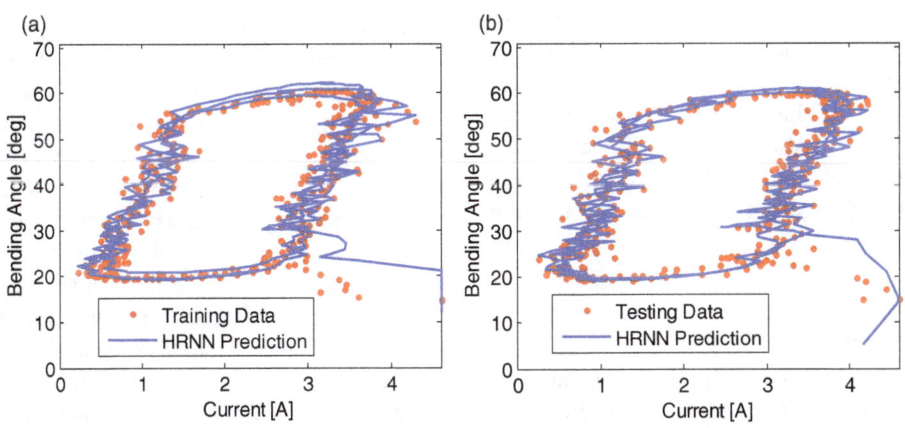

Figure 8. Simulated HRNN-predicted bending angle as a function of SMA temperature compared to (a) training data and (b) testing data obtained from the flexible beam model.

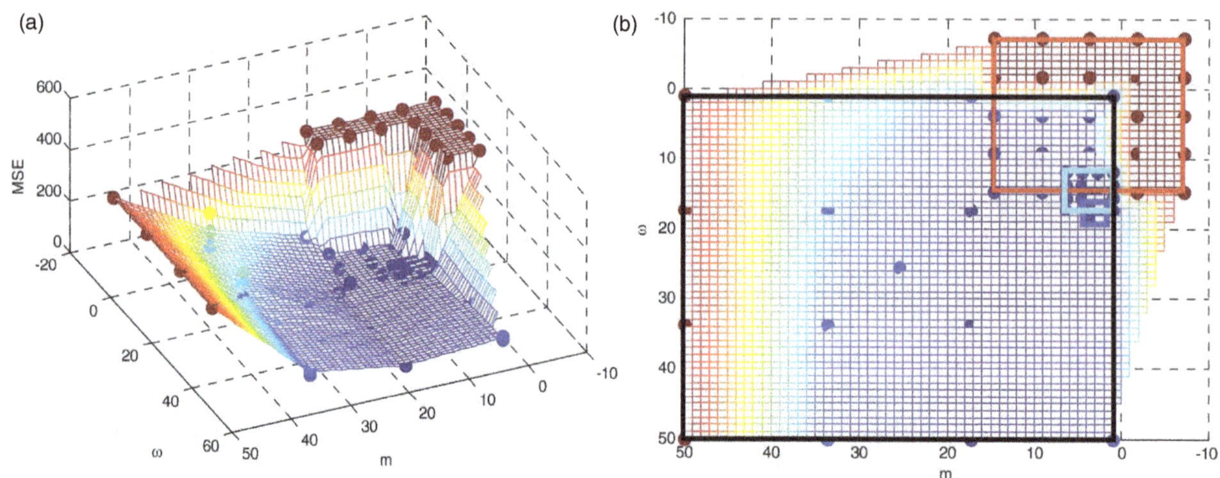

Figure 9. Optimization results of the $\mathcal{L}_1$ control parameters showing (a) the 3D optimization search space and (b) the 2D visualization of the ω and m grid progressions.

Table 1. Simulation and experimental $\mathcal{L}_1$ parameters.

Parameter	Description	Optimization range	Simulation values	Experimental values
m	Reference model	[1–50]	1.94	1.94
ω	Low-pass filter	[1–50]	[16–50]	45
Γ	Adaptive gain	[100–1000]	500	500

Figure 9 shows a 3D representation of the grid search's convergence to the optimal solution.

Table 1 summarizes the resulting $\mathcal{L}_1$ controller parameters. It shows the parameter ranges used in the grid search and the values that were subsequently used in the simulations and experimental implementation. With the reference model m set to the grid search optimum of 1.94, simulation results revealed that tracking performance is relatively insensitive to low pass filter parameters ω in the range of 16–50. This can be seen in the flatness of the surface along the ω axis at around $m = 1.94$ as illustrated in Figure 9(a). This result helps define the minimum low-pass filter bandwidth for the $\mathcal{L}_1$ controller.

Controller performance was simulated using a 0.1 Hz sinusoidal reference, with results shown in Figure 10. Figure 10(a) shows that the plant response $y(t)$ tracks the reference trajectory $r(t)$ reasonably well, but exhibits a lag of approximately 0.55 s due to the SMA's relatively slow heat transfer dynamics. While this phase lag could be significantly reduced using feedforward control, the tracking performance is otherwise quite good. Figure 10(b) shows the tracking error and Figure 10(c) shows the actuator control voltage $u(t)$.

3.2. *Experimental results*

Using the simulation-optimized $\mathcal{L}_1$ control parameters in the experimental system resulted in reasonable tracking performance for a 0.1 Hz sinusoidal reference bending angle. In order to achieve experimental tracking performance that was comparable to the simulation results, Γ and m were set to the simulation-optimized values (500 and 1.94, respectively) while ω had to be at least 45. The controller parameters that resulted in the best experimental tracking performance are presented in Table 1, and corresponding experimental tracking results for a 0.10 Hz sine wave are presented in Figure 11. Similar results for a triangle wave reference are presented in Figure 12. There were no significant changes in the system tracking performance for ω values greater than 45 (within the range that worked well for simulation: 16–50).

4. Discussion

Generally, a very strong correlation was observed between the dynamic behaviour of the simulated plant and the experimental SMA-actuated flexible beam, suggesting that the plant model used in simulation effectively captured the nonlinear characteristics of the flexible beam plant. The experimental tracking performance substantiates the viability of the $\mathcal{L}_1$ adaptive control scheme to effectively control a highly nonlinear hysteretic system.

Although there is still no intuitive way to select the optimal set of $\mathcal{L}_1$ controller parameters, a major a step towards a formal guideline for parameter selection and practical implementation was achieved. The following observations from the grid search parameter selection lay the foundation for the development of $\mathcal{L}_1$ adaptive controller guidelines and considerations for both simulation and implementation.

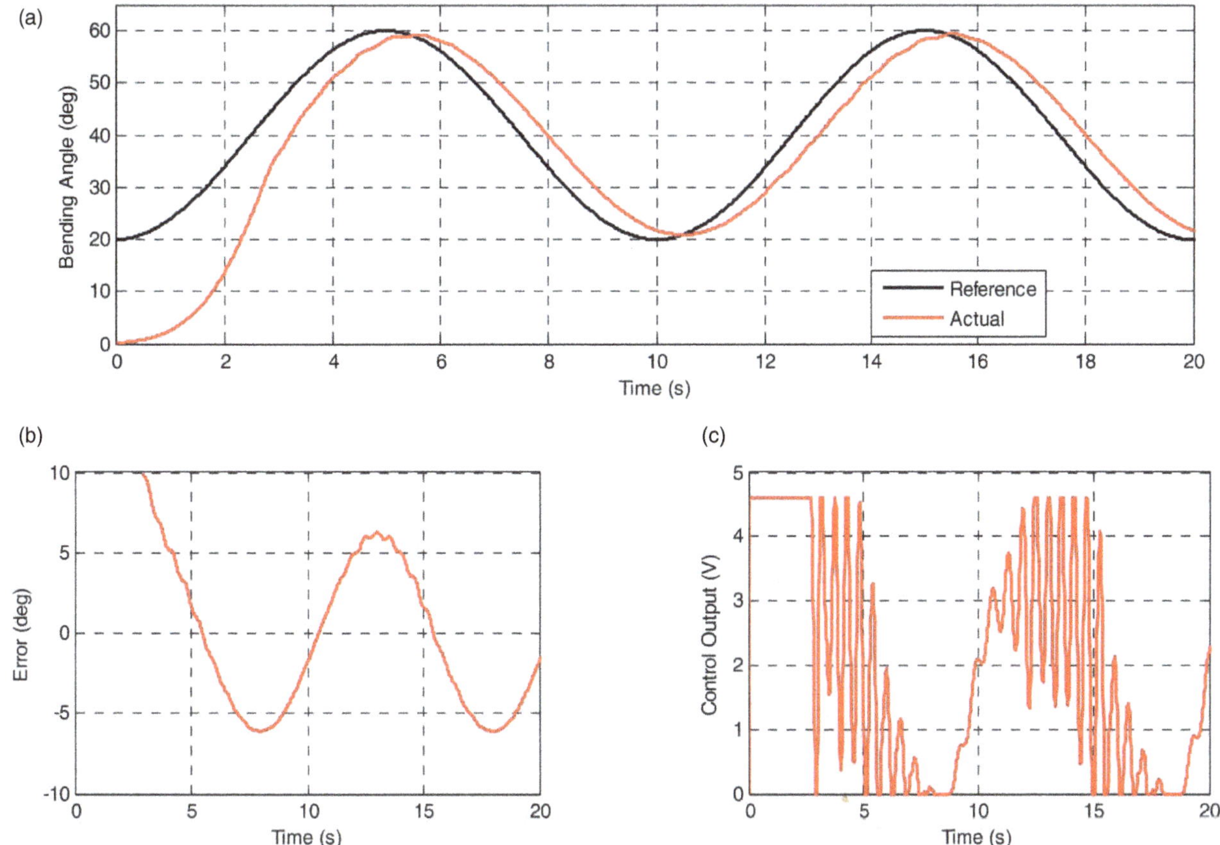

Figure 10. Simulated L_1 controller tracking results for a 0.1 Hz sinusoidal reference trajectory showing (a) bending angle, (b) tracking error and (c) control input.

The reference model, $M(s)$: The reference model defines the desired system behaviour but, unlike those used in MRAC (Astrom and Wittenmark, 1994; Ioannou and Kokotovic, 1982; Krstic, Kanellakopoulos, & Kokotovic, 1995), it serves as a state estimator. The simulated grid search optimization and experimental performance results suggest that, assuming that the reference model $M(s)$ itself is BIBO stable, the choice of the reference model alone cannot cause system instability. Thus, for transient performance enhancement the choice of $M(s)$ can be trivialized to choosing a stable reference model whose response time is at least within the bandwidth of the expected reference trajectories.

Adaptive gain, Γ: Low gains result in slow adaptation and very poor tracking, while high gains yield rapid adaptation. In general, the simulated and experimental evaluations show that by increasing the adaptation gain Γ, the $\mathcal{L}_1$ adaptive controller tracking performance improves. Thus, it can be concluded that an arbitrarily high adaptation gain can be set, reducing the parameter selection problem to the selection of the reference model $M(s)$ and the low pas filter $C(s)$ such that the system has the desired response. However, in almost all real-time implementations that use discretized algorithms running at fixed sample rates, the trade-off between numerical accuracy and fixed sampling interval almost always results in the loss of precision during integration. Therefore, finding the optimal compromise between accurate integration, a feasible sampling time, and numerical stability is of paramount importance. Numerical instability was observed in simulations where the adaptation gain was set excessively high ($\Gamma > 2000$).

Low pass filter, $C(s)$: While the low-pass filter bandwidth is upper-bounded by the actuator bandwidth, and no intuitive lower bound exists, the lower bound was experimentally determined. The simulation results showed slower system response as the filter bandwidth was reduced. The ideal control signal is the one that leads to the desired system response by compensating for the uncertainties exactly. In practice, however, only the uncertainties within the bandwidth of $C(s)$, are cancelled, thus, we want the bandwidth of the low-pass filter to be as high as possible but also making sure it is lower than the actuator limit. The need for higher low-pass filter bandwidth in experiment is attributed to fact that the experimental setup, there is an inherent filter between the digital controller output and the analog electronics that supply the SMA wire with the joule heating current. In addition, there are additional filter and delay dynamics associated with the sensor information being fed back into the digital closed

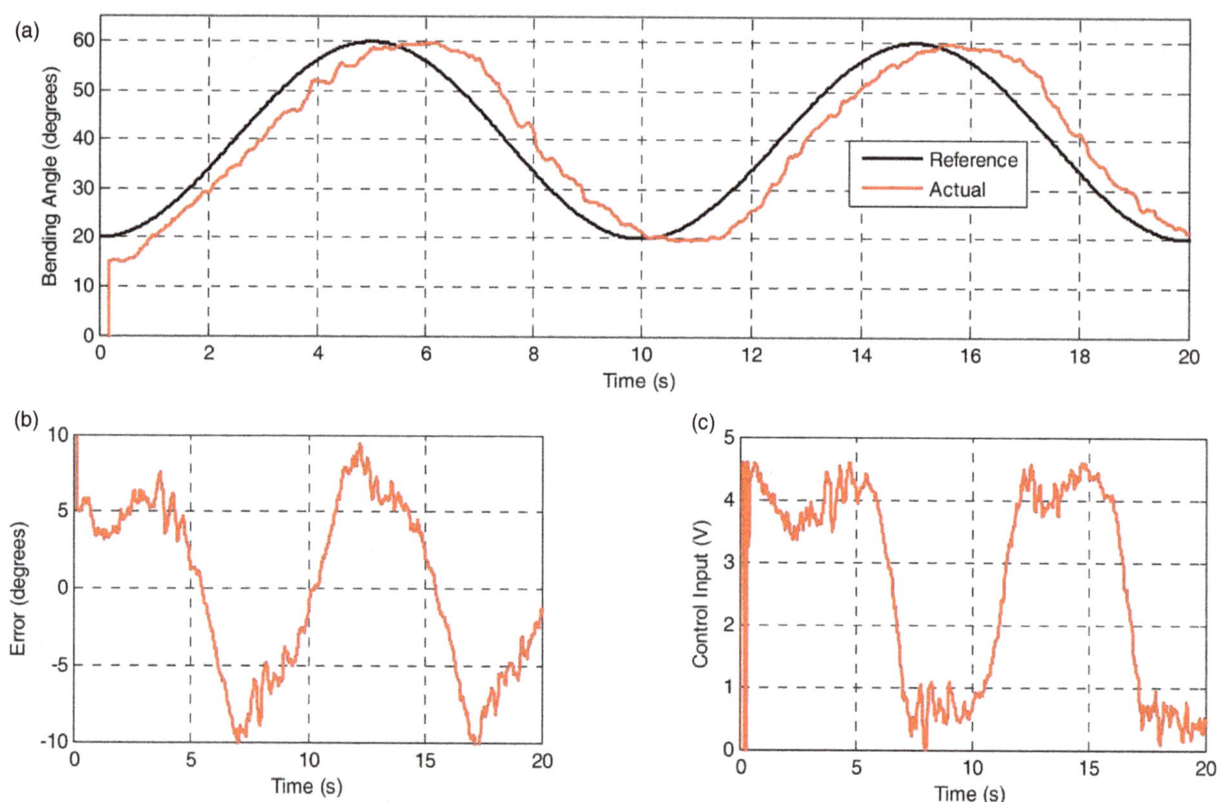

Figure 11. Experimental $\mathcal{L}_1$ control tracking results for a 0.1 Hz sinusoidal reference trajectory showing (a) bending angle, (b) tracking error and (c) controller generated plant input.

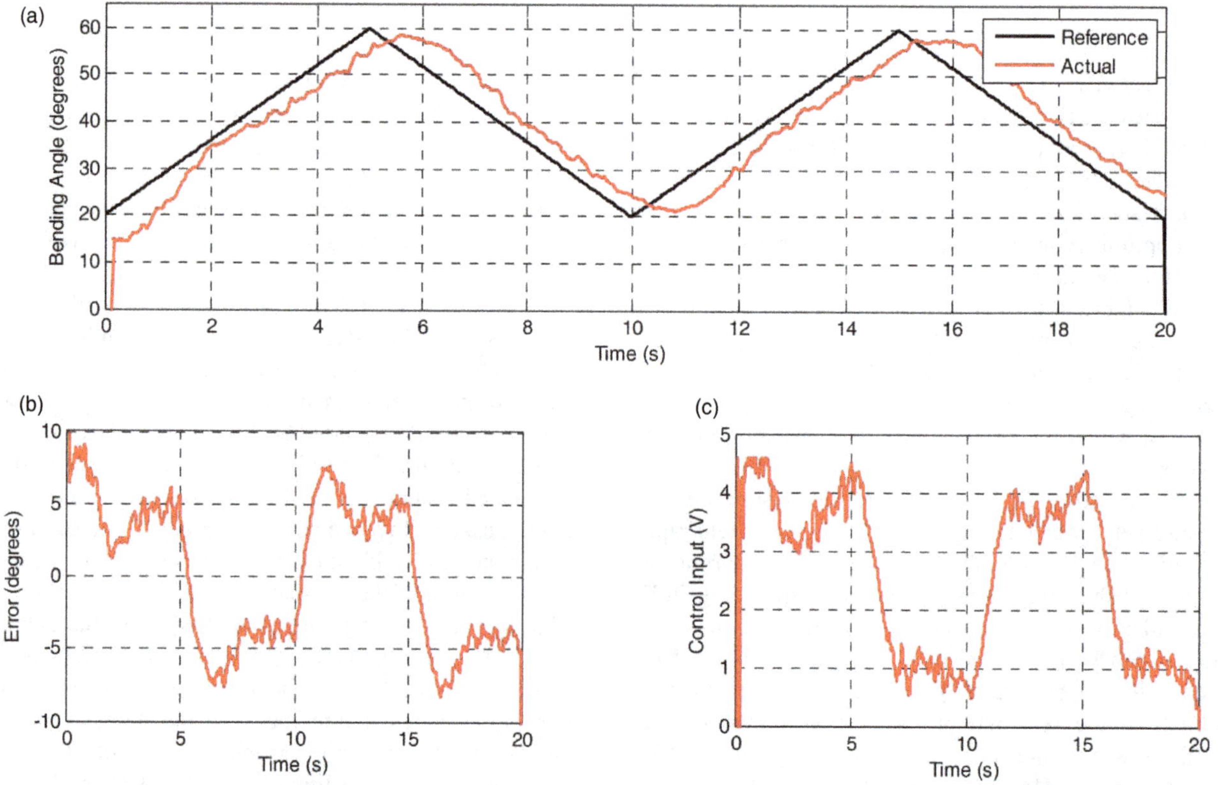

Figure 12. Experimental $\mathcal{L}_1$ control tracking results for a 0.1 Hz ramp reference trajectory showing (a) bending angle, (b) tracking error and (c) controller generated plant input.

loop controller. In experimental implementation, in order to realize a certain actuator bandwidth response, the digital filter bandwidth needs to be set much higher so that it compensates for the total effective feedback path delays and response times. Thus, in order to get similar results that the simulation model shows, it is conceivable that the digital filter in the experimental setup needs to have higher bandwidth.

5. Conclusions and future work

This paper demonstrates the implementation of the $\mathcal{L}_1$ adaptive control law to regulate the bending angle of a flexible beam actuated by a single SMA tendon. The results demonstrate controller feasibility, showing fast response times and precise tracking of a variety of reference trajectories with the controller effectively compensating for hysteresis in the SMA tendon behaviour. Control parameter guidelines and limitations associated with experimental implementation of the output feedback $\mathcal{L}_1$ controller realistic control applications are presented. The proposed parameterization of the $\mathcal{L}_1$ parameters $C(s)$ and $M(s)$ enables the implementation of a manageable grid-search optimization, providing the control designer with an intuitive method of linking performance metrics to the selection of the L_1 parameters. This design process represents a step in the direction of more readily applying $\mathcal{L}_1$ adaptive control to real-world control systems and taking advantage of its potential benefits.

Although SMA-based actuators have been successfully implemented in a number of applications, limitations associated with nonlinear and hysteretic behaviour have presented challenges to the development of robust, high performance, high bandwidth controllers that can be implemented in real-time. While high-performance control can be achieved using high fidelity models that are computationally expensive, such controllers can be difficult to implement in real-time and may have significant bandwidth limitations. This has motivated research into the development of simpler control strategies that can effectively compensate for nonlinearities and hysteresis while providing high performance over a broad frequency range. The performance results presented here suggest an advancement in this direction: a relatively simple but effective adaptive control architecture for systems that exhibit nonlinear and hysteretic behaviour.

A fundamental challenge of $\mathcal{L}_1$ adaptive control theory involves optimizing the design of the bandwidth-limited filter, which defines the trade-off between performance and robustness. The design process requires specifying the filter order, parameterizing the filter, and ensuring system stability with a bounded $\mathcal{L}_1$ norm (a non-convex constrained optimization problem). While this paper presents preliminary design guidelines with the assumed forms of $C(s)$ and $M(s)$, $\mathcal{L}_1$-controller filter design is still an open problem. Another challenge involves specification of the adaptive gain. While increasing this gain can improve the performance bounds and stability margins of the $\mathcal{L}_1$ adaptive controller, it is critical to ensure that high gain implementations do not cause numerical instabilities and that the CPU has enough computational power to robustly execute the real-time integration algorithms.

Future work includes exploring the potential benefits that higher order forms of $C(s)$ and $M(s)$ might provide, while preserving the parameterization simplicity presented in this paper. Additional future work will focus on extending the $\mathcal{L}_1$ controller to bidirectional SMA systems. An antagonistic pair of SMA tendons can increase actuator bandwidth, limited in the single tendon case by slow tendon cooling. This would involve system model reformulation to incorporate the dynamics introduced by the two parallel SMA tendons and improving the actuator response time.

Using the $\mathcal{L}_1$ controller as an adaptive augmentation to an existing base controller is an attractive research topic. This control architecture represents a hybrid system that is more likely to be accepted industry-wide and would provide quantifiable benefits without the additional verification and validation efforts required for certifying new control schemes. Anticipated implementation challenges include analysis of the $\mathcal{L}_1$ adaptive controller performance bounds in the presence of input saturation. This might require appropriate modification of the state predictor to remove the effect of the control deficiency from the adaptation process.

Disclosure statement

No potential conflict of interest was reported by the authors.

References

Achenbach, M., & Müller, I. (1985). Simulation of material behaviour of alloys with shape memory. *Archives of Mechanics*, *35*, 537–585.

Astrom, K., & Wittenmark, B. (1994). *Adaptive control*. Boston, MA: Addison-Wesley Longman.

Brinson, L. C. (1993). One-dimensional constitutive behavior of shape memory alloys: Thermomechanical derivation with non-constant material functions and redefined martensite internal variable. *Journal of Intelligent Material Systems and Structures*, *4*(2), 229–242.

Cao, C., & Hovakimyan, N. (2006). *Design and analysis of a novel L1 adaptive controller architecture with guaranteed transient performance*. Minneapolis, MN: American Control Conference.

Crews, J. H., Smith, R. C., Pender, K. M., Hannen, J. C., & Buckner, G. D. (2012). Data-driven techniques to estimate parameters in the homogenized energy model for shape memory alloys. *Journal of Intelligent Material Systems and Structures*, *23*(17), 1897–1920.

Ensor, K. B., & Glynn, P. W. (1997). Stochastic optimization via grid search. *Lectures in Applied Mathematics, Mathematics of Stochastic Manufacturing Systems*, *33*, 89–100.

Haga, Y., Mineta, T., Makishi, W., Matsunaga, T., & Esashi, M. (2010). Active bending catheter and endoscope using shape memory alloy actuators. In C. Cismasiu, (Ed.), *Shape memory alloys* (pp. 107–126). Rijeka, Croatia: Sciyo.

Hannen, J. C., Crews, J. H., & Buckner, G. D. (2012). Indirect intelligent sliding mode control of a shape memory alloy actuated flexible beam using hysteretic recurrent neural networks. *Smart Materials and Structures*, *21*(8), 921–929.

Ioannou, P. A., & Kokotovic, P. V. (1982). An asymptotic error analysis of identifiers and adaptive observers in the presence of parasitics. *IEEE Trans. on Automatic Control*, *27*(4), 921–927.

Jia, H., Lalande, F., & Rogers, C. A. (1997). *Modeling of strain energy absorption in superelastic shape memory alloys*. Proceedings of SPIE, 3039, San Diego, CA, USA, pp. 548–558.

Kleinstreuer, C., Li, Z., Basciano, C. A., Seelecke, S., & Farber, M. A. (2008). Computational mechanics of nitinol stent grafts. *Journal of Biomechanics*, *41*(11), 2370–2378.

Kohl, M. (2004). *Shape memory microactuators*. New York, NY: Springer.

Krstic, M., Kanellakopoulos, I., & Kokotovic, P. V. (1995). *Nonlinear and adaptive control design*. New York, NY: John Wiley & Sons.

Liang, C., & Rogers, C. A. (1990). One-dimensional thermomechanical constitutive relations for shape memory materials. *Journal of Intelligent Material Systems and Structures*, *1*(2), 207–234.

Marquardt, D. W. (1963). An algorithm for least-squares estimation of nonlinear parameters. *Journal of the Society for Industrial and Applied Mathematics*, *11*(2), 431–441.

Michini, B., & How, J. (2009). *L1 adaptive control for indoor autonomous vehicles: Design process and flight testing.* AIAA Guidance, Navigation, and Control Conference, Chicago.

Moallem, M. (2003). Deflection control of a flexible beam using shape memory alloy actuators. *Smart Materials and Structures*, *12*, 1023–1027.

Nakamura, Y., Matsui, A., Saito, T., & Yoshimoto, K. (1995). *Shape Memory Alloy active forceps for laparoscopic surgery*. IEEE International Conference on Robotics and Automation, 3, Nagoya, Aichi, Japan, pp. 2320–2327.

Seelecke, S., & Muller, I. (2003). Shape memory alloy actuators in smart structures – modeling and simulation. *ASME Applied Mechanics Reviews*, *56*(8), 23–46.

Stoeckel, W. (1991). *Use of NiTi shape memory alloys for thermal sensor actuators*. Proceedings of SPIE, San Diego, CA, USA. pp. 382–387.

Tanaka, K., & Nagaki, S. (1982). A thermomechanical description of materials with internal variables in the process of phase transformation. *Ingenieur-Archiv*, *51*(5), 287–299.

Van Humbeeck, J., Chandrasekaran, M., & Delaey, L. (1991). Shape memory alloys: Materials in action. *Endeavour*, *15*(4), 148–154.

Veeramani, A. S., Buckner, G. D., Owen, S. B., Cook, R. C., & Bolotin, G. (2008). Modeling the dynamic behavior of a shape memory alloy actuated catheter. *Smart Materials and Structures*, *17*(1): 1–14.

Veeramani, A. S., Crews, J. H., & Buckner, G. D. (2009). Hysteretic recurrent neural networks: A tool for modeling hysteretic materials and systems. *Smart Materials and Structures*, *18*(7), 1–15.

Wu, M. H., & Schetky, L. M. (2000). *Industrial applications for shape memory alloys*. Proceedings of the International Conference on Shape Memory and Superelastic, Pacific Grove, CA, USA, pp. 171–182.

Dynamical observer-based fault detection and isolation for linear singular systems

G.-L. Osorio-Gordillo[a,b*], M. Darouach[b], L. Boutat-Baddas[b] and C.-M. Astorga-Zaragoza[a]

[a] *Tecnológico Nacional de México, Centro Nacional de Investigación y Desarrollo Tecnológico, CENIDET, Interior Internado Palmira S/N, Col. Palmira, 62490 Cuernavaca, Mor. Mexico;* [b] *CRAN-CNRS (UMR 7039), Université de Lorraine, IUT de Longwy, 186, Rue de Lorraine, 54400 Cosnes et Romain, France*

This paper is concerned with the fault detection and isolation (FDI) for singular systems by using a dynamical observer having a new structure. The goal set in the task of FDI is to obtain a transfer function from fault to residual of the error dynamics system equal to a diagonal transfer function, to allow multiple faults isolation. Sufficient conditions for the existence and stability of the observer are given. A numerical example is given to illustrate our approach.

Keywords: fault detection and isolation; dynamical observer; singular systems

1. Introduction

This paper concerns the dynamical observer-based fault detection and isolation (FDI) for singular systems. Singular systems also known as descriptor or differential-algebraic systems can be considered as a generalization of dynamical systems. The singular system representation is a powerful modeling tool since they can describe processes governed by both, differential equations (dynamic) and algebraic equations (static). So that represents the physical phenomena that the model by ordinary differential equation cannot describe. These systems were introduced by Luenberger (1977) from a control theory point of view and since, great efforts have been made to investigate singular systems theory and its applications (Araujo, Barros, and Dorea, 2012, Boulkroune, Darouach, Zasadzinski, Gillé, and Fiorelli, 2009, Darouach, 2009, 2012, Liu, Zhang, Yang, and Yang, 2008, Müller, 2005, Müller and Hou, 1993, Zhou and Lu, 2009).

The general goal of fault detection is to determine the fault presence into a system, whereas fault isolation is used to determine the location of the fault, after detection. During the last two decades, FDI has been of considerable interest (Hamdi, Rodriguez, Mechmeche, Theilliol, and Braiek, 2012, Li and Jaimoukha, 2009, Li and Yang, 2011). In Theilliol, Noura, and Ponsart (2002), the fault diagnosis for linear systems is treated. Rodrigues, Theilliol, Adam-Medina, and Sauter (2008) address the FDI for multi-models representation and in Bokor and Szabo (2009), the fault detection for nonlinear systems is presented.

In Liu and Si (1997), the problem of isolating multiples faults in linear systems, is presented by using an approach based on eigenstructure assignment to generate structured residuals, then the observer matrices are determined so that the ith residual represents the ith fault.

In Li and Jaimoukha (2009), an approach which generalizes the results of Liu and Si (1997) is presented. The authors construct an H_∞ FDI observer for linear systems, with the constraint that the transfer function from faults to residual of the error dynamics is equal to a pre-assigned diagonal transfer matrix. All these results use the proportional observer (PO).

By using a singular system representation, it is shown that the remainder kinds of FDI problems, that is, fault detection, FDI, and disturbance decoupled FDI are generally equivalent to the FDI problem for singular systems (Patton, Frank, and Clark, 2000). To our knowledge, few works have considered the approach of FDI based on an observer for singular systems, due to the structural complexity and strong constraints in designing procedure. Fault detection using unknown input observers were presented in Kim, Yeu, and Kawaji (2001), and in Duan, Howe, and Patton (2002) where the residuals are not affected by the unknown input. In Hamdi et al. (2012), Yeu, Kim, and Kawaji (2005), and Astorga, Theilliol, Ponsart, and Rodrigues (2012), fault diagnosis problem is treated by a PO.

In the estimation by a PO, there always exists a static error estimation. In order to deal with the inconveniences

*Corresponding author. Email: gloriaosorio@cenidet.edu.mx

of PO, proportional-integral observers (PIO) were introduced with an integral gain of the output error in their structure. This change in the structure achieve steady-state accuracy in their estimations. Also a new structure of the observers was developed by Goodwin and Middleton (1989) and Marquez (2003), known as dynamic observers. This structure presents an alternative state estimation which can be considered as more general than PO and PIO. These last can be only considered as particular cases of this structure. The idea of including additional dynamics in the observer was presented by Goodwin and Middleton (1989).

In this paper, we consider FDI problem for singular systems with actuator faults. Residual signals are determined from properly weighted output errors between measurements and estimated outputs.

The main contribution of this paper is that the designed observer is presented in a more general form than the existing dynamical observers: the PO and PIO which are only particular cases of the structure of our observer. This observer is used for actuator FDI in singular systems, which are a generalization of the standard systems. The proposed method is based on the directional residual generation in order to locate simultaneous faults. Finally, the effectiveness of this approach is shown through a numerical example simulation.

2. Preliminaries

In the present paper, the set of real matrices $n \times m$ is denoted by $\mathbb{R}^{n\times m}$. A^{T} denotes the transpose of the matrix A, A^{+} is the generalized inverse of A, i.e. $AA^{+}A = A$. I_n denotes an $n \times n$ identity matrix, I denotes an identity matrix with appropriate dimension, 0 denotes a zero scalar or matrix with appropriate dimension. The notation $A = \mathrm{diag}(a_1, \ldots, a_n)$ denotes that A is a diagonal matrix with elements $(a_1, \ldots, a_n)$ in its diagonal, $\mathrm{ones}_{n,m}$ denotes an $n \times m$ matrix with all elements one.

In Section 4.3, we use the following lemma to solve Linear matrix inequalities (LMIs).

LEMMA 1 (Skelton, Iwasaki, and Grigoriadis, 1998). *Let matrices $\mathcal{B}, \mathcal{C}, \mathcal{D} = \mathcal{D}^{\mathrm{T}}$ be given, then the following statements are equivalent:*

(1) There exists a matrix $\mathcal{X}$ satisfying

$$\mathcal{B}\mathcal{X}\mathcal{C} + (\mathcal{B}\mathcal{X}\mathcal{C})^{\mathrm{T}} + \mathcal{D} < 0.$$

(2) The following two conditions hold:

$$\mathcal{B}^{\perp}\mathcal{D}\mathcal{B}^{\perp\mathrm{T}} < 0 \quad \text{or} \quad \mathcal{B}\mathcal{B}^{\mathrm{T}} > 0,$$
$$\mathcal{C}^{\mathrm{T}\perp}\mathcal{D}\mathcal{C}^{\mathrm{T}\perp\mathrm{T}} < 0 \quad \text{or} \quad \mathcal{C}^{\mathrm{T}}\mathcal{C} > 0.$$

Suppose that the statement 2 holds. Let r_b and r_c be the ranks of $\mathcal{B}$ and $\mathcal{C}$, respectively, and $(\mathcal{B}_l, \mathcal{B}_r)$ and $(\mathcal{C}_l, \mathcal{C}_r)$ be any full rank factors of $\mathcal{B}$ and $\mathcal{C}$, i.e. $\mathcal{B} = \mathcal{B}_l\mathcal{B}_r$, $\mathcal{C} = \mathcal{C}_l\mathcal{C}_r$. Then, the matrix $\mathcal{X}$ in statement 1 is given by

$$\mathcal{X} = \mathcal{B}_r^{+}\mathcal{K}\mathcal{C}_l^{+} + \mathcal{Z} - \mathcal{B}_r^{+}\mathcal{B}_r\mathcal{Z}\mathcal{C}_l\mathcal{C}_l^{+},$$

where $\mathcal{Z}$ is an arbitrary matrix and

$$\mathcal{K} = -\mathcal{R}^{-1}\mathcal{B}_l^{\mathrm{T}}\vartheta\mathcal{C}_r^{\mathrm{T}}(\mathcal{C}_r\vartheta\mathcal{C}_r^{\mathrm{T}})^{-1} + \mathcal{S}^{1/2}\mathcal{L}(\mathcal{C}_r\vartheta\mathcal{C}_r^{\mathrm{T}})^{-1/2},$$
$$\mathcal{S} = \mathcal{R}^{-1} - \mathcal{R}^{-1}\mathcal{B}_l^{\mathrm{T}}[\vartheta - \vartheta\mathcal{C}_r^{\mathrm{T}}(\mathcal{C}_r\vartheta\mathcal{C}_r^{\mathrm{T}})^{-1}\mathcal{C}_r\vartheta]\mathcal{B}_l\mathcal{R}^{-1},$$

where $\mathcal{L}$ is an arbitrary matrix such that $\|\mathcal{L}\| < 1$ and $\mathcal{R}$ is an arbitrary positive-definite matrix such that

$$\vartheta = (\mathcal{B}_r\mathcal{R}^{-1}\mathcal{B}_l^{\mathrm{T}} - \mathcal{D})^{-1} > 0.$$

3. Problem formulation

Consider the following singular system subject to actuator fault:

$$E\dot{x}(t) = Ax(t) + Bu(t) + Gf(t),$$
$$y(t) = Cx(t), \tag{1}$$

where $x(t) \in \mathbb{R}^n$ is the semi state vector, $u(t) \in \mathbb{R}^m$ is the input, $f(t) \in \mathbb{R}^{n_f}$ is the fault vector and $y(t) \in \mathbb{R}^{n_y}$ represents the measured output vector. Matrices $E \in \mathbb{R}^{n\times n}$, $A \in \mathbb{R}^{n\times n}$, $B \in \mathbb{R}^{n\times m}$, $G \in \mathbb{R}^{n\times n_f}$ and $C \in \mathbb{R}^{n_y\times n}$. Let $\mathrm{rank}(E) = \varrho \leq n$ and $E^{\perp} \in \mathbb{R}^{\varrho_1\times n}$ be a full row rank matrix such that $E^{\perp}[E \;\; G] = 0$, in this case $\varrho_1 = n - \varrho$.

In the sequel, we assume that

Assumption 1

$$\mathrm{rank}\begin{bmatrix} E \\ E^{\perp}A \\ C \end{bmatrix} = n$$

Remark 1 Assumption 1 is equivalent to the impulse observability for singular systems. This condition is more general than $\mathrm{rank}\left[\begin{smallmatrix} E \\ C \end{smallmatrix}\right] = n$, generally considered, see, for example, Verhaegen and Dooren (1986), Darouach and Boutayeb (1995), and Hou and Müller (1995).

Now, let us consider the following fault isolation dynamical observer for the system (1):

$$\dot{\zeta}(t) = N\zeta(t) + Hv(t) + F\begin{bmatrix} -E^{\perp}Bu(t) \\ y(t) \end{bmatrix} + Ju(t), \tag{2}$$

$$\dot{v}(t) = S\zeta(t) + Lv(t) + M\begin{bmatrix} -E^{\perp}Bu(t) \\ y(t) \end{bmatrix}, \tag{3}$$

$$\hat{x}(t) = P\zeta(t) + Q\begin{bmatrix} -E^{\perp}Bu(t) \\ y(t) \end{bmatrix}, \tag{4}$$

$$r(t) = W(C\hat{x}(t) - y(t)), \tag{5}$$

where $\zeta \in \mathbb{R}^{q_0}$ represents the state vector of the observer, $v(t) \in \mathbb{R}^{q_1}$ is an auxiliary vector, $\hat{x}(t) \in \mathbb{R}^n$ is the estimate of $x(t)$ and $r(t) \in \mathbb{R}^{n_f}$ is the residual vector.

Now, the following lemma is considered.

LEMMA 2 *There exist a fault isolation observer of the form Equations* (2)–(5) *for the system* (1) *if the following two statements hold.*

I. There exist a matrix T of appropriate dimension such that the following conditions are satisfied:
(a) $NTE + F\begin{bmatrix} E^{\perp}A \\ C \end{bmatrix} - TA = 0$,
(b) $J = TB$,
(c) $STE + M\begin{bmatrix} E^{\perp}A \\ C \end{bmatrix} = 0$, and
(d) $[P \;\; Q]\begin{bmatrix} TE \\ E^{\perp}A \\ C \end{bmatrix} = I_n$.

II. The matrix $\begin{bmatrix} N & H \\ S & L \end{bmatrix}$ is a stability matrix, when $f(t) = 0$.

Proof Let $T \in \mathbb{R}^{q_0 \times n}$ be a parameter matrix and define $\varepsilon(t) = \zeta(t) - TEx(t)$, then its dynamic is given by

$$\dot{\varepsilon}(t) = N\varepsilon(t) + Hv(t) + (J - TB)u(t) + \left(NTE + F\begin{bmatrix} E^{\perp}A \\ C \end{bmatrix} - TA\right)x(t) - TGf(t) \quad (6)$$

by using the definition of $\varepsilon(t)$, Equations (3) and (4), can be written as

$$\dot{v}(t) = S\varepsilon(t) + Lv(t) + \left(STE + M\begin{bmatrix} E^{\perp}A \\ C \end{bmatrix}\right)x(t), \quad (7)$$

$$\hat{x}(t) = P\varepsilon(t) + [P \quad Q]\begin{bmatrix} TE \\ E^{\perp}A \\ C \end{bmatrix}x(t). \quad (8)$$

Now, if the conditions (a)–(d) of Lemma 2 are satisfied, the following observer error dynamics is obtained from Equations (6) and (7)

$$\begin{bmatrix} \dot{\varepsilon}(t) \\ \dot{v}(t) \end{bmatrix} = \begin{bmatrix} N & H \\ S & L \end{bmatrix}\begin{bmatrix} \varepsilon(t) \\ v(t) \end{bmatrix} - \begin{bmatrix} TG \\ 0 \end{bmatrix}f(t) \quad (9)$$

and from Equation (4)

$$\hat{x}(t) - x(t) = P\varepsilon(t), \quad (10)$$

$$e(t) = P\varepsilon(t) \quad (11)$$

so, if $f(t) = 0$ and matrix $\begin{bmatrix} N & H \\ S & L \end{bmatrix}$ is a stability matrix, then $\lim_{t\to\infty} e(t) = 0$. ■

The residual equation is obtained from Equation (5)

$$r(t) = WCP\varepsilon(t). \quad (12)$$

Remark 2

- The observer (2)–(4) is in general form and generalizes the existing ones. In fact:
  - For $H = 0$, $S = 0$ and $M = 0$ with L a stability matrix, the observer reduces to the PO for singular systems (see, for example, Darouach, 2012 and references therein).
  - For $S = 0, H = 0, P = I, M = 0$ and L a stability matrix, let matrices F and Q be partitioned according to the partition of $\begin{bmatrix} -E^{\perp}Bu(t) \\ y(t) \end{bmatrix}$ as $F = [0 \;\; F_a]$ and $Q = [0 \;\; Q_a]$, respectively, then we obtain the following observer:

$$\dot{\zeta}(t) = N\zeta(t) + F_a y(t) + Ju(t)$$
$$\hat{x}(t) = \zeta(t) + Q_a y(t)$$

    Which is the form used for the unknown input PO for singular systems (Darouach, Zasadzinski, and Hayar, 1996).
  - For $P = I$, $L = 0$ and let $S = -C$ and $M = -CQ + [0 \;\; I]$, then we obtain the following observer:

$$\dot{\zeta}(t) = N\zeta(t) + Hv(t) + F\begin{bmatrix} -E^{\perp}Bu(t) \\ y(t) \end{bmatrix} + Ju(t),$$
$$\dot{v}(t) = y(t) - C\hat{x}(t),$$
$$\hat{x}(t) = \zeta(t) + Q\begin{bmatrix} -E^{\perp}Bu(t) \\ y(t) \end{bmatrix}.$$

    Which is the form used for the unknown input PIO for singular systems.
- The order of the observer is $q_0 \leq n$, when $q_0 = n - p$, we obtain the reduced order observer and for $q_0 = n$, we obtain the full order one.

Now, the problem of the fault isolation observer is reduced to determine the matrices $N, H, F, S, L, M, P, Q, W$, and T such that conditions (a)–(d) from Lemma 2 are satisfied.

4. Observed-based FDI design

4.1. Parameterization of the observers matrices

Before giving the solution to the dynamical observer design, the parameterization of the solutions to the algebraic constraints (a)–(d) of Lemma 2 is presented. Let $R \in \mathbb{R}^{q_0 \times n}$ be a full row rank matrix such that the matrix $\Sigma = \begin{bmatrix} R \\ E^{\perp}A \\ C \end{bmatrix}$ is of full column rank, and let $\Omega = \begin{bmatrix} E \\ E^{\perp}A \\ C \end{bmatrix}$, and define the following matrices: $T_1 = R\Omega^+\begin{bmatrix} I_n \\ 0 \end{bmatrix}$, $T_2 = (I_{n+\varrho_1+n_y} - \Omega\Omega^+)\begin{bmatrix} I_n \\ 0 \end{bmatrix}$, $N_1 = T_1 A\Sigma^+\begin{bmatrix} I_{q_0} \\ 0 \end{bmatrix}$, $N_2 = T_2 A\Sigma^+\begin{bmatrix} I_{q_0} \\ 0 \end{bmatrix}$, $N_3 = (I_{q_0+\varrho_1+n_y} - \Sigma\Sigma^+)\begin{bmatrix} I_{q_0} \\ 0 \end{bmatrix}$, and $P_1 = \Sigma^+\begin{bmatrix} I_{q_0} \\ 0 \end{bmatrix}$.

The following lemma gives the general form of matrices T, S, M, P, Q, N, and F.

LEMMA 3 *The general form of* T, S, M, P, Q, N, *and* F *are*

$$T = T_1 - Z_1 T_2, \tag{13}$$

$$S = -Y_1 N_3, \tag{14}$$

$$M = -Y_1 F_3, \tag{15}$$

$$P = P_1 - Y_2 N_3, \tag{16}$$

$$Q = Q_1 - Y_2 F_3, \tag{17}$$

$$N = N_1 - Z_1 N_2 - Y_3 N_3, \tag{18}$$

$$F = F_1 - Z_1 F_2 - Y_3 F_3, \tag{19}$$

where Z_1, Y_1, Y_2, *and* Y_3 *are arbitrary matrices of appropriate dimensions.*

Proof Conditions (c) and (d) of Lemma 2 can be rewritten as

$$\begin{bmatrix} S & M \\ P & Q \end{bmatrix} \begin{bmatrix} TE \\ E^{\perp}A \\ C \end{bmatrix} = \begin{bmatrix} 0 \\ I_n \end{bmatrix}. \tag{20}$$

The necessary and sufficient conditions for (20) to have a solution is

$$\text{rank} \begin{bmatrix} TE \\ E^{\perp}A \\ C \end{bmatrix} = \text{rank} \begin{bmatrix} TE \\ E^{\perp}A \\ C \\ 0 \\ I_n \end{bmatrix} = n. \tag{21}$$

Now, since $\text{rank} \begin{bmatrix} TE \\ E^{\perp}A \\ C \end{bmatrix} = \text{rank}(\Omega) = n$, there exist matrices $T \in \mathbb{R}^{q_0 \times n}$, and $K \in \mathbb{R}^{q_0 \times (\varrho_1 + p)}$ such that

$$TE + K \begin{bmatrix} E^{\perp}A \\ C \end{bmatrix} = R, \tag{22}$$

which can also be written as

$$[T \quad K]\Omega = R \tag{23}$$

and since $\text{rank} \begin{bmatrix} \Omega \\ R \end{bmatrix} = \text{rank}(\Omega)$, or equivalently

$$R\Omega^{+}\Omega = R \tag{24}$$

the general solution to Equation (23) is given by

$$[T \quad K] = R\Omega^{+} - Z_1(I_{n_1+\varrho_1+p} - \Omega\Omega^{+}) \tag{25}$$

or equivalently

$$T = T_1 - Z_1 T_2, \tag{26}$$

$$K = K_1 - Z_1 K_2, \tag{27}$$

where $K_1 = R\Omega^{+} \begin{bmatrix} 0 \\ I_{\varrho_1+ny} \end{bmatrix}$, $K_2 = (I_{n+\varrho_1+n_y} - \Omega\Omega^{+}) \begin{bmatrix} 0 \\ I_{\varrho_1+ny} \end{bmatrix}$, and Z_1 is an arbitrary matrix of appropriate dimension.

Now, define the following matrices: $F_1 = T_1 A \Sigma^{+} \begin{bmatrix} K \\ I_{\varrho_1+ny} \end{bmatrix}$, $F_2 = T_2 A \Sigma^{+} \begin{bmatrix} K \\ I_{\varrho_1+ny} \end{bmatrix}$, $F_3 = (I_{q_0+\varrho_1+n_y} - \Sigma\Sigma^{+}) \begin{bmatrix} K \\ I_{\varrho_1+ny} \end{bmatrix}$, $Q_1 = \Sigma^{+} \begin{bmatrix} K \\ I_{\varrho_1+ny} \end{bmatrix}$, $\tilde{K}_1 = T_1 A \Sigma^{+} \begin{bmatrix} 0 \\ I_{\varrho_1+ny} \end{bmatrix}$, $\tilde{K}_2 = T_2 A \Sigma^{+} \begin{bmatrix} 0 \\ I_{\varrho_1+ny} \end{bmatrix}$, and $\tilde{K}_3 = (I_{q_0+\varrho_1+n_y} - \Sigma\Sigma^{+}) \begin{bmatrix} 0 \\ I_{ny} \end{bmatrix}$.

From Equation (22), we obtain

$$\begin{bmatrix} TE \\ E^{\perp}A \\ C \end{bmatrix} = \begin{bmatrix} I_{q_0} & -K \\ 0 & I_p \end{bmatrix} \Sigma \tag{28}$$

by replacing Equation (28) into Equation (20), it leads to

$$\begin{bmatrix} S & M \\ P & Q \end{bmatrix} \begin{bmatrix} I_{q_0} & -K \\ 0 & I_p \end{bmatrix} \Sigma = \begin{bmatrix} 0 \\ I_n \end{bmatrix}. \tag{29}$$

Since Σ is a full column rank matrix, and $\begin{bmatrix} I_{q_0} & -K \\ 0 & I_p \end{bmatrix}^{-1} = \begin{bmatrix} I_{q_0} & K \\ 0 & I_p \end{bmatrix}$, the general solution of Equation (29) is given by

$$\begin{bmatrix} S & M \\ P & Q \end{bmatrix} = \left(\begin{bmatrix} 0 \\ I_n \end{bmatrix} \Sigma^{+} - \begin{bmatrix} Y_1 \\ Y_2 \end{bmatrix} (I_{q_0+\varrho_1+p} - \Sigma\Sigma^{+}) \right) \times \begin{bmatrix} I_{q_0} & K \\ 0 & I_p \end{bmatrix}, \tag{30}$$

where Y_1 and Y_2 are arbitrary matrices of appropriate dimensions. Then matrices S, M, P, and Q can be determined as

$$S = -Y_1 N_3, \tag{31}$$

$$M = -Y_1 F_3, \tag{32}$$

$$P = P_1 - Y_2 N_3, \tag{33}$$

$$Q = Q_1 - Y_2 F_3. \tag{34}$$

By inserting TE from Equation (22) into condition (a) of Lemma 2 leads to

$$N \left(R - K \begin{bmatrix} E^{\perp}A \\ C \end{bmatrix} \right) + F \begin{bmatrix} E^{\perp}A \\ C \end{bmatrix} = TA,$$

$$NR + \tilde{K} \begin{bmatrix} E^{\perp}A \\ C \end{bmatrix} = TA. \tag{35}$$

where $\tilde{K} = F - NK$. Equation (35) can be written as

$$[N \quad \tilde{K}]\Sigma = TA. \tag{36}$$

The general solution of Equation (36) is given by

$$[N \quad \tilde{K}] = TA\Sigma^{+} - Y_3(I_{q_0+\varrho_1+p} - \Sigma\Sigma^{+}). \tag{37}$$

By replacing T from Equation (26) into Equation (37) it gives

$$N = N_1 - Z_1 N_2 - Y_3 N_3, \tag{38}$$

$$\tilde{K} = \tilde{K}_1 - Z_1 \tilde{K}_2 - Y_3 \tilde{K}_3, \tag{39}$$

where Y_3 is an arbitrary matrix of appropriate dimension.

As N, T, K, and $\tilde{K}$ are known, we can deduce the form of F as follows:

$$F = F_1 - Z_1F_2 - Y_3F_3 \tag{40}$$

■

Remark 3 From the above results, we can see that the determination of the matrices of the observer (2)–(4) can be done as follows: Matrices N, T, K, and $\tilde{K}$ are known matrices, matrix $J = TB$, and matrices S, M, P, and Q can be deduced from Equations (31)–(34). On the other hand, parameter matrices H, L, Z_1, Y_1, and Y_3 can be obtained from the stability of Equation (9).

Now, let $\bar{T}_2 = T_2G$ and $Z_1 = Z(I_{n+n_y} - \bar{T}_2\bar{T}_2^+)$, where Z is an arbitrary matrix of appropriate dimension. Then, matrices $T, K, N, \tilde{K}$, and F can be expressed as

$$T = T_1 - Z\mathcal{T}_2, \tag{41}$$

$$K = K_1 - Z\mathcal{K}_2, \tag{42}$$

$$N = N_1 - Z\mathcal{N}_2 - Y_3N_3, \tag{43}$$

$$\tilde{K} = \tilde{K}_1 - Z\tilde{\mathcal{K}}_2 - Y_3\tilde{K}_3, \tag{44}$$

$$F = F_1 - Z\mathcal{F}_2 - Y_3F_3, \tag{45}$$

where $\mathcal{T}_2 = (I_{n+n_y} - \bar{T}_2\bar{T}_2^+)T_2$, $\mathcal{K}_2 = (I_{n+n_y} - \bar{T}_2\bar{T}_2^+)K_2$, $\mathcal{N}_2 = (I_{n+n_y} - \bar{T}_2\bar{T}_2^+)N_2$, $\tilde{\mathcal{K}}_2 = (I_{n+n_y} - \bar{T}_2\bar{T}_2^+)\tilde{K}_2$, and $\mathcal{F}_2 = (I_{n+n_y} - \bar{T}_2\bar{T}_2^+)F_2$.

The observer error dynamics (9) can be rewriting as

$$\underbrace{\begin{bmatrix}\dot{\varepsilon}(t)\\ \dot{v}(t)\end{bmatrix}}_{\dot{\varphi}(t)} = \left(\underbrace{\begin{bmatrix}N_1 - Z\mathcal{N}_2 & 0\\ 0 & 0\end{bmatrix}}_{\mathbb{A}_1} - \underbrace{\begin{bmatrix}Y_3 & H\\ Y_1 & L\end{bmatrix}}_{\mathbb{Y}}\underbrace{\begin{bmatrix}N_3 & 0\\ 0 & -I_{q_1}\end{bmatrix}}_{\mathbb{A}_2}\right) \times \begin{bmatrix}\varepsilon(t)\\ v(t)\end{bmatrix} + \underbrace{\begin{bmatrix}-T_1G\\ 0\end{bmatrix}}_{\mathbb{B}_1} f(t) \tag{46}$$

$$r(t) = W\underbrace{[CP_1 \quad 0]}_{\mathbb{C}_1}\begin{bmatrix}\varepsilon(t)\\ v(t)\end{bmatrix} \tag{47}$$

without lost of generality, $Y_2 = 0$ was taken for simplicity.

4.2. *FDI design*

The objective of the fault detection is to build a residual, which shows the presence of a fault into a system. The mathematical definition of a residual is

$$\lim_{t\to\infty} r(t) = 0 \quad \text{for } f(t) = 0,$$

$$r(t) \neq 0 \quad \text{for } f(t) \neq 0.$$

The fault isolation objective in this paper is to obtain a transfer function from faults to residual equal to a diagonal, in order to deal with faults that may occur simultaneously.

From Equations (46) and (47), the residual dynamics are given by

$$\dot{\varphi}(t) = (\mathbb{A}_1 - \mathbb{Y}\mathbb{A}_2)\varphi(t) + \mathbb{B}_1 f(t), \tag{48}$$

$$r(t) = W\mathbb{C}_1\varphi(t). \tag{49}$$

Let $G_{fr}(s)$ be the transfer function from the fault $f(t)$ to the residual $r(t)$, such that

$$G_{fr}(s) = \left[\begin{array}{c|c}\mathbb{A}_1 - \mathbb{Y}\mathbb{A}_2 & \mathbb{B}_1\\ \hline W\mathbb{C}_1 & 0\end{array}\right]. \tag{50}$$

The objective is to render $G_{fr}(s)$ diagonal, i.e.

$$G_{fr}(s) = \mathrm{diag}\{g_{1,1}(s), \dots, g_{nf,nf}(s)\}, \tag{51}$$

while the stability of the observer is guaranteed. Since $G_{fr}(s)$ has a diagonal structure each residual is affected just by one fault. Considering this, it is possible to isolate simultaneous faults.

PROPOSITION 1 *The transfer function* (50) *can be diagonalized if and only if* $(\mathbb{C}_1\mathbb{B}_1)$ *has full column rank is that* $n_y \geq n_f$.

Proposition 1 is also called output separability condition (White and Speyer, 1987). To isolate n_f faults in Equation (1) the rank of $(\mathbb{C}_1\mathbb{B}_1)$ must be n_f, which in turn requires n_y measured outputs.

The following theorem shows how to design an observer of the form Equations (2)–(4) to perform FDI.

THEOREM 1 *Consider that* $n_y \geq n_f$ *and let*

$$\Lambda = \mathrm{diag}(\lambda_1, \dots, \lambda_{n_f}) \in \mathbb{R}^{n_f \times n_f}, \lambda_i < 0, \tag{52}$$

$$\Gamma = \mathrm{diag}(\gamma_1, \dots, \gamma_{n_f}) \in \mathbb{R}^{n_f \times n_f}, |\gamma_i| > 0, \quad \forall i, \ \{i = 1, \dots, n_f\} \tag{53}$$

be given. Then, there exist matrices $\mathbb{Y}$ *and* W *such that*

$$(\mathbb{A}_1 - \mathbb{Y}\mathbb{A}_2)\mathbb{B}_1 = \mathbb{B}_1\Lambda, \tag{54}$$

$$W\mathbb{C}_1\mathbb{B}_1 = \Gamma. \tag{55}$$

If $(\mathbb{A}_2\mathbb{B}_1)$ *has full column rank, then matrices* $\mathbb{Y}$ *and* W *are given by*

$$\mathbb{Y} = (\mathbb{A}_1\mathbb{B}_1 - \mathbb{B}_1\Lambda)(\mathbb{A}_2\mathbb{B}_1)^+ - \tilde{Z}(I - (\mathbb{A}_2\mathbb{B}_1)(\mathbb{A}_2\mathbb{B}_1)^+), \tag{56}$$

$$W = \Gamma(\mathbb{C}_1\mathbb{B}_1)^+, \tag{57}$$

where $\tilde{Z}$ *is an arbitrary matrix of appropriate dimension. Finally, if there exist matrices* $\mathbb{Y}$ *and* W *satisfying Equations* (54) *and* (55), *then*

$$G_{fr}(s) = \left[\begin{array}{c|c}\Lambda & I\\ \hline \Gamma & 0\end{array}\right] = \mathrm{diag}\left(\frac{\gamma_1}{s - \lambda_1}, \dots, \frac{\gamma_{n_f}}{s - \lambda_{n_f}}\right). \tag{58}$$

Proof Since $\mathbb{B}_1$ has full column rank there exist a matrix completion $\mathbb{B}_1^{\perp} \in \mathbb{R}^{(q_0+q_1)\times(q_0+q_1-n_f)}$ such that $\tilde{\mathcal{B}} = [\mathbb{B}_1\ \ \mathbb{B}_1^{\perp}] \in \mathbb{R}^{(q_0+q_1)\times(q_0+q_1)}$ is nonsingular. Let $\tilde{\mathcal{B}}^{-1} = [\tilde{\mathcal{B}}_1\ \ \tilde{\mathcal{B}}_2]^{\mathrm{T}}$ with $\tilde{\mathcal{B}}_1 \in \mathbb{R}^{(q_0+q_1)\times n_f}$. Then, we obtain

$$
\begin{aligned}
G_{fr}(s) &= \left[\begin{array}{c|c} \tilde{\mathcal{B}}^{-1}(\mathbb{A}_1 - \mathbb{Y}\mathbb{A}_2)\tilde{\mathcal{B}} & \tilde{\mathcal{B}}^{-1}\mathbb{B}_1 \\ \hline W\mathbb{C}_1\tilde{\mathcal{B}} & 0 \end{array}\right] \\
&= \left[\begin{array}{c|c} \begin{bmatrix}\tilde{\mathcal{B}}_1^{\mathrm{T}} \\ \tilde{\mathcal{B}}_2^{\mathrm{T}}\end{bmatrix}(\mathbb{A}_1 - \mathbb{Y}\mathbb{A}_2)\begin{bmatrix}\mathbb{B}_1 & \mathbb{B}_1^{\perp}\end{bmatrix} & \begin{bmatrix}\tilde{\mathcal{B}}_1^{\mathrm{T}} \\ \tilde{\mathcal{B}}_2^{\mathrm{T}}\end{bmatrix}\mathbb{B}_1 \\ \hline W\mathbb{C}_1\begin{bmatrix}\mathbb{B}_1 & \mathbb{B}_1^{\perp}\end{bmatrix} & 0 \end{array}\right] \\
&= \left[\begin{array}{cc|c} \tilde{\mathcal{B}}_1^{\mathrm{T}}(\mathbb{A}_1 - \mathbb{Y}\mathbb{A}_2)\mathbb{B}_1 & \tilde{\mathcal{B}}_1^{\mathrm{T}}(\mathbb{A}_1 - \mathbb{Y}\mathbb{A}_2)\mathbb{B}_1^{\perp} & I \\ \tilde{\mathcal{B}}_2^{\mathrm{T}}(\mathbb{A}_1 - \mathbb{Y}\mathbb{A}_2)\mathbb{B}_1 & \tilde{\mathcal{B}}_2^{\mathrm{T}}(\mathbb{A}_1 - \mathbb{Y}\mathbb{A}_2)\mathbb{B}_1^{\perp} & 0 \\ \hline W\mathbb{C}_1\mathbb{B}_1 & W\mathbb{C}_1\mathbb{B}_1^{\perp} & 0 \end{array}\right]
\end{aligned}
$$

consider $[\tilde{\mathcal{B}}_1\ \ \tilde{\mathcal{B}}_2]^{\mathrm{T}}[\mathbb{B}_1\ \ \mathbb{B}_1^{\perp}] = I$, then we have

$$
G_{fr}(s) = \left[\begin{array}{cc|c} \tilde{\mathcal{B}}_1^{\mathrm{T}}(\mathbb{A}_1 - \mathbb{Y}\mathbb{A}_2)\mathbb{B}_1 & \tilde{\mathcal{B}}_1^{\mathrm{T}}(\mathbb{A}_1 - \mathbb{Y}\mathbb{A}_2)\mathbb{B}_1^{\perp} & I \\ 0 & \tilde{\mathcal{B}}_2^{\mathrm{T}}(\mathbb{A}_1 - \mathbb{Y}\mathbb{A}_2)\mathbb{B}_1^{\perp} & 0 \\ \hline W\mathbb{C}_1\mathbb{B}_1 & W\mathbb{C}_1\mathbb{B}_1^{\perp} & 0 \end{array}\right]
$$

now, removing an uncontrollable subspace, we get

$$
\begin{aligned}
G_{fr}(s) &= \left[\begin{array}{c|c} \Lambda & I \\ \hline \Gamma & 0 \end{array}\right] \\
&= \operatorname{diag}\left(\frac{\gamma_1}{s-\lambda_1}, \ldots, \frac{\gamma_{nf}}{s-\lambda_{nf}}\right). \qquad (59)
\end{aligned}
$$

From Equation (59), we found that

$$(\mathbb{A}_1 - \mathbb{Y}\mathbb{A}_2)\mathbb{B}_1 = \mathbb{B}_1\Lambda, \qquad (60)$$

$$W\mathbb{C}_1\mathbb{B}_1 = \Gamma \qquad (61)$$

the general form of $\mathbb{Y}$ from Equation (60) is given by

$$
\begin{aligned}
\mathbb{Y} &= (\mathbb{A}_1\mathbb{B}_1 - \mathbb{B}_1\Lambda)(\mathbb{A}_2\mathbb{B}_1)^{+} \\
&\quad - \tilde{Z}(I - (\mathbb{A}_2\mathbb{B}_1)(\mathbb{A}_2\mathbb{B}_1)^{+}), \qquad (62)
\end{aligned}
$$

where $\tilde{Z}$ is an arbitrary matrix of appropriate dimension.

And the particular form of W in Equation (61) is

$$W = \Gamma(\mathbb{C}_1\mathbb{B}_1)^{+}, \qquad (63)$$

■

Replacing Equations (62) and (63) in Equations (48) and (49), we obtain

$$
\begin{aligned}
\dot{\varphi}(t) &= \underbrace{\left[\mathbb{A}_1 - (\mathbb{A}_1\mathbb{B}_1 - \mathbb{B}_1\Lambda)(\mathbb{A}_2\mathbb{B}_1)^{+}\mathbb{A}_2\right.}_{\bar{\mathbb{A}}_1} \\
&\quad + \tilde{Z}\underbrace{\left.(I - (\mathbb{A}_2\mathbb{B}_1)(\mathbb{A}_2\mathbb{B}_1)^{+})\mathbb{A}_2\right]}_{\bar{\mathbb{A}}_2}\varphi(t) + \mathbb{B}_1 f(t), \qquad (64)
\end{aligned}
$$

$$r(t) = \Gamma(\mathbb{C}_1\mathbb{B}_1)^{+}\mathbb{C}_1\varphi(t). \qquad (65)$$

Now, is necessary to study the stability of the observer and determine the remainder of matrices of the observer.

4.3. *Observer design*

The following theorem gives the LMI conditions that allow the determination of the dynamical observer matrices.

THEOREM 2 *There exist matrices $\tilde{Z}$ and Z such that system (64)–(65) is asymptotically stable if and only if there exist a matrix $X = X^{\mathrm{T}} > 0$ such that the following LMIs are satisfied:*

$$X = \begin{bmatrix} X_1 & X_2 \\ X_2^{\mathrm{T}} & X_3 \end{bmatrix} > 0, \qquad (66)$$

where $X_1 = X_1^{\mathrm{T}} > 0$, $X_3 = X_3^{\mathrm{T}} > 0$ *and* $X_3 - X_2^{\mathrm{T}}X_1^{-1}X_2 > 0$.

$$
\begin{aligned}
&(N_3 - N_3T_1G(N_3T_1G)^{+}N_3)^{\mathrm{T}\perp}(\Pi), \\
&(N_3 - N_3T_1G(N_2T_1G)^{+}N_3)^{\mathrm{T}\perp\mathrm{T}} < 0,
\end{aligned} \qquad (67)
$$

where

$$\Pi = \Pi_1{}^{\mathrm{T}}X_1 + X_1\Pi_1 - \Pi_2{}^{\mathrm{T}}W_1^{\mathrm{T}} - W_1\Pi_2, \qquad (68)$$

$$\Pi_1 = N_1 + (T_1G\Lambda - N_1T_1G)(N_3T_1G)^{+}N_3, \qquad (69)$$

$$\Pi_2 = \mathcal{N}_2 - \mathcal{N}_2T_1G(N_3T_1G)^{+}N_3. \qquad (70)$$

In this case matrix, $Z = X_1^{-1}W_1$ *and the matrix* $\tilde{Z}$ *is parameterized as follows:*

$$\tilde{Z} = X^{-1}(\mathcal{K}\mathcal{C}_l^{+} + \mathcal{Z}(I - \mathcal{C}_l\mathcal{C}_l^{+})), \qquad (71)$$

where

$$\mathcal{K} = -\mathcal{R}^{-1}\vartheta\mathcal{C}_r^{\mathrm{T}}(\mathcal{C}_r\vartheta\mathcal{C}_r^{\mathrm{T}})^{-1} + \mathcal{S}^{-1/2}\mathcal{L}(\mathcal{C}_r\vartheta\mathcal{C}_r^{\mathrm{T}})^{-1/2}, \qquad (72)$$

$$\vartheta = (\mathcal{R}^{-1} - \mathcal{D})^{-1} > 0, \qquad (73)$$

$$\mathcal{S} = \mathcal{R}^{-1} - \mathcal{R}^{-1}[\vartheta - \vartheta\mathcal{C}_r^{\mathrm{T}}(\mathcal{C}_r\vartheta\mathcal{C}_r^{\mathrm{T}})^{-1}\mathcal{C}_r\vartheta]\mathcal{R}^{-1} \qquad (74)$$

with $\mathcal{C} = \begin{bmatrix} N_3 - N_3T_1G(N_3T_1G)^{+}N_3 & 0 \\ 0 & -I \end{bmatrix}$ *and* $\mathcal{D} = \begin{bmatrix} \Pi & \Pi_1{}^{\mathrm{T}}X_2 - \Pi_2{}^{\mathrm{T}}W_2^{\mathrm{T}} \\ X_2^{\mathrm{T}}\Pi_1 - W_2\Pi_2 & 0 \end{bmatrix}$, *where* Π, Π_1 *and* Π_2 *are defined in Equations* (68)–(70), *respectively, and* $W_2 = X_2^{\mathrm{T}}Z$.

Matrices $\mathcal{R}$, $\mathcal{L}$, and $\mathcal{Z}$ are arbitrary matrices of appropriate dimensions satisfying $\mathcal{R} > 0$ and $\|\mathcal{L}\| < 1$. Matrices $\mathcal{C}_l$ and $\mathcal{C}_r$ are full rank matrices such that $\mathcal{C} = \mathcal{C}_l\mathcal{C}_r$.

Proof Consider a matrix $X = X^{\mathrm{T}} > 0$ such that

$$(\bar{\mathbb{A}}_1 + \tilde{Z}\bar{\mathbb{A}}_2)^{\mathrm{T}}X + X(\bar{\mathbb{A}}_1 + \tilde{Z}\bar{\mathbb{A}}_2) < 0. \qquad (75)$$

This last inequality can be rewritten as

$$\mathcal{B}\mathcal{X}\mathcal{C} + (\mathcal{B}\mathcal{X}\mathcal{C})^{\mathrm{T}} + \mathcal{D} < 0, \qquad (76)$$

where $\mathcal{X} = X\tilde{Z}$, $\mathcal{D} = \bar{\mathbb{A}}_1^{\mathrm{T}}X + X\bar{\mathbb{A}}_1$ its equivalence is defined in Theorem 2. Also matrix $\mathcal{B}$ is taken as $\mathcal{B} = I$, then $\mathcal{B}_l = I$, $\mathcal{B}_r = I$ and $\mathcal{B}^{\perp} = 0$.

The solvability conditions of Lemma 1 applied to Equation (76) are reduced to

$$\mathcal{C}^{\mathrm{T}\perp}\mathcal{D}\mathcal{C}^{\mathrm{T}\perp\mathrm{T}} < 0 \tag{77}$$

with $\mathcal{C}^{\mathrm{T}\perp} = [(N_3 - N_3T_1G(N_3T_1G)^{+}N_3)^{\mathrm{T}\perp}\ 0]$. By using the definition of $\mathcal{D}$ and W_1 we obtain Equation (67).

From Theorem 2 if condition (67) is satisfied, the matrix $\tilde{Z}$ is obtained as in Equations (71)–(74). ■

Remark 4 The dynamical observer application to standard systems can be obtained directly form our results by setting $E = I$, then we have $E^{\perp} = 0$, $\Sigma = \begin{bmatrix} R \\ C \end{bmatrix}$, and $\Omega = \begin{bmatrix} E \\ C \end{bmatrix}$.

The following algorithm summarize the procedure to compute all the observer matrices.

ALGORITHM 1

Step 1. Select the observer order q_0 and a matrix $R \in \mathbb{R}^{q_0 \times n}$ such that $\mathrm{rank}(\Sigma) = n$.
Step 2. Compute the matrices $T_1, \mathcal{T}_2, K_1, \mathcal{K}_2, N1, \mathcal{N}_2, N_3, \tilde{K}_1, \tilde{\mathcal{K}}_2, \tilde{K}_3$, and P_1 defined in Section 4.
Step 3. Select the matrices Λ and Γ as were defined in Theorem 1.
Step 4. Compute matrix W as in Equation (63).
Step 5. Find $\mathcal{R} > 0$ such that Equation (73) be positive definite.
Step 6. Find the matrices $\mathcal{L}$ and $\mathcal{Z}$ such that $\|\mathcal{L}\| < 1$ to solve Equations (66) and (67), then obtain the matrix $\tilde{Z}$ as in (71).
Step 7. Compute the matrices of the dynamical observer (2)–(3): N, H, F, J, S, L, M, P and Q, by using (43) to compute N, (56) to compute H and L, (31)–(34) to compute S, M, P and Q, F is defined in (45) and J is defined by Lemma 2.

5. Illustrative example

In order to illustrate our results, let us consider the following singular system:

$$\begin{bmatrix} 1 & 0 & 0 \\ 0 & 1 & 0 \\ 0 & 0 & 0 \end{bmatrix} \dot{x}(t) = \begin{bmatrix} -2.7 & 0 & 0.3 \\ -0.2 & -3 & 0 \\ -0.11 & 1.74 & -1 \end{bmatrix} x(t) + \begin{bmatrix} 1 \\ 0.5 \\ 1 \end{bmatrix} u(t) + \begin{bmatrix} 1 & 1 \\ 0 & 1 \\ 0 & 0 \end{bmatrix} f(t),$$

$$y(t) = \begin{bmatrix} 1 & 0 & 1 \\ 0 & 0 & 1 \end{bmatrix} x(t).$$

By following Algorithm 1, an observer with order $q_0 = 3$ was selected and $R = I_3$, such that $\mathrm{rank}(\Sigma) = 3$. Matrices $\mathcal{R}$, $\mathcal{L}$ and $\mathcal{Z}$ were selected as $\mathcal{R} = I_6$, $\mathcal{L} = \mathrm{ones}_{6,4} \times 0.1$, and

$$\mathcal{Z} = \begin{bmatrix} 9 & 3 & 2 & 1 & 8 & 9 & 9 & 0 & 3 \\ 9 & 4 & 1 & 3 & 8 & 7 & 8 & 2 & 8 \\ 9 & 2 & 8 & 4 & 1 & 7 & 7 & 4 & 2 \\ 9 & 4 & 4 & 5 & 7 & 8 & 3 & 8 & 3 \\ 9 & 1 & 8 & 4 & 7 & 2 & 8 & 4 & 1 \\ 9 & 4 & 8 & 2 & 8 & 4 & 9 & 3 & 8 \end{bmatrix}.$$

By using the LMI toolbox of MATLAB, we solved the inequalities (66) and (67). The observer gains are constructed using Theorem 2.

$$\begin{aligned} \dot{\zeta}(t) = & \begin{bmatrix} -3.11 & -2.89 & -3.67 \\ -1.10 & -2.04 & -2.08 \\ 0.21 & 0.36 & -0.66 \end{bmatrix} \zeta(t) \\ & + \begin{bmatrix} 0.87 & 0.87 & 0.87 \\ 0.36 & 0.36 & 0.36 \\ 0.20 & 0.20 & 0.20 \end{bmatrix} \times 0.01 v(t) \\ & + \begin{bmatrix} 14.56 & -0.97 & 2.23 \\ 6.13 & 0.10 & -0.10 \\ 2.55 & 0.42 & -0.88 \end{bmatrix} \begin{bmatrix} -E^{\perp}Bu(t) \\ y(t) \end{bmatrix} \\ & + \begin{bmatrix} 14.09 \\ 6.17 \\ 2.75 \end{bmatrix} u(t), \end{aligned}$$

$$\begin{aligned} \dot{v}(t) = & \begin{bmatrix} 0.44 & -0.69 & 1.18 \\ 0.44 & -0.69 & 1.18 \\ 0.44 & -0.69 & 1.18 \end{bmatrix} \times 0.01 \zeta(t), \\ & + \begin{bmatrix} -2.03 & 0.22 & 0.22 \\ 0.22 & -2.03 & 0.22 \\ 0.22 & 0.22 & -2.03 \end{bmatrix} \times 0.01 v(t) \\ & + \begin{bmatrix} 0 & 0 & 0 \\ 0 & 0 & 0 \\ 0 & 0 & 0 \end{bmatrix} \begin{bmatrix} -E^{\perp}Bu(t) \\ y(t) \end{bmatrix} \end{aligned}$$

$$\begin{aligned} \hat{x}(t) = & \begin{bmatrix} 0.59 & -0.05 & -0.19 \\ -0.05 & 0.31 & 0.15 \\ -0.19 & 0.15 & 0.37 \end{bmatrix} \zeta(t) \\ & + \begin{bmatrix} 0.07 & 0.56 & -0.49 \\ 0.49 & 0.16 & 0.33 \\ -0.09 & 0.24 & 0.67 \end{bmatrix} \begin{bmatrix} -E^{\perp}Bu(t) \\ y(t) \end{bmatrix}, \end{aligned}$$

$$r(t) = \begin{bmatrix} -2.20 & -5.85 \\ 6.05 & 4.97 \end{bmatrix} (C\hat{x}(t) - y(t)),$$

In order to evaluate the observer performance a measurement noise $n(t)$ was considered in the measured output, then the noise-corrupted outputs become $y_1(t) = x_1(t) + x_3(t) + n(t)$, and $y_2(t) = x_3(t) + n(t)$.

The results are depicted in Figure 1–5, which show two cases of simulation. The first case shows step faults with different time of apparition, and the second case shows simultaneous faults, and one of these is time variant.

Figure 1 shows the measurement noise, the system input $u(t)$ was considered as a constant $u(t) = 2$.

Case 1 Step actuator faults: In this case, the actuators faults were considered as a step, each one applied at different time, see Figure 2.

Figure 3 gives the residual where each fault can be readily distinguished from the other, which illustrates that the proposed observer satisfies the requirement of FDI.

Once the residuals were generated, the next step is the evaluation of the residuals by assigning a symptom.

symptom	
1	if residue > threshold,
0	if residue < threshold.

With these symptoms we can generate the following signature table.

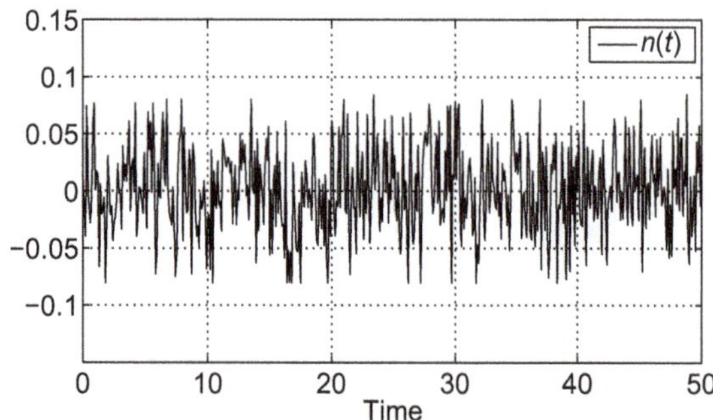

Figure 1. The measurement noise $n(t)$.

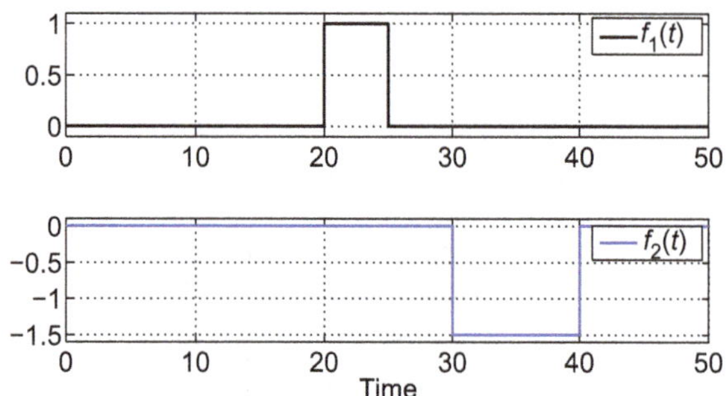

Figure 2. Case 1: step actuator faults.

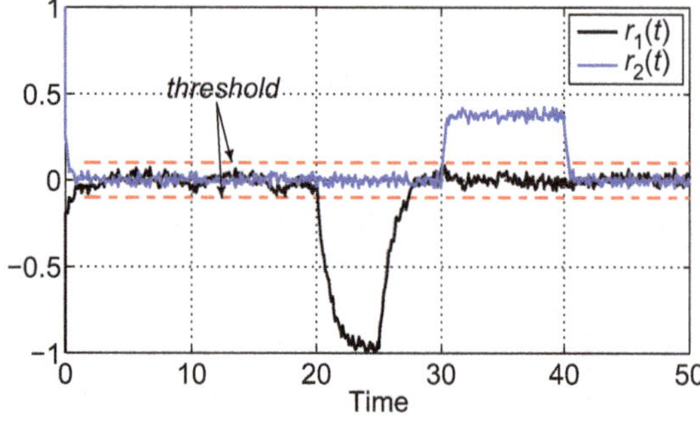

Figure 3. Case 1: residuals fault detection.

Table 1. Case 1: residual evaluation.

	Time		
Residue	20.12–27.37	30.09–40.38	Other time
$r_1(t)$	1	0	0
$r_2(t)$	0	1	0

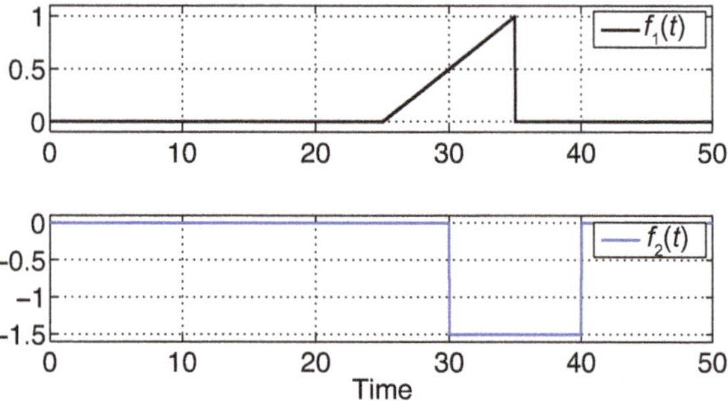

Figure 4. Case 2: actuator faults($f_1(t)$ time variant and $f_2(t)$ step).

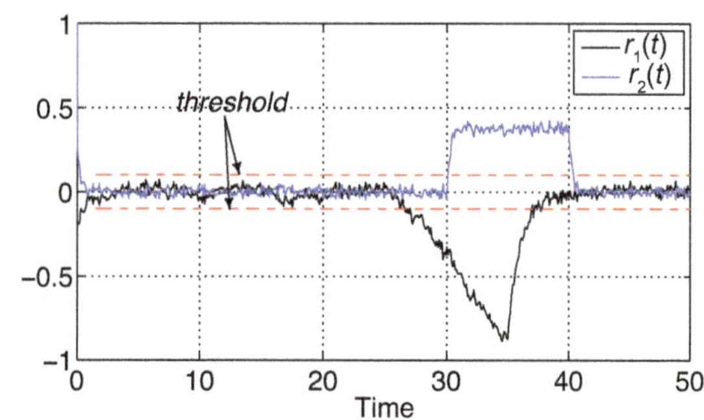

Figure 5. Case 2: residuals simultaneous faults.

Table 2. Case 2: residual evaluation.

	Time			
Residue	26.57–30.09	30.1–37.78	37.79–40.38	Other time
$r_1(t)$	1	1	0	0
$r_2(t)$	0	1	1	0

From Table 1, we observe that the signature to represent the presence of the fault $f_1(t)$ is different from the signature representing the fault $f_2(t)$, so that we can isolate each fault.

Case 2 Simultaneous faults: Figure 4 shows the faults, where $f_1(t)$ is time variant with a ramp behavior. Between 30 and 35 both faults are present into the system.

From Table 2, we observer that the signature in the case of simultaneous faults is different from the other cases, so that even simultaneous faults can be isolated.

The example makes clear that the actual residual response reflects the fault presence in the system. With a good choice of threshold, the observer designed satisfies the performance requirements of FDI.

6. Conclusion

In this paper, a dynamical observer-based FDI for singular systems has been presented. The conditions for the existence of this observer were given in terms of a set of LMIs. The obtained observer satisfies the constraint to obtain a diagonal transfer function between the fault $f(t)$ and the residual $r(t)$ to isolate faults. The approach presented here permits to parameterize the others dynamical observers for fault detection (Remark 2). The case of standard systems can be directly obtained from our results (Remark 4).

Disclosure statement

No potential conflict of interest was reported by the author(s).

References

Araujo J. M., Barros P. R., & Dorea C. E. T. (2012). Design of observers with error limitation in discrete-time descriptor systems: A case study of a hydraulic tank system. *IEEE Transactions on Control Systems Technology*, *20*, 1041–1047.

Astorga-Zaragoza C. M., Theilliol D., Ponsart J. C., & Rodrigues M. (2012). Fault diagnosis for a class of descriptor linear parameter-varying systems. *International Journal of Adaptive Control and Signal Processing*, *26*, 208–223.

Bokor J., & Szabó Z. (2009). Fault detection and isolation in nonlinear systems. *Annual Reviews in Control*, *33*(2), 113–123.

Boulkroune B., Darouach M., Zasadzinski M., Gillé S., & Fiorelli D. (2009). A nonlinear observer design for an activated sludge wastewater treatment process. *Journal of Process control*, *19*, 1558–1565.

Darouach M. (2009). H_∞ unbiased filtering for linear descriptor systems via LMI. *IEEE Transactions on Automatic Control*, *54*, 1966–1972.

Darouach M. (2012). On the functional observers for linear descriptor systems. *Systems & Control Letters*, *61*, 427–434.

Darouach M., & Boutayeb M. (1995). Design of observers for descriptor systems. *IEEE Transactions on Automatic Control*, *40*, 1323–1327.

Darouach M., Zasadzinski M., & Hayar M. (1996). Reduced-order observer design for descriptor systems with unknown inputs. *IEEE Transaction on Automatic Control*, (41), 1068–1072.

Duan G. R., Howe D., & Patton R. J. (2002). Robust fault detection in descriptor linear systems via generalized unknown input observers. *International Journal of Systems Science*, *33*, 369–377.

Goodwin G. C., & Middleton R. H. (1989). The class of all stable unbiased state estimators. *Systems & Control Letters*, *13*, 161–163.

Hamdi H., Rodriguez M., Mechmeche C., Theilliol D., & Benhadj Braiek N. (2012). Fault detection and isolation in linear parameter-varying descriptor systems via proportional integral observer. *International Journal of Adaptive Control and Signal Processing*, *26*, 208–223.

Hou M., & Müller P. C. (1995). Design of a class of Luenberger observers for descriptor systems. *IEEE Transactions on Automatic Control*, *40*, 133–136.

Kim H. S., Yeu T. K., & Kawaji S. (2001). Fault detection in linear descriptor systems via unknown input PI observer. *Transactions on Control, Automation and Systems Engineering*, *3*, 77–82.

Li Z., & Jaimoukha I. M. (2009). Observer-based fault detection and isolation filter design for linear time-invariant systems. *International Journal of Control*, *82*, 171–182.

Li X. J., & Yang G. H. (2011). Dynamic observer-based robust control and fault detection for linear systems. *IET Control Theory and Applications*, *6*, 2657–2666.

Liu B., & Si J. (1997). Fault isolation filter design for linear time-invariant systems. *IEEE Transaction on Automatic Control*, *42*, 704–707.

Liu P., Zhang Q., Yang X., & Yang L. (2008). Passivity and optimal control of descriptor biological complex systems. *IEEE Transactions on Automatic Control*, *53*, 122–125.

Luenberger D. G. (1977). Dynamic equations in descriptor form. *IEEE Transactions on Automatic Control*, *AC-22*, 312–321.

Marquez H. J. (2003). A frequency domain approach to state estimation. *Journal of the Franklin Institute*, *340*, 147–157.

Müller P. C. (2005). Modelling and control of mechatronic systems by the descriptor approach. *Journal of Theoretical and Applied Mechanics*, *43*, 593–607.

Müller P. C., & Hou M. (1993). On the observer design for descriptor systems. *IEEE Transactions on Automatic Control*, *38*, 1666–1671.

Patton R. J., Frank P. M., & Clark R. N. (2000). *Issues of fault detection for dynamic systems*. London: Springer.

Rodrigues M., Theilliol D., Adam-Medina M., & Sauter D. (2008). A fault detection and isolation scheme for industrial systems based on multiple operating models. *Control Engineering Practice*, (16), 225–239.

Skelton R. E., Iwasaki T., & Grigoriadis K. (1998). *A unified algebraic approach to linear control design*. London: Taylor & Francis.

Theilliol D., Noura H., & Ponsart J. C. (2002). Fault diagnosis and accommodation of three-tank system based on analytical redundancy. *ISA Transactions*, (41), 365–382.

Verhaegen M. H., & Dooren P. V. (1986). A reduced order observer for descriptor systems. *Systems & Control Letters*, *8*, 29–37.

White J. E., & Speyer J. L. (1987). Detection filter design: Spectral theory and algorithms. *IEEE Transactions on Automatic Control*, *AC-32*, 593–603.

Yeu T. K., Kim H. S., & Kawaji S. (2005). Fault detection, isolation and reconstruction for descriptor systems. *Asian Journal of Control*, *7*, 356–367.

Zhou L., & Lu G. (2009). Detection and stabilization for discrete-time descriptor systems via a limited capacity communication channel. *Automatica*, *45*, 2272–2277.

11

The optimal extended warranty length of durable-goods-based preventive maintenance behaviour

Na Tao[a] and Sheng Zhang[b]*

[a]*School of Management, Xi'an Jiaotong University, Xi'an 710049, People's Republic of China* [b]*School of Public Policy and Administration, Xi'an Jiaotong University, Xi'an, 710049, People's Republic of China*

The extended warranty is an important strategy for the manufacturers to expand their market share. In this work, we studied the optimal extended warranty length considering preventive maintenance behaviour, which mainly avoids the producers and consumers' moral hazard. Based on the profit function maximization, the optimal extended warranty length under different situations can be achieved. We conducted a simulation analysis to see the influence on preventive maintenance behaviour during the warranty period to the optimal extended warranty length.

Keywords: base warranty; extended warrant; preventive maintenance behaviour

1. Introduction

Product performances during the warranty period are determined by product characteristics and its use. From the manufacturer point of view, the consumers' abuse may lead to an increase in the total numbers of failures, which further generate more maintenance cost during the warranty period. If the maintenance cost is greater than the income derived, the manufacturer would not provide the extended warranty. Since the income derived and the extended warranty is fixed, the manufacturer should control the absolute numbers of failures to manage the maintenance cost. Preventive maintenance behaviour is acted, on the one hand, to guide the consumers to use the product correctly, and, on the other hand, to detect the quality of the products to reduce the total numbers of failures.

The relation between warranty and preventive maintenance behaviour is closeness. These include (i) the preventive maintenance behaviour models classes and policy (Jain & Maheshwari, 2006; Rai & Allada, 2006; Wu & Zuo, 2010), (ii) Preventive maintenance behaviour during the warranty period has an important impact on the warranty servicing cost (Chien, 2008; Kleyner & Sandborn, 2008; Maronick, 2007), (iii) the relation between preventive maintenance behaviour and age replacement (Chien, 2010; Jack & Iskandar, 2009; Jung & Park, 2010) and (iv) the optimal warranty policy (Wu & Li, 2007; Wu & Xie, 2008; Wua & Choub, 2009). Wu and Longhurst (2011) studied the lifecycle cost of a product protected by the extended warranty policies from the consumers' perspective. Jack and Murthy (2007) proposed that the method describing the degree of preventive maintenance behaviour is failure-rate reduction and age-reduction. Jun, Min, and Tsan (2011) studied a general periodic preventive maintenance behaviour policy. The research conclusions show that the time of the first preventive maintenance behaviour and its corresponding maintenance level determined the optimal preventive maintenance strategies for the consumers.

The paper integrates the base warranty and extended warranty breaking through the orthodox research separating the base warranty and extended warranty. Manufacturers carry preventive maintenance behaviour during the base warranty period and the extended warranty period to avoid the risk of increasing maintenance cost as the products' quality and the consumers' abuse. We seek to obtain the influence on preventive maintenance behaviour during the warranty period to the extended warranty length.

Due to difficulty of the empirical study of preventive maintenance and the extended warranty, at present, most scholars adopt the mathematical model method to study the topic on the extended warranty and preventive maintenance behaviour. Chien (2005) determined the optimal warranty period from the perspective of the seller and out-of-warranty replacement age from that of the buyer respectively, minimizing the corresponding cost functions by a concise numerical example. Lin, Wang, and Chin (2009) established a model-driven Decision Support System to determine the dynamic optimization of price, warranty length and production rate. Chang and Lin (2012) took the numerical examples to illustrate the influences of

*Corresponding author. Email: zhangsheng@xjtu.edu.cn

the optimal extended warranty length. Therefore, this study establishes mathematical model and adopts the simulation analysis method to study the problem.

Preventive maintenance behaviour is quantized as the number of carrying preventive maintenance behaviour during the base warranty period and the extended warranty period. The numbers of failures during the warranty period can be expressed as the function of the number of carrying preventive maintenance behaviour. Combining the maintenance cost and the preventive maintenance behaviour cost, the profit function is constructed. Therefore, we can get the optimal extended warranty length under different situations by maximizing the profit function. The simulation analysis tells us the influence on preventive maintenance behaviour during the warranty period to the extended warranty length. Ultimately the influencing mechanism is helpful to increase consumers' purchase of hardware products and expanded market share with indirect network effects. The rest of the paper is outlined as follows. The mathematical model is constructed based on some assumption, in Section 2. Section 3 elaborates the optimal extended warranty length under different situations. The numerical examples are analysed in Section 4. Finally, the conclusions are presented at the end of the paper.

2. Model formulation

Consider a new durable product sold with the base warranty period t_{bw} and the extended warranty period t_{ew}. Assume that the relation between the base warranty and the extended warranty periods satisfies the following equation:

$$t_{bw} = t_{ew} + kl, k = 1, 2, \dots \tag{1}$$

where l means a fixed time length unit of the extended warranty, k presents purchasing the number of a fixed time length unit of the extended warranty.

The preventive maintenance behaviour can be measured with the maintenance degree of a preventive maintenance behaviour x, the time epochs for carrying out the first preventive maintenance behaviour t_1 and the marginal cost of carrying the preventive maintenance behaviour ρ. Assuming the relation between the maintenance degree of a preventive maintenance behaviour x and the time epochs for carrying out preventive maintenance behaviour t_1 satisfies the following relation:

$$t_i = t_1 + (i-1)x,\ i = 1, 2, \dots \tag{2}$$

Let $r(t)$ be the failure rate of the product. Assuming that the failure-rate function follows square function, the failure-rate function is

$$r(t) = t^2 \tag{3}$$

The expected total number of failures within the interval $(0, t)$ is

$$R(t) = \int_0^{\infty} r(t)dt \tag{4}$$

The probability intensity follows $r(t - ix)$. k is maximum when the extended warranty periods cover the life cycle. That is

$$t_{ew} = t_d \tag{5}$$

Combing the Equation (1), the maximum k is

$$k_{\max} = \frac{t_d - t_{bw}}{l} \tag{6}$$

When consumers do not buy the extended warranty, the minimum k is

$$k_{\min} = 0 \tag{7}$$

Therefore k is located during $[0, (t_d - t_{bw}/l)]$.

2.1. *Cost to manufacturer*

The total cost to the manufacturer during the product sold includes two parts: the cost C_{bw} during the base warranty period and the cost C_{ew} during the extended warranty period. The cost during the warranty period mainly is the maintenance cost and preventive maintenance behaviour cost. The maintenance cost equals to each maintenance cost plus the total numbers of failures. Preventive maintenance behaviour cost is each preventive maintenance behaviour cost $C_{pm}(x)$ plus the total number of preventive maintenance behaviour. Each preventive maintenance behaviour cost $C_{pm}(x)$ is non-negative and non-decreasing function of maintenance degree x.

$$C_{pm}(x) \geq 0, C'_{pm}(x) \geq 0 \text{ for all } x \geq 0 \tag{8}$$

During the base warranty period, the numbers of failures can be expressed as

$$\begin{aligned}\sum_{i=1}^{n} \int_{t_{i-1}}^{t_i} r(t - ix)\mathrm{d}t + \int_{t_n}^{t_{bw}} r(t - nx)\mathrm{d}t \\ = \sum_{i=1}^{n} [R(t_i - ix) - R(t_{i-1} - ix)] \\ + R(t_{bw} - nx) - R(t_n - nx) \\ = nR(t_1) + (n-1)R(t_1 - x) + R(t_{bw} - nx) - R(t_n - nx)\end{aligned} \tag{9}$$

The maintenance cost during the base warranty period equals to r_p plus the numbers of failures during the base warranty period, which is

$$\begin{aligned}C_{bm} = r_p[nR(t_1) + (n-1)R(t_1 - x) \\ + R(t_{bw} - nx) - R(t_n - nx)]\end{aligned} \tag{10}$$

The preventive maintenance behaviour cost during the base warranty period is equal to each preventive maintenance behaviour cost plus the numbers of preventive maintenance behaviour during the base warranty period, which denotes

$$C_{bp} = nC_{pm}(x) \tag{11}$$

Let n denote the total number of preventive maintenance behaviour within base warranty period $(0, t_{bw})$. Combining Equations (10) and (11), we have the cost during the base warranty given by

$$C_{bw} = C_{bm} + C_{bp} = r_p[nR(t_1) + (n-1)R(t_1 - x) + R(t_{bw} - nx) - R(t_n - nx)] + nC_{pm}(x) \tag{12}$$

During the extended warranty period, the numbers of failures can be expressed as

$$\begin{aligned}\sum_{i=1}^{m}\int_{t_{bw+i-1}}^{t_{bw+i}} r(t-ix)\mathrm{d}t + \int_{t_m}^{t_{bw+kl}} r(t-mx)\mathrm{d}t &= \sum_{i=1}^{m}[R(t_{bw+i} - ix) - R(t_{bw+i-1} - ix)] \\ &\quad + R(t_{bw} + kl - mx) - R(t_m - mx) \\ &= m\{R[t_1 + (t_{bw} - 1)x] - R[t_1 + (t_{bw} - 2)x]\} \\ &\quad + R(t_{bw} + kl - mx) - R(t_m - mx)\end{aligned} \tag{13}$$

The maintenance cost during the extended warranty period equals to r_p plus the numbers of failures during the extended warranty period, which is

$$C_{em} = r_p\{m\{R[t_1 + (t_{bw} - 1)x] - R[t_1 + (t_{bw} - 2)x]\} + R(t_{bw} + kl - mx) - R(t_m - mx)\} \tag{14}$$

The preventive maintenance behaviour cost in the extended warranty period is equal to each preventive maintenance behaviour cost plus the numbers of preventive maintenance behaviour during the extended warranty period, which denotes

$$C_{ep} = mC_{pm}(x) \tag{15}$$

Let m denote total number of preventive maintenance behaviour within extended warranty period (t_{bw}, t_{ew}). Combining Equations (14) and (15), we can get the cost during the extended warranty period given by

$$\begin{aligned}C_{ew} = C_{ep} + C_{em} &= r_p\{m\{R[t_1 + (t_{bw} - 1)x] - R[t_1 + (t_{bw} - 2)x]\} \\ &\quad + R(t_{bw} + kl - mx) - R(t_m - mx)\} + mC_{pm}(x)\end{aligned} \tag{16}$$

From Equations (12) and (16), we can obtain the expected total cost given by

$$\begin{aligned}\mathrm{TC} &= C_{bw} + C_{ew} \\ &= r_p[nR(t_1) + (n-1)R(t_1 - x) \\ &\quad + R(t_{bw} - nx) - R(t_n - nx)] + nC_{pm}(x) \\ &\quad + r_p\{m\{R[t_1 + (t_{bw} - 1)x] - R[t_1 + (t_{bw} - 2)x]\} \\ &\quad + R(t_{bw} + kl - mx) - R(t_m - mx)\} + mC_{pm}(x)\end{aligned} \tag{17}$$

2.2. *Profit to manufacturer*

When the consumer purchased the product, the total income to manufacturer derives two aspects: the first is the selling price of the extended warranty C_E, the second is the manufacturer's total maintenance income. It equals to each maintenance income plus the absolute numbers of failures. After the extended warranty period expires, the consumer must pay maintenance cost r_c to the manufacturer when the product failed. The manufacturer must exclude the maintenance cost r_p, then each maintenance income denotes $r_c - r_p$. The total numbers of failures after the extended warranty period expires are

$$\int_{t_{bw}+kl}^{t_d} r[t - (n+m)x]\mathrm{d}t = R[t_d - (n+m)x] - R[t_{bw} + kl - (n+m)x] \tag{18}$$

The manufacturer's total maintenance income is

$$(r_c - r_p)\{R[t_d - (n+m)x] - R[t_{bw} + kl - (n+m)x]\} \tag{19}$$

The manufacturer's total income is

$$C_E + (r_c - r_p)\{R[t_d - (n+m)x] - R[t_{bw} + kl - (n+m)x]\} \tag{20}$$

Combining the total cost and income yield the total profit, which is

$$\begin{aligned}\mathrm{TP} = \mathrm{TR} - \mathrm{TC} &= C_E + (r_c - r_p)\{R[t_d - (n+m)x] \\ &\quad - R[t_{bw} + kl - (n+m)x]\} \\ &\quad - r_p[nR(t_1) + (n-1)R(t_1 - x) + R(t_{bw} - nx) \\ &\quad - R(t_n - nx)] - (n+m)C_{pm}(x) \\ &\quad - r_p\{m\{R[t_1 + (t_{bw} - 1)x] - R[t_1 + (t_{bw} - 2)x]\} \\ &\quad + R(t_{bw} + kl - mx) - R(t_m - mx)\}\end{aligned} \tag{21}$$

3. Optimal policy

Our objective is to determine an optimal extended warranty period such that the expected total profit in Equation (13) is maximized. To assess the properties of the optimal extended warranty period, we take the first and second partial derivatives of Equation (21) with respect to k. We have

$$\frac{\partial \mathrm{TP}}{\partial k} = (r_c - r_p)l\{r[t_{bw} + kl - (n+m)x]\} - r_p l[r(t_{bw} + kl - mx)] \tag{22}$$

And

$$\frac{\partial^2 \mathrm{TP}}{\partial k^2} = r_p l^2\{r'[t_{bw} + kl - (n+m)x] - r'(t_{bw} + kl - mx)\} - r_c l^2\{r'[t_{bw} + kl - (n+m)x]\} \tag{23}$$

Analysing the first and second partial derivatives, if $(\partial^2\mathrm{TP}/\partial k^2) < 0, (r_p/r_c) < (r'[t_{bw} + kl - (n+m)x]/r'[t_{bw} + kl - (n+m)x] - r'(t_{bw} + kl - mx))$, it is shown that $\partial\mathrm{TP}/\partial k$ is the decreasing function of k.

$$\left.\frac{\partial \mathrm{TP}}{\partial k}\right|_{k=0} = r_p l\{r[t_{bw} - (n+m)x] - r(t_{bw} - mx)\} - r_c l\{r[t_{bw} - (n+m)x]\}$$

$$\left.\frac{\partial \mathrm{TP}}{\partial k}\right|_{k=(t_d - t_{bw})/l} = r_p l\{r[t_d - (n+m)x] - r(t_d - mx)\} - r_c l\{r[t_d - (n+m)x]\}$$

We arrive at the following Lemma 1–3.

LEMMA 1 *If*

$$\frac{r_p}{r_c} < \frac{r'[t_{bw} + kl - (n+m)x]}{r'[t_{bw} + kl - (n+m)x] - r'(t_{bw} + kl - mx)},$$

and

$$\frac{r_p}{r_c} > \frac{r[t_{bw} - (n+m)x]}{r[t_{bw} - (n+m)x - r(t_{bw} - mx)]}$$

$$\frac{r_p}{r_c} > \frac{r[t_d - (n+m)x]}{r[t_d - (n+m)x - r(t_d - mx)]}, \text{ then } k = \frac{t_d - t_{bw}}{l}.$$

If

$$\left.\frac{\partial \mathrm{TP}}{\partial k}\right|_{k=0} > 0,$$

that is

$$\frac{r_p}{r_c} > \frac{r[t_{bw} - (n+m)x]}{r[t_{bw} - (n+m)x - r(t_{bw} - mx)]}$$

$$\left.\frac{\partial \mathrm{TP}}{\partial k}\right|_{k=\frac{t_d - t_{bw}}{l}} > 0,$$

that is

$$\frac{r_p}{r_c} > \frac{r[t_d - (n+m)x]}{r[t_d - (n+m)x - r(t_d - mx)]}.$$

The profit expression is the increasing function of k. Based on the profit maximum; the optimal extended warranty period is $k = (t_d - t_{bw}/l)$. This proves the Lemma 1.

LEMMA 2 *If*

$$\frac{r_p}{r_c} < \frac{r'[t_{bw} + kl - (n+m)x]}{r'[t_{bw} + kl - (n+m)x] - r'(t_{bw} + kl - mx)}$$

and

$$\frac{r_p}{r_c} < \frac{r[t_{bw} - (n+m)x]}{r[t_{bw} - (n+m)x - r(t_{bw} - mx)]}$$

$$\frac{r_p}{r_c} < \frac{r[t_d - (n+m)x]}{r[t_d - (n+m)x - r(t_d - mx)]} \text{ then } k = 0.$$

If

$$\left.\frac{\partial \mathrm{TP}}{\partial k}\right|_{k=0} < 0,$$

that is

$$\frac{r_p}{r_c} < \frac{r[t_{bw} - (n+m)x]}{r[t_{bw} - (n+m)x - r(t_{bw} - mx)]};$$

$$\left.\frac{\partial \mathrm{TP}}{\partial k}\right|_{k=\frac{t_d - t_{bw}}{l}} < 0,$$

that is

$$\frac{r_p}{r_c} < \frac{r[t_d - (n+m)x]}{r[t_d - (n+m)x - r(t_d - mx)]}$$

The profit expression is the decreasing function of k. Based on the profit maximum, the optimal extended warranty period is $k = 0$. This proves Lemma 2.

LEMMA 3 *If*

$$\frac{r_p}{r_c} < \frac{r'[t_{bw} + kl - (n+m)x]}{r'[t_{bw} + kl - (n+m)x] - r'(t_{bw} + kl - mx)}$$

and

$$\frac{r_p}{r_c} > \frac{r[t_{bw} - (n+m)x]}{r[t_{bw} - (n+m)x - r(t_{bw} - mx)]}$$

$$\frac{r_p}{r_c} < \frac{r[t_d - (n+m)x]}{r[t_d - (n+m)x - r(t_d - mx)]},$$

then there exists a unique solution $k \in [0, (t_d - t_{bw}/l)]$.

If

$$\left.\frac{\partial \mathrm{TP}}{\partial k}\right|_{k=0} > 0,$$

that is

$$\frac{r_p}{r_c} > \frac{r[t_{bw} - (n+m)x]}{r[t_{bw} - (n+m)x - r(t_{bw} - mx)]};$$

$$\left.\frac{\partial \mathrm{TP}}{\partial k}\right|_{k=\frac{t_d - t_{bw}}{l}} < 0,$$

that is

$$\frac{r_p}{r_c} < \frac{r[t_d - (n+m)x]}{r[t_d - (n+m)x - r(t_d - mx)]}.$$

The profit expression is not monotonous function of k. Therefore, the optimal extended warranty period exists between 0 and $(t_d - t_{bw}/l)$. *That is* $k \in [0, (t_d - t_{bw}/l)]$. *This proves Lemma 3.*

In order to seek the optimal extended warranty length during $[0,(t_d - t_{bw}/l)]$, we let Equation (22) be zero,

$$\frac{\partial \mathrm{TP}}{\partial k} = (r_c - r_p)l\{r[t_{bw} + kl - (n+m)x]\} - r_p l[r(t_{bw} + kl - mx)] = 0 \quad (24)$$

That is,

$$\frac{r_c}{r_p} = 1 + \frac{r(t_{bw} + kl - mx)}{r[t_{bw} + kl - (n+m)x]} \quad (25)$$

Combing specific failure-rate function, Equation (25) will become

$$\frac{r_c}{r_p} = 1 + \frac{(t_{bw} + kl - mx)^2}{[t_{bw} + kl - (n+m)x]^2}. \quad (26)$$

4. Numerical examples

4.1. *Example 1*

Example 1 mainly analyses the influence of preventive maintenance behaviour during the base warranty to the optimal extended warranty length. When analysing the influence of preventive maintenance behaviour during the base warranty to the optimal extended warranty length, the paper assumes that preventive maintenance behaviour during the extended warranty period is fixed. The value of preventive maintenance behaviour during the base warranty period is from 0 to 1.5, and the corresponding step is 0.03. The total number of the value of preventive maintenance behaviour during the base warranty period is 51.

The other parameters in the simulation are as follows:

$$x = 1, \quad l = 0.5, \quad t_1 = 0, \quad r_c = 100, \quad r_p = 40, \quad m = 1, \quad t_{bw} = 1$$

The simulation result is observed in Figure 1. Specially, preventive maintenance behaviour during the base warranty period positively influences the optimal extended warranty length. The increasing extent keeps steady.

4.2. *Example 2*

Example 2 mainly analyses the influence of preventive maintenance behaviour during the extended warranty period to the optimal extended warranty length. When analysing the influence of preventive maintenance behaviour during the extended warranty period to the optimal extended warranty length, we assume that preventive maintenance behaviour during the base warranty period is fixed. The value of preventive maintenance behaviour during the extended warranty period is from 0.5 to 2.5, and the

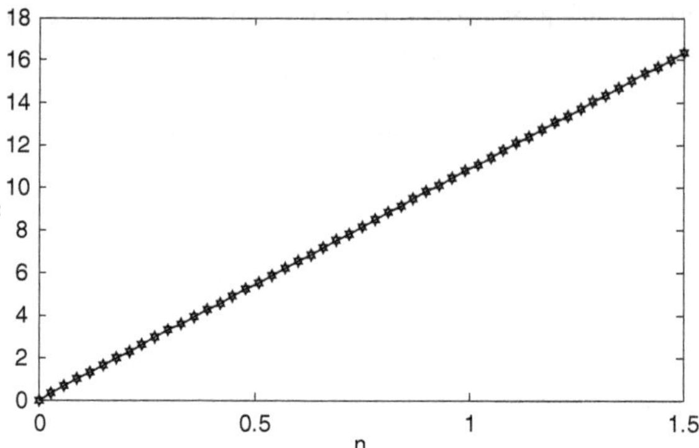

Figure 1. The influencing relation on n to k.

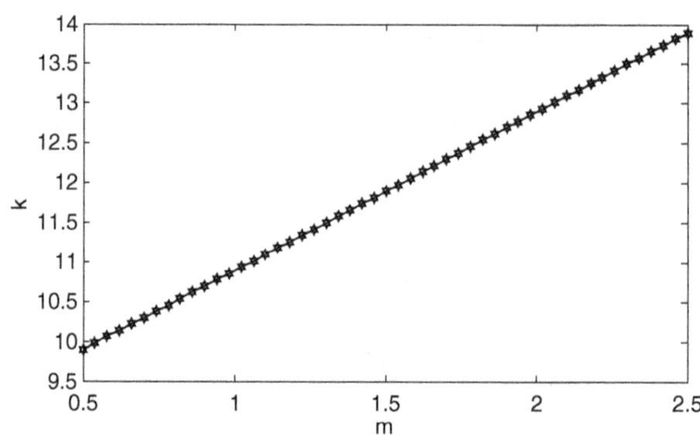

Figure 2. The influencing relation on m to k.

corresponding step is 0.04. The total number of the value of preventive maintenance behaviour during the extended warranty period is 51.

The other parameters in the simulation are as follows:

$$x = 1, \quad l = 0.5, \quad t_1 = 0, \quad r_c = 100, \quad r_p = 40, \quad n = 1, \quad t_{bw} = 1$$

The simulation result is seen from Figure 2. Specially, preventive maintenance behaviour during the extended warranty period positively influences the optimal extended warranty length. The increasing extent keeps steady.

Figures 1 and 2 describe the properties of the extended warranty optimal length influenced preventive maintenance behaviour during the base warranty period and extended warranty period. From Figures 1 and 2, we can see that preventive maintenance behaviour during the base warranty period and extended warranty period positively influences the optimal extended warranty length. The differences embody two aspects: firstly, the influence timeliness of preventive maintenance behaviour during the extended warranty period is ahead of that during the base warranty period. Secondly, the influence degree of preventive maintenance behaviour during the base warranty period is greater than during the extended warranty period. The conclusions are helpful for manufacturers to make the optimal extended warranty length. On the one hand, the manufacturers should notice the positive influence relation of preventive maintenance behaviour during the base warranty period and the extended warranty period to the extended warranty optimal length. On the other hand, as the influence degree of preventive maintenance behaviour during the base warranty period is greater, the manufacturers should pay attention to preventive maintenance behaviour during the base warranty period making the optimal extended warranty length.

5. Conclusion

The manufacturers optimize the extended warranty to increase the consumers' purchase of hardware products and expand market share with indirect network effects. The manufacturers carry preventive maintenance behaviour during the base warranty period and the extended warranty period to avoid the risk of increasing the warranty cost as the products' quality and the consumers' abuse, reducing the failure rate.

Therefore, the paper breaks the past method for studying the extended warranty period which combines the relation between the failure rate and the maintenance cost. Preventive maintenance behaviour is embedded in the conventional profit function to control the maintenance cost during the extended warranty period from the perspective of in-process control. Preventive maintenance behaviour is divided into the BWPM behaviour and EWPM behaviour. The paper analyses the influence relation on BWPM behaviour and EWPM behaviour to the extended warranty length and compares the influence difference to propose the new mechanism of designing the extended warranty length for the manufacturers.

This study constructs the model introducing preventive maintenance behaviour to get the optimal extended warranty length under different situations by maximizing the profit function. The simulation analysis can be done to observe the influence on preventive maintenance behaviour during the base warranty period and the extended warranty period to the optimal extended warranty length. The following conclusions can be obtained: firstly, preventive maintenance behaviour during the base warranty period and the extended warranty period positively influences the optimal extended warranty length. Secondly, the influence degree of preventive maintenance behaviour during the base warranty period is later and greater than during the extended warranty period.

Disclosure statement

No potential conflict of interest was reported by the authors.

References

Chang, W. L., & Lin, J. H. (2012). Optimal maintenance policy and length of extended warranty within the life cycle of products. *Computers and Mathematics with Applications, 63*, 144–150.

Chien, Y. H. (2005). Determining optimal warranty periods from the seller's perspective and optimal out-of-warranty replacement age from the buyer's perspective. *International Journal of Systems Science, 36*(10), 631–637.

Chien, Y. H. (2008). A general age-replacement model with minimal repair under renewing free-replacement warranty. *European Journal of Operational Research, 186*, 1046–1058.

Chien, Y. H. (2010). Optimal age for preventive replacement under a combined fully renewable free replacement with a pro-rata warranty. *International Journal of Production Economics, 124*, 198–205.

Jack, N., & Iskandar, B. P. (2009). A repair-replace strategy based on usage rate for items sold with a two-dimensional warranty. *Reliability Engineering & System Safety, 94*, 611–617.

Jack, N., & Murthy, D. N. P. (2007). A flexible extended warranty and related optimal strategies. *Journal of the Operational Research Society, 58*, 1612–1620.

Jain, M., & Maheshwari, S. (2006). Discounted costs for repairable units under hybrid warranty. *Applied Mathematics and Computation, 173*, 887–901.

Jun, W., Min, X., & Tsan, S. A. N. (2011). On a general periodic preventive maintenance policy incorporating warranty contracts and system ageing losses. *International Journal of Production Economics, 129*, 102–110.

Jung, K. M., & Park, M. (2010). System maintenance cost dependent on life cycle under renewing warranty policy. *Reliability Engineering & System Safety, 95*, 816–821.

Kleyner, A., & Sandborn, P. (2008). Minimizing life cycle cost by managing product reliability via validation plan and warranty return cost. *International Journal of Production Economics, 112*, 796–807.

Lin, P. C., Wang, J., & Chin, S. S. (2009). Dynamic optimisation of price, warranty length and production rate. *International Journal of Systems Science, 40*(4), 411–420.

Maronick, T. J. (2007). Consumer perceptions of extended warranties. *Journal of Retailing and Consumer Services, 14*, 224–231.

Rai, R., & Allada, V. (2006). Product warranty and reliability. *Annals of Operations Research, 143*, 133–146.

Wu, S., & Li, H. (2007). Warranty cost analysis for products with a dormant state. *European Journal of Operational Research, 182*, 1285–1293.

Wu, S., & Longhurst, P. (2011). Optimizing age-replacement and extended non-renewing warranty policies in lifecycle costing. *International Journal of Production Economics, 130*, 262–267.

Wu, S., & Xie, M. (2008). Warranty cost analysis for nonrepairable services products. *International Journal of Systems Science, 39*, 279–288.

Wu, S., & Zuo, M. J. (2010). Linear and nonlinear preventive maintenance models. *IEEE Transactions on Reliability, 59*, 242–249.

Wua, C., & Choub, C. Y. (2009). Optimal Price, warranty length and production rate for free replacement policy in the static demand market. *The International Journal of Management Science, 37*, 29–39.

12

Modeling and model predictive control of dividing wall column for separation of Benzene–Toluene-*o*-Xylene

Rajeev Kumar Dohare[a]*, Kailash Singh[a] and Rajesh Kumar[b]

[a]*Department of Chemical Engineering, Malaviya National Institute of Technology, Jaipur 302017, Rajasthan, India;* [b]*Department of Electrical Engineering, Malaviya National Institute of Technology, Jaipur 302017, Rajasthan, India*

In this paper, dividing wall column (DWC) has been chosen for a BTX (Benzene–Toluene–*o*-Xylene) system. A MATLAB® program has been written for nonlinear unsteady-state DWC, which is used in Simulink environment for control of the system by Model Predictive Control (MPC). Compositions of the three products (benzene, toluene, and *o*-xylene) are indirectly controlled by controlling the corresponding temperatures of the respective tray due to requirement of online analyzer. The temperature of uppermost tray in the rectifying section, stage temperature in the main column corresponding to the side stream withdrawn, and the bottom stage temperature in the stripping section have been chosen in order to maintain the compositions of the three products. The manipulated variables are reflux rate (L_0), side-stream flow rate (SSRF), and reboiler heat duty (Q_B). It has been observed that MPC shows good performance even in the presence of $\pm 10\%$ change in the feed flow rate, feed composition, and liquid split factor in comparison with conventional controllers. The MPC has less settling time (almost 1.5 h) compared with the PI controller (approximately 3–4 h).

Keywords: BTX separation; dividing wall column; multivariable control; model predictive control

1. Introduction

Distillation is the most desirable process for the separation of multi-component liquid mixture. However, when separation of high-purity products is required, a single distillation column is not sufficient; for the separation of *n*-component mixture, $n - 1$ columns are required (Sotudeh & Shahraki, 2007). An important effort has been focused on the development of a new design, optimization, and control method for thermally coupled distillation columns, which provide saving up to 30–40% of the total annual cost for the separation of some multi-component liquid mixtures as compared with classical distillation sequences (Triantafyllou & Smith, 1992). The use of fully thermally coupled distillation arrangements, such as the dividing wall column (DWC), leads to significant reductions in both energy and capital costs when compared with conventional two-column arrangements in the separation of ternary mixtures.

Although some authors have studied control of DWC to improve the process operation, yet model-based control finds its good applicability on this type of the distillation column. DWC is a nonlinear system due to its inherent complex dynamics because of the middle wall. Therefore, this system is challenging to control at its optimum operating parameters. Rewagad and Kiss (2012) showed the model predictive control (MPC) results for composition control; however, in this study, we have focused the control of temperatures. Besides disturbance in feed flow rate and feed composition, the disturbance in liquid split factor has also been considered. The parameter-based controllers, like proportional-integral-derivative controllers, are characterized by a short development time and smaller development efforts. The development of MPC for a DWC is a promising approach as the advantages of this control methodology can be intensified in view of the operational difficulties of the column. For maintaining the product purity at its desirable conditions, proper control is required. Therefore, model predictive control was selected to analyze its suitability.

The huge impact of distillation processes in both operation and investment costs has motivated the development of various types of fully thermally coupled distillation columns that can be used in saving energy and capital cost (Halvorsen & Skogestad, 2004). Conventionally, a ternary mixture can be separated via direct sequence (most volatile component is separated first), indirect sequence (heaviest component is separated first), or distributed sequence (mid-split) consisting of two to three distillation columns (Van Diggelen, Kiss, & Heemink, 2010). Eventually, this led to the concept known today as DWC that integrates in fact the two columns of a Petlyuk system into one column shell (Kaibel, 1987; Kolbe & Wenzel, 2004; Schultz et al.,

*Corresponding author. Email: rajeevdohare@gmail.com

2002). Mostly, the position of the wall in the DWC is in the middle, but off-center positions of the dividing wall may also be possible. This might be useful in situations, when the concentration of the medium boiling component is small as compared with the overhead and bottom (Asprion & Kaibel, 2010). Woinaroschy and Isopescu (2010) investigated the ability of computational dynamic programming to solve time optimal control of DWC. They focused on start-up control of the DWC.

In DWC, a single column is divided into four sections by inserting a wall in it. DWC is especially advantageous for separating ternary mixtures. The schematic diagram of the DWC is shown in Figure 1. It is divided into four sections: top section of the column is known as a rectifying section, left section as prefractionator, right section as main column, and the bottom section as stripping section. Feed consisting of 30 mol% benzene, 30 mol% toluene, and 40 mol% *o*-xylene is introduced on the 12th stage in the prefractionator. Almost pure toluene is withdrawn from the 11th stage of the main column. Benzene and *o*-xylene are obtained as top and bottom products, respectively.

Van Diggelen et al. (2010) proposed a model of DWC with the assumptions of constant pressure, no vapor flow dynamics, linearized liquid dynamics, and neglecting energy balance and changes in enthalpy. The researchers used this model to compare various control strategies. Hiller, Buck, Ehlers, and Fieg (2010) developed a non-equilibrium stage model by assuming heat and mass transfer between the liquid and vapor phases for the ideal component system. Ignat and Woinaroschy (2011) used a dynamic model for minimizing the distillation start-up time for the separation of an ideal benzene–toluene–ethylbenzene ternary mixture and the separation of a non-ideal methanol–ethanol–1-propanol mixture. The effect of

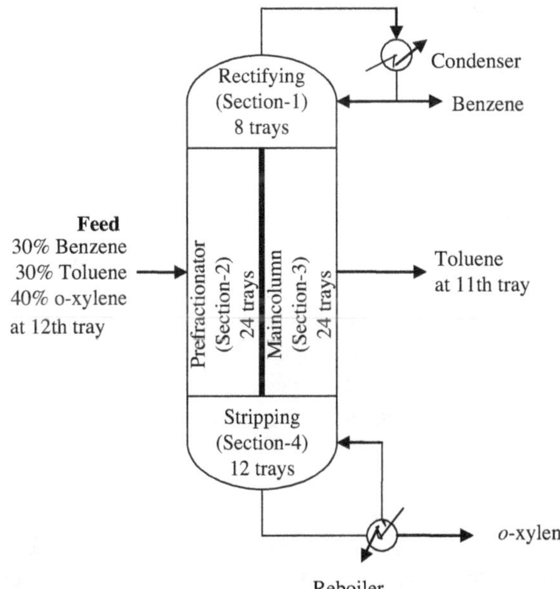

Figure 1. Schematic diagram of the DWC.

the liquid split ratio and vapor split ratio is responsible for internal disturbance and the dynamics of the DWC. Ignat and Woinaroschy (2011) also assumed the hold-up volume in the column. The vapor split ratio is assumed to be constant and liquid split ratio as a load change in this study.

Recently, Buck, Hiller, and Fieg (2011) also reported experimentally, temperature profile of the column by implementing MPC. Their study proves the real-life practicability of MPC; they did not provide an analysis of the transient behavior of DWC under desired disturbances. Kvernland, Halvorsen, and Skogestad (2010) applied MPC on the Kaibel column that separates a feed stream into four product streams using only a single column. The main objective for optimal operation was to minimize the total impurity flow. Finally, they concluded that MPC obtained typically less total impurity flow as compared with conventional decentralized control.

In this work, a mathematical model of DWC has been developed assuming non-constant volatility. Several simulation runs of the model have been used to investigate the effects of several parameters and dynamics of the system. BTX (Benzene–Toluene–Xylene) has been chosen as a component system. A MATLAB® program was written for unsteady-state DWC for investigating the MPC methodology. The controlled variables selected are: temperature of uppermost tray in the rectifying section, stage temperature corresponding to the side stream withdrawn, and the bottom stage temperature in the stripping section in order to maintain the compositions of the three products (Ling & Luyben, 2009). The manipulated variables are reflux rate, side-stream flow rate (SSRF), and reboiler heat duty. The performance of MPC controller has been verified by giving load changes in the feed flow rate and feed composition. The performance of MPC in the presence of disturbances variables has been compared with that of PI controller.

2. Mathematical model

The following assumptions have been taken into account for model development:

(1) Constant volume holdup of condenser/reflux
(2) Fast energy dynamics on trays (molar enthalpy change on trays with respect to time is negligible)
(3) Vapor split ratio is constant as it is fixed by column design.
(4) Raoult's law is assumed as BTX is close to ideal solution.

For the rectifying section, prefractionator, main column, and stripping section:

Mass balance for component i at tray j,

$$\frac{\mathrm{d}(M_j x_{j,i})}{\mathrm{d}t} = V_{j+1}, y_{j+1,i} + L_{j-1} x_{j-1,i} - V_j y_{j,i} - L_j x_{j,i} + F_j z_{j,i} - S_j x_{j,i}, \quad (1)$$

where n_c denotes the number of components and M_j denotes the total liquid holdup on the jth stage.

Summation Equations:

$$\sum_i x_{j,i} = 1; \quad \sum_i y_{j,i} = 1, \tag{2}$$

where $x_{j,i}$ and $y_{j,i}$ represent the mole fraction of the ith component in liquid and vapor phase at the jth stage, respectively.

Energy Balance at tray j:

$$\frac{d(H_{Lj} M_j)}{dt} = V_{j+1} H_{V_{j+1}} + L_{j-1} H_{L_{j-1}} - V_i H_{V_j} - L_j H_{L_j} + F_j H_{F_j} - S_j H_{L_j}. \tag{3}$$

In energy balance equation, H_{L_j} and H_{V_j} represent the liquid and vapor enthalpy on the jth stage, respectively.

Equilibrium relationship:

$$y_{j,i} = K_{j,i} x_{j,i} \text{ where } K_{j,i} = \frac{\gamma_{j,i} P^{sat}_{j,i}}{P_j}, \tag{4}$$

where $\gamma_{j,i}$ denotes the activity coefficient of the ith component in liquid phase at the jth stage.

For the condenser,

Material Balance:

$$\frac{d(M_0 x_{D,i})}{dt} = V_1 y_{1,i} + L_0 x_{D,i} - D x_{D,i}. \tag{5}$$

Energy Balance:

$$\frac{d(M_0 H_D)}{dt} = V_1 H_{V,1} - L_0 H_{L,0} - D H_D - q_C. \tag{6}$$

Summation equation:

$$\sum_i x_{D,i} = 1. \tag{7}$$

For the reboiler,

material Balance:

$$\frac{d(M_R x_{w,i})}{dt} = L_{n_4} x_{n_4,i} - V_{n_4+1} y_{n_4+1,i} - w x_{n_4+1,i}, \tag{8}$$

where M_R denotes liquid holdup in reboiler

Energy Balance:

$$\frac{d(M_{n_4+1} H_{L,n_4+1})}{dt} = L_{n_4} H_{L,n_4} - H_{V,n_4+1} V_{n_4+1} - w H_{L,n_4+1} + q_R. \tag{9}$$

Equilibrium Relationship:

$$y_{n_4+1,i} = K_{n_4+1,i} x_{w,i}. \tag{10}$$

Summation Equations:

$$\sum_i x_{n_4+1,i} = 1 \quad \text{and} \quad \sum_i y_{n_4+1,i} = 1. \tag{11}$$

At the intersection of rectifying section (Section 1) with prefractionator (Section 2) and main column (Section 3):

Vapor Mixing:

$$V^{(1)}_{n_1+1} = V^{(2)}_1 + V^{(3)}_1, \tag{12}$$

$$V^{(1)}_{n_1+1} y^{(1)}_{n_1+1,i} = V^{(2)}_1 y^{(2)}_{1,i} + V^{(3)}_1 y^{(3)}_{1,i}. \tag{13}$$

Liquid Splitting:

$$L^{(2)}_0 = \alpha L^{(1)}_{n_1}, \quad L^{(3)}_0 = (1-\alpha) L^{(1)}_{n_1}, \tag{14}$$

where α is liquid split factor.

$$x^{(2)}_{0,i} = x^{(1)}_{n_1,i}, \quad x^{(3)}_{0,i} = x^{(1)}_{n_1,i}.$$

At the intersection of Sections 2 and 3 with Section 4 (Stripping section):

Vapor splitting:

$$V^{(2)}_{n_2+1} = \beta V^{(4)}_1; \quad V^{(3)}_{n_2+1} = (1-\beta) V^{(4)}_1; \tag{15}$$

where β is a vapor splitting factor.

$$y^{(2)}_{n_2+1,i} = y^{(4)}_{1,i}; \quad y^{(3)}_{n_2+1,i} = y^{(4)}_{1,i}.$$

Liquid mixing:

$$L^{(4)}_0 = L^{(2)}_{n_2} + L^{(3)}_{n_2}, \tag{16}$$

$$L^{(2)}_{n_2} x^{(2)}_{n_2,i} + L^{(3)}_{n_2} x^{(3)}_{n_2,i} = L^{(4)}_0 x^{(4)}_{0,i}. \tag{17}$$

These model equations are set of ordinary differential equations-initial value problems, which were solved by ode15s solver (an inbuilt function in MATLAB® to solve ordinary differential equations). For simulation, physical properties and the nominal operating conditions of the investigated system (Benzene, Toluene and *o*-Xylene) are given in Table 1.

3. Model predictive control of a DWC

The controllability indicator such as relative gain array (RGA) is a useful method to understand the behavior of the system (Segovia-Hernandez, Hernandez-Vargas, & Marquez-Munoz, 2007; Skogestad & Postlethwaite, 2005). The RGA provides information about the interactions among the controlled and manipulated variables. The RGA element has been calculated as the ratio of open-loop gain to the closed-loop gain for a pair of variables.

For a selected pair of variables, values of the RGA element close to unity are preferred as the best combination. Due to complex interactions among the manipulated and controlled variables, decentralized controller's performance is not very good.

Table 1. Properties of the system and nominal operating parameters.

Properties of the system	Benzene	Toluene	Xylene
Molecular formula	C_6H_6	C_7H_8	C_8H_{10}
Molecular weight	78.11	92.14	106.17
Boiling point	80.2 °C	110.7 °C	144.5 °C
Density (mol/cm^3)	0.01133	0.00941	0.00829
Critical temperature	289.1 °C	318.7 °C	357.2 °C
Antoine equation and its constants			
$T_{sat} = B/(A - \log P) - C$, units of T in K and P in atm			
Components	A	B	C
Benzene	15.9008	2788.51	220.79
Toluene	16.0137	3096.52	219.48
Xylene	16.1156	3395.57	213.69
Nominal operating parameters	Value		
Feed flow rate	1 kmol/s		
Feed temperature	85 °C		
Feed composition (mol%)	30% B, 30% T, 40% X		
Reflux rate	0.860 kmol/s		
Vapor split factor	0.627		
Liquid split factor	0.353		
Side stream flow rate	0.296 kmol/s		
Reboiler heat duty	40.544 MW		
Bottoms flow rate	0.401 kmol/s		
Reflux ratio	2.84		

The DWC is a multivariable process that gives motivation for use of a MPC. The MPC offers a large number of operational advantages for diverse processes like reaction, separation, etc. (Backx, Bosgra, & Marquardt, 2000; De Temmerman, Dufour, Nicolaï, & Ramon, 2009). In view of the multi-input–multi-output system, the MPC is characterized by many manipulated and controlled variables and it may be performed for a complex system (Buck et al., 2011). It is an appropriate controller to control a multivariable process with considerable interactions. The MPC shows a good result in eliminating loop interaction when it is compared with a decentralized control scheme. Model predictive controller is good for creating a linear model at any operating point. Total operating cost and settling time can be reduced by controlling the temperature of the different sections in spite of the compositions of the products (Rodríguez Hernández & Chinea-Herranz, 2012). The manipulated variables, reflux ratio, side stream flow rate, and reboiler heat duty are used to control the temperatures (and therefore product purities). For investigating the controllability of MPCs, MPC Toolbox in MATLAB® was used. The cost function in MPC methodology is

$$\text{Min}\,F = \sum_{j=1}^{N_y}\sum_{i=1}^{P}(y_{\text{sp}_{i,i}} - y_{i,j}^{*})^2 w_{y,j} + \sum_{j=1}^{N_u}\sum_{i=1}^{M}(\Delta u_{i,j})^2 w_{u,j}, \tag{18}$$

where N_u and N_y denote the number of manipulated and controlled variables, respectively. p denotes the length of prediction horizon and M denotes the length of control horizon with respect to the sequence of input increments Δu.
subject to:

$$\Delta u_{\text{LB}} \le \Delta u \le \Delta u_{\text{UB}}, \tag{19}$$

$$u_{\text{LB}} \le u \le u_{\text{UB}}, \tag{20}$$

where u_{LB} and u_{UB} represent the lower and upper bounds to control the oscillations of the manipulated variables.

Model predictive control works on prediction horizon, control horizon and robustness factor parameters. Prediction horizon is the number of sampling intervals over which the cost function is minimized. Control horizon is the number of sampling intervals over which the control moves are estimated by optimization routine. The best suitable MPC parameters based on several runs by selecting different values are given in Table 2. Robustness factor controls the sluggishness and the speed of the controller. The working scheme of MPC is shown in Figure 2.

These parameters have a high impact on the total controller performance. The lower and upper limits on the manipulated variables (reflux rate, side stream flow rate,

Table 2. Best suitable model predictive control parameters.

Parameter	Value
Control interval (sampling interval)	10 s
Prediction horizon	20
Control horizon	6
Robustness factor	0.25

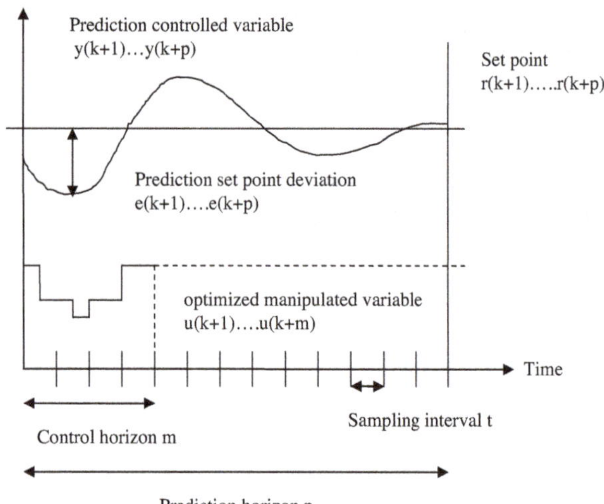

Figure 2. Working scheme of model predictive control.

and reboiler heat duty) have been set as [360.5 96 20.54] and [2000 496 60.54], respectively. The prediction horizon is coupled with the system's time constants and the chosen sampling rate (Agachi, Nagy, Cristea, & Imre-Lucaci, 2006).

In this study, temperature of tray 6 in Section 1 (Tsec1 (6)), temperature of tray 18 in Section 3 (Tsec3 (18)), and temperature of tray 6 in Section 4 (Tsec4 (6)) have been controlled by the reflux rate (L_0), side stream flow rate (SSFR) and reboiler heat duty (Q_B).

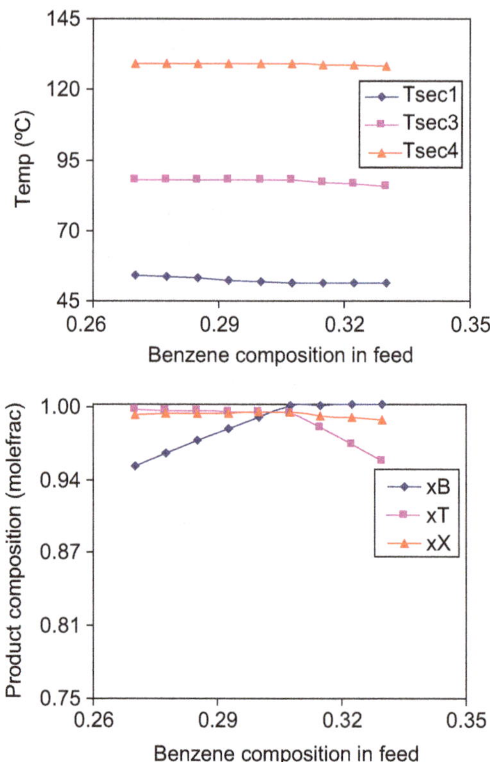

Figure 3. Effect of benzene composition in feed on temperatures and product composition.

4. Effect of parameters on temperature and composition

Some important parameters of DWC such as feed composition of benzene, toluene, and xylene, feed flow rate, reflux rate, SSRF, and reboiler heat duty were selected for investigation of the effect on temperatures and the resulting product compositions.

4.1. *Effect of benzene composition in the feed*

The effect of benzene composition of the feed is shown in Figure 3. While changing the benzene composition, the other two components were assumed to be in the same proportion. As can be seen, on increasing benzene composition in the feed, there is no significant change in *o*-xylene composition in the bottoms product. Toluene composition in the middle product does not change much until 30 mol% of benzene composition in the feed, after which there is a declining trend because extra benzene is mixed up in the middle product. The benzene composition of the top product increases up to 30 mol% of benzene in the feed, beyond which, there is no significant change. The temperatures also follow the trend accordingly. Therefore, it is evident that 30 mol% composition of benzene in feed is an optimum value giving the maximum purity of all the products.

4.2. *Effect of toluene composition in the feed*

The effect of toluene composition of the feed is shown in Figure 4. On increasing the feed composition of toluene, the benzene composition in the top product and *o*-xylene composition of the bottom product do not show significant effect up to 30 mol% of toluene composition in the feed but decreases afterwards. However, toluene composition in the middle product increases and becomes almost constant after 30 mol% of toluene composition in the feed. Therefore, the toluene composition of 30% seems to be an optimum feed composition as it gives the highest purity of all the products. The temperatures also follow the trend accordingly in equilibrium with the product composition.

4.3. *Effect of o-xylene composition in the feed*

The effect of *o*-xylene composition in the feed is shown in Figure 5. While changing the *o*-xylene composition, the other two components are supposed to be in the same proportion. Purity of benzene is constant up to 40 mol% *o*-xylene in the feed after which, it suddenly decreases. Moreover, *o*-xylene purity of the product increases up to 40% and then it becomes constant on further increasing xylene composition. Toluene purity decreases beyond 40

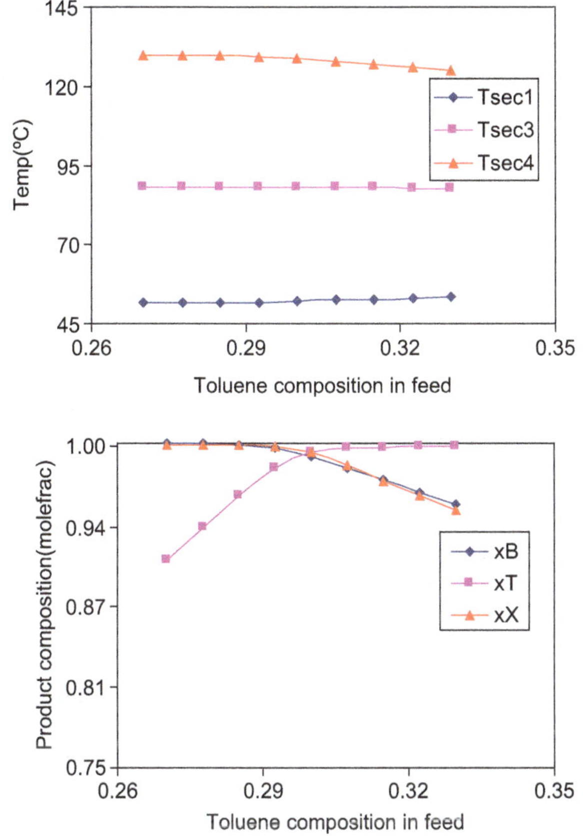

Figure 4. Effect of toluene composition in feed on temperature and product composition.

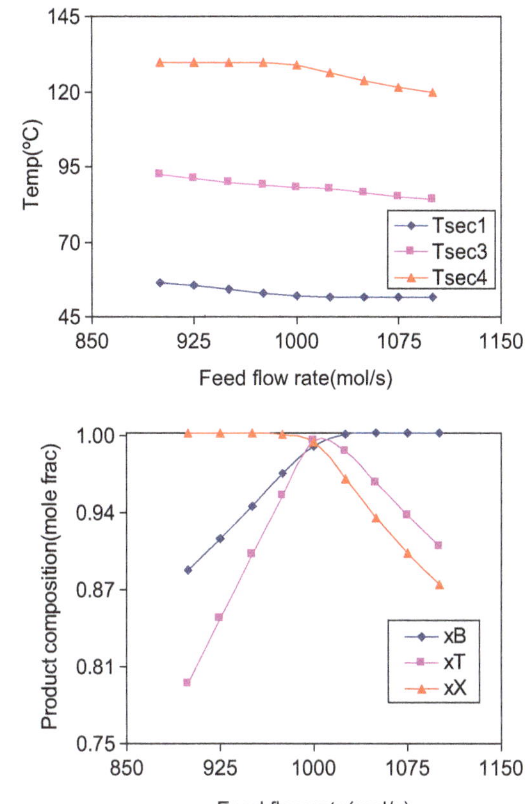

Figure 6. Effect of feed flow rate on temperature and product composition.

mol% composition of the *o*-xylene in the feed. Therefore, the temperature in the bottom section (stripping section) increases and the temperatures in the top section (rectifying section) and middle section decrease slightly.

4.4. *Effect of feed flow rate*

As the feed flow rate increases up to 1000 mol/s, benzene and toluene compositions in the product increase, but the xylene composition remains maximum at almost 100%. On increasing the feed flow rate further, the benzene composition remains constant; however, other two product compositions decrease as shown in Figure 6. Accordingly, there is a small decrease in the temperature of the rectifying section, while the temperatures of the main column and stripping section decrease significantly.

4.5. *Effect of reflux rate*

The reflux rate plays an important role in the product purities as shown in Figure 7. Below reflux rate of 860 mol/s, benzene and toluene compositions increase and *o*-xylene composition in the bottom product remains constant at almost 100%. For more than reflux rate of 860 mol/s, benzene composition of the top product remains

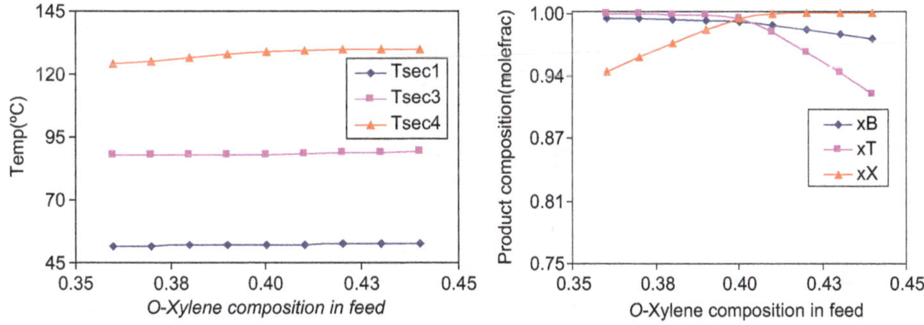

Figure 5. Effect of *o*-xylene composition in feed on temperature and product composition.

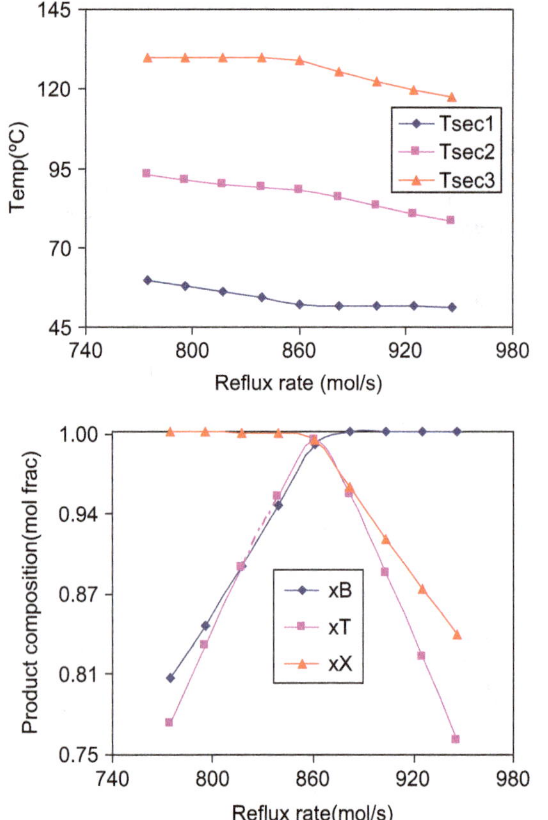

Figure 7. Effect of reflux rate on temperature and product composition.

constant, but *o*-xylene and toluene compositions decrease sharply. Accordingly, the temperature in the rectifying section (Tsec1) initially decreases and then becomes constant after 860 mol/s of reflux rate, while the temperature in the stripping section (Tsec4) decreases after 860 mol/s of reflux rate.

4.6. *Effect of side stream flow rate*

For side stream flow rate less than 296 mol/s, benzene composition of the top product does not change much for a change in the side stream flow rate; however, toluene composition in the middle product decreases after 296 mol/s as shown in Figure 8. The *o*-xylene composition in the bottom product increases and becomes constant after 296 mol/s of side stream flow rate. Therefore, side stream flow rate of 296 mol/s is the optimum flow rate as it gives the highest purity of all the products.

4.7. *Effect of reboiler heat duty*

The effect of the reboiler heat duty is shown in Figure 9. As can be seen, benzene composition of the top product does not show significant change until 41 MW, beyond which it starts decreasing. Toluene composition first increases and then decreases, and *o*-xylene in the bottom product increases up to 41 MW after which it remains almost constant. Therefore, 41 MW is the optimum reboiler heat duty

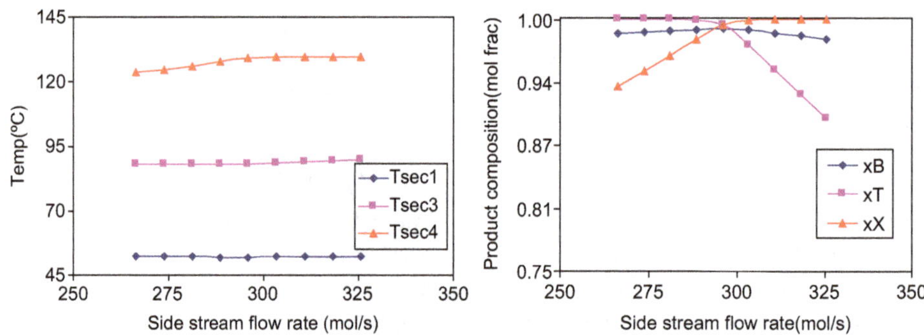

Figure 8. Effect of side stream flow rate on temperature product composition.

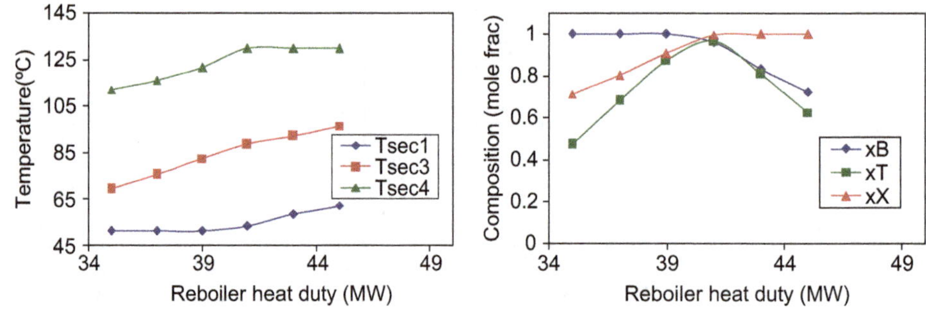

Figure 9. Effect of reboiler duty on temperature and product composition.

at which all the three products have highest purity at 99.9%. The temperatures increase on increasing reboiler heat duty.

5. Results for model predictive control of DWC

5.1. Load change in benzene composition of feed

± 10% load changes were given in the feed flow rate to observe the performance of the MPC as shown in Figure 10. Temperature overshoots in Section 1 are approximately 0.3 °C and 0.2 °C for − 10% and + 10% change, respectively. Similarly, there is only very small overshoot (only 0.5%) in the product composition. Moreover, the offset is negligibly very small in both cases.

To compare the performance of MPC with conventional controller, Ling and Luyben's (2010) results have been considered for the study. The PI controller stabilized the controlled variables in about 3–4 h; however, time taken by MPC controller is one-fourth time of that by PI controller. Temperature overshoot in Section 1 is 0.35 and 0.4 °C for − 10% and + 10% change, respectively, in case of PI controller; in case of MPC, this is marginally lower, that is, 0.3 °C and 0.2 °C, respectively. Similarly, in Section 3, temperature overshoot is 0.6 in PI control; however, in case of MPC, there is very small overshoot, that is, 0.13 °C and 0.12 °C for − 10% and + 10% load change, respectively. In Section 4 also, the overshoot is marginally better for MPC performance.

5.2. Load change in toluene composition of feed

The MPC performance for ± 10% load change in toluene feed composition is also studied as shown in Figure 11. The maximum overshoot was 0.1 °C in the temperatures. The controller brought back the temperature to their set points within approximately half an hour. Accordingly, the composition was also controlled with negligible offset.

PI control achieved stabilization in 4–5 h as reported by Ling and Luyben (2010); however, MPC stabilized in 0.6 h upon ± 10% disturbance in toluene feed composition. Temperature overshoot in Section 1 is 0.5 °C in PI and 0.1 °C in MPC Controller. Temperature overshoot in Section 3 is 0.7 in case of PI controller and 0.1 in MPC. Similarly, in Section 4, the overshoot is marginally better.

5.3. Load change in o-xylene composition of feed

The MPC performance for load change of ± 10% in *o*-xylene composition of feed is shown in Figure 12. The maximum overshoot is approximately 0.2 °C in temperature with almost no offset. The product compositions were also controlled according to their initial values. PI control achieved stabilization in 4–5 h as reported by Ling and Luyben (2010); however, MPC stabilized in 1.5 h upon ± 10% disturbance in the *o*-xylene feed composition. Temperature overshoot in Sections 1 and 3 are a little bit better in the case of MPC; however, in Section 4, MPC shows more overshoot than PI control (a difference of 0.5 °C).

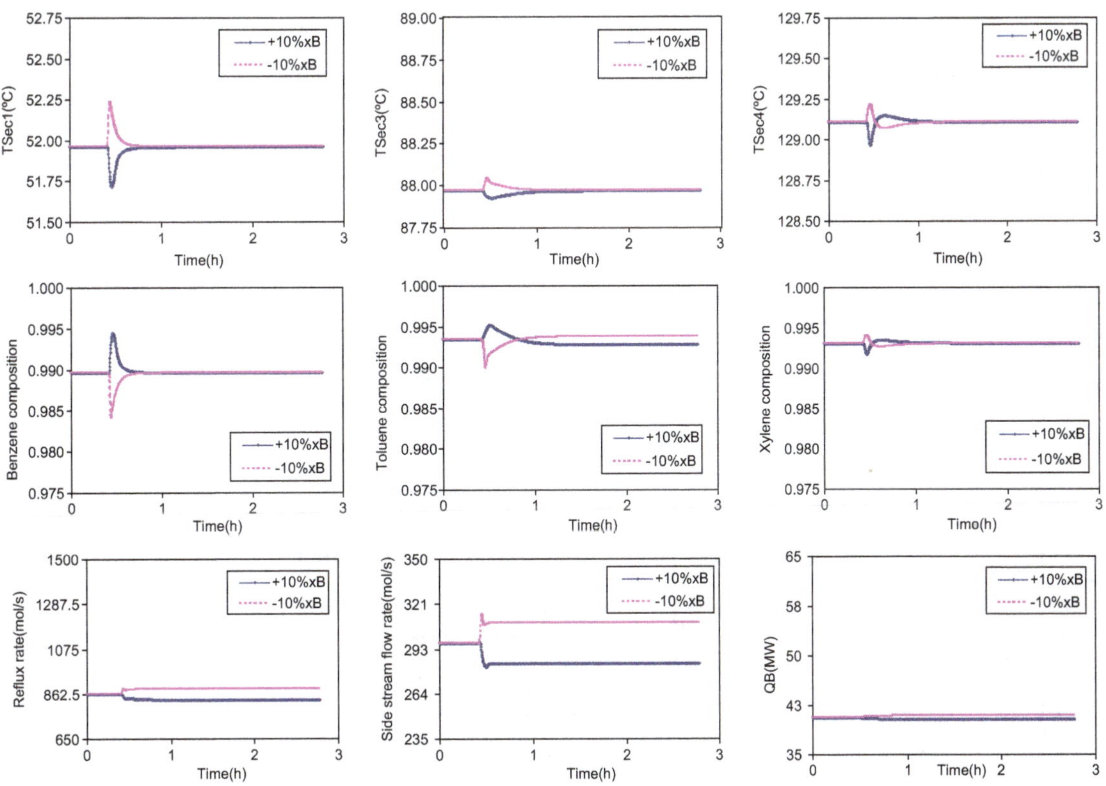

Figure 10. MPC performance for ± 10% disturbance in benzene feed composition.

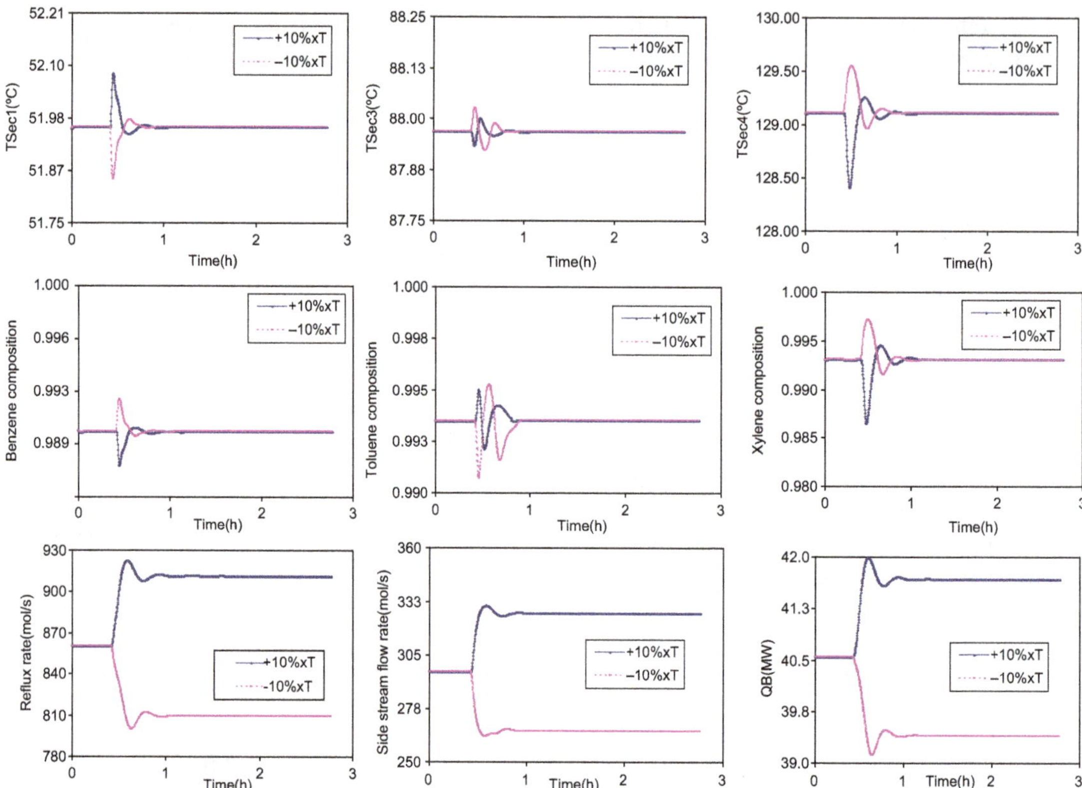

Figure 11. MPC performance for ± 10% disturbance in toluene feed composition.

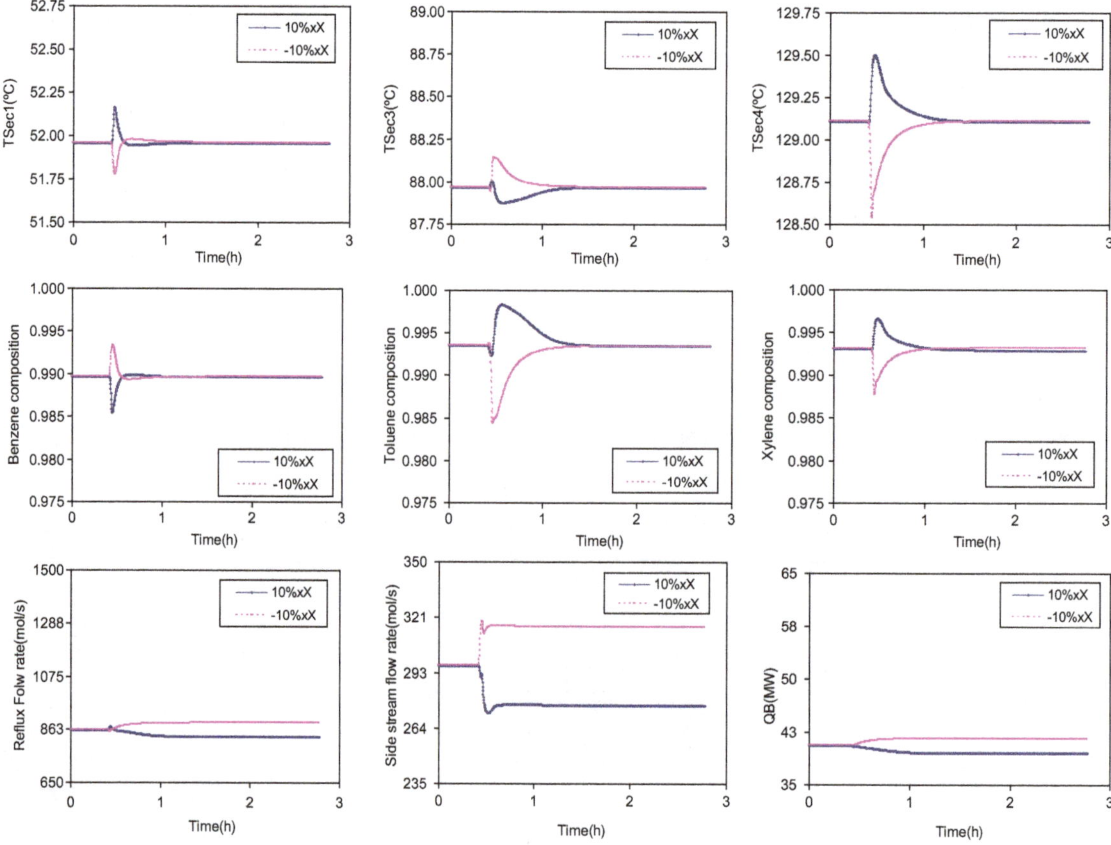

Figure 12. MPC performance for ± 10% disturbance in *o*-xylene feed composition.

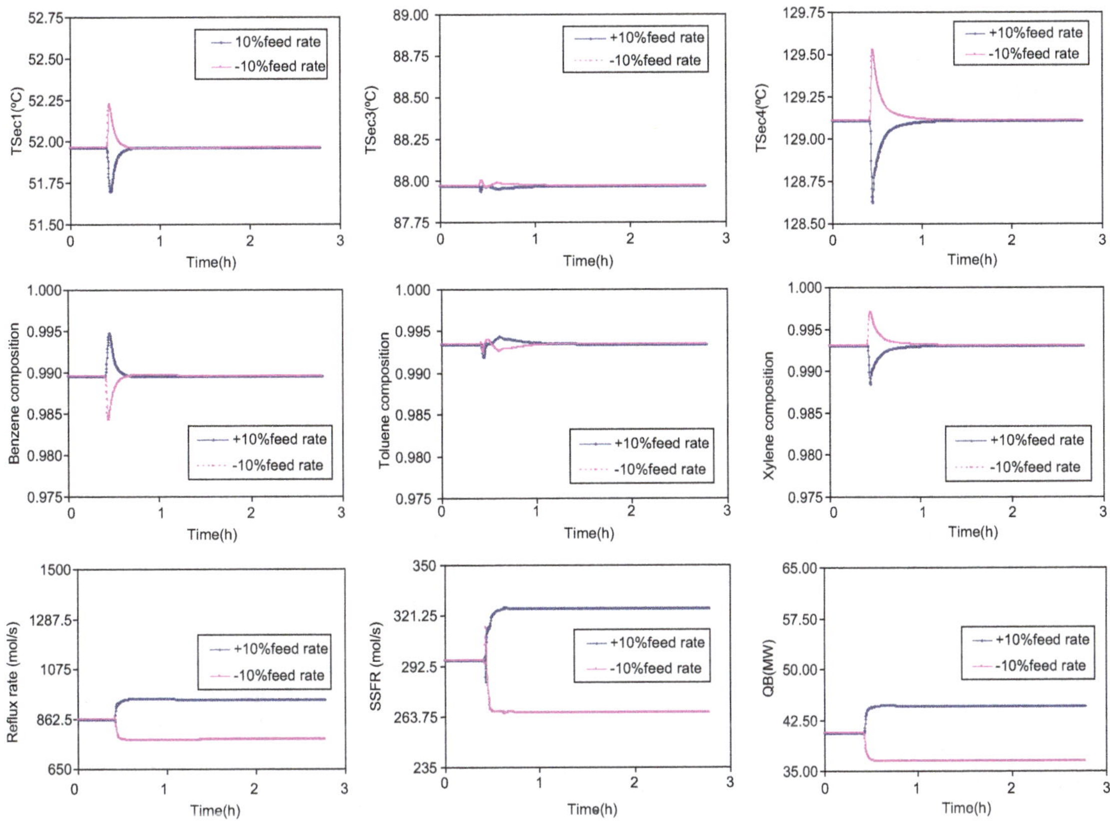

Figure 13. MPC performance for ± 10% disturbance in feed flow rate.

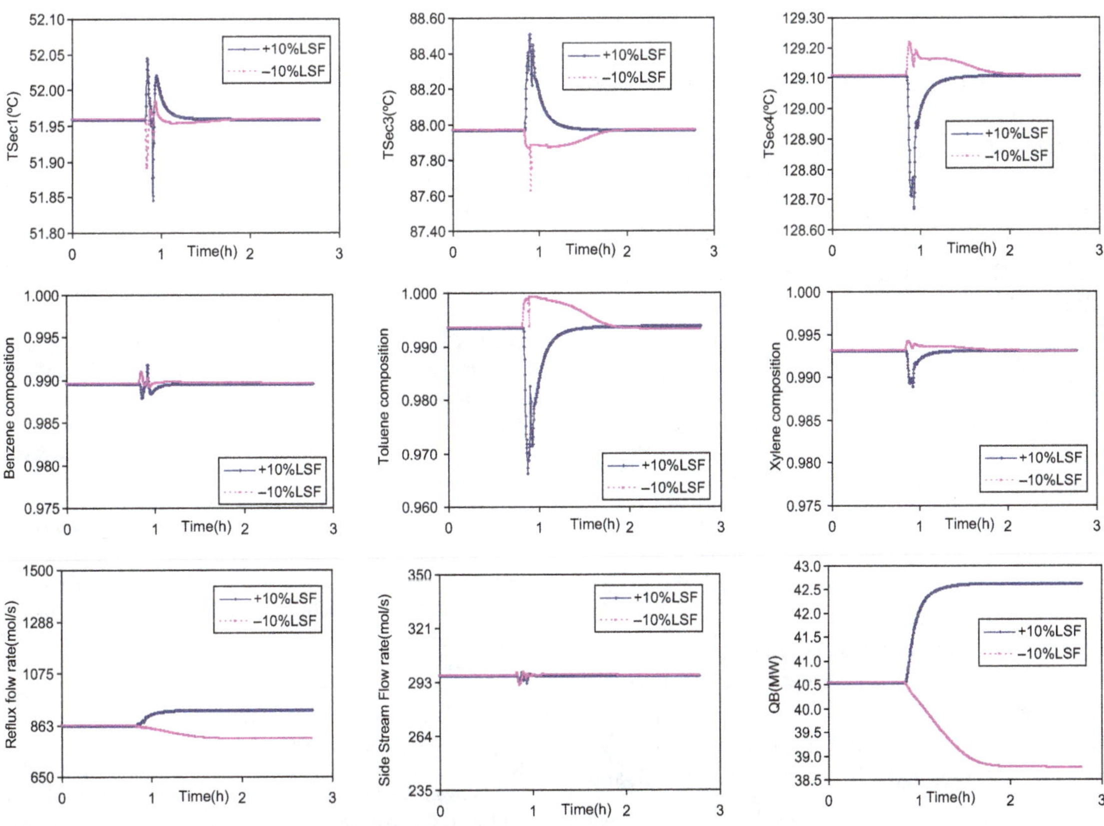

Figure 14. MPC performance for ± 10% disturbance in liquid split factor.

5.4. *Load change in feed flow rate*

The MPC response for $\pm 10\%$ load change in feed flow rate is shown in Figure 13. The temperature of all the three sections was controlled at their respective set points after maximum overshoot of 0.4 °C. The compositions were also maintained at their initial values with maximum deviation of 0.5%. PI controller stabilized in about 4–5 h (Ling and Luyben, 2010); however, MPC achieved the control performance in 1.5 h.

5.5. *Load change in liquid split factor*

The response for $\pm 10\%$ load change in liquid split factor is shown in Figure 14. The temperatures in Sections 1, 3, and 4 are controlled at their respective set points with maximum overshoot of 0.1, 0.4, and 0.4, respectively. Toluene composition shows a maximum overshoot of 0.02; other two compositions have a smaller overshoot in comparison with the toluene.

6. Conclusions

A mathematical model was developed for DWC, which was used as a case study for studying the control behavior of model predictive control. The controlled variables were selected as the temperatures of the 6th tray in the rectifying section, the 18th tray in the main column, and the 6th tray in the stripping section to ensure the maximum purity of benzene, toluene, and xylene, respectively. The manipulated variables selected are reflux rate, side stream flow rate, and reboiler heat duty. Model predictive controller was able to control the temperatures (and therefore product purities) of all the three sections in the presence of $\pm 10\%$ load changes in feed composition, flow rate, and liquid split factor. The performance of MPC was also compared with conventional controller and it is concluded that being a multivariable controller, MPC performs better than PI controller.

Nomenclature

M_j	Total liquid holdup on the jth stage, moles
$x_{j,i}$	Mole fraction of the ith component in the liquid phase at the jth stage
V_j	Vapor flow rate from the jth stage, mole/s
L_j	Liquid flow rate from the jth stage, mole/s
$y_{j,i}$	Mole fraction of the ith component in the vapor phase at the jth stage
F_j	Feed flow rate on the jth stage, mole/s
$Z_{j,i}$	Mole fraction of the ith component in feed at the jth stage
S_j	Side stream flow rate at the jth stage, mole/s
n_k	Number of stages in the kth section ($k = 1,2,3$, and 4)
n_c	Number of components $= 3$
H_{L_j}	Liquid enthalpy on the jth stage, J/mole
H_{V_j}	Vapor enthalpy on the jth stage, J/mole
$P^{sat}_{j,i}$	Saturation pressure of the ith component at the jth stage, Pa
P_j	Pressure on the jth stage, Pa
M_0	Liquid holdup in condenser, moles
$x_{D,i}$	Mole fraction of the ith component in distillate
L_0	Reflux rate, mole/s
D	Distillate rate, mole/s
H_D	Liquid enthalpy of distillate, joule/mol
q_C	Condenser duty, J/s
M	Control horizon
M_R	Liquid holdup in reboiler, mole
N_u	Number of manipulated variables
N_y	Number of controlled variables
p	Prediction horizon
$x_{w,i}$	Mole fraction of the ith component in bottom product
w	Bottom product flow rate, mole/s
$w_{u,j}$	Weight on control move
$w_{y,j}$	Weight on controlled variable deviation
y_{sp}	Set point of controlled variables
Δu	control move of manipulated variables
$y*$	Reference value of controlled variables

Disclosure statement

No potential conflict of interest was reported by the authors.

References

Agachi, P. S., Nagy, Z. K., Cristea, M. V., & Imre-Lucaci, A. (2006). *Model based control. Case studies in process engineering*. Weinheim: Wiley-VCH.

Asprion, N., & Kaibel, G. (2010). Dividing wall columns: Fundamentals and recent advances. *Chemical Engineering and Processing: Process Intensification, 49*(2), 139–146.

Backx, T., Bosgra, O., & Marquardt, W. (2000). *Integration of model predictive control and optimization of processes*. Aachen: LPT, RWTH.

Buck, C., Hiller, C., & Fieg, G. (2011). Applying model predictive control to dividing wall columns. *Chemical Engineering & Technology, 34*(5), 663–672.

De Temmerman, J., Dufour, P., Nicolaï, B., & Ramon, H. (2009). MPC as control strategy for pasta drying processes. *Computers & Chemical Engineering, 33*(1), 50–57.

Halvorsen, I. J., & Skogestad, S. (2004). Shortcut analysis of optimal operation of Petlyuk distillation. *Industrial & Engineering Chemistry Research, 43*(14), 3994–3999.

Hiller, C., Buck, C., Ehlers, C., & Fieg, G. (2010). Nonequilibrium stage modeling of dividing wall columns and experimental validation. *Heat and Mass Transfer, 46*(10), 1209–1220.

Ignat, R., & Woinaroschy, A. (2011). Dynamic analysis and controllability of dividing-wall distillation columns. *Chemical Engineering, 25*, 647–652.

Kaibel, G. (1987). Distillation columns with vertical partitions. *Chemical Engineering & Technology, 10*(1), 92–98.

Kolbe, B., & Wenzel, S. (2004). Novel distillation concepts using one-shell columns. *Chemical Engineering and Processing: Process Intensification, 43*(3), 339–346.

Kvernland, M., Halvorsen, I., & Skogestad, S. (2010). Model predictive control of a Kaibel distillation column. *In: Proceedings of the 9th International Symposium on Dynamics and Control of Process Systems (DYCOPS)*, 539–544.

Ling, H., & Luyben, W. L. (2009). New control structure for divided-wall columns. *Industrial & Engineering Chemistry Research*, *48*(13), 6034–6049.

Ling, H., & Luyben, W. L. (2010). Temperature control of the BTX divided-wall column. *Industrial & Engineering Chemistry Research, 49*(1), 189–203.

Rodríguez Hernández, M., & Chinea-Herranz, J. A. (2012). Decentralized control and identified-model predictive control of divided wall columns. *Journal of Process Control, 22*(9), 1582–1592.

Rewagad, R. R., & Kiss, A. A. (2012). Dynamic optimization of a dividing-wall column using model predictive control. *Chemical Engineering Science*, *68*(1), 132–142.

Schultz, M. A., Stewart, D. G., Harris, J. M., Rosenblum, S. P., Shakur, M. S., & O'Brien, D. E. (2002). Reduce costs with dividing-wall columns. *Chemical Engineering Progress*, *98*(5), 64–71.

Segovia-Hernandez, J. G., Hernandez-Vargas, E. A., & Marquez-Munoz, J. A. (2007). Control properties of thermally coupled distillation sequences for different operating conditions. *Computers & Chemical Engineering*, *31*(7), 867–874.

Skogestad, S., & Postlethwaite, I. (2005). *Multivariable feedback control. Analysis and design*. 2nd ed. Chichester: John Wiley.

Sotudeh, N., & Shahraki, B. H. (2007). A method for the design of divided wall columns. *Chemical Engineering & Technology*, *30*(9), 1284–1291.

Triantafyllou, C., & Smith, R. (1992). The design and optimisation of fully thermally coupled distillation columns: Process design. *Chemical Engineering Research & Design*, *70*(A2), 118–132.

Van Diggelen, R. C., Kiss, A. A., & Heemink, A. W. (2010). Comparison of control strategies for dividing-wall columns. *Industrial & Engineering Chemistry Research*, *49*(1), 288–307.

Woinaroschy, A., & Isopescu, R. (2010). Time-optimal control of dividing-wall distillation columns. *Industrial & Engineering Chemistry Research*, *49*(19), 9195–9208.

13

Robust fault diagnosis for an exothermic semi-batch polymerization reactor under open-loop

Abdelkarim M. Ertiame[a*], Dingli Yu[a], Feng Yu[b] and J.B. Gomm[a]

[a]*Control System Research Group, School of Engineering, Liverpool John Moores University, Byrom Street, Liverpool L3 3AF, UK;* [b]*School of Electronic Information, Changchun Architecture &Civil Engineering College, Changchun, Jilin Province, People's Republic of China*

An independent radial basis function neural network (RBFNN) is developed and employed here for an online diagnosis of actuator and sensor faults. In this research, a robust fault detection and isolation scheme is developed for an open-loop exothermic semi-batch polymerization reactor described by Chylla–Haase. The independent RBFNN is employed here for online diagnosis of faults when the system is subjected to system uncertainties and disturbances. Two different techniques to employ RBFNNs are investigated. Firstly, an independent neural network (NN) is used to model the reactor dynamics and generate residuals. Secondly, an additional RBFNN is developed as a classifier to isolate faults from the generated residuals. Three sensor faults and one actuator fault are simulated on the reactor. Moreover, many practical disturbances and system uncertainties, such as monomer feed rate, fouling factor, impurity factor, ambient temperature and measurement noise, are modelled. The simulation results are presented to illustrate the effectiveness and robustness of the proposed method.

Keywords: robust fault detection; independent RBF model; RBF neural networks; open-loop Chylla–Haase reactor

1. Introduction

In recent years, the task of monitoring complex non-linear processeslts has been intensively studied. Fault detection and isolation (FDI) techniques have attracted much interest due to the increasing demand for good performance and higher standards of safety and reliability of technical plants for improving the supervision and monitoring as part of the overall control of processes (Isermann, 1984). FDI has become a critical issue in the operation of high-performance chemical plants, nuclear plants, airplanes, ships, submarines and space vehicles (Gertler, 1988). In the chemical industry, faults can occur due to sensor failures, equipment failures or changes in process parameters. The occurrence of a fault may cause process performance degradation, or in the worst cases, may cause disastrous accidents. However, FDI can help avoid all these major consequences (Deibert & Isermann, 1992).

Due to several non-linearity and time-varying feature of the reactor dynamics, the observer methods, parity space methods and other first-principle model-based methods cannot be successfully applied for FDI of the Chylla–Haase reactor.

Many research works have been carried out to study neural networks (NNs) for FDI. Yu, Gomm, and Williams (1999) studied sensor fault diagnosis in a chemical process via radial basis function neural networks (RBFNNs); a semi-independent NN was used for sensor fault diagnosis. Moreover, the thins-plate-spline function was used for the neural model and the Gaussian function was used for the neural classifier. Another study was conducted by Gomm and Yu (2000) that introduced the selection of RBF network centres with recursive orthogonal least squares training. Frank and Köppen-Seliger (1997) and Koppen-Seliger and Frank (1995) studied fuzzy logic and NN applications for fault diagnosis. Their paper introduced fuzzy logic for residual evaluation, a dependent NN for residual generation and a NN for residual evaluation by using another dependent NN for generating residuals. All those authors used dependent and semi-dependent mode of NN for FDI. As the residual of these methods is affected by the plant output, the residual is made insensitive to the faults. Although a partial dependent mode is used to enhance the residual to fault sensitivity, the fault detect threshold is still high such that fault with small amplitude cannot be detected.

Patton, Chen, and Siew (1994) proposed an approach for detecting and isolating faults in a non-linear dynamic process using NNs. Firstly, a multi-layer perceptron (MLP) network was trained to predict the future system states, then the residual was generated using the differences between the actual and predicted states. Secondly, another NN was used as a classifier to isolate faults from these

*Corresponding author. Email: A.M.Ertiame@2011.ljmu.ac.uk

state prediction errors. However, this method used the NN model in its so-called dependent mode.

Ferrari, Parisini, and Polycarpou (2008), Xiaodong (2011), Xiaodong, Polycarpou, and Parisini (2002) and Zhang, Polycarpou, and Parisini (2010) studied the design and analysis of a robust FDI scheme for non-linear uncertain dynamic systems, the proposed architecture consists of a bank of non-linear adaptive estimator, one of the estimators is used for the detection and approximation of a fault, whereas the rest are used for an online fault isolation decision scheme which is based on adaptive threshold functions. In their method, they used a state-space non-linear model and then used a simple NN as an estimator for online learning. The model output must be equal to the plant output; however, this method needs to have a plant non-linear dynamic model and sometimes the model needs to be very accurate; this accurate model is difficult to produce. Most of the recent investigations of fault diagnosis for chemical reactors using an independent RBFNN have been studied by Ertiame, Dingli, Feng, and Gomm (2013).

In this research, a new robust FDI scheme is developed for an open-loop Chylla–Haase polymerization reactor using an independent RBFNN. The independent RBFNN is employed here for online diagnosis of faults on the actuator and sensors when the system is subjected to system uncertainties and disturbances. The independent NN mode is developed to generate enhanced residuals for diagnosing faults in the reactor. Then, a second NN is developed as a classifier to isolate these faults. The basis Gaussian function is used for the NN model and for the NN classifier. The K-means clustering algorithm is used to choose the centres of the RBF networks, and a p-nearest-neighbours algorithm is used to choose the widths. Moreover, a recursive least squares (RLS) algorithm is used to update the weights.

2. The Chylla–Haase benchmark reactor

Batch and semi-batch reactors have been widely used in the chemical industry. In this research, a semi-batch polymerization reactor benchmark is considered which is described by Chylla and Haase (1993) and used as a benchmark for process control applications. The schematic diagram of the semi-batch polymerization reactor is shown in Figure 1 (Chylla & Haase, 1993). It consists of a stirred tank reactor with a cooling jacket and a coolant recirculation. The reactor temperature is controlled by manipulating the temperature of the coolant, which is recirculated through the cooling jacket of the reactor. The heat released through the reaction must be removed by circulating cold water through the jacket, where both hot and cold jacket streams are available. When the jacket temperature controller output is between 0% and 50%, the valve is opened and cold water is inserted, and when the jacket controller output is between 50% and 100%, the valve is opened and steam is inserted (Beyer, Grote, & Reinig, 2008; Graichen, Hagenmeyer, & Zeitz, 2005).

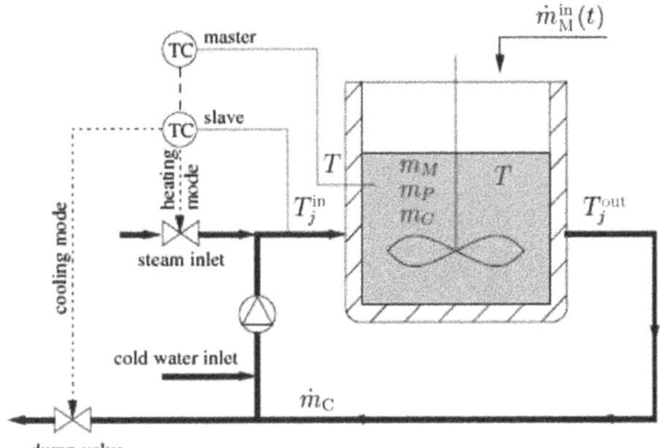

Figure 1. Chylla–Haase reactor schematic.

2.1. *Polymerization reactor model*

The mathematical model of the Chylla–Haase reactor is described by a set of five ordinary differential equations (ODEs) which come from material and heat balances inside the reactor:

$$\frac{dm_M}{dt} = \dot{m}_M^{in}(t) + \frac{Q_{rea}}{\Delta H}, \tag{1}$$

$$\frac{dm_P}{dt} = -\frac{Q_{rea}}{\Delta H}, \tag{2}$$

$$\frac{dT}{dt} = \frac{1}{\sum_i m_i C_{p,i}}[\dot{m}_M^{in}(t)C_{p,M}(T_{amb} - T) - UA(T - T_j) - (UA)_{loss}(T - T_{amb}) + Q_{rea}], \tag{3}$$

$$\frac{dT_{j\,out}}{dt} = \frac{1}{m_C\, C_{p,C}}[\dot{m}_C C_{p,C}(T_{j\,in}(t-\theta_1) - T_{j\,out}) + UA(T - T_j)], \tag{4}$$

$$\frac{dT_{j\,in}}{dt} = \frac{dT_{j\,out}(t-\theta_2)}{dt} + \frac{T_{j\,out}(t-\theta_2) - T_{j\,in}}{\tau_p} + \frac{K_p(c)}{\tau_p}. \tag{5}$$

The reactor model includes the material balances (1) and (2) for the monomer mass $m_M(t)$ and the polymer mass $m_P(t)$, the energy balance (3) with the reactor temperature $T(t)$, and the energy balances (4) and (5) of the cooling jacket and the recirculation loop with the outlet and inlet temperatures $T_{j\,in}(t)$ and $T_{j\,out}(t)$ of the coolant. The available measurements of the process are the temperature of the reactor and the cooling circuitry (Graichen, Hagenmeyer, & Zeitz, 2006):

The heating/cooling function $K_p(c)$ is influenced by an equal-percentage valve with valve position $c(t)$ as shown

Table 1. Variables and parameters of the reactor model.

$\dot{m}_{\mathrm{M}}^{\mathrm{in}}(t)$	Monomer feed rate (kg s^{-1})
$Q_{\mathrm{rea}} = -\Delta H \cdot R_{\mathrm{P}}$	Reaction heat (kW)
R_{P}	Rate of polymerization (kg s^{-1})
$-\Delta H$	Reaction enthalpy (kJ kg^{-1})
U	Overall heat transfer coefficient (kW m^{-2} K^{-1})
A	Jacket heat transfer area (m^2)
$(UA)_{\mathrm{loss}}$	Heat loss coefficient (kW K^{-1})
$C_{p,\mathrm{M}}, C_{p,\mathrm{P}}, C_{p,\mathrm{C}}$	Specific heat at constant pressure (kJ kg^{-1} K^{-1})
θ_1, θ_2	Transport delay (s)
$T_j = (T_{j\,\mathrm{in}} + T_{j\,\mathrm{out}})/2$	Average cooling jacket temperature (K)
$K_{\mathrm{p}}(c)$	Heating/cooling function (K)
τ_{p}	Heating/cooling time constant (s)

in the following equation:

$$K_p(c) = \begin{cases} 0.8 \times 30^{-c/50}(T_{\mathrm{inlet}} - T_{j\,\mathrm{in}}(t)), & c < 50\% \\ 0, & c = 50\%, \\ 0.15 \times 30^{(c/50-2)}(T_{\mathrm{steam}} - T_{j\,\mathrm{in}}(t)), & c > 50\% \end{cases} \tag{6}$$

For $c < 50\%$, cold water with inlet temperature T_{inlet} is injected into the cooling jacket, whereas a valve position $c > 50\%$ leads to a heating of the coolant by injecting steam with temperature T_{steam} into the recirculating water steam. Moreover, the variables and the parameters of the reactor model are listed in Table 1 (Graichen et al., 2006).

2.2. *Uncertainties and disturbances in the process*

In order to model the following practical issues of the control of polymerization reactors, various disturbances and uncertainties are identified:

- The impurity factor $i \in [0.8, 1.2]$ in the polymerization rate R_{P} is random but constant during one batch, which tries to simulate fluctuations in monomer kinetics caused by batch-to-batch variations in reactive impurity.
- The fouling factor $1/h_{\mathrm{f}}$ in the overall heat transfer coefficient U increases with each batch and accounts for the fact that during successive batches a polymer film builds up on the wall resulting in a decrease of U.
- The delay times θ_1 and θ_2 of the cooling jacket and the recirculation loop may vary by $\pm 25\%$ compared to nominal values.
- The ambient temperature T_{amb} is different during summer and winter. This affects the temperature of the monomer feed $\dot{m}_{\mathrm{M}}^{\mathrm{in}}$, as well as the initial conditions $T(0)$, $T_{j\,\mathrm{in}}(0)$ and $T_{j\,\mathrm{out}}(0)$ given by T_{amb} (Graichen et al., 2006).

Table 2 describes the empirical relations for the polymerization rate, the jacket heat transfer area and the overall heat transfer coefficient (Graichen et al., 2006).

3. Residual generation with RBF model

3.1. *Independent model of RBF modelling*

Using RBFNN for modelling, a non-linear dynamic system can be modelled in two modes: a dependent mode and an independent mode as shown in Figures 2 and 3, respectively. The first model referred to is a dependent mode, since the past system output is used as a network input. Thus, the model is dependent on the system output and cannot operate independently from the system. In the independent mode, the past model output is used as a network input. Therefore, the model is not dependent on the system output and can operate independently from the system. The independent model has an advantage in that the model can be used to simulate the system to obtain long-range prediction. In contrast, the dependent model performs as one-step-ahead predication.

Table 2. Empirical relations.

$R_{\mathrm{P}} = ikm_{\mathrm{M}}$	i	Impurity factor (–)
	$k = k_0 \exp(-E/RT) \cdot (k_1\mu)^{k_2}$	First-order kinetic constant (s^{-1})
	$\mu = c_0 \exp(c_1 f) \times 10^{c_2(a_0/T - c_3)}$	Batch viscosity (kg m^{-1} s^{-1})
	$f = m_{\mathrm{P}}/(m_{\mathrm{M}} + m_{\mathrm{P}} + m_{\mathrm{C}})$	Auxiliary variable (–)
	$k_0, k_1, E, R, a_0, c_0, c_1, c_2, c_3$	Constants
	R	Natural gas (kJ kmol^{-1} K^{-1})
$A = \left(\frac{m_{\mathrm{M}}}{\rho_{\mathrm{M}}} + \frac{m_{\mathrm{P}}}{\rho_{\mathrm{P}}} + \frac{m_{\mathrm{W}}}{\rho_{\mathrm{W}}}\right)\frac{P}{B_1} + B_2$	$\rho_{\mathrm{M}}, \rho_{\mathrm{P}}, \rho_{\mathrm{W}}$	Densities (kg m^{-3})
	B_1	Reactor bottom area (m)
	P	Jacket perimeter (m)
	B_2	Jacket bottom area (m^2)
$U = 1/(h^{-1} + h_{\mathrm{f}}^{-1})$	$h = d_0 \exp(d_1 \mu_{\mathrm{wall}})$	heat transfer coefficient (kW m^{-2} K^{-1})
	$\mu_{\mathrm{wall}} = c_0 \exp(c_1 f) \times 10^{c_2(a_0/T_{\mathrm{wall}} - c_3)}$	Wall viscosity (kg m^{-1} s^{-1})
	$T_{\mathrm{wall}} = (T + T_j)/2$	Wall temperature (K)
	h_{f}^{-1}	Fouling factor (m^2 K kW^{-1})
	d_0, d_1	Constants

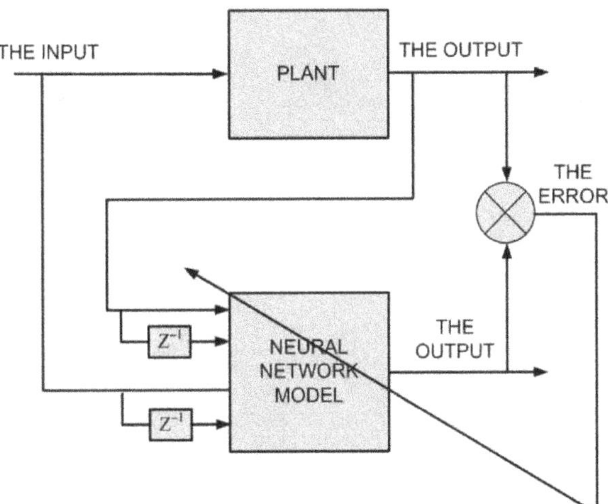

Figure 2. Dependent mode.

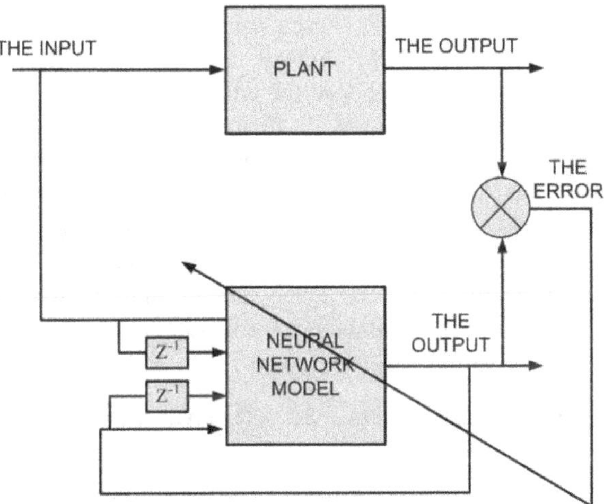

Figure 3. Independent mode.

The non-linear dynamic plant to be modelled is presented by the non-linear autoregressive with exogenous inputs (NARX) model as follows:

$$y(t) = f\,[y(t-1), \ldots, y(t-n_y), u(t-1-d), \ldots, u(t-n_u-d)] + e(t), \tag{7}$$

where $u \in \Re^m$ and $y \in \Re^p$ are the plant input and output, respectively. $e \in \Re^p$ is the random noise, m and $p$are the number of plant inputs and outputs, respectively, n_y and n_u are the maximum lags in the model output and input, respectively, $d$is the time delay in inputs and $f(*)$ is a vector-valued non-linear function.

The dependent mode of the network model can be represented by Equation (8), which is referred to dependent mode as the prediction uses the process output and, therefore the model cannot run independent of the process

$$\hat{y}(t) = \hat{f}\,[y(t-1), \ldots, y(t-n_y),\ u(t-1-d), \ldots, u(t-n_u-d)], \tag{8}$$

where $\hat{f}(*)$ is a function approximation of $f(*)$. If the past process outputs in the network input are replaced by the network outputs as in Equation (9), then the model is referred to an independent model

$$\hat{y}(t) = \hat{f}\,[\hat{y}(t-1), \ldots, \hat{y}(t-n_y),\ u(t-1-d), \ldots, u(t-n_u-d)]. \tag{9}$$

The RBF network performs non-linear mapping, and is used because of its advantages over the MLP network of short training time. The RBFNN in this research consists of three layers: an input layer, a hidden layer and an output layer. The hidden layer contains a number of RBF neurons; each of them represents a single RBF, with associated centre and width, and calculates the Euclidean distance between centre c and RBF network input vector x defined by $\| x(t) - c_j(t) \|$, where $c_j(t)$ is jth centre and $x(t)$ is the NN input vector which is given as follows:

$$x(t) = f\,[y(t-1), \ldots, y(t-n_y),\ u(t-1-d), \ldots, u(t-n_u-d)]. \tag{10}$$

Then, the output of the hidden layer nodes is produced by a so-called non-linear activation function $\varphi_j(t)$. In this work, the Gaussian basis function is chosen as the non-linear activation function

$$\varphi_j(t) = \exp\left(-\frac{\| x(t) - c_j(t) \|^2}{\sigma_j^2}\right), \quad j = 1, \ldots, n_h, \tag{11}$$

where σ_j is a positive scalar called a width and n_h is the number of centres. The network outputs are then computed as a linear weighted sum of the hidden node outputs and bias as follows:

$$\hat{y}_i(t) = \sum_{j}^{n_h} \varphi_j(t)^T w_{ji}, \quad i = 1, \ldots, q, \tag{12}$$

where w_{ji} is the output layers weight connecting the jth centre output and ith network output, and q is the number of outputs.

3.2. *Input–output determination of RBF model*

The first step towards developing a NN model of the process is to obtain training data. The training data are obtained by designing a set of random amplitude signals (RAS) for the five inputs to the reactor: monomer feed rate, fouling factor, ambient temperature, impurity factor and valve position, as shown in Figure 4. These five inputs are defined as, the system inputs (monomer feed rate and manipulated variable) including the uncertainties and disturbances in the process (fouling factor, ambient temperature and impurity factor). The second step towards developing a NN model of the process is to determine the

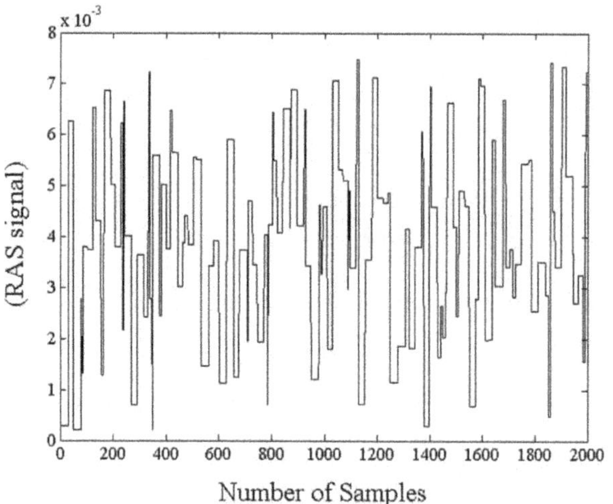

Figure 4. Random amplitude signal.

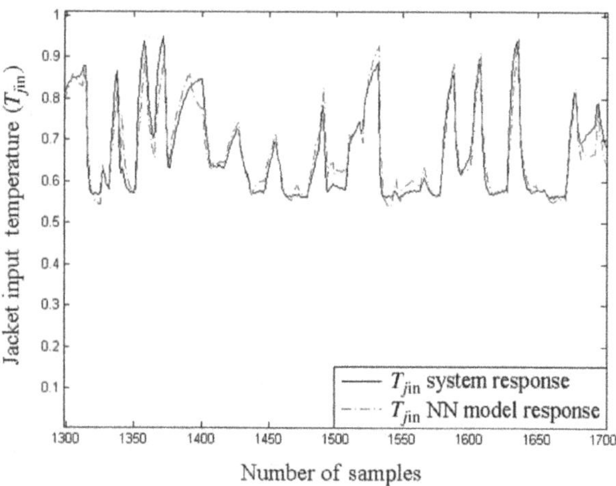

Figure 5. Simulation result of jacket input temperature system output and RBF model.

network input variables and the input vector and the output vector. The network input vector consists of the past values of the five system inputs and the past values of the three system outputs. The determination of the inputs and outputs of the system is based on Equations (1)–(5). A total data set of 2000 samples is collected from the system Simulink model, and 4 s are used as the sampling time. The first 1500 samples are used for training the network model and the remaining 500 samples are used for testing the network model. Before training and testing, the raw data are scaled linearly into the range of [0 1] using the following formulae:

$$u = \begin{bmatrix} m_M \\ 1/h_f \\ T_{amb} \\ i \\ c \end{bmatrix}, \quad y = \begin{bmatrix} T_{j\,in} \\ T_{j\,out} \\ T \end{bmatrix}, \tag{13}$$

$$u_{scaled}(k) = \frac{u(k) - u_{min}}{u_{max} - u_{min}}, \quad y_{scaled}(k) = \frac{y(k) - y_{min}}{y_{max} - y_{min}}. \tag{14}$$

3.3. *RBF model training data acquisition for open-loop and validation*

In this research, an independent RBF network is used to represent the NARX model in Equation (9). Thus, in order to get a good training result with a minimum modelling error, several numbers of maximum lags in the outputs and inputs and several numbers of the maximum time delay in the inputs are tried. The maximum lags in the output were selected as 3, the maximum lags in the input were selected as 2 and the maximum time delay in the inputs is selected as 2, as described in Equation (15). Thus, the RBF model is designed to have 19 inputs and 3 outputs, as shown in Figure 12. The hidden layer nodes are selected as 21. The centres are chosen using a K-means clustering algorithm as 21. Moreover, a p-nearest-neighbours algorithm is used to choose the widths. In the training of the network model, the RLS algorithm is used to update the weight matrix since the weights are linearly related to the output, and the parameters of the RLS algorithm are selected as follows: $\mu = 0.999$, $w(0) = 10^{-6} * U(n_h, 3)$ and $p(0) = 10^6 * I(n_h)$, where μ is the forgetting factor, I is an identity matrix, U is the element unity matrix and n_h is the number of hidden layer nodes:

$$x(t) = [y(t-1) \quad y(t-2) \quad y(t-3) \quad u(t-k-1) \quad u(t-k-2)]^T. \tag{15}$$

Figures 5–7 show the last 200 sample intervals in the training data set and the first 200 sample intervals in the testing data set. It can be clearly seen that the model outputs track the system output with a small modelling error. The mean absolute error (MAE) for the jacket input temperature, jacket output temperature and reactor temperature are 0.004, 0.0054 and 0.0072, respectively.

4. Fault detection

4.1. *Simulating faults*

In this study, after training the independent RBF network model with healthy data, the model will be tested with faulty data. The faulty data are obtained by simulating different faults in the proposed reactor. These faults are classified as three sensor faults and one actuator fault. The sensor faults are jacket input temperature sensor fault, jacket output temperature sensor fault and reactor temperature sensor fault, and the actuator fault is the inlet temperature. These faults are simulated in the following sections.

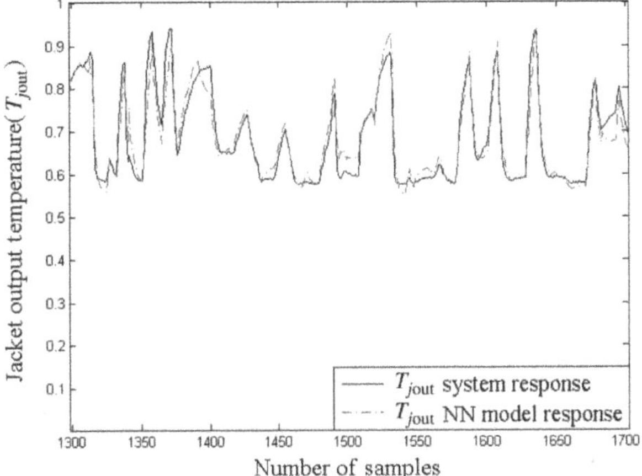

Figure 6. Simulation result of jacket output temperature system output and RBF model output.

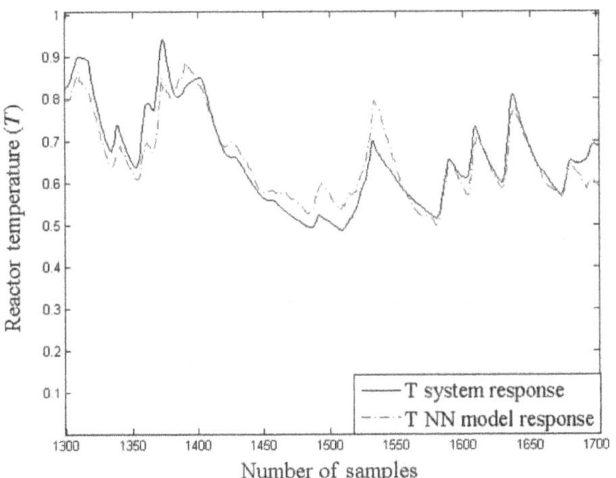

Figure 7. Simulation result of reactor temperature system output and RBF model output.

4.1.1. *Simulating sensor faults*

The jacket input temperature sensor fault is superimposed with 10% change in the measured jacket input temperature, and simulated from the sample number 400 to 500, as shown in Figure 8. Additionally, the jacket output temperature sensor fault is superimposed with 10% change in the measured jacket output temperature, and simulated from the sample number 600 to 700, as shown in Figure 8. Furthermore, the sensor fault of the reactor temperature is superimposed with 10% change in the measured temperature, and simulated from the sample number 800 to 900, as shown in Figure 8.

4.1.2. *Simulating actuator fault*

The heating–cooling function is influenced by an equal-percentage valve with valve position. When the valve position $c < 50\%$, cooling water with inlet temperature (278.71 k) is inserted into the cooling jacket. When the valve position $c > 50\%$, steam with temperature (449.82

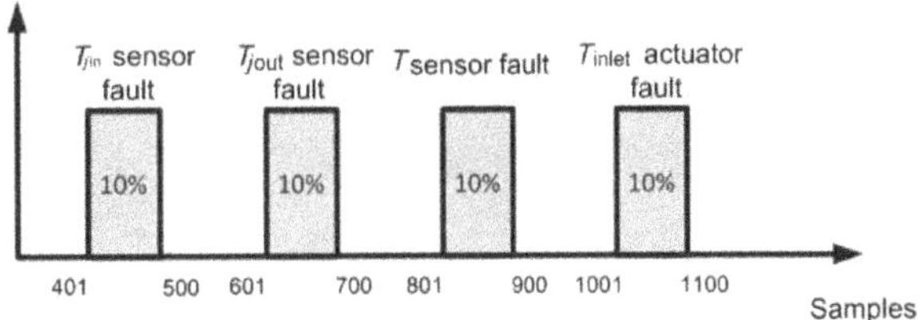

Figure 8. Fault structure with respect to the number of samples.

k) is injected into the recirculating water stream, which will lead to heating up of the coolant. Consequently, it is assumed here that a failure in the pump position of cooling mode has occurred, which leads to an increase in the temperature by 10% change in the measured inlet temperature. This inlet temperature fault is simulated from the sample number 1000 to 1100, as shown in Figure 8.

4.2. *Residual generation*

Figure 9 demonstrates the fault detection approach. An independent model is implemented in parallel with the system to generate the residuals for detecting the sensor and actuator faults in the reactor. After training the network model with healthy random data, as described in the previous section, all four faults were simulated to the reactor model. Then, with another set of 2000 samples, faulty square data are collected. These faulty data are collected by designing a set of square waves for all inputs.

These five inputs are the system inputs (monomer feed rate and manipulated variable) including the uncertainties and disturbances in the process (fouling factor, ambient temperature and impurity factor). The second step towards developing a NN model of the process is to determine the network input variables and the input vector and the output vector. The network input vector consists of the past values of the five system inputs and the past values of the three system outputs, where $m_M(t)$, $1/h_f$, T_{amb}, i and $c(t)$ are the inputs of the system; and jacket input temperature $T_{j\,in}(t)$, jacket output temperature $T_{j\,out}(t)$ and reactor temperature $T(t)$ are the outputs of the system. Moreover, the collected data are scaled linearly. After determining and scaling the input and output vectors of the system, the multivariable NARX is used to represent the non-linear dynamics of the reactor, The maximum lags in the output were selected as 3, the maximum lags in the input were selected as 2 and the maximum time delay in the inputs is selected as 2, as described in Equation (14). Here again the NN is realized by a RBF network with Gaussian basis functions. Moreover, the centres are chosen again using a K-means clustering algorithm and the widths are chosen using p-nearest-neighbours. Different numbers of hidden nodes, such as 21, 31 and 51, are used in order to get good results. The RLS algorithm is used to update the weight matrix. The parameters of the RLS algorithm

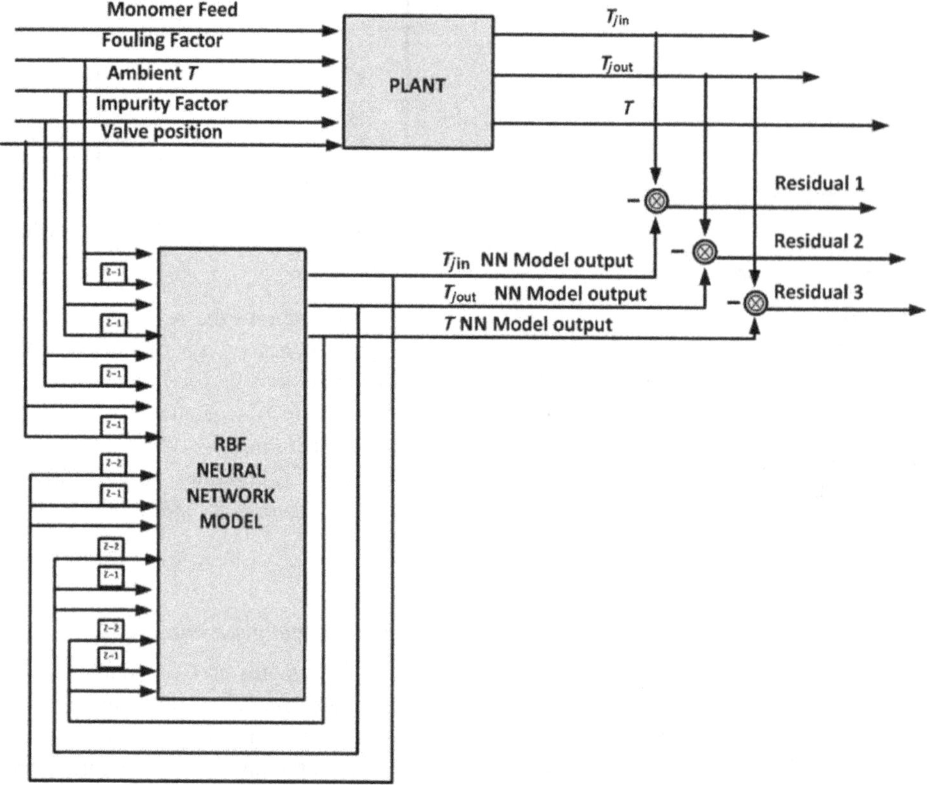

Figure 9. The structure of FD using an independent RBFNN.

are selected as follows: $\mu = 0.999$, $w(0) = 10^{-5} * U(n_h, 5)$ and $p(0) = 10^5 * I(n_h)$, where μ is the forgetting factor, I is an identity matrix, U is the element unity matrix and n_h is the number of hidden layer nodes. The RBF network model is tested with these faulty square data to generate fault-detection residuals. The filtered model prediction errors are shown in Figures 10–12. The first model prediction error of jacket input temperature is shown in Figure 10 and that for jacket output temperature and reactor temperature are shown in Figures 11 and 12, respectively. In this study, the residual ε is generated as the sum-squared filtered modelling error as follows:

$$e(t) = [y(t) - \hat{y}(t)],$$

$$\varepsilon(t) = \sqrt{(e_{T_{j\text{in}}})^2 + (e_{T_{j\text{out}}})^2 + (e_T)^2}.$$

The residuals of testing the neural model are slightly bigger than the residuals of training the neural model. The MAE index is used to evaluate the modelling effects. The MAE for the jacket input temperature, jacket output temperature and reactor temperature are 0.004, 0.0054 and 0.0072, respectively. Figure 10 demonstrates the residuals after using a low-pass filter. It can be observed that the independent network model output is not influenced by any type of fault, because an independent model does not use past faulty measurements as inputs. Thus, it can be clearly noticed that all faults have been clearly detected.

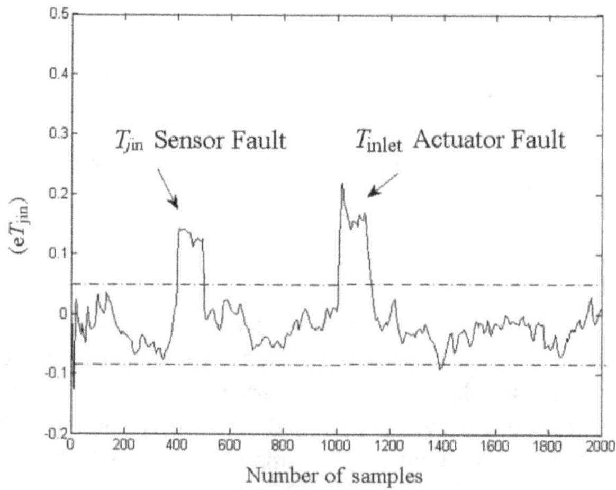

Figure 10. Residual filtered model prediction error of T_{jin}.

Moreover, no false alarms are thereby produced, so this verifies that the proposed scheme has shown an excellent diagnostic performance.

5. Fault isolation

Figure 13 illustrates the fault isolation strategy; an additional NN is applied as a classifier for fault isolation. The application of NNs for fault isolation has been used by many researchers, such as Patton et al. (1994) and Yu et al.

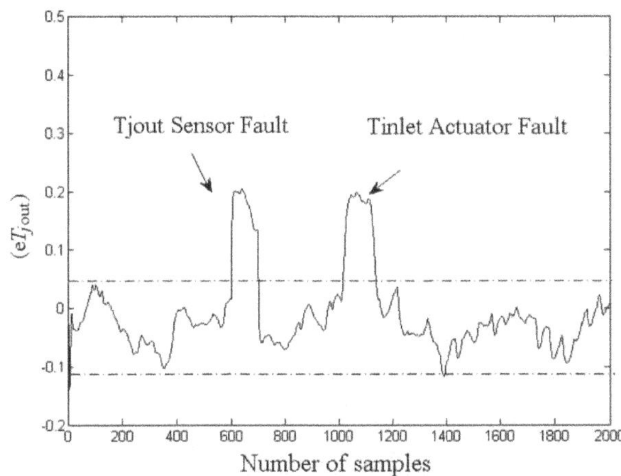

Figure 11. Residual filtered model prediction error of $T_{j\text{out}}$.

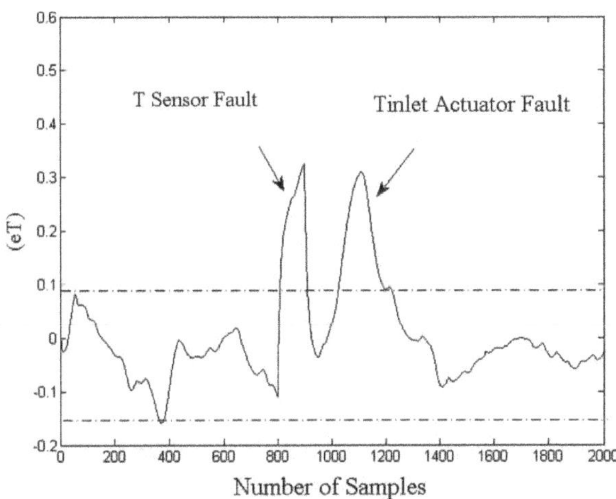

Figure 12. Residual filtered model prediction error of T.

(1999) used an RBF network. In the fault detection, a residual is generated to report a fault occurrence. However, it is difficult to identify which fault has occurred among all pre-specified possible faults using the residual, due to the fact that the residual is a scalar and carries little information about fault types. In this work, it is proposed to isolate faults according to model prediction errors. The model prediction errors are multi-dimensional, three-dimension in this case, and different faults will have different impacts on these vectors in three-dimensional vector space. Classification of these features of different faults on the model prediction error vectors will lead to classification of different faults. Therefore, the faults that have occurred can be isolated. In this work, the neural classifier is developed by an RBF network with Gaussian basis functions. The residuals shown in Figures 10–12 which are the difference between the real system output and the tested neural output were used as inputs for the RBF network classifier. Moreover, the neural classifier was developed with five outputs, with four outputs associated with the four faults and one output for no-fault case. The centres are chosen

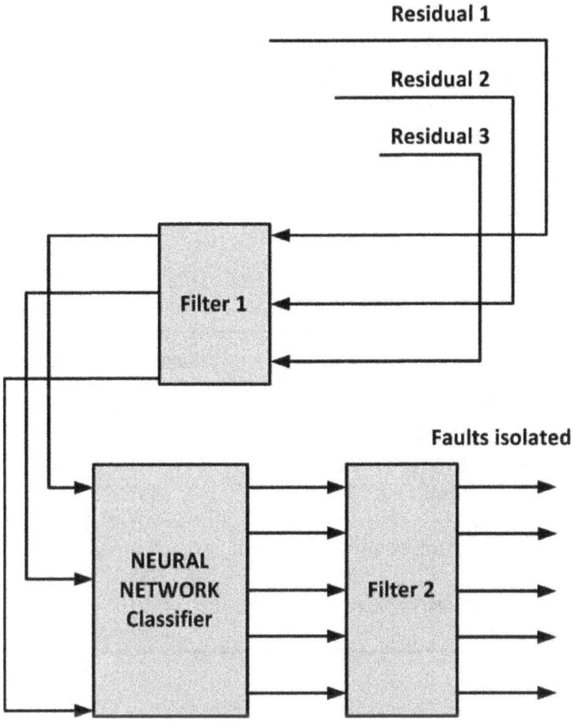

Figure 13. Block diagram for fault isolation.

again using a K-means clustering algorithm and the widths are chosen using p-nearest-neighbours. Different numbers of hidden nodes, such as 51, 151 and 251, are used in order to get good results. Finally, 51 hidden layer nodes are selected and the centres are chosen as 51. The parameters of the RLS algorithm are selected as follows: $\mu = 0.9999$, $w(0) = 10^{-6} * U(n_h, 5)$ and $p(0) = 10^6 * I(n_h)$. The samples arranged for fault occurrence are illustrated in Table 3. Moreover, the target is set such that all four outputs are set as zero for the healthy condition data, and one output is set as 1 for a specific fault, with the others remaining at zero. Thus, once the first output is 1 and the other outputs are zero, this means that the jacket input temperature sensor fault with 10% change has occurred. In the same way, the jacket output temperature sensor fault with 10% is believed to have occurred when the second output is 1, while the others remain at zero. Similarly, the reactor temperature sensor fault and the inlet temperature actuator fault with 10% changes will have occurred when the third and the fourth outputs are 1. After training, the RBF network classifier is tested with another set of faulty data with the same arrangement of training data. The samples arranged for fault occurrence can be different from those of the training data. Table 1 shows the classification of faults with respect to the number of samples. The four outputs of the neural classifier after use of a filter are displayed in Figures 14–17. It can be clearly noticed that all faults have been clearly detected and isolated.

Robust is that the fault detection always see residual sensitive to the fault but insensitive to the disturbances.

When the disturbances come in it will not affect the report of the fault, and will not increase false alarm reading. False alarm reading is that when there is no fault but fault is reported and when there is a fault but is not reported. False alarm reading should be zero percentage when all faults are reported, if there is no fault but report affected by disturbances this should be zero, but if not zero then should

Table 3. Classification of faults with respect to the number of samples.

Faults	Number of samples
No fault	0–400
$T_{j\,\mathrm{in}}$ sensor fault	401–400
No fault	501–600
$T_{j\,\mathrm{out}}$ sensor fault	601–700
No fault	701–800
Reactor temperature sensor fault	801–900
No fault	901–1000
Inlet temperature actuator fault	1001–1100
No fault	1101–2000

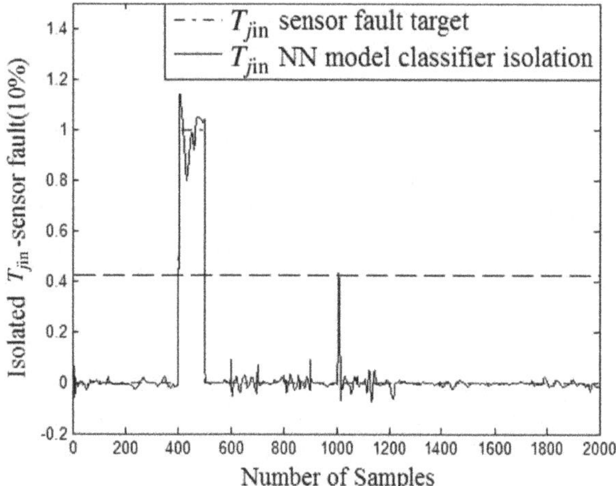

Figure 14. Classifier output 1.

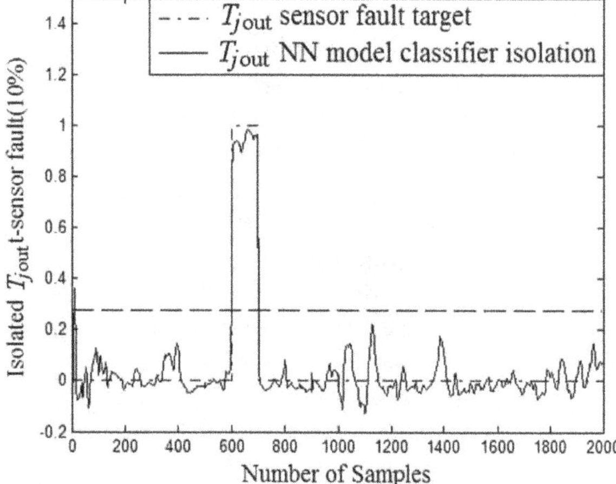

Figure 15. Classifier output 2.

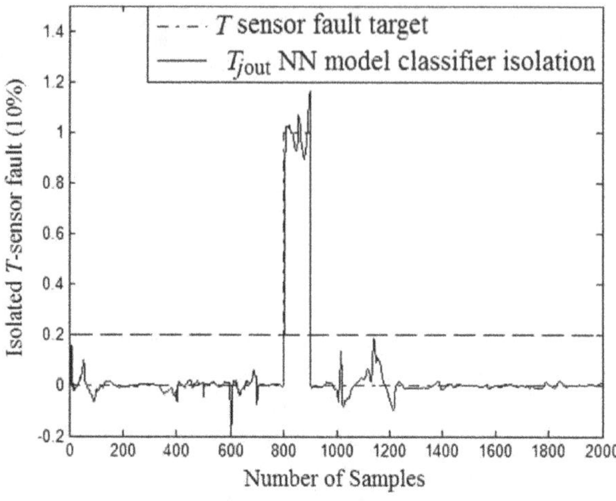

Figure 16. Classifier output 3.

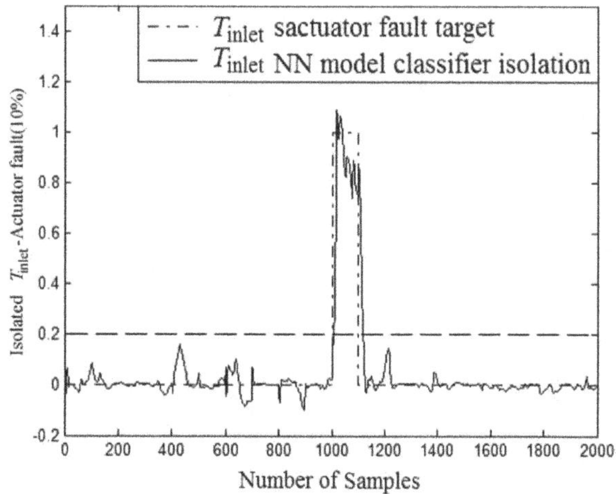

Figure 17. Classifier output 4.

be reduced as small as possible. In this research work, when collecting training data all disturbances are simulated, because of this the model is trained considering the disturbances. When disturbances occurred will not affect the residual that because the disturbances in this system are not big enough to make the residual high, in this process the disturbances just change the non-linear function of the system and that are big enough from the control point of view. It is observed from simulation results that all faults have been clearly detected and isolated, and no false alarm was thereby produced, so this verifies that the proposed scheme has shown an excellent performance. Note that the outputs are not zero when no faults occur, as a result of the effects of the disturbances.

6. Conclusion

A new robust fault diagnosis scheme has been developed for an open-loop Chylla–Haase reactor using an independent RBFNN. Three sensor faults and one actuator fault have been simulated on the Chylla–Haase reactor. All the simulated faults are superimposed with 10% changes in the measured temperatures, and simulated for different numbers of samples. Moreover, the uncertainties and disturbances in the process, such as fouling factor, impurity factor and measurement noise, have been simulated. Two different techniques to employ RBF NNs for fault diagnosis have been investigated. The first technique is implementing an independent RBNN for residual generation. Moreover, the generated residuals were used for detecting actuator and sensor faults. The second technique is applying an additional RBFNN as a classifier to perform the classification task for residual evaluation and therefore to diagnose and isolate the actuator and sensor faults from the generated residuals. The simulation results show that all faults were clearly detected and isolated. Moreover, no false alarms are thereby produced, so this verifies that the proposed scheme has shown an excellent diagnosis performance. The main contribution of this work is to show how to apply an independent RBFNN to open-loop Chylla–Haase benchmark polymerization reactor fault diagnosis. Therefore, this proposed method can contribute to the safety of chemical reactors. In future work, we propose to develop a new robust FDI method for closed-loop reactor using an independent RBFNN. Due to the high non-linearity and the noise in the process, we will propose to develop a new FDI method for closed-loop reactor using the extended Kalman filter.

Disclosure statement

No potential conflict of interest was reported by the author(s).

References

Beyer, M.-A., Grote, W., & Reinig, G. (2008). Adaptive exact linearization control of batch polymerization reactors using a sigma-point Kalman filter. *Journal of Process Control*, *18*(7–8), 663–675.

Chylla, R. W., & Haase, D. R. (1993). Temperature control of semibatch polymerization reactors. *Computers & Chemical Engineering*, *17*(3), 257–264.

Deibert, R., & Isermann, R. (1992). Examples for fault detection in closed loops. *Annual Review in Automatic Programming*, *17*, 235–240.

Ertiame, A. M., Dingli, Y., Feng, Y., & Gomm, J. B. (2013). *Fault detection and isolation for open-loop Chylla–Haase polymerization reactor*. 19th international conference on automation and computing (ICAC), Brunel University, London, UK (pp. 1–6).

Ferrari, R. M. G., Parisini, T., & Polycarpou, M. M. (2008). *A robust fault detection and isolation scheme for a class of uncertain input-output discrete-time nonlinear systems*. American control conference, Seattle, Washington, USA (pp. 2804–2809).

Frank, P. M., & Köppen-Seliger, B. (1997). Fuzzy logic and neural network applications to fault diagnosis. *International Journal of Approximate Reasoning*, *16*(1), 67–88.

Gertler, J. J. (1988). Survey of model-based failure detection and isolation in complex plants. *IEEE Control Systems Magazine*, *8*(6), 3–11.

Gomm, J. B., & Yu, D. L. (2000). Selecting radial basis function network centers with recursive orthogonal least squares training. *IEEE Transactions on Neural Networks*, *11*(2), 306–314.

Graichen, K., Hagenmeyer, V., & Zeitz, M. (2005). *Adaptive feedforward control with parameter estimation for the Chylla–Haase polymerization reactor*. 44th IEEE conference on decision and control, and the European Control Conference, Seville, Spain (pp. 3049–3054).

Graichen, K., Hagenmeyer, V., & Zeitz, M. (2006). Feedforward control with online parameter estimation applied to the Chylla–Haase reactor benchmark. *Journal of Process Control*, *16*(7), 733–745.

Isermann, R. (1984). Process fault detection based on modeling and estimation methods—A survey. *Automatica*, *20*(4), 387–404.

Koppen-Seliger, B., & Frank, P. M. (1995). *Fault detection and isolation in technical processes with neural networks*. 34th IEEE conference on decision and control, New Orleans, LA (Vol. 3, pp. 2414–2419).

Patton, R. J., Chen, J., & Siew, T. M. (1994). *Fault diagnosis in nonlinear dynamic systems via neural networks*. International conference on control, Coventry, UK (Vol. 2, pp. 1346–1351).

Xiaodong, Z. (2011). Sensor bias fault detection and isolation in a class of nonlinear uncertain systems using adaptive estimation. *IEEE Transactions on Automatic Control*, *56*(5), 1220–1226.

Xiaodong, Z., Polycarpou, M. M., & Parisini, T. (2002). A robust detection and isolation scheme for abrupt and incipient faults in nonlinear systems. *IEEE Transactions on Automatic Control*, *47*(4), 576–593.

Yu, D. L., Gomm, J. B., & Williams, D. (1999). Sensor fault diagnosis in a chemical process via RBF neural networks. *Control Engineering Practice*, *7*(1), 49–55.

Zhang, X., Polycarpou, M. M., & Parisini, T. (2010). Fault diagnosis of a class of nonlinear uncertain systems with Lipschitz nonlinearities using adaptive estimation. *Automatica*, *46*(2), 290–299.

14

An EOQ model for deteriorating items with inflation and time value of money considering time-dependent deteriorating rate and delay payments

P. Muniappan[a]*, R. Uthayakumar[b] and S. Ganesh[a]

[a]*Department of Mathematics, Sathyabama University, Chennai – 600 119, Tamil Nadu, India;* [b]*Department of Mathematics, Gandhigram Rural Institute – Deemed University, Dindigul 624 302, Tamil Nadu, India*

A finite time horizon inventory problem for a deteriorating item is developed with constant demand and time-varying deterioration rate under inflation and time value of money. Some of the items may deteriorate in the course of time. In this regard, the authors develop an EOQ model for time-varying deterioration rate. Shortages are allowed in each cycle and backlogged them completely. In this model, the fixed credit period is offered by the supplier to the retailer. During the credit period, the retailers are allowed a trade-credit offer by the suppliers to buy more items and earn more by selling their products. The interest on purchasing cost is charged for the delay of payment by the retailers. The objective of the study is to find optimal decision variables and order quantities of the products so that the net present value of total system cost over a finite planning horizon is minimized. Also, the profit function of the model is maximized. Finally, numerical results are presented to analyse the sensitivity of the optimal policies with respect to changes in some parameters of the system.

Keywords: inventory; order quantity; deteriorating items; shortages; credit periods; inflation

1. Introduction

In recent years, many researchers have investigated on inventory models for deteriorating items. The deterioration of items becomes a common factor in daily life. In reality, many products such as fruits, vegetables, medicines, volatile liquids, blood banks, high-tech products and others deteriorate continuously due to evaporation, spoilage, obsolescence, etc.

The retailer must pay off as soon as the items are received. It is tacitly assumed in the classical economic order quantity inventory model. In the actual business world, that would not be true always in today's competitive business environment. A supplier frequently offers his retailers a delay of payment for settling the amount owed to him. The permissible delay in payments is an effective method of attracting new customers and increasing sales. It may be applied as an alternative to price discount additionally, because it does not provoke competitors to reduce their prices and thus introduce lasting price reductions. Hu and Liu (2010) analysed an optimal replenishment policy for the EPQ model with permissible delay in payments and allowable shortages. Huang (2007) developed economic order quantity under conditionally permissible delay in payments. Ouyang, Wu, and Yang (2006) developed a study on an inventory model for non-instantaneous deteriorating items with permissible delay in payments. Uthayakumar and Parvathi (2006) developed a deterministic inventory model for deteriorating items with time-dependent demand, backlogged partially when delay in payments is permissible. Chung and Huang (2009) analysed an ordering policy with allowable shortage and permissible delay in payments. Chang, Wu, and Chen (2009) studied optimal payment time with deteriorating items under inflation and permissible delay in payments.

Inflation also plays an important role for the optimal order policy and influences the demand of certain products. The value of money goes down and erodes the future worth of saving and forces one for more current spending as inflation increases. These spending are on peripherals and luxury items that give rise to demand of these items usually. As a result, the effect of inflation and time value of the money cannot be ignored for determining the optimal inventory policy.

Hou (2006) studied an inventory model for deteriorating items with stock-dependent consumption rate and shortages under inflation and time discounting. Hou and Lin (2006) developed an EOQ model for deteriorating items with price- and stock-dependent selling rates under inflation and time value of money. Mirzazadeh, Seyyed-Esfahani, and Fatemi-Ghomi (2009) analysed an inventory model under uncertain inflationary conditions, finite production rate and inflation-dependent demand rate for deteriorating items with shortages. Roy, Pal, and Maiti (2009) analysed a production inventory model with

*Corresponding author. Email: a.munichandru@yahoo.com

stock-dependent demand incorporating learning and inflationary effect in a random planning horizon. Sarkar and Moon (2011) developed an EPQ model with inflation in an imperfect production system. Sarkar, Sana, and Chaudhuri (2011) analysed an imperfect production process for time-varying demand with inflation and time value of money – an EMQ model. Taheri-Tolgari, Mirzazadeh, and Jolai (2012) developed an inventory model for imperfect items under inflationary conditions with considering inspection errors. Guria, Das, Mondal, and Maiti (2013) considered the inventory policy for an item with inflation-induced purchasing price, selling price and demand with immediate part payment. Guan and Liang (2014) studied optimal reinsurance and investment strategies for insurer under interest rate and inflation risks. Sarkar, Mandal, and Sarkar (2014) analysed an EMQ model with price- and time-dependent demand under the effect of reliability and inflation. Ftiti and Hichri (2014) considered price stability under an inflation targeting regime: an analysis with a new intermediate approach. Gilding (2014) studied inflation and the optimal inventory replenishment schedule within a finite planning horizon.

Combining the above arguments, so for not all the EOQ models considered all situations such as shortages, deterioration, delayed payments, inflation and time value of money. In the proposed paper, we develop an inventory model for time-varying deteriorating items with constant demand, shortages, delayed payment, inflation and time value of money. The total cost per inventory cycle depends on two decision variables N (number of replenishment) and F (fraction of replenishment cycle). An algorithm is proposed to derive optimal decision variables N^*, F^* so as to minimize the total cost and maximize the profit.

The rest of the paper is organized as follows. In the next section, assumptions, notations and mathematical formulation are given. In Section 3, a numerical example and sensitivity analysis are given in detail to illustrate the models. Finally, conclusion and summary are presented.

2. Model formulaton

The following notations and assumptions are used to develop the model.

2.1. Notations

D	the constant demand rate per unit time
L	the length of the finite planning horizon
θ	the deterioration rate per unit time
r	the fixed ordering cost, \$ per order
h	the fixed holding cost, \$ per order
p_0	the unit purchase cost at time zero, \$ per order
V	the unit selling price at time zero, \$/item
p_t	the unit purchase cost at time t, $p_t = p_0 e^{-Rt}$
s	shortage cost, per unit time \$ per order
s_t	shortage cost, per unit per unit time at time t, $s_t = se^{-Rt}$
d	the discount rate, representing the time value of money
R	$d - i$, representing the net constant discount rate of inflation
M	permissible delay in settling the accounts and $0 < M < 1$
Q	the order quantity per replenishment
$I(t)$	the inventory level at time t
T	the length of replenishment cycle
t_j	the total time that is elapsed upto and including the jth replenishment cycle where $t_0 = 0, t_1 = T, t_N = L$
T_j	the time at which the inventory level in the jth replenishment cycle drops to zeros

Dependent variables

I_b	the maximum shortage quantity
I_m	the maximum inventory level
Q_m	the maximum order quantity
TC_r	the total ordering cost during (0, L)
TC_h	the total holding cost during (0, L)
TC_p	the total purchasing cost during (0, L)
TC_s	the total shortage cost during (0, L)
$\varphi(N, F)$	the total relevant cost during (0, L)

Decision variables

F	the fraction of replenishment cycle where the net stock is positive
N	the number of replenishment during the planning horizon $N = (L/T)$

2.2. Assumptions

(1) The inventory system considers a single item and the demand rate is known and constant.
(2) Deterioration rate is depending on time and there is no replacement or repair of deteriorated units.
(3) Shortages are allowed and completely backlogged.
(4) The time horizon of the inventory system is finite.
(5) The fixed credit period is offered by the supplier to the retailer.
(6) Inflation rate is constant and time value of money is considered.

2.3. Mathematical formulaton

$$\frac{dI_1(t)}{dt} + \theta t I_1(t) = -D, \quad 0 \le t \le T_1, \tag{1}$$

$$\frac{dI_2(t)}{dt} = -D, \quad T_1 \leq t \leq T, \tag{2}$$

with the boundary condition $I(T_1) = 0$. The solutions of Equations (1) and (2) are

$$I_1(t) = De^{-\theta t^2/2}[(T_1 - t) + \frac{\theta}{6}(T_1^3 - t^3)], \quad 0 \leq t \leq T_1 \tag{3}$$

$$I_2(t) = -D(t - T_1), \quad T_1 \leq t \leq T. \tag{4}$$

According to Equations (3) and (4), the maximum inventory quantity at the beginning of each period and the maximum shortage quantity at the end of each period are

$$\begin{aligned} I_m &= I_1(0) \\ &= D\left[T_1 + \frac{\theta}{6}T_1^3\right], \quad T_1 = \frac{FL}{N} \\ &= D\left[\frac{FL}{N} + \frac{\theta}{6}\left(\frac{FL}{N}\right)^3\right], \end{aligned} \tag{5}$$

$$I_b = D(T - T_1) = D\left(T - \frac{FL}{N}\right). \tag{6}$$

The total minimum cost can include the following elements:

(i) Fixed ordering cost

Since the number of replenishment or period is N, the fixed ordering cost over the planning horizon under net present value and inflation is

$$TC_r = \sum_{j=0}^{N} r e^{-jRT} = r\left[\frac{e^{-(N+1)RT} - 1}{e^{-RT} - 1}\right], \tag{7}$$

$$= r\left[\frac{e^{-((N+1)/N)RL} - 1}{e^{-(RL/N)} - 1}\right] \quad \text{since} T = \frac{L}{N}. \tag{8}$$

(ii) Holding cost

In order to determine the holding cost, firstly the average inventory quantity should be determined. Using Equation (3), the average inventory is

$$\begin{aligned} \bar{I} &= \int_0^{T_1} I_1(t)dt = \int_0^{T_1} De^{-\theta t^2/2}\left[(T_1 - t) + \frac{\theta}{6}(T_1^3 - t^3)\right]dt \\ &= D\left[\frac{T_1^2}{2} + \frac{\theta T_1^4}{12} - \frac{\theta^2 T_1^6}{72}\right]. \end{aligned} \tag{9}$$

Then, using Equation (9), the holding cost over the planning horizon under net present value and inflation is

$$\begin{aligned} TC_h &= \sum_{j=0}^{N-1} hp_0 e^{-jRT}\bar{I} \\ &= hDp_0\left[\frac{T_1^2}{2} + \frac{\theta T_1^4}{12} - \frac{\theta^2 T_1^6}{72}\right]\left[\frac{e^{-NRT} - 1}{e^{-RT} - 1}\right]. \end{aligned} \tag{10}$$

Put $T = L/N$ in Equation (10)

$$TC_h = hDp_0\left[\frac{(FL/N)^2}{2} + \frac{\theta(FL/N)^4}{12} - \frac{\theta^2(FL/N)^6}{72}\right]\left[\frac{e^{-RL} - 1}{e^{-RL/N} - 1}\right]. \tag{11}$$

(iii) Shortage cost

$$\begin{aligned} \bar{B} &= \int_{T_1}^{T} I_2(t)dt = \int_{T_1}^{T} D(t - T_1)dt \\ &= \frac{D}{2}[T - T_1]^2 = \frac{D}{2}\left[\frac{L}{N} - \frac{FL}{N}\right]^2. \end{aligned} \tag{12}$$

Therefore, the shortage cost over the planning horizon under net present value and inflation is

$$\begin{aligned} TC_s &= \sum_{j=0}^{N-1} se^{-jRT}\bar{B} = s\left[\frac{e^{-NRT} - 1}{e^{-RT} - 1}\right]\bar{B} \\ &= \frac{sD}{2}\left[\frac{L}{N} - \frac{FL}{N}\right]^2\left[\frac{e^{-RL} - 1}{e^{-(RL/N)} - 1}\right]. \end{aligned} \tag{13}$$

(iv) Purchasing cost

Purchasing cost of the jth cycle is

$$\begin{aligned} c_p(j) &= c_{(j)}I_m + c_{(j+1)T}I_b \\ &= c_{(j)}D\left[\frac{FL}{N} + \frac{\theta}{6}\left(\frac{FL}{N}\right)^3\right] + c_{(j+1)T}D\left(\frac{L}{N} - \frac{FL}{N}\right). \end{aligned} \tag{14}$$

Therefore, the total purchasing cost over the planning horizon with $T = L/N$ is

$$\begin{aligned} TC_p &= \sum_{j=0}^{N-1} c_p(j) \\ &= p_0D\left[\frac{FL}{N} + \frac{\theta}{6}\left(\frac{FL}{N}\right)^3\right]\left[\frac{e^{-RL} - 1}{e^{-RL/N} - 1}\right] \\ &\quad + p_0De^{-RL/N}\left(\frac{L}{N} - \frac{FL}{N}\right)\left[\frac{e^{-RL} - 1}{e^{-RL/N} - 1}\right]. \end{aligned} \tag{15}$$

(v) Interest charged and earned

Regarding interests charged and earned, we have the following two possible cases based on the value of $M \leq T_1$ and $M > T_1$ which are described as follows.

Case i: $M \leq T_1$

Interest earned

As items are sold and before the replenishment account is settled, the sales revenue is used to earn interest. At the beginning of the time interval, the backordered quantity which is I_b, should be replenished first and the maximum accumulated is sold until M is equal to $\int_0^M Dt\,dt$.

Therefore, the interest earned in the first cycle is

$$\begin{aligned} \mathrm{IE}_1 &= I_e V(0)\left[I_b M + \int_0^M Dt\,dt\right] \\ &= I_e V(0)\left[MD\left(\frac{L}{N} - \frac{FL}{N}\right) + \frac{DM^2}{2}\right]. \end{aligned} \tag{16}$$

And total interest earned over the horizon planning using $T = L/N$ will be

$$\begin{aligned} TIE_1(N,F) &= \sum_{j=0}^{N-1} IE_1(j) = \sum_{j=0}^{N-1} IE_1 e^{-jRT} \\ &= I_e V\left[MD\left(\frac{L}{N} - \frac{FL}{N}\right) + \frac{DM^2}{2}\right] \\ &\quad \times \left[\frac{e^{-RL} - 1}{e^{-RL/N} - 1}\right] \end{aligned} \tag{17}$$

Interest charged

When the replenishment account is settled, the situation is reversed and effectively the items still in stock, which is equal to $\int_M^{T_1} I(t)\,dt$, have to be financed at interest rate I_r.

Therefore, the interest payable in the first cycle is

$$\begin{aligned} I_p &= p_0 I_r \int_M^{T_1} I(t)\,dt = p_0 I_r \int_M^{T_1} De^{(-\theta t^2/2)} \\ &\quad \left[(T_1 - t) + \frac{\theta}{6}(T_1^3 - t^3)\right] dt \\ &= p_0 I_r D\left[\frac{T_1^2}{2} - M\left(T_1 - \frac{M}{2}\right)\right. \\ &\quad -\theta\left[\frac{MT_1}{6}(T_1^2 - 2M^2) + \frac{T_1^4 + M^4}{12}\right] \\ &\quad \left. -\theta^2\left[\frac{T_1^6 + M^6}{72} - \frac{T_1^3 M^3}{36}\right]\right]. \end{aligned} \tag{18}$$

And total interest payable over the horizon planning using $T = L/N$ will be

$$\begin{aligned} TI_p(N,F) &= \sum_{j=0}^{N-1} I_p(j) = \sum_{j=0}^{N-1} I_p e^{-jRT} \\ &= p_0 I_r D\left[\frac{(FL)^2}{2N^2} - M\left(\frac{FL}{N} - \frac{M}{2}\right)\right. \\ &\quad - \theta\left[\frac{MT_1}{6}\left(\frac{(FL)^2}{N^2} - 2M^2\right) + \frac{((FL)^4/N^4) + M^4}{12}\right] \\ &\quad \left. - \theta^2\left[\frac{((FL)^6/N^6) + M^6}{72} - \frac{(FL)^3 M^3}{36N^3}\right]\right] \\ &\quad \times \left[\frac{e^{-RL} - 1}{e^{-RL/N} - 1}\right], \end{aligned} \tag{19}$$

$$\begin{aligned} \varphi_1(N,F) &= \text{ordering cost} + \text{holding cost} \\ &\quad + \text{shortage cost} \\ &\quad + \text{purchasing cost} + \text{interest payable} \\ &\quad - \text{interest earned} \end{aligned}$$

$$\begin{aligned} \varphi_1(N,F) &= r\left[\frac{e^{-((N+1)/N)RL} - 1}{e^{-RL/N} - 1}\right] + \left[\frac{e^{-RL} - 1}{e^{-RL/N} - 1}\right] \\ &\left\{hDp_0\left[\frac{(FL/N)^2}{2} + \frac{\theta(FL/N)^4}{12} - \frac{\theta^2(FL/N)^6}{72}\right]\right. \\ &+ \frac{sD}{2}\left[\frac{L}{N} - \frac{FL}{N}\right]^2 + p_0 D\left[\frac{FL}{N} + \frac{\theta}{6}\left(\frac{FL}{N}\right)^3\right. \\ &\left.+ e^{-RL/N}\left(\frac{L}{N} - \frac{FL}{N}\right)\right] + p_0 I_r D\left[\frac{T_1^2}{2} - M\left(T_1 - \frac{M}{2}\right)\right. \\ &-\theta\left[\frac{MT_1}{6}(T_1^2 - 2M^2) + \frac{T_1^4 + M^4}{12}\right] \\ &\left.-\theta^2\left[\frac{T_1^6 + M^6}{72} - \frac{T_1^3 M^3}{36}\right]\right] \\ &\left.-I_e V\left[MD\left(\frac{L}{N} - \frac{FL}{N}\right) + \frac{DM^2}{2}\right]\right\}. \end{aligned} \tag{20}$$

For optimality, $\partial\varphi_1/\partial N = 0$ and $\partial\varphi_1/\partial F = 0$

$$\begin{aligned} \frac{\partial\varphi_1}{\partial N} &= RL(1 - e^{-RL})\left\{re^{-RL/N} + \frac{e^{-RL/N} - 1}{RL}\right. \\ &\times \left[\frac{hDp_0}{N}(FL)^2\left(\frac{\theta^2(FL)^4}{12N^3} - \frac{\theta(FL)^2}{3N^2} - 1\right)\right. \\ &- Dp_0\left(FL + \frac{\theta(FL)^3}{2N^2} + L(1-F)e^{(-RL/N)}\left(1 - \frac{1}{N}\right)\right) \\ &- \frac{sDL^2(1-F)^2}{N} - \frac{I_e VDMLF}{N^2} + e^{-RL/N} \\ &\times \left[hDp_0\left(\frac{(FL)^2}{2N^2} + \frac{\theta(FL)^4}{12N^4} - \frac{\theta^2(FL)^6}{72N^6}\right)\right. \\ &+ Dp_0\left(\frac{FL}{N} + \frac{\theta(FL)^3}{6N^3} + \frac{e^{-RL/N}L(1-F)}{N}\right) \end{aligned}$$

$$+\frac{sDL^2(1-F)^2}{2N^2}+I_r p_0 D\left[\frac{(FL)^2}{2N^2}-M\left(\frac{FL}{N}-\frac{M}{2}\right)\right.$$
$$-\theta\left(\frac{M(FL)^3}{6N^3}-M^3\frac{FL}{3N}+\frac{(FL)^4}{12N^4}\right)$$
$$\left.+\theta^2\left(\frac{M^6}{72}+\frac{(FL)^6}{72N^6}-\frac{M^3(FL)^3}{36N^3}\right)\right]$$
$$\left.\left.-\ I_e VD\left(\frac{ML}{N}(1-F)-\frac{M^2}{2}\right)\right]\right\}=0. \quad (21)$$

$$\frac{\partial\varphi_1}{\partial F}=hDp_0\left[\frac{FL^2}{N^2}+\frac{\theta}{3}\left(\frac{L}{N}\right)^4F^3-\frac{\theta^2}{12}\left(\frac{L}{N}\right)^6F^5\right]$$
$$+Dp_0\left[\frac{L}{N}+\frac{\theta}{2}\left(\frac{L}{N}\right)^3F^2-\mathrm{e}^{-RL/N}\left(\frac{L}{N}\right)\right]$$
$$-sD\left(\frac{L}{N}\right)^2(1-F)+I_e VD\frac{ML}{N}=0. \quad (22)$$

Case ii: $M > T_1$

Interest earned

At the beginning of the time interval, the backordered quantity, I_b, should be replenished first and interest earned for the first cycle will be $I_e v(0)MI_b$. Then, the maximum accumulated is sold until M is equal to $\int_0^{T_1} Dt\,\mathrm{d}t$, while the interest earned will be $I_e V(0)[(M-T_1)DT_1+(DT_1^2/2)]$.

Therefore, the interest earned in the first cycle is

$$IE_2=I_e V(0)\left[I_b M+(M-T_1)DT_1+\int_0^m DT_1\,\mathrm{d}t\right]$$
$$=I_e V(0)\left[MD\left(\frac{L}{N}-\frac{FL}{N}\right)+\left(M-\frac{FL}{N}\right)D\frac{FL}{N}\right.$$
$$\left.+\frac{(FL)^2}{2N^2}\right]. \quad (23)$$

And total interest earned over the planning horizon using $T=L/N$ will be

$$TIE_2(N,F)=\sum_{j=0}^{N-1}IE_2(j)=\sum_{j=0}^{N-1}IE_2\mathrm{e}^{-jRT}$$
$$=I_e V\left[MD\left(\frac{L}{N}-\frac{FL}{N}\right)+\left(M-\frac{FL}{N}\right)D\frac{FL}{N}\right.$$
$$\left.+\frac{(FL)^2}{2N^2}\right]\left[\frac{\mathrm{e}^{-RL}-1}{\mathrm{e}^{-RL/N}-1}\right]. \quad (24)$$

Interest charged

In this case when the replenishment accounts is settled, the number of items which are in stock is zero, so the interest payable will be zero.

Therefore, the total inventory cost function is

$$\varphi_2(N,F)=\text{ordering cost}+\text{holding cost}+\text{shortage cost}$$
$$+\text{purchasing cost}-\text{interest earned},$$

$$\varphi_2(N,F)=r\left[\frac{\mathrm{e}^{-((N+1)/N)RL}-1}{\mathrm{e}^{-RL/N}-1}\right]$$
$$+\left\{hDp_0\left[\frac{(FL/N)^2}{2}+\frac{\theta(FL/N)^4}{12}-\frac{\theta^2(FL/N)^6}{72}\right]\right.$$
$$+\frac{sD}{2}\left[\frac{L}{N}-\frac{FL}{N}\right]^2+p_0D\left[\frac{FL}{N}+\frac{\theta}{6}\left(\frac{FL}{N}\right)^3\right]$$
$$+p_0D\mathrm{e}^{-RL/N}\left(\frac{L}{N}-\frac{FL}{N}\right)-I_eV\left[MD\left(\frac{L}{N}-\frac{FL}{N}\right)\right.$$
$$\left.\left.+\left(M-\frac{FL}{N}\right)D\frac{FL}{N}+\frac{(FL)^2}{2N^2}\right]\right\}\left[\frac{\mathrm{e}^{-RL}-1}{\mathrm{e}^{-RL/N}-1}\right]. \quad (25)$$

For optimality, $\partial\varphi_2/\partial N=0$ and $\partial\varphi_2/\partial F=0$

$$\frac{\partial\varphi_2}{\partial N}=RL(1-\mathrm{e}^{-RL})\left\{r\mathrm{e}^{-RL/N}+\frac{\mathrm{e}^{-RL/N}-1}{RL}\right.$$
$$\left[\frac{hDp_0}{N}(FL)^2\left(\frac{\theta^2(FL)^4}{12N^3}-\frac{\theta(FL)^2}{3N^2}-1\right)\right.$$
$$-Dp_0\left(FL+\frac{\theta(FL)^3}{2N^2}+L(1-F)\mathrm{e}^{-RL/N}\left(1-\frac{1}{N}\right)\right)$$
$$\left.-\frac{sDL^2(1-F)^2}{N}+I_eVD\left(ML-\frac{(FL)^2}{N}\right)\right]$$
$$+\mathrm{e}^{-RL/N}\left[hDp_0\left(\frac{(FL)^2}{2N^2}+\frac{\theta(FL)^4}{12N^4}-\frac{\theta^2(FL)^6}{72N^6}\right)\right.$$
$$+Dp_0\left(\frac{FL}{N}+\frac{\theta(FL)^3}{6N^3}+\frac{\mathrm{e}^{-RL/N}L(1-F)}{N}\right)$$
$$\left.\left.+\frac{sDL^2(1-F)^2}{2N^2}-I_eVD\left(\frac{ML}{N}-\frac{(FL)^2}{2N^2}\right)\right]\right\}=0, \quad (26)$$

$$\frac{\partial\varphi_2}{\partial F}=hDp_0\left[\frac{FL^2}{N^2}+\frac{\theta}{3}\left(\frac{L}{N}\right)^4F^3-\frac{\theta^2}{12}\left(\frac{L}{N}\right)^6F^5\right]$$
$$+Dp_0\left[\frac{L}{N}+\frac{\theta}{2}\left(\frac{L}{N}\right)^3F^2-\mathrm{e}^{-RL/N}\left(\frac{L}{N}\right)\right] \quad (27)$$
$$-sD\left(\frac{L}{N}\right)^2(1-F)+I_eVD\left(F\left(\frac{L}{N}\right)^2\right)=0.$$

Calculation of profit

$$\text{Revenue}=\text{price}\times\text{order quantity}=p_0(I_m+I_b),$$

where $I_m=D[FL/N+(\theta/6)(FL/N)^3]$ and $I_b=D(L/N-FL/N)$.

The total maximum order quantity

$$Q_{\mathrm{m}} = I_{\mathrm{m}} + I_{\mathrm{b}} = D\left[\frac{L}{N} + \frac{\theta}{6}\left(\frac{FL}{N}\right)^3\right].$$

Therefore, the revenue cost over the planning horizon under net present value and inflation consideration is

$$TR_{\mathrm{c}} = \sum_{j=0}^{N-1} r_{\mathrm{c}} p_0 \overline{Q_{\mathrm{m}}}$$

$$= rp_0 D\left[\frac{L}{N} + \frac{\theta}{6}\left(\frac{FL}{N}\right)^3\right]\left[\frac{\mathrm{e}^{-RL} - 1}{\mathrm{e}^{-RL/N} - 1}\right],$$

$$\text{Profit} = \text{revenuecost} - \text{total inventory},$$

$$\phi(N^*, F^*) = TR_{\mathrm{c}} - \varphi(N^*, F^*). \tag{28}$$

By using the following algorithm, we have to find optimal values of N^* and F^* to minimize the total cost $\varphi(N^*, F^*)$ and maximize the profit $\phi(N^*, F^*)$

Algorithm 1 **Step 1 Perform (i)–(v)**

(i) Input the values
(ii) Substituting the values into Equation (21) and find $N_{(1)}$
(iii) Using $N_{(1)}$ determine $F_{(1)}$ from Equation (22)
(iv) Using Equation (20) determine $\varphi_1(N_{(1)}, F_{(1)})$
(v) Repeat (ii) and (v) until $\varphi_1(N_{(n)-1}, F_{(n)-1}) \leq \varphi_1(N_{(n)}, F_{(n)})$. Consider $N_1 = N_{(n)-1}$, $F_1 = F_{(n)-1}$ and go to step 2

Step 2 Perform (i)–(v)

(i) Input the values
(ii) Substituting the values into Equation (26) and find $N_{(1)}$
(iii) Using $N_{(1)}$ determine $F_{(1)}$ from Equation (27)
(iv) Using Equation (25) determine $\varphi_2(N_{(1)}, F_{(1)})$
(v) Repeat (ii) and (v) until $\varphi_2(N_{(n)-1}, F_{(n)-1}) \leq \varphi_2(N_{(n)}, F_{(n)})$. Consider $N_2 = N_{(n)-1}$, $F_2 = F_{(n)-1}$ and go to step 3

Step 3 Case (i) $M \leq F_1L/N_1$ and $M > F_2L/N_2$

(i) If $\varphi_1(N_1, F_1) \leq \varphi_2(N_2, F_2)$ then $N^* = N_1$, $F^* = F_1$ and $\varphi(N^*, F^*) = \varphi_1(N_1, F_1)$
(ii) If $\varphi_1(N_1, F_1) > \varphi_2(N_2, F_2)$ then $N^* = N_2$, $F^* = F_2$ and $\varphi(N^*, F^*) = \varphi_2(N_2, F_2)$
(ii) Go to step 4

Case (ii) $M > F_1L/N_1$ and $M > F_2L/N_2$

(i) Input the values
(ii) Substituting the values into Equation (21) and find $N_{(1)}$
(iii) Using $N_{(1)}$ determine $F_{(1)}$ by using $F_{(1)} = N_{(1)}M/L$
(iv) Using Equation (20) determine $\varphi_1(N_{(1)}, F_{(1)})$
(v) Repeat (ii) and (v) until $\varphi_1(N_{(n)-1}, F_{(n)-1}) \leq \varphi_1(N_{(n)}, F_{(n)})$. Consider $N_3 = N_{(n)-1}$, $F_3 = F_{(n)-1}$
(vi) If $\varphi_1(N_3, F_3) \leq \varphi_2(N_2, F_2)$ then $N^* = N_3$, $F^* = F_3$ and $\varphi(N^*, F^*) = \varphi_1(N_3, F_3)$
(vii) If $\varphi_1(N_3, F_3) > \varphi_2(N_2, F_2)$ then $N^* = N_2$, $F^* = F_2$ and $\varphi(N^*, F^*) = \varphi_2(N_2, F_2)$
(viii) Go to step 4

Case (iii) $M \leq F_1L/N_1$ and $M \leq F_2L/N_2$

(i) Input the values
(ii) Substituting the values into Equation (26) and find $N_{(1)}$
(iii) Using $N_{(1)}$ determine $F_{(1)}$ by using $F_{(1)} = N_{(1)}M/L$
(iv) Using Equation (25) determine $\varphi_2(N_{(1)}, F_{(1)})$
(v) Repeat (ii) and (v) until $\varphi_2(N_{(n)-1}, F_{(n)-1}) \leq \varphi_2(N_{(n)}, F_{(n)})$. Consider $N_4 = N_{(n)-1}$, $F_4 = F_{(n)-1}$
(vi) If $\varphi_1(N_1, F_1) \leq \varphi_2(N_4, F_4)$ then $N^* = N_1$, $F^* = F_1$ and $\varphi(N^*, F^*) = \varphi_1(N_1, F_1)$
(vii) If $\varphi_1(N_1, F_1) > \varphi_2(N_4, F_4)$ then $N^* = N_4$, $F^* = F_4$ and $\varphi(N^*, F^*) = \varphi_2(N_4, F_4)$
(viii) Go to step 4

Case (iv) $M > F_1L/N_1$ and $M \leq F_2L/N_2$

(i) If $\varphi_1(N_3, F_3) \leq \varphi_2(N_4, F_4)$ then $N^* = N_3$, $F^* = F_3$ and $\varphi(N^*, F^*) = \varphi_1(N_3, F_3)$
(ii) If $\varphi_1(N_3, F_3) > \varphi_2(N_4, F_4)$ then $N^* = N_4$, $F^* = F_4$ and $\varphi(N^*, F^*) = \varphi_2(N_4, F_4)$
(iii) Go to step 4

Step 4 Determination of I_m, I_b and $\phi(N^*, F^*)$

(i) Determine I_m by using $I_m = D[(F^*L/N^*) + (\theta/6)(F^*L/N^*)^3]$
(ii) Determine I_b by using $I_b = D((L/N^*) - (F^*L/N^*))$
(iii) Determine order quantity $Q_m = I_m + I_b$
(iv) Determine maximum profit $\phi(N^*, F^*)$ by using Equation (28)

3. Numerical examples

Example 1 Let $r = 100$, $D = 10{,}000$, $R = 0.05$, $L = 30/365$, $F = 0.001$, $\theta = 0.01$, $h = 0.5$, $p_0 = 8$, $s = 2$, $I_{\mathrm{r}} = 0.06$, $I_{\mathrm{e}} = 0.05$, $N = 1$, $V = 10$, $M = 5/365$. The computational result shows the following optimal values by using the above algorithm:

$$N^* = 1.8309,\ F^* = 0.2413,\ \in T^* = 0.0449,\ Q^*_m = 449,$$

$$\varphi(N^*, F^*) = 112.3436,\ \phi(N^*, F^*) = 2214.7.$$

Example 2 Let $r = 500$, $D = 10{,}000$, $R = 0.05$, $L = 60/365$, $F = 0.001$, $\theta = 0.01$, $h = 0.8$, $p_0 = 8$, $s = 4$, $I_{\mathrm{r}} = 0.06$, $I_{\mathrm{e}} = 0.05$, $N = 1$, $V = 10$, $M = 10/365$. The computational result shows the following optimal values by using above algorithm.

$$N^* = 1.6577,\ F^* = 0.2763,\ T^* = 0.0992,\ Q^*_m = 992,$$

$$\varphi(N^*, F^*) = 560.9096,\ \phi(N^*, F^*) = 50,657.$$

3.1. Sensitivity analysis

We now study the effects of changes in the value of system parameters $R, L, \theta, D, r, I_r, I_e, s, p_0, M, V$ on the optimal length of order cycle T^*, the optimal number of replenishment N^*, the optimal order quantity per cycle Q^*_m, the minimum total relevant cost per unit time $\varphi(N^*, F^*)$ and the total profit $\phi(N^*, F^*)$ of Example 1. The analysis is carried out by changing the value of only one parameter at a time keeping the rest of the parameters at their initial values. The results are shown in Table 1.

From Table 1, the following inferences can be observed:

(1) An increase in the value of demand D, the relevant total costs $\varphi(N^*, F^*)$, optimum order size Q^*_m and profit $\phi(N^*, F^*)$ will increase, that is,if the retailer's demand

Table 1. Sensitivity of the optimal solution with respect to change in values of the model parameters.

Parameters		N^*	F^*	Q_m^*	$\varphi(N^*,F^*)$	$\phi(N^*,F^*)$
r	100	1.8309	.2413	449	112.3436	2214.7
	150	2.0261	.2386	406	160.8830	2993.8
	200	2.1769	.2365	378	209.7621	3705.3
	250	2.3016	.2348	357	258.7958	4370.3
R	0.05	1.8309	.2413	449	112.3436	2214.7
	0.1	1.8024	.1753	456	123.0813	4588.8
	0.15	1.7616	.1096	467	132.4330	7074.9
	0.2	1.7127	.0442	480	140.4481	9711.0
D	10,000	1.8309	.2413	449	112.3436	2214.7
	15,000	1.6544	.2438	745	120.6680	3741.8
	20,000	1.5395	.2454	1068	129.6569	5403.8
	25,000	1.4549	.2425	1412	139.0332	7179.5
L	30/365	1.8309	.2413	449	112.3436	2214.7
	45/365	1.4658	.1629	841	132.0780	6393.8
	60/365	1.2423	.1035	1323	162.1879	13,494
	75/365	1.0970	.0731	1873	204.8207	23,895
p_0	6	1.8781	.3088	438	109.7437	1591.7
	7	1.8517	.2720	444	111.0706	1902.2
	8	1.8309	.2413	449	112.3436	2214.7
	9	1.8139	.2153	453	113.5750	2528.8
θ	0.01	1.8309	.2413	449	112.3436	2214.7
	0.05	1.8308	.2413	449	112.3443	2214.8
	0.1	1.8307	.2413	449	112.3453	2215.0
	0.15	1.8305	.2413	449	112.3462	2215.2
I_r	0.06	1.8309	.2413	449	112.3436	2214.7
	0.08	1.8310	.2413	449	112.3430	2214.6
	0.10	1.8311	.2413	449	112.3425	2214.5
	0.12	1.8312	.2413	449	112.3420	2214.3
I_e	0.05	1.8309	.2413	449	112.3436	2214.7
	0.07	1.8248	.2313	450	112.2320	2222.6
	0.09	1.8187	.2213	452	112.1213	2230.5
	0.11	1.8126	.2144	454	112.0112	2238.5
s	2	1.8309	.2413	449	112.3436	2214.7
	3	1.7002	.2834	483	114.1531	2391.5
	4	1.5901	.2650	517	115.0037	2563.9
	5	1.5038	.2506	547	115.7620	2716.6
V	10	1.8309	.2413	449	112.3436	2214.7
	12	1.8278	.2363	450	112.2877	2218.6
	14	1.8248	.2313	450	112.2320	2222.6
	16	1.8217	.2263	451	112.1766	2226.5
M	5/365	1.8309	.2413	449	112.3436	2214.7
	10/365	1.8176	.2163	452	112.0525	2532.8
	15/365	1.8059	.1915	455	111.7568	2547.9
	20/365	1.7956	.1670	458	111.4577	2561.2

increases, then the retailer orders a high amount of quantity with the supplier, so that the costs automatically increase.

(2) An increase in the value of deterioration rate θ, the relevant total costs $\varphi(N^*,F^*)$ and profit $\phi(N^*,F^*)$ increases without affecting the optimum order size Q_m^*, that is, θ is independent of optimum order size Q_m^*.

(3) An increase in the value of ordering cost r, the relevant total costs $\varphi(N^*,F^*)$, profit $\phi(N^*,F^*)$ will increase and the optimum order size Q_m^* will decrease.

(4) An increase in the value of the parameters R, L, s, p_0 the optimum order size Q_m^*, the relevant total costs $\varphi(N^*,F^*)$ and profit $\phi(N^*,F^*)$ will increase.

(5) An increase in the value of the parameters I_e, V, h and M the optimum order size Q_m^*, the profit $\phi(N^*,F^*)$ will increase and the relevant total costs $\varphi(N^*,F^*)$ will decrease.

(6) An increase in value of the parameter I_r, the relevant total costs $\varphi(N^*,F^*)$ will decrease, the profit $\phi(N^*,F^*)$ will increase and the optimum order size Q_m^* will remain same.

4. Conclusion

This paper deals with the EOQ model for deteriorating items with inflation and time value of money over the finite horizon. The model assumes constant demand, time-varying deteriorating rate and finite planning horizon. Shortages are allowed in the inventory system and are fully backlogged and also supplier offers the delayed payment strategy. Our aim is to find the optimal decision variables to maximize the net present value of the profit and minimize the net present value of the total inventory system cost over a finite horizon. Numerical examples are also provided to illustrate the proposed model. Moreover, sensitivity analysis of the optimal solutions with respect to major parameters is carried out. The model can be extended in several ways. For instance, we may generalize the models by allowing price- or stock-dependent demand, fuzzy demand, two-level trade-credit policy, temporary discounts, quantity discounts, etc.

Disclosure statement

No potential conflict of interest was reported by the authors.

References

Chang, C. T., Wu, S. J., & Chen, L. C. (2009). Optimal payment time with deteriorating items under inflation and permissible delay in payments. *International Journal of System Sciences, 40*, 985–993.

Chung, K. J., & Huang, Y. F. (2009). An ordering policy with allowable shortage and permissible delay in payments. *Applied Mathematical Modelling, 33*, 2518–2525.

Ftiti, Z., & Hichri, W. (2014). The price stability under inflation targeting regime: An analysis with a new intermediate approach. *Economic Modelling, 38*, 23–32.

Gilding, B. H. (2014). Inflation and the optimal inventory replenishment schedule within a finite planning horizon. *European Journal of Operational Research, 234*, 683–693.

Guan, G., & Liang, Z. (2014). Optimal reinsurance and investment strategies for insurer under interest rate and inflation risks. *Insurance: Mathematics and Economics, 55*, 105–115.

Guria, A., Das, B., Mondal, S., & Maiti, M. (2013). Inventory policy for an item with inflation induced purchasing price, selling price and demand with immediate part payment. *Applied Mathematical Modelling, 31*, 240–257.

Hou, K. L. (2006). An inventory model for deteriorating items with stock-dependent consumption rate and shortages under inflation and time discounting. *European Journal of Operational Research, 168*, 463–474.

Hou, K. L., & Lin, L. C. (2006). An EOQ model for deteriorating items with price- and stock-dependent selling rates under inflation and time value of money. *International Journal of System Sciences, 37*, 1131–1139.

Hu, F., & Liu, D. (2010). Optimal replenishment policy for the EPQ model with permissible delay in payments and allowable shortages. *Applied Mathematical Modelling, 34*, 3108–3117.

Huang, Y. F. (2007). Economic order quantity under conditionally permissible delay in payments. *European Journal of Operational Research, 176*, 911–924.

Mirzazadeh, A., Seyyed-Esfahani, M., & Fatemi-Ghomi, M. T. (2009). An inventory model under uncertain inflationary conditions, finite production rate and inflation-dependent demand rate for deteriorating items with shortages. *International Journal of System Sciences, 40*, 21–31.

Ouyang, L. Y., Wu, K. S., & Yang, C. T. (2006). A study on an inventory model for non-instantaneous deteriorating items with permissible delay in payments. *Computers and Industrial Engineering, 51*, 637–651.

Roy, A., Pal, S., & Maiti, M. K. (2009). A production inventory model with stock dependent demand incorporating learning and inflationary effect in a random planning horizon: A fuzzy genetic algorithm with varying population size approach. *Computers and Industrial Engineering, 57*, 1324–1335.

Sarkar, B., Mandal, P., & Sarkar, S. (2014). An EMQ model with price and time dependent demand under the effect of reliability and inflation. *Applied Mathematics and Computation, 231*, 414–421.

Sarkar, B., & Moon, I. (2011). An EPQ model with inflation in an imperfect production system. *Applied Mathematics and Computation, 217*, 6159–6167.

Sarkar, B., Sana, S. S., & Chaudhuri, K. (2011). An imperfect production process for time varying demand with inflation and time value of money – An EMQ model. *Expert Systems with Applications, 38*, 13543–13548.

Taheri-Tolgari, J., Mirzazadeh, A., & Jolai, F. (2012). An inventory model for imperfect items under inflationary conditions with considering inspection errors. *Computers and Mathematics with Applications, 63*, 1007–1019.

Uthayakumar, R., & Parvathi, P. (2006). A deterministic inventory model for deteriorating items with time-dependent demand, backlogged partially when delay in payments is permissible. *International Journal of Mathematical Sciences, 5*, 201–219.

15

Dynamic modelling and experimental validation of an automotive windshield wiper system for hardware in the loop simulation

Mark Dooner[a], Jihong Wang[a]* and Alexandros Mouzakitis[b]

[a]*School of Engineering, The University of Warwick, Coventry, UK;* [b]*Jaguar Land Rover Product Development Centre, Gaydon, Warwickshire, UK*

In order to remain competitive, automotive companies use advanced simulation methods to assist in product development. Hardware in the loop (HIL) simulation is one such technique. To use HIL in the development of automotive electronic control units (ECU), accurate simulation models of the ECU's sensors and actuators are needed. In this work, a full dynamic mathematical model of an automotive windshield wiper system is developed and validated. In the modelling phase, the wiper motor is analysed and a unique mathematical model is developed to capture the devices two speed operation. A multi-body dynamic model of the linkages is implemented using the MathWorks' SimMechanics software. The model is validated experimentally and its parameters are identified using genetic algorithms. The model is then simplified to allow it to be simulated in real time, making it suitable for HIL simulation. The HIL compatible model is used in the development of Automotive ECUs.

Keywords: windshield-wiper; hardware-in-the-loop; real-time simulation; genetic algorithm

1. Introduction

The Global Automotive Report 2013 compiled by Clearwater Corporate Finance LLP estimated the value of the automotive industry at $800bn (Clearwater Corporate Finance, 2013). In addition, worldwide passenger car sales in 2012 exceeded 60 million units and sales have tended to increase over the last decade (http://www.oica.net). The total number of passenger cars produced in 2012 was also greater than 60 million – representing an increase of 5.3% in production from 2011 (http://www.oica.net). New vehicles must meet stringent safety and environmental requirements, with industry standards such as ISO26262 being widely adhered to (Jeon, Cho, Jung, Park, & Han, 2011) whilst maintaining acceptable comfort and performance standards to make the product viable. It follows that automotive companies capable of producing quality and safe products in relatively short production times will benefit from the highly lucrative global automotive industry.

The development process for new products in the automotive industry follows the classic V model (Robert Bosch GmbH, 2007). Figure 1 shows an adapted version of this, highlighting the use of models in the production process. Modern product development processes use model-based design, development and testing tools to improve the tractability of requirements, the speed of development and the validity of early testing. A particularly important simulation-based testing procedure is hardware in the loop (HIL) simulation.

HIL is an advanced real-time simulation technique in which a purely simulated system has certain aspects replaced by hardware components (Hu & Azarnasab, 2013). In the automotive industry, HIL simulation is used extensively in the development and testing of electronic control units (ECU) (Ganesh, 2005; Schuette & Ploeger, 2007) because they allow extensive tests to be carried out before the final design and manufacture of components interfacing with the ECUs. Most of the innovations in modern luxury vehicles are in the electronic/software domains, with electronic systems replacing traditionally mechanical and pneumatic systems (Von Tils, 2006) and there are in excess of 100 ECUs in a modern luxury car (Waltermann, 2009). The purpose of the model developed here is to provide a simulation model of a windscreen wiper system to be used in the HIL testing of an ECU by Jaguar Land Rover.

The benefits of using a simulation model such as the one developed in this paper are as follows: (1) reduced need for a hardware prototype which takes up space and resources, (2) models can be quickly updated to incorporate design changes, whereas up to date prototypes are often unavailable, (3) tests done with simulation models have higher repeatability because all variables can be controlled and (4) testing for the development of ECUs

*Corresponding author. Email: jihong.wang@warwick.ac.uk

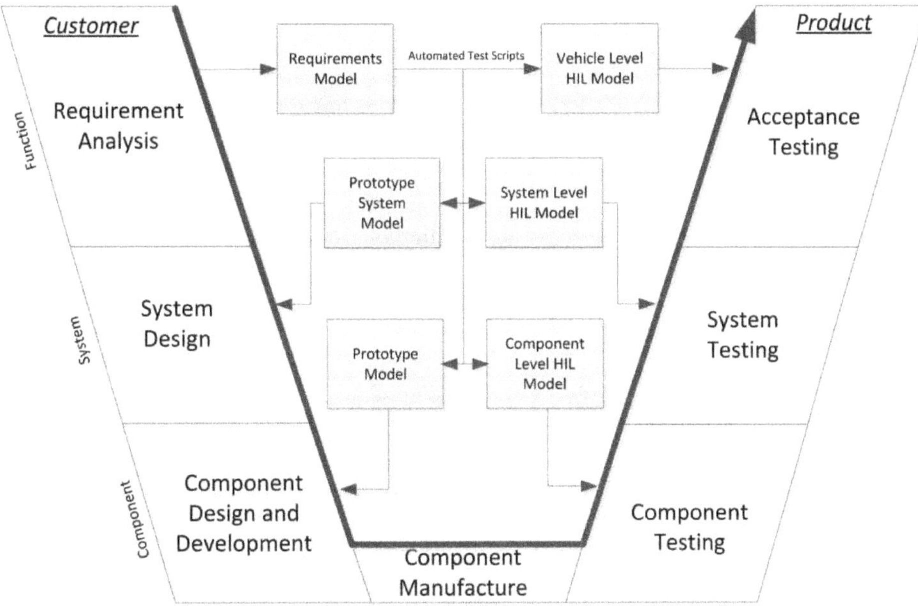

Figure 1. V model for product development.

can be carried out sooner than tests relying on hardware prototypes.

This paper first describes the wiper system to be modelled by showing its physical structure and operation principles. Then the modelling process of each element of the system is shown. Genetic algorithms (GA) are then used to identify the unknown parameters in the model. The model is then validated by comparing its performance against real data. Once this is complete, the model is simplified to allow it to be simulated in real time, making it suitable for HIL simulation.

2. Existing windscreen wiper models

Models of windscreen wiper systems in the existing literature tend to concentrated on one particular aspect of the system and are not suitable for real-time simulation. Some examples of previous work are given to illustrate this point: In Chang and Lin (2004) a mathematical model of a wiper system is developed in order to investigate, and attempt to control, chaotic behaviour present the system. A mathematical model of the mechanics of a wiper blade is developed in Okura, Sekiguchi, and Oya (2000) to investigate the vibrational behaviour of the system during reversal (i.e. when the blades change direction). Significant noise can be heard at this point and the model was used to identify design parameters which could be modified to reduce this noise. Chatter vibration was investigated in Grenouillat and Leblanc (2002) using a simulation model of the blade on the windscreen. It was found that the mechanical configuration of the wipers had a large effect on the chatter noise. Models such as the three identified above which deal with vibrations are generally too complicated for HIL simulation and are unsuitable for developing ECUs.

More similar research to that presented here is given in Xiaoyu, Yanfeng, and Yengjie (2011) where physical modelling techniques are used to create a mechanical model of a wiper system to assist in the design of future components; however the model was not validated. Finally, previous publications of this work Wei, Mouzakitis, Wang, and Sun (2011) and Dooner, Wang, and Mouzakitis (2013) develop a model for HIL simulation which includes the motor and linkages.

3. Wiper system modelling

The model developed in the section simulates different behaviours from the models discussed in Section 2. The important elements of the wiper system that this model captures are (1) the dynamic torque load of the linkages driven by the motor and how this affects the motor current, (2) the switching transients on the motor current when switching between ON and OFF modes as well as FAST and SLOW, (3) such behaviour must be modelled across the systems entire operating speed. The model does not take into account effects such as vibration of the mechanical element.

In addition to capturing the system dynamics and coupling between the motor and linkages, the model must be capable of being simulated in real time (for use in HIL simulation) and be easy to update to incorporate design changes since the model will be used in the early stages of product development. To meet these goals a physical modelling approach has been employed using the MathWorks physical modelling tool, Simscape.

The system can be split into two elements for modelling: the wiper motor and the linkage system, as shown in

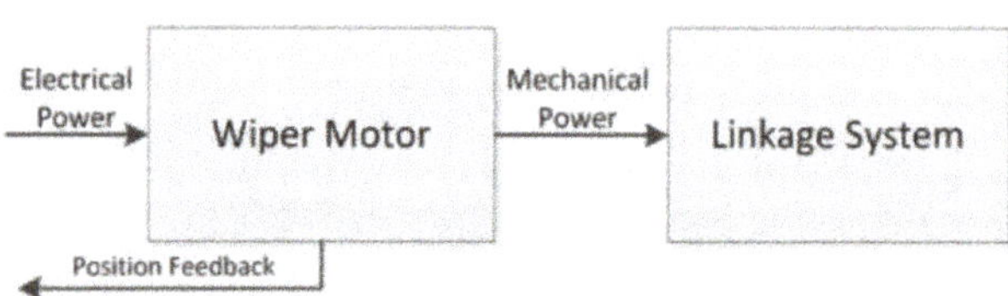

Figure 2. System block diagram.

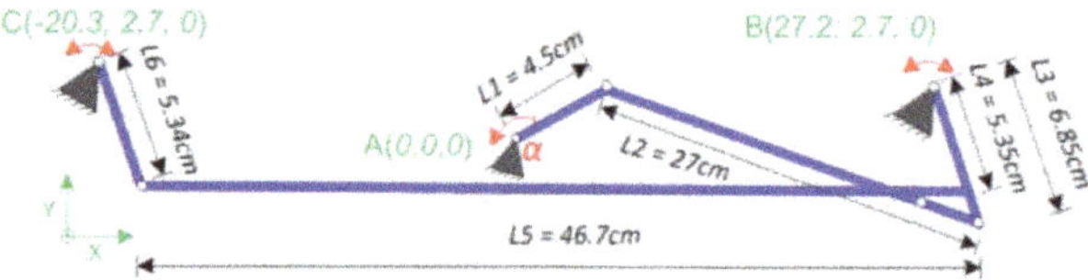

Figure 3. Linkages operation.

Figure 2. These are both initially modelled separately and then coupled together into one model.

3.1. *Linkage system modelling*

The linkages used in a windshield wiper system are designed such that unidirectional rotational motion from a Permanent Magnet Direct Current (PMDC) motor can be translated into the oscillatory motion of the two wipers. The design and operation of the linkage system is demonstrated in Figure 3. The linkages have initially been modelled using the MathWorks tool SimMechanics. This was chosen because it simplifies the modelling and simulation process of Multi-body dynamic systems by using embedded equations and modular design (Wood & Kennedy, 2003). SimMechanics has also been successfully employed in similar projects (Xiaoyu et al., 2011).

The linkages are defined in the SimMechanics modelling environment as shown in Figure 4. Each individual linkage is represented by a 'Solid' block which defines the shape and material of the linkage. Two transformation blocks are used to define the length of the linkage and assign a co-ordinate system to either end. Each end is then attached to a revolute joint. The full five bar system is modelled using this method and is approximated as being in a plane. The complete model of the linkages is shown in Figure 5.

The measured parameters of the model are shown in Figure 3, adapted from Wei et al. (2011). The model is designed to allow any of the parameters to be easily changed so that the model can represent any physically viable design of the linkage system shown in Figure 3, allowing for design updates to be easily implemented. Also included in the parameters is the density of the material, 2.7 g/cm^3, and the initial angle, (α), of the crank, determining the park position of the wipers. The park position refers to the wipers rest position when not in operation, that is, at the bottom of the windscreen.

3.2. *Wiper motor modelling*

The brushed PMDC wiper motor is connected directly to the battery meaning that the speed of the motor, and thus

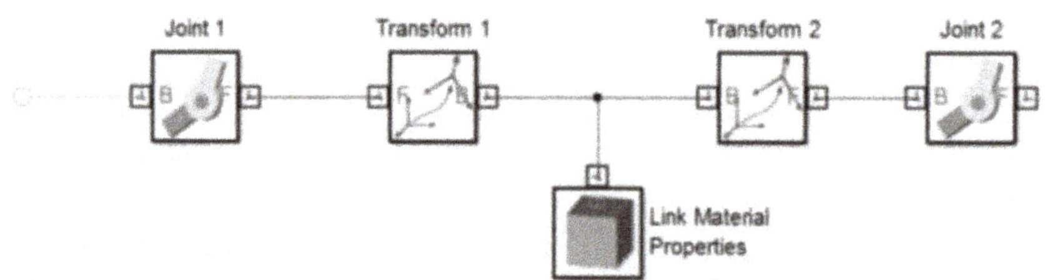

Figure 4. SimMechanics tool.

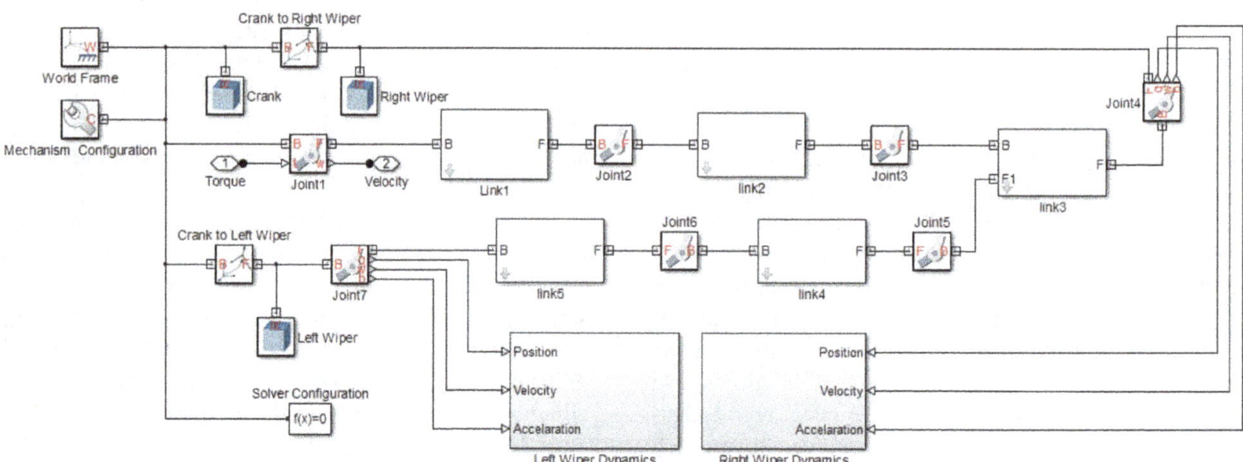

Figure 5. Complete linkages model.

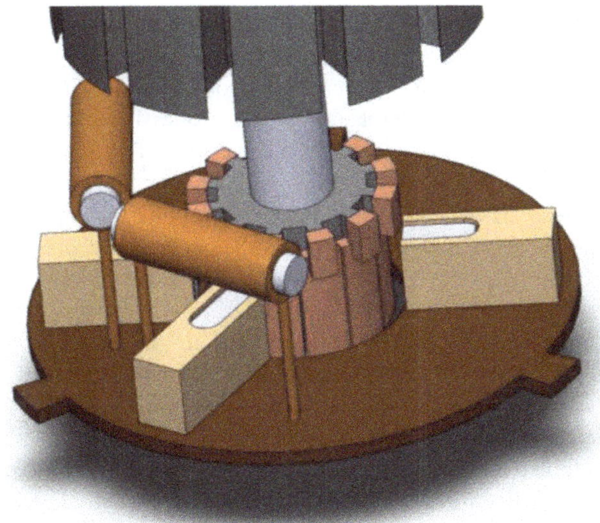

Figure 6. Wiper motor structure.

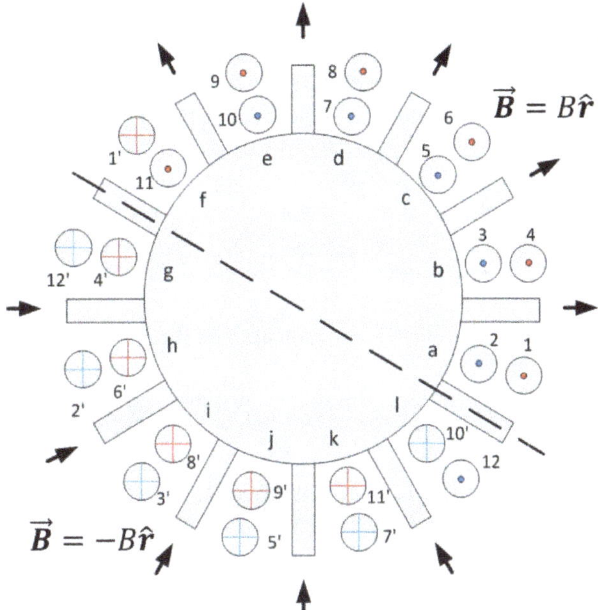

Figure 7. Slow speed armature winding.

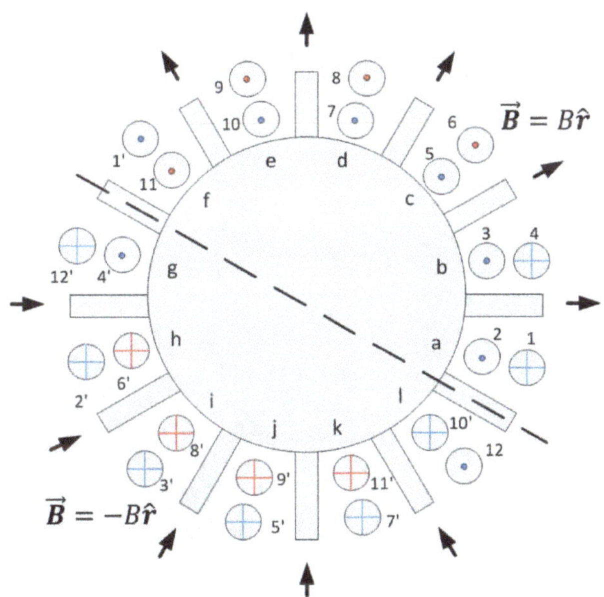

Figure 8. Fast speed armature winding.

the wipers, cannot be controlled by changing the input voltage. In order to control the speed of the motor a third brush is added, offset from the magnetic neutral line, giving the motor two electrical inputs and one common ground output (Hameyer & Belmans, 1996). The structure can be seen in Figure 6.

The brush directly opposite to the common brush in the magnetic neutral line is connected to the slow speed input. Applying a voltage to this brush will cause the motor to operate as a standard two brush PMDC motor. Applying a voltage to the brush offset from the magnetic neutral line will cause the motor to operate in its fast mode. This increases the rotational velocity of the motor shaft and also the armature current, with a reduction in efficiency (Hillier, 1987).

No model of a wiper motor capturing the fast and slow speed operation, and the switching between them, could be found. Therefore a unique model of the wiper motor is developed here using classical electromagnetic theory.

To understand the two brush operation of the wiper motor the difference in the current paths around the armature windings in the two different states are considered. The current paths in the slow and fast operations are shown in Figures 7 and 8, respectively where the 12 armature slots are labelled '*a*' to '*l*' and contain two windings each. The 12 windings are labelled '1' to '12' with a tick representing the return path of the winding. The instantaneous direction of the current is represented using a cross for current into the page and a dot for current out of the page and the current paths are colour coded, with blue representing current path 1 and red representing current path 2. The arrows show the direction of the magnetic field caused by the permanent magnet. The magnetic field can be described in terms of the flux density vector, $\vec{B}$, or the scalar denoting the magnitude of the magnetic flux density, B, and the vector, $\vec{r}$, which points radially away from the centre of the armature.

It can be seen that for the slow operation the current paths are symmetrical around the magnetic neutral line, as in a normal PMDC motor. However, in fast operation the current paths are not symmetrical, that is, they are unbalanced. The value of the current in each current path the same in slow mode, however it is different in fast mode due to the different physical lengths of the current paths.

An analysis is carried out based on Chapter 1 in Chiasson (2005) on the torque developed in each armature slot.

The magnitude of the force in each slot is a function of the dimensions of the armature, the magnitude and direction of the $\vec{B}$ field, and the magnitude and direction of the current. The unbalanced nature of the current paths in the wiper motor complicates the determination of the total torque (which is usually the torque in one armature slot multiplied by the number of slots. There are three cases to consider:

- The currents in the slot are equal and in the same direction.
- The currents in the slot are equal and in opposite directions.
- The currents in the slot are unequal and in the same direction.

In the first case, that is, the normal case as in slot 'a' Figure 6, the torque generated per slot can be shown to be,

$$\vec{T}_a = l_1 l_2 B i \hat{z} = K_t i \hat{z},$$

where $\vec{T}_a$ is the torque in slot 'a', l_1 is the length of the armature, l_2 is the diameter of the armature, i is the current, B magnetic flux density and K_t is the torque constant of the motor. Unit vector $\hat{z}$ points along the axis of rotation.

In the second case (as in slot 'f' in Figure 7), the torque produced by the two windings in the slot cancel each other out. In the case of slot 'f':

$$\vec{T}_{11} = l_1 \left(\frac{l_2}{2}\right) B i \hat{z},$$

$$\vec{T}_{1'} = -l_1 \left(\frac{l_2}{2}\right) B i \hat{z},$$

$$\vec{T}_{11} + \vec{T}_{1'} = \vec{T}_f = 0.$$

In the third case (as in slot 'c' of Figure 8), the torques produced by the two windings sum, but are of different magnitudes. In the case of slot 'c':

$$\vec{T}_5 = l_1 \left(\frac{l_2}{2}\right) B i_1 \hat{z},$$

$$\vec{T}_6 = l_1 \left(\frac{l_2}{2}\right) B i_2 \hat{z},$$

$$\vec{T}_5 + \vec{T}_6 = \vec{T}_c = l_1 l_2 B (i_1 + i_2)\hat{z} = K_t (i_1 + i_2)\hat{z},$$

where i_1 and i_2 are the currents through the separate current paths.

Expressions for the torque produced by the motor in slow and fast modes can now be derived by summing the torque produced in each individual slot, depending on current pattern,

$$T_{\text{slow}} = T_a + T_b + \cdots + T_l = 10 K_t i_{\text{slow}}, \quad (1)$$

$$T_{\text{fast}} = T_a + T_b + \cdots + T_l = 8 K_t i_{\text{fast}} \quad (2)$$

in scalar form where T_{slow} and T_{fast} are the torques produced in the motor's slow and fast modes, respectively. Likewise, i_{slow} and i_{fast} are the input currents in its slow and fast operations, respectively, with i_{fast} equalling $(i_1 + i_2)$.

When analysing the back electromotive force (EMF) produced by the motor, it is simpler to analyse armature winding loops, rather than armature slots. The following shows an analysis of the EMF produced by a single loop, which will then be applied to the wiper motor under investigation; the analysis is based on Chapter 1 in Chiasson (2005).

Figure 9 represents a single coil of wire in the armature. The transparent section represents the air gap and thus the flux surface of the magnetic field, S, in the motor. Lengths l_1 and l_2 are the height and width of the armature, respectively, meaning that the flux surface is approximately a half cylinder of radius $l_2/2$ and length l_1. Figure 9 shows that the positive direction of travel around the coil is taken to be anti-clockwise, in accordance with the vector $\vec{r}$ pointing radially away from the centre of the armature. The magnetic field in the air gap generated by the permanent magnet is known to be approximately radially directed with a constant magnitude of B. Therefore an expression of the vector $\vec{B}$ is given as:

$$\vec{B} = \begin{cases} +B\hat{r} & \text{for} \quad 0 < \theta < \pi, \\ -B\hat{r} & \text{for} \quad \pi < \theta < 2\pi. \end{cases}$$

A surface element, $d\vec{S}$, of the flux surface defined in Figure 9 is shown in Figure 10; its direction is always outward of the cylinder. From Figure 10, an expression for $d\vec{S}$ can be derived as:

$$d\vec{S} = \left(\frac{l_2}{2}\right) d\theta \, dz \vec{r}, \quad (3)$$

where θ is the position of the coil, with $\theta = 0$ on the magnetic neutral line between slots g and f in Figures 7 and 8. The flux can be derived by integrating the flux density over the air gap shown in Figure 9, making use of the expression

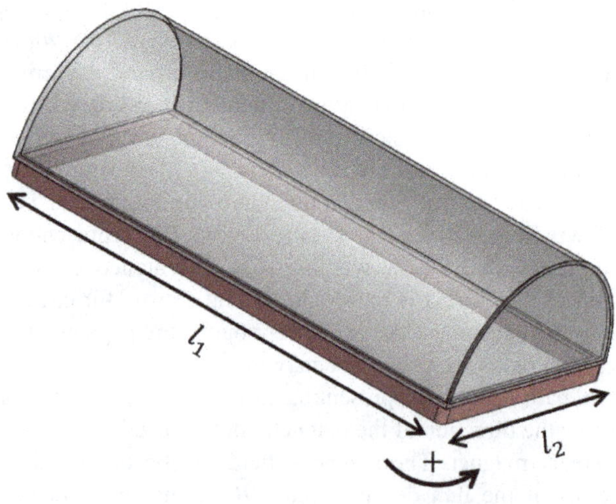

Figure 9. Flux surface of a single coil.

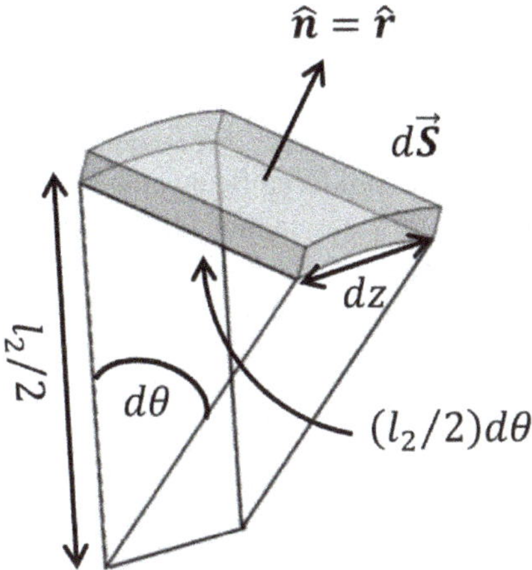

Figure 10. Surface element.

for the surface element area in Equation (3). Thus, using Equation (3), an expression for the flux, ϕ, can be derived:

$$\phi(\theta_R) = \int_S \vec{B} \cdot d\vec{S},$$

$$\phi(\theta_R) = \int_0^{l_1} \int_{\theta=\theta_R}^{\theta=\pi} (B\hat{r}) \cdot \left(\frac{l_2}{2} d\theta\, dz\hat{r}\right) + \int_0^{l_1} \int_{\theta=\pi}^{\theta=\pi+\theta_R} (-B\hat{r}) \cdot \left(\frac{l_2}{2} d\theta\, dz\hat{r}\right),$$

$$\phi(\theta_R) = -l_1 l_2 B\left(\theta_R - \frac{\pi}{2}\right) \quad \text{for } 0 < \theta_R < \pi.$$

Similarly, it can be shown that the flux for $\pi < \theta_R < 2\pi$ is:

$$\phi(\theta_R) = -l_1 l_2 B\left(\theta_R - \frac{\pi}{2} - \pi\right).$$

The induced EMF in the rotor loop can therefore be calculated as:

$$\xi = -\frac{d\phi}{dt} = (l_1 l_2 B)\frac{d\theta_R}{dt} = K_e \omega_R,$$

where ξ is the EMF, K_e is the back EMF constant and equals $l_1 l_2 B$ and ω_R is the rotor's angular velocity.

Using the above expression, an expression for the EMF produced by each coil can be derived for both the slow and fast modes of operation. These are then summed to get an expression for the total EMF produced by the motor in each mode. Referring to Figure 7, it can be seen that the forward and return paths of coil 1 both lie in the positive direction of $\vec{B}$ when the motor is operating in its slow mode. Likewise, coil 12's forward and return paths both lie in the negative direction of $\vec{B}$. This means that these two coils will produce no back EMF. This is also true for the fast mode. Figure 8 shows that the current in coil 4 in the fast mode is in the reverse direction to the so called positive direction defined in Figure 9. This means that the EMF produced by the will be negative. Therefore, the total EMF in the motors slow and fast modes are:

$$\xi_{slow} = 10 K_e \omega_r, \tag{4}$$

$$\xi_{fast} = 8 K_e \omega_r. \tag{5}$$

When a voltage is applied to either of the inputs, the current will see two paths to ground. In the slow operation the resistance of the paths is virtually equal. However in the fast operation, one path is physically shorter than the other and thus its resistance is lower. Considering the two current paths as two resistors in parallel, the overall armature resistance of the motor in its fast mode is smaller than in its slow mode. A similar analysis can be done for the inductance, with the same conclusion.

The eight parameter dynamic simulation model of the wiper motor is implemented as shown in Figure 11. Each separate DC motor represents a separate speed and implements the model with the following equations (Chiasson, 2005):

$$V - R_a I_a - L\frac{dI_a}{dt} - K_e\frac{d\theta}{dt} = 0,$$

$$K_t I_a - J\frac{d^2\theta}{dt^2} - b\frac{d\theta}{dt} - T_L = 0,$$

where V is input voltage, R_a is armature resistance, I_a is armature current, L is armature inductance, K_e is the EMF constant, θ is the rotor position, K_t is the torque constant, J is the motor inertia, b is the damping and T_L is the torque load.

The using Equations (1)–(5), along with the analysis of the resistance and inductance of the motor, it is concluded that to represent the fast mode, the resistance, inductance and K parameters will be lower than the slow mode. The mechanical parameters remain the same for both modes.

The model also includes a gear box to model the worm and wheel gears built into the motor. The park switch is

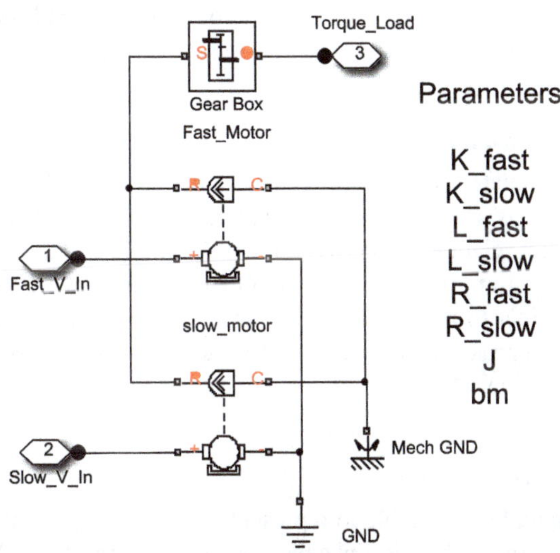

Figure 11. Wiper motor model implementation.

generated by measuring the position of the crank shaft and outputting a pulse when the wipers are in their park position.

4. Modelling for real-time simulation

The model presented in Section 3 is not optimized for real-time simulation. A powerful enough machine could simulate the model in real time; however there are steps that can be taken to make its simulation less computationally intensive.

The off-line model was simulated on a Desk-top PC with 4 GB of RAM and a 3.4 GHz processor. The solver used was a fixed step (ode14x) solver with a step size of 1 ms. Simulating 60 s of data took 86 s, giving a time per simulation step of 1.43 ms, that is, one simulation step of 1 ms takes 1.43 ms to simulate. This means that to simulate the model in real time for HIL, significant increases in code efficiency would be required.

To move from off-line to online real-time simulation there are four areas in which changes can be made (Miller & Wendlandt 2013):

- Increase the step size of the simulation; however the step size is fixed at 0.001 s in this case.
- Utilize the local solvers supplied by SimScape. The local solvers greatly increases the simulation speed but cannot be used when a SimMechanics model is connected to the physical network.
- Reduce the number of iterations the solver makes, reducing accuracy but increasing speed.
- Decrease the overall model fidelity by removing insignificant or irrelevant elements.

After considering the available options, the decision was made to remove the SimMechanics element from the model and replace it with a mathematical model so that local solvers could be used. Therefore the linkage model needed to be replaced, without a significant loss of accuracy.

4.1. Position of the wipers

To simulate the position of the wipers, a simple look-up table was used. This method was chosen because there is a direct relationship between the position of the crank and the position of the wipers. Two look-up tables were used per wiper to represent the forward and backward sweeps. This approximation introduced virtually no error into the simulation.

4.2. Torque load

Using a look-up table to represent the torque load applied to the motor by the linkages would have been unwieldy because, not only is the instantaneous torque value dependent of the position of the rotor, it is also dependent on the motor's angular velocity.

The fundamental shape of the torque produced by the linkages is the same for each revolution, or 'wipe'. The shape of the torque produced by a forward and backward sweep of the wipers has been modelled separately with two polynomials, $f(x)$ and $g(x)$, respectively, where x is the angle of the motor's rotor and is between 0 and 2π. The polynomials are:

$$\begin{aligned} f(x) &= -0.19224x^4 + 0.021796x^3 - 0.15435x^2 \\ &\quad - 0.077812x + 0.68626, \\ g(x) &= 0.4184x^7 - 0.30315x^6 - 0.42441x^5 + 0.12634x^4 \\ &\quad + 0.21484x^3 - 0.10093x^2 - 0.29859x + 0.63276. \end{aligned}$$

Both the average and peak-to-peak values of the torque increase with velocity. This was modelled using Simulink to represent the following equation:

$$\tau = \begin{cases} (C_2 * v) * (f(x) + (C_1 * v)) & \text{if } \cos x \text{ is } + \\ (C_2 * v) * (g(x) + (C_1 * v)) & \text{if } \cos x \text{ is } - \end{cases}, \tag{6}$$

where τ is the torque, v is the velocity of the motor and C_1 and C_2 are unknown unit-less constants to be identified. C_1 determines the average value of the torque and C_2 determines the peak-to-peak value of the fundamental shape.

To estimate C_1 and C_2 a GA was used. A GA is an advanced optimization tool based on the principles of biological evolution and natural selection. Typically, a GA will attempt to minimize a cost function by changing the values of its population using the following operators: selection, mating and mutation. In this case the population consists of the two constants to be identified, C_1 and C_2. For more information on GAs refer to Goldberg (1989) and Haupt and Haupt (2004).

The cost function used in the GA is defined below.

$$\sigma_i = \sum_1^N |\tau - \tau'|, \tag{7}$$

$$\text{cost} = \frac{1}{N}\sum_{i=1}^{5} \sigma_i, \tag{8}$$

where σ_i is the result of the absolute error between the torque, τ, calculated from the off-line SimMechanics model and the torque, τ', calculated by the simplified approximated model described in Equation (6) at a specific speed. N is the number of sampled points, in this case $N = 1001$. The value of *cost* is the sum of σ_i for $i = (1, 2, \ldots, 5)$, that is, the torque load at five different speeds, representing a range of speeds of the wiper system. The result of *cost* is the value to be minimized by the GA.

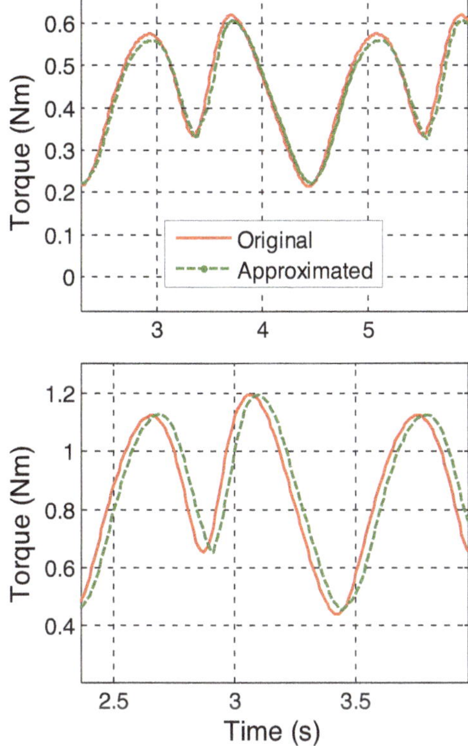

Figure 12. Torque load approximation.

The GA identified the constants as $C_1 = 0.0035$ and $C_2 = 0.2738$. The performance of the torque approximation at the slowest and fastest speeds is shown in Figure 12. It can be seen that the approximated model accurately matches the original torque.

The look-up tables and torque approximation model replace the SimMechanics linkages in the model.

4.3. *Performance*

With the SimMechanics linkages removed and replaced with a simplified model, local solvers can be used. The modified model was simulated under the same conditions as the off-line model and was measured as being able to simulate 60 s of simulation time in 4 s of real time, meaning each simulation step of 1 ms can be simulated in 0.067 ms. The model is now suitable for simulation in real time and ready for implementation in an HIL test facility.

5. Identifying the model parameters

Motor parameterization could be achieved in a number of ways, for example: Measuring the parameters directly, datasheets or by using optimization algorithms. A GA was chosen for the following reasons: (1) Although some parameters can be measured directly, many cannot because the motor and linkages are a single unit and thus no load or constant load tests (including step inputs) are difficult to do. Also, data used in a GA may be available where a physical prototype is not, (2) Because the motor model is unique, datasheets do not contain the relevant information and (3) For off-line parameterization GA is ideal because it is superior at finding a global maximum solution over local optimizers but takes more time, which is not a problem for off-line parameterization.

The eight parameters to identify are: K_fast, K_slow, L_fast, L_slow, R_fast. R_slow, J and bm. The cost function for this GA is more complicated than Equation (7) because it is attempting to minimize the error in both the angle of the wipers and the motor current, hence the cost function is a multi-objective function. The cost function used is:

$$\mathrm{Cost}_{\mathrm{cur}} = W_{\mathrm{cur}} \sum_{j=1}^{N} \sum_{i=1}^{4} \{\mu(j,i) - \mu'(j,i)\}^2, \tag{9}$$

$$\mathrm{Cost}_{\mathrm{pos}} = W_{\mathrm{pos}} \sum_{j=1}^{N} \sum_{i=1}^{4} \{\theta(j,i) - \theta'(j,i)\}^2, \tag{10}$$

$$\mathrm{Cost}_{\mathrm{total}} = \frac{1}{N}(\mathrm{Cost}_{\mathrm{cur}} + \mathrm{Cost}_{\mathrm{pos}}). \tag{11}$$

Equations (9) and (10) compare the simulated current, μ', and simulated wiper angle, θ', with measured current, μ, and position, θ, data, respectively. W_{cur} and W_{pos} are the weights applied to Equations (9) and (10), respectively, to bring the errors of each equation into a similar range. $W_{\mathrm{cur}} = 8$ and W_{pos} is 1, which were chosen to bring the errors in the current and position to with a similar absolute value so one cost function does not dominate in Equation (11). In this case N, which is the number of data points sampled, is 501. Equation (11) sums the value of Equations (9) and (10) and divides it by N to calculate the accumulated cost of the current and angle. In around 50 generations the algorithm found an optimal set of parameters. Table 1 shows the values of the parameters identified by the GA. The parameters were identified using four data sets, representing the full operating range of the system. The results were then validated against a fifth data set which was not used for parameter identification.

The performance of the model in representing the motor current and wiper angle of the real system is shown in Figures 13 and 14 at the slowest and fastest speeds,

Table 1. Wiper motor equivalent parameters.

Parameter	Value	Unit
K_fast	0.0423	V/rpm
K_slow	0.065	V/rpm
L_fast	9.41×10^{-10}	H
L_slow	9.6×10^{-10}	H
R_fast	0.58	Ω
R_slow	0.6	Ω
J	0.004	kgm^2
bm	0.0185	Nm/(rad/s)

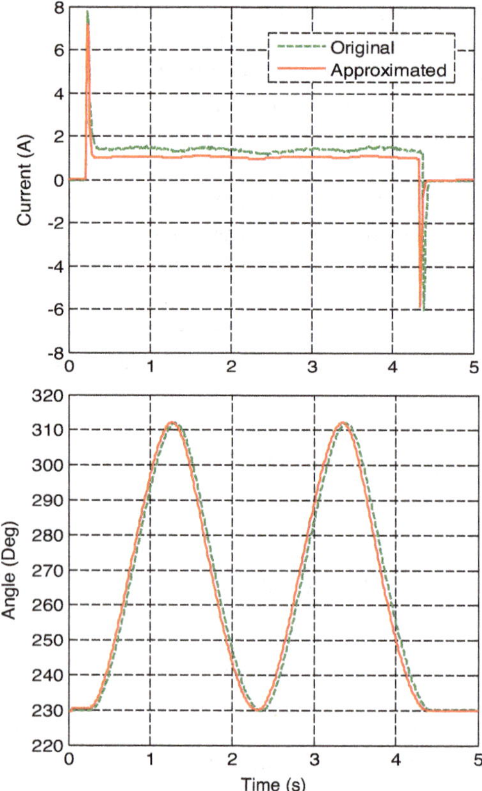

Figure 13. Model performance (slow).

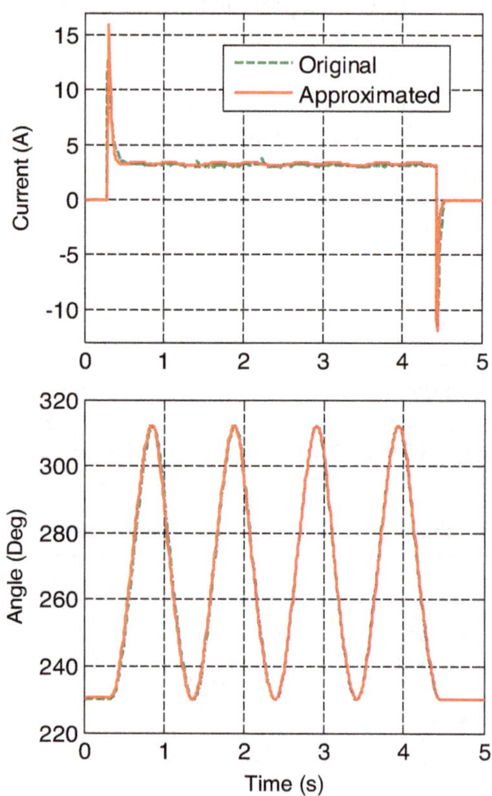

Figure 14. Model performance (fast).

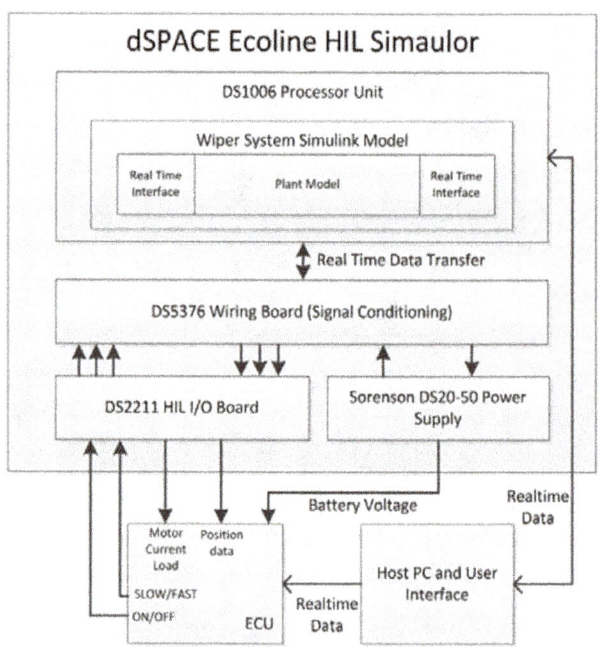

Figure 15. HIL test rig configuration.

respectively. It can be seen that the model accurately recreates the output of the real system, particularly in the fast mode of operation. The largest error, and thus the worst-case performance of this model, is the steady state current error in the slow operation. This error is likely to be due to the simplified friction model implemented in the motor and linkages. The accuracy model could be improved by increasing the complexity and fidelity of the friction model, however this will add extra parameters to be identified by the GA.

6. Implementing the model in an HIL facility

A diagram of the HIL facility in which the wiper model is implemented is shown in Figure 15.

The model is integrated into an existing control system for the wipers and real-time executable code is generated using the MathWorks real-time workshop tool. This code is transferred to the DS1006 processor board and can be simulated in real time whilst interacting with hardware. The results of the simulation can now be used to test the ECU's control system.

7. Conclusion

In the automotive industry ECUs are developed with the aid of HIL simulation. Accurate, real-time capable simulation models of the loads and actuators connected to the ECUs are required in order to carry out this testing. An off-line physical model of a windscreen wiper motor and linkage system has been developed whose parameters can be updated to capture design changes in the development process. The physical model has then been simplified using system identification methods to decrease its simulation time, making it suitable for HIL simulation. GAs were used to identify the unknown parameters of the model. The model has been implements in an HIL rig used for the testing of ECUs.

Further work on this project will be to add the wiper blades and windscreen elements to the model.

Acknowledgements

The authors would like to thank the support in test facilities from the Advantage West Midlands and the European Regional Development Fund for Birmingham Science City Energy Efficiency and Demand Reduction project.

Disclosure statement

No potential conflict of interest was reported by the authors.

References

Chang, S. C., & Lin, H. P. (2004). Chaos attitude motion and chaos control in an automotive wiper system. *International Journal of Solids and Structures*, *41*, 3491–3504. doi:10.1016/j.ijsolstr.2004.02.005

Chiasson, J. (2005). *Modeling and high-performance control of electric machines*. Hoboken, New Jersey: John Wiley & Sons.

Clearwater Corporate Finance, Global Automotive Report. (2013). Retrieved from http://www.clearwatercf.com/

Dooner, M., Wang, J., & Mouzakitis, A. (2013). *Development of a simulation model of a windshield wiper system for Hardware in the Loop simulation*. 19th International Conference on Automation and Computing (ICAC), London.

Ganesh, B. (2005). *Hardware in the loop simulation (HIL) for vehicle electronics systems testing and validation*. SAE Technical Paper 2005-26-304. doi:10.4271/2005-26-304

Goldberg, G. (1989). *Genetic algorithms in search, optimization, and machine learning*. Addison-Wesley Publishing Company.

Grenouillat, R., & Leblanc, C. (2002). *Simulation of chatter vibrations for wiper systems*. SAE Technical Paper 2002-01-1239. doi:10.4271/2002-01-1239

Hameyer, K., & Belmans, R. J. M. (1996). Permanent magnet excited brushed DC motors. *IEEE Transactions on Industrial Electronics*, *43*, 247–255. doi:10.1109/41.491348

Haupt, R., & Haupt, S. (2004). *Practical genetic algorithms*. Hoboken, New Jersey: John Wiley & Sons.

Hillier, V. (1987). *Fundamentals of automotive electronics*. London: Hutchinson Education.

Hu, X., & Azarnasab, E. (2013). Progressive simulation-based design for networked real-time embedded systems. In K. Popovici, & P. J. Mosterman (Eds.), *Real-time simulation technologies: Principles, methodologies and applications* (pp. 181–198). Boca Raton, FL: Taylor Francis Group.

Jeon, S. H., Cho, J. H., Jung, Y., Park, S., & Han, T. M. (2011). *Automotive hardware development according to ISO 26262*. 13th International Conference on Advanced Communication Technology (ICACT), Seoul.

Miller, S., & Wendlandt, J. (2013). Real-time simulation of physical systems using simscape™. In K. Popovici & P. J. Mosterman (Eds.), *Real-time simulation technologies: Principles, methodologies and applications* (pp. 581–597). Boca Raton, FL: Taylor Francis Group.

Okura, S., Sekiguchi, T., & Oya, T. (2000). *Dynamic analysis of blade reversal behavior in a windshield wiper system*. SAE Technical Paper 2000-01-0127. doi:10.4271/2000-01-0127

Robert Bosch GmbH. (2007). *Automotive electrics automotive electronics*. Sussex: John Wily and Sons Ltd.

Schuette, H., & Ploeger, M. (2007). *Hardware-in-the-loop testing of engine control units – a technical survey*. SAE Technical Paper 2007-01-0500. doi:10.4271/2007-01-0500

Von Tils, V. (2006). *Trends and challenges in automotive electronics*. IEEE International Symposium on Power Semiconductor Devices and IC's, Naples.

Waltermann, P. (2009). *Hardware-in-the-loop: The technology for testing electronic controls in automotive engineering*. Translation of 6th Paderborn Workshop 'Designing Mechatronic Systems', Paderborn.

Wei, J., Mouzakitis, A., Wang, J., & Sun, H. (2011). *Vehicle windscreen wiper mathematical model development and optimisation for model based hardware-in-the-loop simulation and control*. 17th International Conference on Automation and Computing (ICAC), Huddersfield.

Wood, G., & Kennedy, D. (2003). *Simulating mechanical systems in Simulink with SimMechanics*. Retrieved from http://www.mathworks.com

Xiaoyu, Z., Yanfeng, X., & Yengjie, L. (2011). *Based on Matllab electrically operated windshield wiper systems design method research*. Third International Conference on Measuring Technology and Mechatronics Automation, Shanghai.

16

Robust H_∞ control of time delayed power systems

Mohammed Jamal Alden and Xin Wang*

Department of Electrical and Computer Engineering, Southern Illinois University Edwardsville, Edwardsville, IL 62026, USA

Power system is the backbone of our society. The purpose of this work is to design a stable, robust and efficient controller for a power-generation system with time delays, model uncertainties and disturbances. Based on the practical dynamics of generator, prime mover, exciter and automatic voltage regulator, a mathematical power-generation system model is developed with state space dynamical equations involving time delays in the feedback. A novel robust H_∞ control framework based on linear matrix inequalities is proposed in the paper, which controls the energy system effectively. Computer simulations are used to show the efficacy of the proposed control algorithm.

Keywords: time delay; linear matrix inequalities; power system; H_∞

Nomenclature

δ	rotor angle
ω_s	synchronous speed
ν	normalized frequency $\nu = \omega/\omega_s$
T'_{do}	equivalent transient rotor time constant
P	total number of poles
S_B	rated three phase voltage ampere
ω_B	rated speed in electrical radians per second, $\omega_B = \omega_s$
H	the shaft inertia constant is scaled by defining $H = \frac{1}{2}J(\omega_B(2/P))^2/S_B$
K_d	the damping factor
I_d	direct axis current
I_q	quadrature axis current
X_d	direct axis reactance
X_q	quadrature axis reactance
X'_d	direct axis transient reactance
E_q	quadrature axis voltage
E'_q	quadrature axis transient voltage
E_{fd}	excitation voltage
E'_{fd}	transient excitation voltage
T_{mech}	mechanical torque
R_s	stator resistance
V_T	terminal bus voltage
V_{ref}	reference voltage
R_e	transmission line resistance
X_e	transmission linear reactance
z_1, z_2, z_3	state variables of power system stabilizer
U_{pss}	power system stabilizer control signal
$\star$	transposed value of the corresponding element

1. Introduction

Power systems are the basic infrastructure of modern civilization in our society. Stability analysis and control system development of the smart power grid are becoming more and more important, due to the rapid deployment of the distributed energy resources. In practical engineering applications, time delay plays a significant role in performance and stability of the overall power systems. Severe delays can even lead to catastrophic breakdown of the entire energy system due to instability (Alrifai, Zribi, Rayan, & Mahmoud, 2013; Bayrak & Tatlicioglu, 2012; Mahmoud, 2000; Okuno & Nakabayashi, 2006; Scorletti & Fromion, 1998; Wu, Ni, & Heydt, 2002; Zribi, Mahmoud, Karkoub, & Lie, 2000). For this reason, extensive research of transient and steady-state stability analysis and controller design have been conducted during the past decade (Jiang, 2007). System design engineers should consider time delays in designing and implementing practical power systems due to their significance (Alrifai et al., 2013; Jiang, Cai, Dorsey, & Qu, 1997; Mahmoud & Zribi, 1999; Wu et al., 2002; Yu, Jia, & Zhao, 2008).

Many different control approaches have been studied in the literature for effectively controlling the time delayed power systems Zhang et al. (2012). Based on an optimal control approach, the effect of time delays on the region of stability for small signals variation is studied in Jia, Yu, Yu, and Wang (2008). Zhang, Jiang, Wu, and Wu (2013) present a robust control method to design a PID type load frequency control of power systems considering time delays. Snyder, Ivanescu, Hadjsaid, Georges,

*Corresponding author. Email: xwang@siue.edu

and Margotin (2000) introduce a robust controller for a wide-area power system involving input time delays. The controller is developed based on the model reduction and linear matrix inequalities (LMIs). An adaptive wide-area damping controller based on generalized predictive control and model identification for time delayed power system is proposed in Yao, Jiang, Wen, Cheng, and Wu (2009). Yu, Zhang, Xie, and Wang (2007) propose a nonlinear robust control algorithm for power system considering signal delays and measurement incompleteness. Yu et al. (2008) discuss the maximal allowable time delay margin for a stable power systems based on the Lyapunov method involving three generators and nine buses. Chowdhury, Kulhare, and Raina (2011) present the nonlinear limit cycle effect of time delays on local stability of the single machine infinite bus system. In Liu, Zhu, and Jiang (2008), the cluster treatment of eigenvalues is introduced to analyze the stability of a power system with time delays in the feedback loop.

This paper presents a general robust control framework based on linear matrix inequality for time delayed power systems. The mathematical dynamics is modeled as a seventh-order nonlinear system with bounded model uncertainties and external disturbances. By formulating the nonlinear control as a convex optimization problem, the linear matrix inequality can provide the optimal and robust solution satisfying the Lyapunov stability and the robust H_∞ performance objective.

This paper is organized as follows: Section 2 discusses the mathematical modeling of the power-generation system. Section 3 presents the novel H_∞ controller design with linear matrix inequality. Computer simulations conducted with MATLAB is given in Section 4. In Section 5, the conclusion is reached and future work is discussed.

2. Mathematical model of power-generation system

In this section, the mathematical model of an infinite bus power system involving a synchronous generator is developed.

2.1. Synchronous generator model

A model for the synchronous generator is given as follows (Wang & Gu, 2014):

$$\dot{E}'_\mathrm{q} = -\frac{1}{T'_\mathrm{do}}(E'_\mathrm{q} - (X_\mathrm{d} - X'_\mathrm{d})I_\mathrm{d} - E_\mathrm{fd}), \tag{1}$$

$$\dot{\delta} = \omega - \omega_\mathrm{s}, \tag{2}$$

$$\dot{\omega} = \frac{\omega_\mathrm{s}}{2H}[T_\mathrm{mech} - (E'_\mathrm{q}I_\mathrm{q} + (X_\mathrm{q} - X'_\mathrm{d})I_\mathrm{d}I_\mathrm{q} + K_\mathrm{d}(\omega - \omega_\mathrm{s}))]. \tag{3}$$

The stator algebraic equations are

$$V_\mathrm{T}\sin(\delta-\theta) + R_\mathrm{s}I_\mathrm{d} - X_\mathrm{q}I_\mathrm{q} = 0, \tag{4}$$

$$E'_\mathrm{q} - V_\mathrm{T}\cos(\delta-\theta) - R_\mathrm{s}I_\mathrm{q} - X'_\mathrm{d}I_\mathrm{d} = 0. \tag{5}$$

Neglecting stator resistance by assuming $R_\mathrm{s} = 0$, we can write the stator dynamical equations as

$$V_\mathrm{T}\sin(\delta-\theta) - X_\mathrm{q}I_\mathrm{q} = 0, \tag{6}$$

$$E'_\mathrm{q} - V_\mathrm{T}\cos(\delta-\theta) - X'_\mathrm{d}I_\mathrm{d} = 0. \tag{7}$$

Since we have

$$(V_\mathrm{d} + jV_\mathrm{q})\,\mathrm{e}^{j(\delta-\pi/2)} = V_\mathrm{T}\,\mathrm{e}^{j\theta}, \tag{8}$$

therefore,

$$V_\mathrm{d} = V_\mathrm{T}\sin(\delta-\theta), \tag{9}$$

$$V_\mathrm{q} = V_\mathrm{T}\cos(\delta-\theta). \tag{10}$$

Substituting in Equations (6) and (7), we get

$$V_\mathrm{d} - X_\mathrm{q}I_\mathrm{q} = 0, \tag{11}$$

$$E'_\mathrm{q} - V_\mathrm{q} - X'_\mathrm{d}I_\mathrm{d} = 0. \tag{12}$$

Assuming zero degree phase angle for the infinite bus voltage, we have

$$(I_\mathrm{d} + \mathrm{j}I_\mathrm{q})\,\mathrm{e}^{\mathrm{j}(\delta-\pi/2)} = \frac{(V_\mathrm{d} + \mathrm{j}V_\mathrm{q})\,\mathrm{e}^{\mathrm{j}(\delta-\pi/2)} - V_\infty\angle 0^\circ}{R_e + \mathrm{j}X_e}, \tag{13}$$

By separating the imaginary and real parts of Equation (13), we get

$$\begin{aligned} R_\mathrm{d}I_\mathrm{d} - X_eI_\mathrm{q} &= V_\mathrm{d} - V_\infty\sin(\delta), \\ X_eI_\mathrm{d} - R_eI_\mathrm{q} &= V_\mathrm{q} - V_\infty\cos(\delta). \end{aligned} \tag{14}$$

By linearizing Equations (11) and (12), we get

$$\begin{bmatrix}\Delta V_\mathrm{d}\\ \Delta V_\mathrm{q}\end{bmatrix} = \begin{bmatrix}0 & X_\mathrm{q}\\ -X'_\mathrm{d} & 0\end{bmatrix}\begin{bmatrix}\Delta I_\mathrm{q}\\ \Delta I_\mathrm{q}\end{bmatrix} + \begin{bmatrix}0\\ \Delta E'_\mathrm{q}\end{bmatrix}. \tag{15}$$

And by linearizing Equation (14), we get

$$\begin{bmatrix}\Delta V_\mathrm{d}\\ \Delta V_\mathrm{q}\end{bmatrix} = \begin{bmatrix}R_e & -X_e\\ X_e & R_e\end{bmatrix}\begin{bmatrix}\Delta I_\mathrm{q}\\ \Delta I_\mathrm{q}\end{bmatrix} + \begin{bmatrix}V_\infty\sin(\delta)\\ -V_\infty\cos(\delta)\end{bmatrix}\Delta\delta. \tag{16}$$

By equating the right-hand sides of Equations (15) and (16), we have

$$\begin{bmatrix}R_e & -(X_e + X_\mathrm{q})\\ (X_e + X'_\mathrm{d}) & R_e\end{bmatrix}\begin{bmatrix}\Delta I_\mathrm{d}\\ \Delta I_\mathrm{q}\end{bmatrix} = \begin{bmatrix}0\\ \Delta E'_\mathrm{q}\end{bmatrix} + \begin{bmatrix}-V_\infty\cos\delta\\ V_\infty\sin\delta\end{bmatrix}\Delta\delta, \tag{17}$$

$\Delta I_d, \Delta I_q$ can be obtained from Equations (15) and (16) as

$$\begin{bmatrix}\Delta I_d\\ \Delta I_q\end{bmatrix} = \frac{1}{\Delta}\begin{bmatrix}(X_e+X_q) & -R_eV_\infty\cos\delta+V_\infty\sin\delta(X_q+X_e)\\ R_e & R_eV_\infty\sin\delta+V_\infty\cos\delta(X'_d+X_e)\end{bmatrix} \times \begin{bmatrix}\Delta E'_q\\ \Delta\delta\end{bmatrix}, \tag{18}$$

where

$$\Delta = R_e^2+(X_e+X_q)(X_e+X'_d). \tag{19}$$

Denote the normalized frequency $\nu=\omega/\omega_s$. The linearized synchronous generator model of Equations (1)–(3) is given as follows:

$$\begin{bmatrix}\Delta\dot{E}'_q\\ \Delta\dot{\delta}\\ \Delta\dot{\nu}\end{bmatrix} = \begin{bmatrix}-\frac{1}{T'_{do}} & 0 & 0\\ 0 & 0 & \omega_s\\ -\frac{I_q^o}{2H} & 0 & -\frac{K_d\omega_s}{2H}\end{bmatrix}\begin{bmatrix}\Delta E'_q\\ \Delta\delta\\ \Delta\nu\end{bmatrix} + \begin{bmatrix}-\frac{1}{T'_d}(X_d-X'_d) & 0\\ 0 & 0\\ \frac{1}{2H}(X'_d-X_q)I_q^o & \frac{1}{2H}(X'_d-X_q)I_d^o-\frac{1}{2H}E'^o_q\end{bmatrix} \times\begin{bmatrix}\Delta I_d\\ \Delta I_q\end{bmatrix} + \begin{bmatrix}\frac{1}{T'_{do}} & 0\\ 0 & 0\\ 0 & \frac{1}{2H}\end{bmatrix}\begin{bmatrix}\Delta E_{fd}\\ \Delta T_{mech}\end{bmatrix}. \tag{20}$$

Substitute for $\Delta I_d, \Delta I_q$, we have

$$\Delta\dot{E}'_q = -\frac{1}{K_3T'_{do}}\Delta E'_q - \frac{K_4}{T'_{do}}\Delta\delta + \frac{1}{T'_{do}}\Delta E_{fd}, \tag{21}$$

$$\Delta\dot{\delta} = \omega_s\Delta\nu, \tag{22}$$

$$\Delta\dot{\nu} = -\frac{K_2}{2H}\Delta E'_q - \frac{K_1}{2H}\Delta\delta - \frac{K_d\omega_s}{2H}\Delta\nu + \frac{1}{2H}\Delta T_{mech}. \tag{23}$$

where

$$\frac{1}{K_3} = 1+\frac{(X_d-X'_d)(X_q+X_e)}{\Delta}, \tag{24}$$

$$K_4 = \frac{V_\infty(X_d-X'_d)}{\Delta}[(X_q+X_e)\sin\delta - R_e\cos\delta], \tag{25}$$

$$K_2 = \frac{1}{\Delta}[I_q^o\Delta - I_q^o(X'_d-X_q)(X_q+X_e) - R_e(X'_d-X_q)I_d^o + R_eE'^o_q], \tag{26}$$

$$K_1 = -\frac{1}{\Delta}[I_q^oV_\infty(X'_d-X_q)\{(X_q+X_e)\sin\delta - R_e\cos\delta\} + V_\infty\{(X'_d-X_q)I_d^o - E'^o_q\}\{(X'_d+X_e)\cos\delta + R_e\sin\delta\}]. \tag{27}$$

Since

$$V_T^2 = V_d^2+V_q^2,$$

the differential terms is given as follows:

$$\Delta V_T = \frac{V_d^o}{V_T}\Delta V_d + \frac{V_q^o}{V_T}\Delta V_q, \tag{28}$$

Substituting Equation (18) into Equation (15), we obtain

$$\begin{pmatrix}\Delta V_d\\ \Delta V_q\end{pmatrix} = \frac{1}{\Delta}\begin{pmatrix}X_qR_e & X_q(R_eV_\infty\sin\delta+V_\infty\cos\delta(X'_d+X_e))\\ -X'_d(X_q+X_e) & -X'_d(-R_eV_\infty\cos\delta+V_\infty(X_q+X_e)\sin\delta)\end{pmatrix}\begin{pmatrix}\Delta E'_q\\ \Delta\delta\end{pmatrix} + \begin{pmatrix}0\\ \Delta E'_q\end{pmatrix}, \tag{29}$$

Based on Equations (28) and (29), we get

$$\Delta V_T = K_5\Delta\delta + K_6\Delta E'_q, \tag{30}$$

where

$$K_5 = \frac{1}{\Delta}\left\{\frac{V_d^o}{V_T}X_q[R_eV_\infty\sin\delta+V_\infty\cos\delta(X'_d+X_e)] + \frac{V_q^o}{V_T}[X'_d(R_eV_\infty\cos\delta) - V_\infty(X_q+X_e)\sin\delta]\right\}, \tag{31}$$

$$K_6 = \frac{1}{\Delta}\left\{\frac{V_d^o}{V_T}X_qR_e - \frac{V_q^o}{V_T}X'_d(X_q+X_e)\right\} + \frac{V_q^o}{V_T}. \tag{32}$$

2.2. *Automatic voltage regulator (AVR) and exciter circuit dynamics*

The following dynamical equations for AVR and excitation control system are adopted:

$$\dot{E}_{fd} = \frac{K_A}{T_A}\left(V_{ref} - V_T + U_{pss} - \frac{E_{fd}}{T_A}\right). \tag{33}$$

By linearizing Equation (33), we get

$$\Delta\dot{E}_{fd} = \frac{K_A}{T_A}(\Delta V_{ref} - \Delta V_T + \Delta U_{pss}) - \frac{\Delta E_{fd}}{T_A}. \tag{34}$$

Based on (30), (34) can be rewritten as

$$\Delta\dot{E}_{fd} = -\frac{\Delta E_{fd}}{T_A} - \frac{K_AK_5}{T_A}\Delta\delta - \frac{K_AK_6}{T_A}\Delta E'_q + \frac{K_A}{T_A}\Delta U_{pss} + \frac{K_A}{T_A}\Delta V_{ref}. \tag{35}$$

A typical power system stabilizer (PSS) control scheme include a washout filter and two lead-lag blocks. The retarded measure of ν propagates in the PSS equations. The linearized form of PSS can be modeled as

$$\begin{aligned}
\Delta\dot{z}_1 &= \frac{-(K_w\Delta\nu(t-\tau)+\Delta z_1)}{T_w},\\
\Delta\dot{z}_2 &= \frac{[(1-T_1/T_2)(K_w\Delta\nu(t-\tau)+\Delta z_1)-\Delta z_2]}{T_2},\\
\Delta\dot{z}_3 &= \frac{\{(1-T_3/T_4)[\Delta z_2+(T_1/T_2)(K_w\Delta\nu(t-\tau)+\Delta z_1)]-\Delta z_3\}}{T_4},\\
\Delta U_{\text{pss}} &= \Delta z_3+\frac{T_3}{T_4}\left[\Delta z_2+\frac{T_1}{T_2}(K_w\Delta\nu(t-\tau)+\Delta z_1)\right].
\end{aligned} \tag{36}$$

Hence, the overall linearized model of the power-generation system is given as follows:

$$\Delta\dot{\delta} = \omega_s\Delta\nu, \tag{37}$$

$$\Delta\dot{\nu} = -\frac{K_2}{2H}\Delta E_q' - \frac{K_1}{2H}\Delta\delta - \frac{K_d\omega_s}{2H}\Delta\nu + \frac{1}{2H}\Delta T_{\text{mech}}, \tag{38}$$

$$\Delta\dot{E}_q' = -\frac{1}{K_3T_{\text{do}}'}\Delta E_q' - \frac{K_4}{T_{\text{do}}'}\Delta\delta + \frac{1}{T_{\text{do}}'}\Delta E_{\text{fd}}, \tag{39}$$

$$\begin{aligned}
\Delta\dot{E}_{\text{fd}} = &-\frac{\Delta E_{\text{fd}}}{T_A} - \frac{K_AK_5}{T_A}\Delta\delta - \frac{K_AK_6}{T_A}\Delta E_q' + \frac{K_A}{T_A}\Delta V_{\text{ref}}\\
&+\frac{K_A}{T_A}\left\{\Delta z_3+\frac{T_3}{T_4}\left[\Delta z_2 +\frac{T_1}{T_2}(K_w\Delta\nu(t-\tau)+\Delta z_1)\right]\right\},
\end{aligned} \tag{40}$$

$$\Delta\dot{z}_1 = \frac{-(K_w\Delta\nu(t-\tau)+\Delta z_1)}{T_w}, \tag{41}$$

$$\Delta\dot{z}_2 = \frac{[(1-T_1/T_2)(K_w\Delta\nu(t-\tau)+\Delta z_1)-\Delta z_2]}{T_2}, \tag{42}$$

$$\Delta\dot{z}_3 = \frac{\{(1-T_3/T_4)[\Delta z_2+((T_1/T_2)(K_w\Delta\nu(t-\tau)+\Delta z_1)]-\Delta z_3\}}{T_4}. \tag{43}$$

Denote $x=[\Delta\delta,\Delta\nu,\Delta E_q',\Delta E_{\text{fd}},\Delta z_1,\Delta z_2,\Delta z_3]^t$ and $u=[\Delta T_{\text{mech}},\Delta V_{\text{ref}}]^t$, the linearized model becomes

$$\dot{x} = Ax(t) + A_dx(t-\tau) + Bu(t), \tag{44}$$

where

$$A = \begin{pmatrix}
0 & \omega_s & 0 & 0 & 0 & 0 & 0\\
-\frac{K_1}{2H} & -\frac{K_d\omega_s}{2H} & -\frac{K_2}{2H} & 0 & 0 & 0 & 0\\
-\frac{K_4}{T_{\text{do}}'} & 0 & -\frac{1}{K_3T_{\text{do}}'} & -\frac{1}{T_{\text{do}}'} & 0 & 0 & 0\\
-\frac{K_AK_5}{T_A} & 0 & -\frac{K_AK_6}{T_A} & -\frac{1}{T_A} & \frac{K_AT_3T_1}{T_AT_4T_2} & \frac{K_AT_3}{T_AT_4} & \frac{K_A}{T_A}\\
0 & 0 & 0 & 0 & -\frac{1}{T_w} & 0 & 0\\
0 & 0 & 0 & 0 & \left(1-\frac{T_1}{T_2}\right)\frac{1}{T_2} & -\frac{1}{T_2} & 0\\
0 & 0 & 0 & 0 & \left(1-\frac{T_3}{T_4}\right)\frac{T_1}{T_2}\frac{1}{T_4} & \left(1-\frac{T_3}{T_4}\right)\frac{1}{T_4} & -\frac{1}{T_4}
\end{pmatrix}, \tag{45}$$

$$A_d = \begin{pmatrix}
0 & 0 & 0 & 0 & 0 & 0 & 0\\
0 & 0 & 0 & 0 & 0 & 0 & 0\\
0 & 0 & 0 & 0 & 0 & 0 & 0\\
0 & \frac{T_3T_1K_w}{T_4T_2} & 0 & 0 & 0 & 0 & 0\\
0 & -\frac{K_w}{T_w} & 0 & 0 & 0 & 0 & 0\\
0 & \left(1-\frac{T_1}{T_2}\right)K_w\frac{1}{T_2} & 0 & 0 & 0 & 0 & 0\\
0 & \left(1-\frac{T_3}{T_4}\right)\frac{T_1}{T_2}K_w\frac{1}{T_4} & 0 & 0 & 0 & 0 & 0
\end{pmatrix}, \tag{46}$$

$$B = \begin{pmatrix}
0 & 0\\
\frac{1}{2H} & 0\\
0 & 0\\
0 & \frac{K_A}{T_A}\\
0 & 0\\
0 & 0\\
0 & 0
\end{pmatrix}. \tag{47}$$

3. Robust H_∞ controller design

In this section, we propose the novel design of the robust controller satisfying H_∞ performance objective. The system with state and input delays, uncertainties and disturbances is considered. The system is of the form:

$$\dot{x}(t) = (A + \delta A)x(t) + (A_d + \delta A_d)x(t - \tau_s) + (B + \delta B)u(t) + (B_d + \delta B_d)u(t - \tau_i) + Dw(t), \tag{48}$$

the performance output is chosen as

$$z(t) = Ex(t) \tag{49}$$

and

$$x(t) = \phi(t) \quad \text{for } t \in [-\tau_s, 0]. \tag{50}$$

Assume that the state variables are available for feedback. Otherwise, estimators can be developed for state estimation purposes. Then, we have state feedback control input as

$$u(t) = Kx(t).$$

Therefore, the closed-loop system becomes:

$$\dot{x}(t) = (A + \delta A + BK + \delta B)x(t) + (A_d + \delta A_d)x(t - \tau_s) + (B_d + \delta B_d)Kx(t - \tau_i) + Dw(t). \tag{51}$$

Rearranging Equation (51), we get

$$\dot{x}(t) = A_c x(t) + \delta A_c x(t) + (A_d + \delta A_d)x(t - \tau_s) + (B_d + \delta B_d)Kx(t - \tau_i) + Dw(t), \tag{52}$$

where

$$A_c = A + BK,$$
$$\delta A_c = \delta A + \delta BK.$$

Before proceeding to the theorem derivation, Assumption 1 and Lemma 1 are introduced (Wang, Yaz, & Yaz, 2010).

ASSUMPTION 1 *The general form of unstructured L_2 bounded uncertainties is used in this work*:

$$\delta A \delta A^t \leq \gamma_A I,$$
$$\delta A_d \delta A_d^t \leq \gamma_{A_d} I,$$
$$\delta B \delta B^t \leq \gamma_B I,$$
$$\delta B_d \delta B_d^t \leq \gamma_{B_d} I.$$

LEMMA 1

$$AB^t + BA^t \leq \alpha AA^t + \alpha^{-1} BB^t.$$

To prove this inequality, we can consider the following equivalent inequality which always holds, given arbitrary $\alpha > 0$:

$$(\alpha^{1/2}A - \alpha^{-1/2}B)(\alpha^{1/2}A - \alpha^{-1/2}B)^t \geq 0.$$

Furthermore, if A and B are chosen to be $\left[\begin{smallmatrix} a^t \\ 0 \end{smallmatrix}\right]$ and $\left[\begin{smallmatrix} 0 \\ b^t \end{smallmatrix}\right]$, respectively, we get

$$\begin{bmatrix} 0 & a^t b \\ b^t a & 0 \end{bmatrix} \leq \begin{bmatrix} \zeta a^t a & 0 \\ 0 & \zeta^{-1} b^t b \end{bmatrix}.$$

Based on Assumption 1 and Lemma 1, the main theorem of the paper is summarized as follows:

THEOREM 1 *Under the feedback control law $u(t) = Kx(t)$, the system of Equation (52) is asymptotically stable for all delays satisfying $\tau_s, \tau_i \geq 0$. And the H_∞ performance objective $\|T_{zw}\|_\infty \leq \gamma, \gamma > 0$ can be satisfied. If there exist matrices $Y, Q_t^t = Q_t > 0$ and $Q_s^t = Q_s > 0$ satisfying the following LMI*:

$$\begin{bmatrix} m_1 & A_d X & B_d Y & D & X & Y^t \\ XA_d^t & m_2 & 0 & 0 & 0 & 0 \\ Y^t B_d^t & 0 & m_3 & 0 & 0 & 0 \\ D^t & 0 & 0 & m_4 & 0 & 0 \\ X & 0 & 0 & 0 & m_5 & 0 \\ Y & 0 & 0 & 0 & 0 & m_6 \end{bmatrix} < 0, \tag{53}$$

where

$$\begin{aligned} m_1 &= AX + BY + XA^t + Y^t B + Q_t + Q_s + \alpha_1(\gamma_A + \gamma_B)I, \\ m_2 &= \alpha_2^{-1} I - Q_t, \\ m_3 &= \alpha_3^{-1} I - Q_s, \\ m_4 &= -\gamma^2 I, \\ m_5 &= -[\alpha_1^{-1} I + \alpha_2 \gamma_{A_d} I + E^t E]^{-1}, \\ m_6 &= -[\alpha_1^{-1} I + \alpha_3 \gamma_{B_d} I]^{-1}. \end{aligned} \tag{54}$$

Proof A Lyapunov–Krasovskii function is chosen as follows:

$$V(x,t) = x^t(t)Px(t) + \int_{t-\tau_s}^{t} x^t(v)Q_1 x(v)\,dv + \int_{t-\tau_i}^{t} x^t(v)Q_2 x(v)\,dv \tag{55}$$

where $V(x, t)$ is a positive semi-definite functional and the matrices P, Q_1, Q_2 are all positive definite.

By taking derivative, we have

$$\dot{V}(x,t) = \dot{x}^t(t)Px(t) + x^t(t)P\dot{x}(t) + x^t(t)Q_1 x(t) - x^t(t - \tau_s)Q_1 x(t - \tau_s) + x^t(t)Q_2 x(t) - x^t(t - \tau_i)Q_2 x(t - \tau_i). \tag{56}$$

Based on LaSalle's theorem, in order to achieve the asymptotic stability, the conditions $V > 0$ and $\dot{V} < 0$ need to be satisfied.

In order to satisfy H_∞ performance objective, the following H_∞ performance inequality needs be employed.

$$J = \int_0^\infty (z^t z - \gamma^2 \omega^t \omega)\,dt < 0. \tag{57}$$

The sufficient condition to achieve both the asymptotic stability and H_∞ performance objective is

$$J = \int_0^\infty (z^t z - \gamma^2 \omega^t \omega + \dot{V})\,dt < 0. \tag{58}$$

Condition (58) implies

$$z^t z - \gamma^2 \omega^t \omega + \dot{V} < 0. \tag{59}$$

By substituting Equations (52) into (56), we get

$$\begin{aligned}\dot{V}(x,t) = {} & [A_c x(t) + \delta A_c x(t) + (A_d + \delta A_d) x(t-d) \\ & + (B_d + \delta B_d) K x(t-\tau_i) + D w(t)]^t P x \\ & + x^t P [A_c x(t) + \delta A_c x(t) + (A_d + \delta A_d) x(t-d) \\ & + (B_d + \delta B_d) K x(t-\tau_i) + D w(t)] + x^t(t) Q_1 x(t) \\ & - x^t(t-\tau_s) Q_1 x(t-\tau_s) + x^t(t) Q_2 x(t) \\ & - x^t(t-\tau_i) Q_2 x(t-\tau_i) < 0. \end{aligned} \tag{60}$$

Based on condition (58), we have

$$\begin{aligned} & [A_c x(t) + \delta A_c x(t) + (A_d + \delta A_d) x(t-d) \\ & + (B_d + \delta B_d) K x(t-\tau_i) + D w(t)]^t P x \\ & + x^t P [A_c x(t) + \delta A_c x(t) + (A_d + \delta A_d) x(t-d) \\ & + (B_d + \delta B_d) K x(t-\tau_i) + D w(t)] + x^t(t) Q_1 x(t) \\ & - x^t(t-\tau_s) Q_1 x(t-\tau_s) + x^t(t) Q_2 x(t) \\ & - x^t(t-\tau_i) Q_2 x(t-\tau_i) + z^t z - \gamma^2 I < 0. \end{aligned} \tag{61}$$

Denote $\zeta(t) = [x^t(t) \quad x^t(t-\tau_d) \quad x^t(t-\tau_i) \quad \omega^t(t)]^t$, then Equation (61) can be written as

$$\dot{V}(x,t) = \zeta^t(t) W_o \zeta(t),$$

where

$$W_o = \begin{bmatrix} P(A_c+\delta A_c) + (A_c+\delta A_c)^t P + Q_1 + Q_2 + E^t E & P(A_d+\delta A_d) & P(B_d+\delta B_d)K & PD \\ (A_d+\delta A_d)^t P & -Q_1 & 0 & 0 \\ K^t (B_d+\delta B_d)^t P & 0 & -Q_2 & 0 \\ D^t P & 0 & 0 & -\gamma^2 I \end{bmatrix} < 0. \tag{62}$$

By pre- and post-multiplying Equation (62) with the diagonal matrix $\mathrm{diag}(X, I, I, I)$ and denote

$$X = P^{-1}, \quad Y = KX, \quad Q_t = XQ_1X, \quad Q_s = XQ_2X,$$

we get

$$\begin{bmatrix} \phi_{11} & \phi_{12} & \phi_{13} & D \\ \star & \phi_{22} & 0 & 0 \\ \star & 0 & \phi_{33} & 0 \\ \star & 0 & 0 & \phi_{44} \end{bmatrix} < 0, \tag{63}$$

where

$$\begin{aligned} \phi_{11} &= (A + BK + \delta A + \delta BK)X + X(A + BK + \delta A + \\ & \quad \delta BK)^t + Q_s + Q_t + XE^t EX, \\ \phi_{12} &= (A_d + \delta A_d)X, \\ \phi_{13} &= (B_d + \delta B_d)Y, \\ \phi_{22} &= -Q_t, \\ \phi_{33} &= -Q_s, \\ \phi_{44} &= -\gamma^2 I. \end{aligned}$$

Applying Lemma 1, we get

$$\begin{aligned} & (\delta A + \delta BK)X + X(\delta A + \delta BK)^t \\ & = X[I \quad K^t]^t \begin{bmatrix} \delta A^t \\ \delta B^t \end{bmatrix} + [\delta A \quad \delta B] \begin{bmatrix} I \\ K \end{bmatrix} X \\ & \leq \alpha_1 [\delta A \quad \delta B] \begin{bmatrix} \delta A^t \\ \delta B^t \end{bmatrix} + \alpha_1^{-1} X [I \quad K^t] \begin{bmatrix} I \\ K \end{bmatrix} X. \end{aligned}$$

By applying Assumption 1, we have

$$\begin{aligned} (\delta A + \delta BK)X + X(\delta A + \delta BK)^t & \leq \alpha_1 (\gamma_A I + \gamma_B I) \\ & + \alpha_1^{-1} X [I \quad K^t] \begin{bmatrix} I \\ K \end{bmatrix} X. \end{aligned} \tag{64}$$

Based on Lemma 1 and Assumption 1, the following matrix inequality is reached:

$$\begin{bmatrix} 0 & \delta A_d X & \delta B_d Y & 0 \\ X^t \delta A_d^t & 0 & 0 & 0 \\ Y^t \delta B_d^t & 0 & 0 & 0 \\ 0 & 0 & 0 & 0 \end{bmatrix} \leq \begin{bmatrix} \alpha_2 \gamma_{A_d} X^t X + \alpha_3 \gamma_{B_d} Y^t Y & 0 & 0 & 0 \\ 0 & \alpha_2^{-1} I & 0 & 0 \\ 0 & 0 & \alpha_3^{-1} I & 0 \\ 0 & 0 & 0 & 0 \end{bmatrix}. \tag{65}$$

Now, by substituting Equations (64) and (65) into Equation (63) and applying Schur complement, we obtain the following LMI result:

$$\begin{bmatrix} \zeta_{11} & \zeta_{12} & \zeta_{13} & D & X & Y^t \\ \star & \zeta_{22} & 0 & 0 & 0 & 0 \\ \star & 0 & \zeta_{33} & 0 & 0 & 0 \\ \star & 0 & 0 & \zeta_{44} & 0 & 0 \\ \star & 0 & 0 & 0 & \zeta_{55} & 0 \\ \star & 0 & 0 & 0 & 0 & \zeta_{66} \end{bmatrix} < 0, \tag{66}$$

where

$$\zeta_{11} = AX + BY + XA^t + Y^tB + Q_t + Q_s + \alpha_1(\gamma_A + \gamma_B)I,$$
$$\zeta_{12} = A_dX,$$
$$\zeta_{13} = B_dY,$$
$$\zeta_{22} = \alpha_2^{-1}I - Q_t,$$
$$\zeta_{33} = \alpha_3^{-1}I - Q_s,$$
$$\zeta_{44} = -\gamma^2 I,$$
$$\zeta_{55} = -[\alpha_1^{-1}I + \alpha_2\gamma_{A_d}I + E^tE]^{-1},$$
$$\zeta_{66} = -[\alpha_1^{-1}I + \alpha_3\gamma_{B_d}I]^{-1}.$$

■

4. Simulation and results

The following parameters are used for simulations. Assuming that $R_e = 0, X_e = 0.5pu$, $V_T\angle\theta = 1\angle 15^\circ pu$, and $V_\infty\angle 0^\circ = 1.05\angle 0^\circ pu$.

The generator, automatic voltage regulator and exciter parameters are $H = 3.2$ s, $T'_{do} = 9.6$ s, $K_A = 400$, $T_A = 0.2$ s, $R_s = 0$ pu, $X_q = 2.1$ pu, $X_d = 2.5$ pu, $X'_d = 0.39$ pu, $K_d = 0$, and $\omega_s = 377$.

The PSS parameters are $K_w = 0.5, T_1 = 0.5, T_2 = 0.01, T_3 = 1, T_4 = 0.1, T_w = 10$.

Based on Equations (24)–(27), (31) and (32), we can calculate the values: $K_1 = 0.9224$, $K_2 = 1.0739$, $K_3 = 0.296667, K_4 = 2.26555, K_5 = 0.005, K_6 = 0.3572$.

The L_2 of disturbance is chosen as $w(t) = 5 \times 0.9^t$, notice that the disturbance energy is finite.

MATLAB robust control toolbox provide the capability to design the optimal control feedback with LMI. Computer simulation shows that our proposed controller effectively stabilizes the time response of rotor angle in Figure 1, normalized frequency in Figure 2, quadrature axis transient voltage in Figure 3 and excitation voltage in Figure 4. The control input is shown in Figure 5. Simulation results have demonstrated the effectiveness and robustness of our proposed approach (Figure 5).

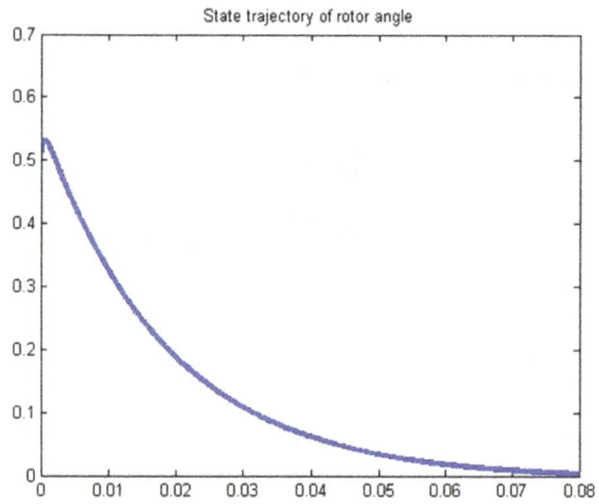

Figure 1. Time response of rotor angle.

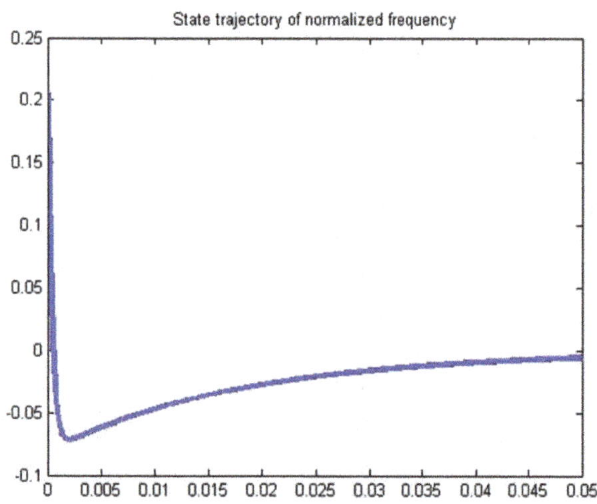

Figure 2. Time response of normalized frequency.

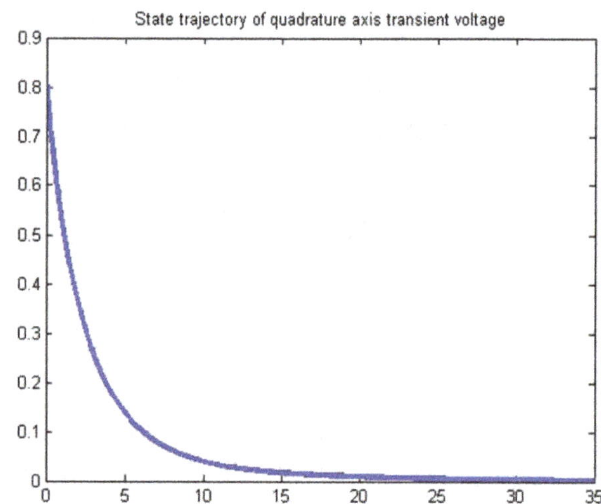

Figure 3. Time response of quadrature axis transient voltage.

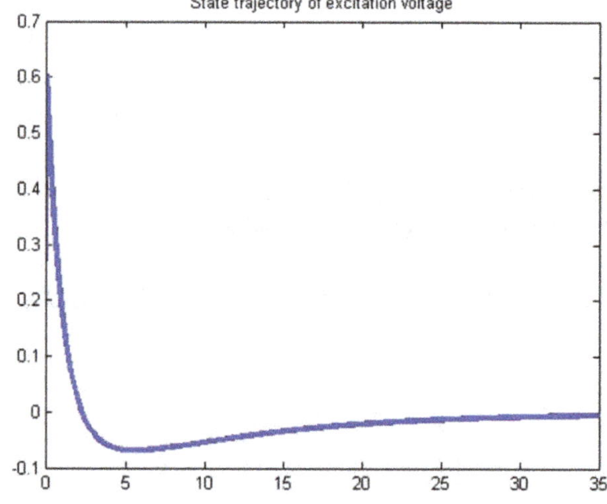

Figure 4. Time response of excitation voltage.

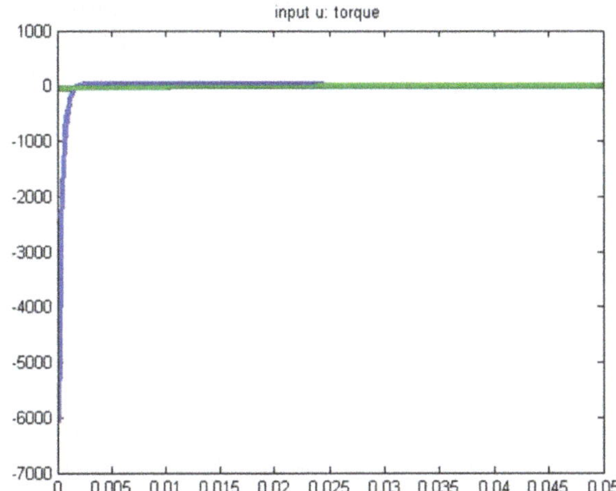

Figure 5. Control input.

5. Conclusion

A general robust H_∞ control approach is proposed in this paper for a power-generation system with state and input delays, disturbances and model uncertainties. The uncertainties are assumed to be bounded and unstructured. A novel seventh-order state space model for the power-generation system is developed. Computer simulation studies conducted through the use of MATLAB show the robustness and effectiveness of the novel approach. Notice that our LMI control solution only applies for delay independent power system control applications. In the future development, we would investigate the delay-dependent cases based on the Discretized Lyapunov–Krasovskii functional method.

Disclosure statement

No potential conflict of interest was reported by the authors.

References

Alrifai M., Zribi M., Rayan M., & Mahmoud M. (2013). On the control of time delay power systems. *International Journal of Innovative Computing, Information and Control*, *9*(2), 762–792.

Bayrak A., & Tatlicioglu E. (2012). *Online time delay identification and control for general classes of nonlinear systems.* Proceedings of the IEEE 51st annual conference on decision and control, Maui, HI, USA (pp. 1591–1596).

Chowdhury B., Kulhare A., & Raina G. (2011). *A study of the SMIB power system model with delayed feedback.* Proc. of international conference on power and energy systems, Chennai, India (pp. 1–6).

Jia H., Yu X., Yu Y., & Wang C. (2008). Power system small signal stability region with time delay. *International Journal of Electrical Power and Energy Systems*, *30*(1), 16–22.

Jiang Z. (2007). *Design of power system stabilizers using synergetic control theory.* Proceedings of IEEE power engineering society general meeting, Tampa, FL, USA (pp. 1–8).

Jiang H., Cai H., Dorsey J. F., & Qu Z. (1997). Toward a globally robust decentralized control for large-scale power systems. *IEEE Transactions on Control Systems Technology*, *5*(3), 309–319.

Liu Z., Zhu C., & Jiang Q. (2008). *Stability analysis of time delayed power system based on cluster treatment of characteristic roots method.* Proceeding of IEEE power and energy society general meeting, Pittsburgh, PA, USA (pp. 1–6).

Mahmoud M. (2000). *Robust control and filtering for time-delay systems.* New York, NY: Marcel Dekker Control Engineering.

Mahmoud M. S., & Zribi M. (1999). H_∞-controllers for time-delay systems using linear matrix inequalities. *Journal of Optimization Theory and Applications*, *100*(1), 89–122.

Okuno H., & Nakabayashi T. (2006). *Basin of attraction and controlling chaos of five-synchronous-generator infinite-bus system.* Proc. of the IEEE international symposium on industrial electronics, Montreal, Quebec, Canada (pp. 1818–1823).

Saad M. S., Hassouneh M. A., Abed E. H., & Edris A. (2005). *Delaying instability and voltage collapse in power systems using SVCs with washout filter-aided feedback.* Proceedings of the 2005 American control conference, Portland, OR, USA (pp. 4357–4362).

Scorletti G., & Fromion V. (1998). *A unified approach to time-delay system control: Robust and gain-scheduled.* Proceedings of the American control conference, Philadelphia, PA (pp. 2391–2395).

Snyder A. F., Ivanescu D., Hadjsaid N., Georges D., & Margotin T. (2000). *Delayed-input wide-area stability control with synchronized phasor measurements and linear matrix inequalities.* Proc. of IEEE power engineering society summer meeting, Seattle, WA (pp. 1009–1014).

Wang X., & Gu K. (2014). *Time-delay power systems control and stability with discretized Lyapunov functional method.* Proceeding of the 2014 Chinese control conference. Invited Session Paper, Nanjing, Jiangsu, China.

Wang X., Yaz E. E., & Yaz Y. I. (2010). *Robust and resilient state dependent control of continuous-time nonlinear systems with general performance criteria.* Proceedings of the 49th IEEE conference on decision and control (CDC) (pp. 603–608).

Wu H., Ni H., & Heydt G. T. (2002). *The impact of time delay on robust control design in power systems.* Proc. of power engineering society winter meeting, New York, NY, USA (pp. 1511–1516).

Yao W., Jiang L., Wen J. Y., Cheng S. J., & Wu Q. H. (2009). *An adaptive wide-area damping controller based on generalized predictive control and model identification.* Proc. of IEEE power and energy society general meeting, Calgary, AB, Canada (pp. 1–7).

Yu X., Jia H., & Zhao J. (2008). *A LMI based approach to power system stability analysis with time delay.* Proceedings of IEEE region 10 conference (TENCON), Hyderabad, India (pp. 1–6).

Yu G. L., Zhang B. H., Xie H., & Wang C. G. (2007). *Wide-area measurement-based nonlinear robust control of power system considering signals delay and incompleteness.* Proceedings of IEEE power engineering society general meeting, Tampa, FL, USA (pp. 1–8) .

Zhang C., Jiang L., Wu Q. H., & Wu M. (2013). Delay-dependent robust load frequency control for time delay power systems. *IEEE Transactions on Power Systems*, *28*(3), 2192–2201.

Zhang W., Xu F., Hu W., Li M., Ge W., & Wang Z. (2012). *Research of coordination control system between nonlinear robust excitation control and governor power system stabilizer in multi-machine power system*. Proceedings of the IEEE international conference on power system technology, Auckland, New Zealand (pp. 1–5).

Zribi M., Mahmoud M. S., Karkoub M., & Lie T. T. (2000). H_∞-controllers for linearised time-delay power systems. *IEEE Transaction on Generation, Transmission and Distribution, 147*(6), 401–408.

17

Dynamic modelling and simulation of IGCC process with Texaco gasifier using different coal

Yue Wang[a], Jihong Wang[a]*, Xing Luo[a], Shen Guo[a], Junfu Lv[b] and Qirui Gao[b]

[a]*School of Engineering, The University of Warwick, Coventry CV4 7AL, UK;* [b]*Department of Thermal Engineering, Tsinghua University, Beijing 100084, People's Republic of China*

Integrated gasification combined cycle (IGCC) is considered as a viable option for low emission power generation and carbon dioxide sequestration. As a part of the process of IGCC plant design and development, modelling and simulation study of the whole IGCC process is important for thermodynamic performance evaluation, study of carbon capture readiness and economic analysis. The work presented in the paper is to develop such a whole system model and simulation platform. A simplified dynamic model for the IGCC process is developed, in which Texaco gasifier is chosen to give the basic representation for the IGCC process. The chemical equilibriums principle is used to predict the syngas contents in the modelling procedure. The influences of key parameters to regulate the input such as oxygen/coal ratio and water/coal ratio to syngas generation are studied. The simulation results are validated by comparing with the industry data provided by the Lu-nan fertilizer factory. Water-shift reactor, gas turbine and heat recovery steam-generation modules are modelled to study the dynamic performance with respect to the variation from the input of syngas stream. The simulation results reveal the dynamic changes in the plant outputs, including gas temperature, power output and mole percentages of hydrogen and carbon dioxide in the syngas. The process dynamic responses with three types of coal inputs are studied in the paper and their dynamic variation trends are presented via the simulation results.

Keywords: chemical equilibrium; dynamic performance; IGCC; syngas; Texaco gasifier

Nomenclature

A	carbon conversion rate (–)
C_p	specific heat capacity (kJ/(kmol·K))
N	molar flow rate (kmol/s)
D	derivative
H	enthalpy (kJ/kmol)
ΔH_{f}	enthalpy of formation (kJ/kmol)
T	Kelvin temperature (K)
X	mole faction
S	simulation
R	reference
P	pressure
Ar	Argon
C	carbon
CH_4	methane
CO	carbon monoxide
CO_2	carbon dioxide
COS	cabonyl sulphide
g	gas
H	hydrogen
H_2	hydrogen element
H_2O	water, vapour
H_2S	hydrogen sulphide
i	gas species
N_2	nitrogen
O_2	oxygen
S	sulphur element
SO_2	sulphur dioxide
0	original value

Abbreviation

CC	combined cycle
HRSG	heat recovery steam generator
WGS	water gas shift reactor

1. Introduction

Integrated gasification combined cycle (IGCC) offers the benefits over conventional coal-fired power plants, especially with regard to the environment and feedstock flexibility (Casclla & Colonna, 2012; Yang et al., 2011). The gasification process of solid fuel, such as coal and biomass, generates much less pollutants than the direct burning process in traditional coal-fired power plants (Maurstad, 2005; Sun, Liu, Chen, Zhou, & Su 2011). IGCC shows its own

*Corresponding author. Email: jihong.wang@warwick.ac.uk

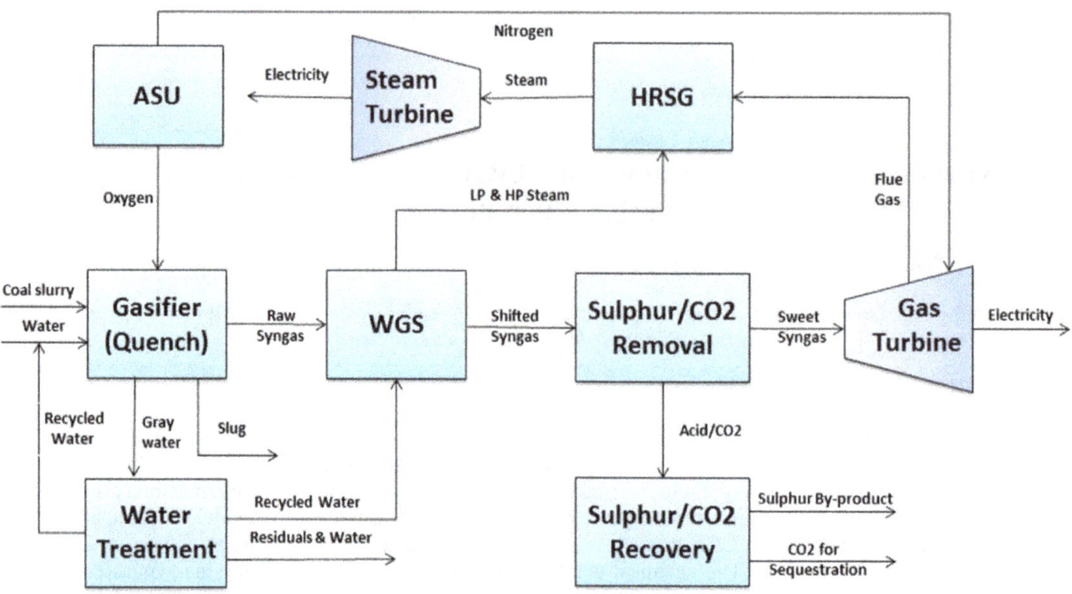

Figure 1. Simplified schematic diagram of an IGCC system (Casella & Colonna, 2012).

merits for integration with carbon capture and storage units. Figure 1 shows a schematic diagram of the whole IGCC process which is, no doubt, a complicated energy conversion process formed by many interconnected subsystem modules. In the IGCC process, the coal slurry and oxygen react in the gasifier and generate syngas primarily made of H_2 and CO. The WGS raises the H_2 and CO_2 concentration while reducing the CO content. The sweet syngas combustion in the gas turbine will generate power and hot flue gas, which is used in HRSG to generate power as well.

The work on IGCC process modelling can be dated back to the 1970s. Researchers in the chemical engineering field studied the coal gasification process and developed models based on mass and energy balances (Beér, 2000; Brown, Smoot, & Hedman, 1986; Buskies, 1996; Chen et al., 2004; Govind & Shah, 1984; Ni & Williams, 1995; Smoot & Smith, 1979; Ubhayakar, Stickler, & Gannon, 1977; Watkinson, Lucas, & Lim, 1991; Wen & Chaung, 1979). Researchers in thermal engineering developed gas turbine and steam power plant models (Colonna & van Putten, 2007; Lu & Hogg, 2000; van Putten & Colonna, 2007). However, dynamic modelling of the whole process is still not mature and requires further study.

Among all the sub-modules of the IGCC process, the most important and complicated one is the one for the gasifier. The challenge of modelling the gasification process will need to deal with complicated chemical reactions involved in the process. The earliest report on modelling this process was published in the 1970s (Ubhayakar et al., 1977), in which a one-dimensional model was reported with consideration of fluid mix in the axial direction. Smoot and Smith (1979) provided an approach to evaluate different chemical kinetics data and to estimate the input parameters; this method has laid the foundation of many subsequent works. Wen and Chung (1979) built a model of Texaco gasifier which divided the furnace into three zones to describe the processes from pyrolytic cracking to gasification; mass balance and energy balance equations are built for each zone. Govind and Shah (1984) introduced momentum conservation to the former work and calculated the temperature, concentration and fluid field in the axial direction.

Most of the models reported in the published literature are based on experimental data using a data-driven approach, which limits the suitability of a model for industry use as its working conditions vary in a wide range. To provide good prediction for syngas output, a generic gasifier model is developed based on the process engineering operation principles discussed in this paper and its steady-state prediction is validated. The syngas output stream from the gasifier varies depending on the reactions in the gasifier and the coal slurry feeding speed. It increases from its initial rate of 0.1 mol/s to settle at 100 mol/s in 100 s.

The auxiliary modules including shift reactor, gas turbine, and HRSG are built with Matlab and a Simulink-based toolbox – Thermolib. The syngas generated by the gasifier will first enter a water quencher and will then be further cooled by a syngas cooler. After hydrolysis reaction and desulphuration reaction, the COS and H_2S contents in the syngas will be removed. The sweet syngas will then pass the shift reactor where the CO contents will be shifted to hydrogen and CO_2, which not only can enhance the fuel gas quality but also raise the CO_2 concentration. After the shift reaction, the shifted syngas will be compressed and heated again and injected to the gas turbine to generate electricity; the heat carried by the exhaust gas will be utilized to generate superheated steam in HRSG and produce more electricity, thus improving the overall system efficiency.

2. Description of Texaco gasifier and mathematical model

Texaco gasification technology, also known as coal slurry gasification technology, is developed by Texaco Company, initially for heavy oil gasification applications. Texaco gasifier structure is shown in Figure 2. It gasifies coal slurry, which is mixed by pulverized coal particles and water, as raw material, and uses oxygen as the gasification agent. Coal slurry is injected into the gasifier furnace through nozzles; the moisture content of coal slurry droplets will evaporate rapidly and the pulverized coal particles will devolatilize and yield coal tar, gaseous hydrocarbons and oil. The gaseous components and volatiles will be consumed rapidly with steam and oxygen. The combustion of carbon char will react with oxygen, carbon dioxide and hydrogen while the reaction products react with each other as well. The whole gasification process involves complicated physical and chemical reactions. The gasifier generates wet syngas composed of CO, CO_2, H_2 and steam. Syngas will leave the gasification zone with slag and enter the water quench zone where the slag will be deposited in the slag tank. Raw syngas will be cooled and cleaned after the quenching process.

The mathematic model for this process is developed by following the work of Watkinson et al. (1991). To simplify the modelling procedure, the following assumptions are made:

(1) The flow in the gasifier furnace is uniformed laminar flow, and the differences in temperature, concentration, pressure and material exchanges in the radial direction are not taken into consideration. Actually, laminar flow only exists in the lower part of the gasifier; the flow between nozzles and the lower part should be jet flow surrounded by a strong back flow zone. The eddy turbulent is not considered in the modelling process as the syngas content will not be affected by this flow type. Thus, it is acceptable for the global laminar assumption.

(2) Preheating of slurry droplets, moisture evaporation and coal devolatilization will be complete as soon as the coal slurry is injected into the gasifier. The nozzles are surrounded by high-pressure high-temperature gas flow.

(3) The released volatile combustion and carbon pyrolytic and char combustion reactions reach chemical equilibrium as soon as the slurry enters the furnace. The chemical equilibrium constants of homogeneous reactions inside the gasifier are used to describe the reactions.

(4) Nitrogen and argon are assumed to be steady and will not participate in any chemical reaction. It is assumed that all oxygen is consumed, and the carbon conversion is 99.5% in the entire gasifier.

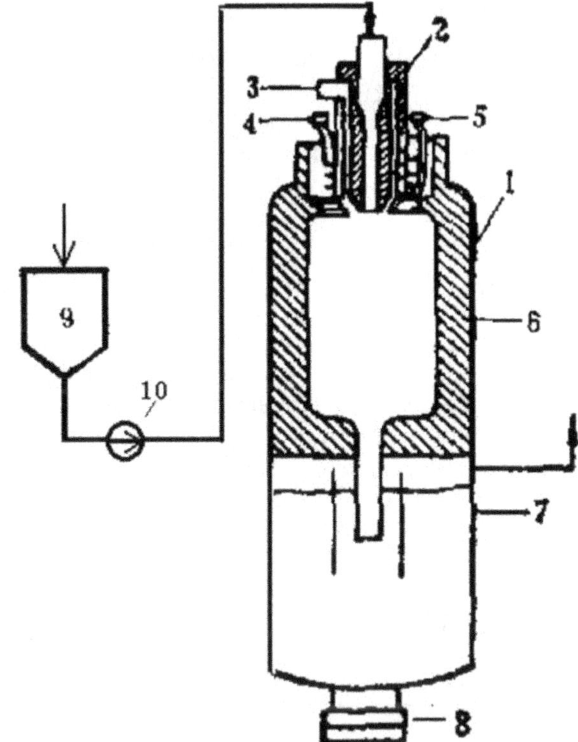

Figure 2. Structure of Texaco gaisifer. 1 – gasifier, 2 – Nozzle, 3 – oxygen input, 4 – cooling water input, 5 – cooling water output, 6 – refractory bricks liner, 7 – quenching water input, 8 – slag output, 9 – coal slurry tank, 10 – coal slurry pump.

Chemical reactions considered in this paper are as follows:

$$C + O_2 \rightarrow CO_2, \quad (1)$$

$$C + \tfrac{1}{2}O_2 \rightarrow CO, \quad (2)$$

$$C + H_2O \rightleftarrows H_2 + CO, \quad (3)$$

$$C + CO_2 \rightleftarrows 2CO, \quad (4)$$

$$CO + H_2O \rightleftarrows H_2 + CO_2, \quad (5)$$

$$CO + 3H_2 \rightleftarrows CH_4 + H_2O, \quad (6)$$

$$SO_2 + 3H_2 \rightleftarrows H_2S + 2H_2O, \quad (7)$$

$$COS + H_2O \rightleftarrows H_2S + CO_2. \quad (8)$$

For the aforementioned chemical reactions, mass balance equations of carbon, oxygen, hydrogen, nitrogen and sulphur can be derived as follows:

$$N_{C,0}A = N_g(X_{CO} + X_{CO_2} + X_{CH} + X_{COS}), \quad (9)$$

$$N_{O_2,0} = N_g(0.5X_{CO} + X_{CO_2} + X_{SO_2} + 0.5X_{COS} + 0.5X_{H_2O}),^y \quad (10)$$

$$N_{H_2,0} = N_g(2X_{CH_4} + X_{H_2S} + X_{H_2} + X_{H_2O}), \quad (11)$$

$$N_{N_2,0} = N_g(X_{N_2}), \tag{12}$$

$$N_{S,0} = N_g(X_{SO_2} + X_{H_2S} + X_{COS}), \tag{13}$$

$$N_{Ar,0} = N_g(X_{Ar}). \tag{14}$$

According to Dalton's law (Watkinson et al., 1991), we have

$$X_{CO} + X_{CO_2} + X_{CH_4} + X_{H_2} + X_{H_2O} + X_{H_2S} + X_{SO_2} + X_{COS} + X_{N_2} + X_{Ar} = 1. \tag{15}$$

Equations (9)–(14) are derived based on the mass conservation of C, O, H, N, S and Ar. From Dalton's law (Watkinson et al., 1991), Equation (15) means that the sum of all the syngas contents equals 1.

Based on chemical equilibriums:

$$\frac{X_{H_2}X_{CO_2}}{X_{CO}X_{H_2O}} = 0.0265\,e^{3956/T_g}, \tag{16}$$

$$\frac{X_{CH_4}X_{H_2O}}{X_{CO}X_{H_2}^3P^2} = 6.7125 \times 10^{-14}\,e^{27020/T_g}, \tag{17}$$

$$\frac{X_{H_2S}X_{H_2O}^2}{X_{SO_2}X_{H_2}^3P} = 4.3554 \times 10^{-4}\,e^{26281/T_g}, \tag{18}$$

$$\frac{X_{H_2S}X_{CO_2}}{X_{COS}X_{H_2O}P} = 0.75314\,e^{4083/T_g}. \tag{19}$$

Equations (16)–(19) are derived based on the chemical equilibrium of reactions (5)–(8) listed earlier; the reaction temperature and pressure will affect the equilibrium and change the concentration of syngas contents. There are 11 variables in Equations (9)–(19); this nonlinear equation system can be solved using the Newton–Raphson method (Wang, Wang, Guo, Lv, & Gao, 2013). Energy balance can be expressed as {Total enthalpy input} − {Total enthalpy output} = {Heat loss to environment}; the enthalpy input includes raw material enthalpy, the chemical reaction formation enthalpy and enthalpy carried by recycled gas. The enthalpy output includes enthalpy carried by output syngas, tar and char. Gaseous enthalpies are calculated as

$$H_{g\,i} = N_{g\,i}\left[\sum_i X_i\left(\Delta H_{f\,i} + \int_{298}^{T_g} C_{p\,i}\,dT\right)\right]. \tag{20}$$

3. Governing equations

In the Thermolib toolbox, the syngas stream is organized as a vector formed by the data flow of mole flow, contents concentration, temperature, pressure, enthalpy flow, entropy flow, Gibbs energy rate, heat capacity rate and vapour fraction of all compounds, which is described as combined flow bus. The gas phase contents are considered as the real gas form, and their enthalpy, entropy and heat capacity rate will be calculated by using the Peng–Robinson real gas equations of state (Peng & Robinson, 1976).

The most important governing equations are formed of mass balance and energy balance; the mass balance for a normal (standard) block is as follows:

Mass balance:

$$\frac{dM_i}{dt} = \sum Y_{i,in}\dot{M}_{in} - \sum Y_{i,out}\dot{M}_{out} + \sum R_i, \tag{21}$$

where $Y_{i,in}$ denotes the mass concentration of content i in inlet flow, $\dot{M}_{in}$ denotes the inlet mass flow rate, $Y_{i,out}$ denotes the mass concentration of content i in outlet flow, $Y_{i,out}$ denotes the mass concentration of content i in outlet flow, $\dot{M}_{out}$ denotes the outlet mass flow rate and R_i denotes the net production rate of i by chemical reactions.

When no chemical reaction occurs in the simulated block, the factor R_i will equal zero. In addition, mass accumulation is not considered in all the blocks simulated in this paper.

The governing equation of energy balance is derived by following the first law of thermodynamics:

$$\frac{dU}{dt} = \sum \dot{H}_{i,in} - \sum \dot{H}_{j,out} + \sum \dot{Q}_k + \sum P_m, \tag{22}$$

where U denotes the internal energy in block, $\dot{H}_{i,in}$ denotes the enthalpy flow rate of content i in the inlet flow, $\dot{H}_{j,out}$ denotes the enthalpy flow rate of content j in the outlet flow, $\dot{Q}_k$ denotes the heat flow and P_m denotes the mechanical power.

All the auxiliary blocks built with Thermolib in this paper will follow Equations (21) and (22). The individual blocks will be explained one by one in the following.

4. Description of the WGS reactor

The syngas generated by the gasifier will be cooled in the quench water pool. The wet syngas will then enter the WGS and the exothermic water gas shift reaction will take place:

$$CO + H_2O \rightleftarrows H_2 + CO_2. \tag{23}$$

In the industry, a shift catalyst is used in the reactor; the catalyst can convert most of the CO to CO_2 with the evolution of heat. There are a number of specific advantages from incorporating a shift reactor into the coal gasification flow scheme in real world; it will improve the H_2 extraction while decreasing the CO concentration in the syngas stream. The shift reactor with heat recovery is able to increase the power output for the same gas turbine investment as well. Moreover, the WGS is important for the preparation of CO_2 sequestration because the CO_2 concentration rises as well, which is ideal for PSA (pressure swing adsorption) adsorption capture (Karmarkar, 2005).

The shift reactor in this study is used to prepare high H_2 and CO_2 content-shifted syngas stream for the hydrogen gas turbine and future carbon capture module. The complicated catalysis or equilibrium process is hard to be developed and validated for this simulation; thus, a reaction rate-controlled reactor block developed using the Thermolib toolbox is adopted to simulate the WGS reaction process, and the heat exchange with the environment is considered.

In this mode, two shift reactors are connected in series and the reaction rates are defined, respectively, as 0.8 and 0.9 (Karmarkar, 2005); meanwhile, the pressure loss and heat exchange with the environment are considered. By defining the reaction rate, the conversion of CO can be controlled; thus, the partial pressure of CO_2 in the shifted syngas will satisfy the demand for further PSA carbon capture simulation. The development of the carbon capture model and its integration with the current IGCC model will be studied in the future.

Based on the energy balance, the reactor model in the WGS process can be described by

$$\sum_{\text{reac}} \dot{m}_{\text{in}} h_{\text{in}} = \sum_{\text{prod}} \dot{m}_{\text{out}} h_{\text{out}} + \dot{Q}. \tag{24}$$

The syngas temperature, pressure, mass flow rate and contents dynamic change are obtained. In this paper, the contents of the syngas entering the WGS are simplified as a mixture of CO, CO_2, H_2 and H_2O only. This simplification will not cause big error because the total portion of CH_4, N_2, H_2S, SO_2, COS and Ar is less than 2% in the syngas, and they will not affect the model to represent the process correctly in terms of the dynamic features concerned in the analysis.

5. Description of the gas turbine

The gas turbine-centred power-generation process is based on Brayton cycle (Lichty, 1967). It has a rotating compressor coupled to the shaft of the turbine, and a combustion chamber in between. The shifted syngas is compressed, mixed with compressed air in the mixer and then ignited in the combustion chamber. The combustion of H_2 under a constant pressure will generate the hot flue gas. The flue gas expands through the turbine to perform work.

An isentropic compressor developed in Thermolib is adopted to model the compression process; this block can increase the pressure of an incoming flow to a given outlet pressure. To achieve more accurate simulation results, a certain isentropic efficiency is given in this process modelling, which can decrease the error between simulation results and practical working conditions. The isentropic efficiency is defined by Equation (25):

$$\eta_s = \frac{h_{\text{out},s} - h_{\text{in}}}{h_{\text{out}} - h_{\text{in}}}, \tag{25}$$

where $h_{\text{out},s}$ is the enthalpy in ideal isentropic state, while h_{out} is the enthalpy in actual state. Thus, the mechanical power consumption can be calculated as

$$W_{mch} = \frac{\dot{m}(h_{\text{out},s} - h_{\text{in}})}{\eta_s}. \tag{26}$$

The compressed syngas flow is then mixed with compressed air in a mixer. It is assumed that there is no mass loss in the mixer. According to the energy balance, the following equation can be obtained:

$$\sum_{\text{input}} \dot{m}_{\text{in}} h_{\text{in}} = \dot{m}_{\text{out}} h_{\text{out}}. \tag{27}$$

The mixed syngas and air will then be ignited in the combustion chamber, where the fuel gas combustion will take place, and the chemical reactions considered include

$$2H_2 + O_2 \rightleftarrows 2H_2O, \tag{28}$$

$$2CO + O_2 \rightleftarrows 2CO_2. \tag{29}$$

The dominant reaction in the chamber is hydrogen combustion. The remaining carbon monoxide combustion is also considered. After passing the WGS, the CO mole content in shifted syngas will drop to less than 2.47% (Karmarkar, 2005); thus, the influence of CO combustion is tenuous.

The heat generation of the combustion will be calculated by hydrogen and carbon monoxide lower heating value (LHV) as 120 and 10.112 MJ/kg. In this model, a reactor block like WGS reactor is adopted to simulate the combustion, and the reaction rate is defined as 1, which means complete combustion happens inside the chamber.

The high-temperature and high-pressure flue gas is then injected into the turbine block. The turbine can decrease the pressure of the incoming flow to a given outlet pressure. It determines the thermodynamic state of the outgoing flow along with the produced mechanical power at a given isentropic efficiency, which is similar to that of the isentropic compressor. The mechanical power generated by the turbine can be calculated by the following equations:

$$\eta_s = \frac{h_{\text{in}} - h_{\text{out}}}{h_{\text{in}} - h_{\text{out},s}}, \tag{30}$$

$$W_{mch} = \dot{m}(h_{\text{in}} - h_{\text{out},s})\eta_s. \tag{31}$$

The parameters with subscript s are isentropic state change. This turbine block is defined as passive, which means the mass flow rate remains unchanged during the calculation. The isentropic efficiency used in this block is retrieved from a lookup table as a function of the mass flow. It is defined as the ratio of actual enthalpy difference to enthalpy difference for isentropic change of state with the same pressure drop. In this study, the value of isentropic efficiency is set as 0.8.

The thermodynamic characteristics of the syngas flow and flue gas flow and the mechanical power-generation results are obtained. The thermal efficiency is calculated based on the fuel heat value and gas turbine net power generation.

6. The HRSG description

The HRSG unit is based on the Rankine cycle (Wong, 2012). It has a heat exchanger to pick up the waste heat of the flue gas to heat the feed water and generate steam. Then, the vapour will drive the turbine and then the generator to produce electricity. Using the HRSG, the thermal efficiency of the power cycle can be improved.

A pump module is adopted to provide feed water to the heat exchanger. The pump increases the pressure of the incoming water flow to target pressure. It determines the thermodynamic state of the outgoing flow along with the required mechanical power consumption. The energy balance of the compression process in the pump is calculated using the following equation (32):

$$\dot{m}\left(h_{\text{in}} + \frac{1}{2}v_{\text{in}}^2\right) = \dot{m}\left(h_{\text{out}} + \frac{1}{2}v_{\text{out}}^2\right) + \dot{W}_{\text{mch}}, \quad (32)$$

while

$$\dot{m}(h_{\text{out}} - h_{\text{in}}) = \frac{P_{\text{out}} - P_{\text{in}}}{\rho}, \quad (33)$$

where P_{out} and P_{in} are the pressure of the pump outlet and inlet. ρ is density of the working flow. The power consumption is calculated by

$$W_{\text{mch}} = \frac{\dot{m}\Delta P}{\rho\eta_{\text{pump}}}, \quad (34)$$

where η_{pump} is the pump efficiency and is set as 0.8. The power consumption of the pump is considered in the total HRSG output calculation. With the mechanical power, the enthalpy of the water outlet is equal to the sum of input enthalpy and electric power.

$$\dot{H}_{\text{out}} = \dot{H}_{\text{in}} + W_{\text{mch}}. \quad (35)$$

The pumped feed water is then transported to the heat exchanger, where its temperature will rise to the target by adsorbing heat from the flue gas generated by the gas turbine. A heat exchanger module based on Thermolib is used for calculating the heat transfer between the flue gas and the water stream. It is important to notice that the heat transfer between these two streams is indirect and they are treated as counter flow. The two media state dynamic change will be simulated by using the number of transfer units method (Incropera & DeWitt, 1985). The actual heat transfer rate can be determined by calculating the effectiveness ε, which is the actual heat transfer divided by the maximum possible heat transfer.

$$\varepsilon = \left(\frac{\dot{Q}}{\dot{Q}_{\text{max}}}\right). \quad (36)$$

For counter flow:

$$\varepsilon = \left(\frac{1 - \exp(-M(1 - C))}{1 - C \times \exp(-M(1 - C))}\right), \quad (37)$$

where

$$M = \frac{UA}{\dot{C}_{\text{min}}}, \quad (38)$$

$$C = \frac{\dot{C}_{\text{min}}}{\dot{C}_{\text{max}}}, \quad (39)$$

$$\dot{C}_{\text{min}} = \min(\dot{m}_1 c_{p1}, \dot{m}_2 c_{p2}), \quad (40)$$

$$\dot{C}_{\text{max}} = \max(\dot{m}_1 c_{p1}, \dot{m}_2 c_{p2}), \quad (41)$$

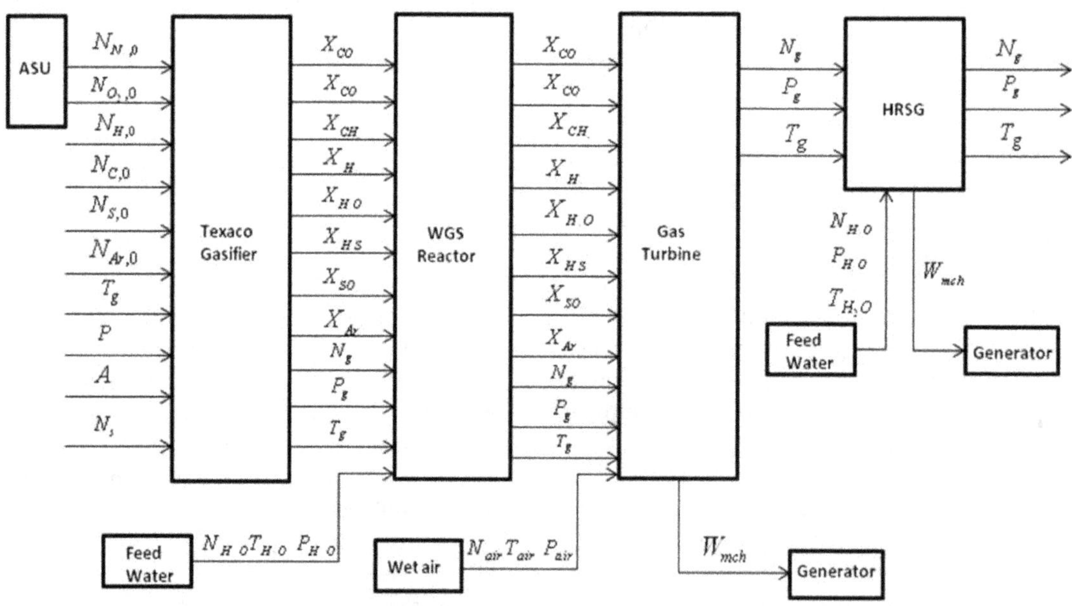

Figure 3. Schematic diagram of process simulation.

where U is the overall heat transfer coefficient, A is the surface area available for the heat transfer and thus UA is the overall heat transfer rate. It represents the heat transfer between the flow and wall as well as the heat conduction in the wall. This is the product of heat transfer coefficient and effective heat exchange area between the flows.

Using the aforementioned equations (36–41), we can have

$$Q_{\max} = \dot{C}_{\min}(T_{\mathrm{hi}} - T_{\mathrm{ci}}), \tag{42}$$

where T_{hi} is the temperature of the hot fluid input to the heat exchanger and T_{ci} is the cold fluid input temperature. In the calculation of this heat exchanger block, the heat exchanges of both the flow streams with environment and pressure loss of the flows are considered.

The heated water steam is then injected to the steam turbine block, where the isentropic expansion process happens. The same block as described in the gas turbine system introduction is adopted for the steam turbine. The net power output of the whole HRSG system will be the difference between steam turbine power generation and pump power consumption. The dynamic change in net power of HRSG and the gas turbine system is obtained and analysed as the power output of the combined cycle.

7. Air separation unit

The air separation unit (ASU) is an important part in the IGCC power plants using oxygen-feed gasifiers (Wang et al., 2013). Air is liquefied and separated into oxygen and nitrogen along with some by-products such as helium. The generated oxygen is then compressed to 40 bar and injected into the gasifier for raw syngas production. Some plants integrate ASU with gas turbine, using nitrogen for combustion diluent aiming to reduce NO_x emission (Karmarkar, 2005). In this paper, only hydrogen and carbon monoxide combustion are considered in the gas turbine module, the integration of ASU and gas turbine will not be included.

The working principle of ASU in this study is based on the Linde–Hampson cycle (Timmerhaus & Reed, 2007). Air is isothermally compressed first and then passes the main heat exchanger; the isobaric heat transfer between the air phase and liquid phase results in huge temperature drop. After the isenthalpic expansion process in the throttle valve, the air will be liquefied and transported to the distillation tower for further separation. In the separation process, liquid nitrogen will reach boiling point first and be separated, and pure nitrogen and oxygen will be generated for the IGCC process.

A polytropic compressor in Thermolib is adopted to simulate the isothermal process. The power consumption is calculated by using the following equation:

$$W_{\mathrm{mch}} = \frac{n(p_{\mathrm{out}}v_{\mathrm{out}} - p_{\mathrm{in}}v_{\mathrm{in}})}{1-n} \tag{43}$$

Table 1. Model input.

Input	Unit	Illinois No. 6	Australia	Fluid coke
Slurry flow rate	kg/s	1	1	1
Slurry concentration	kg coal/kg slurry	0.665	0.621	0.606
O_2 purity	Vol%	98	99.6	100
O_2/coal	kg O_2/kg dry coal (no ash)	0.86	0.87	1.03
Ar/O_2	kg Ar/kg O_2	0	0	0
Gasifier pressure	MPa	4.083	4.083	4.083
Temp.	°C	1141	1044	1060
Heat loss	H.H.V.%	2	2	2
Ultimate analysis(dry)	Mass %			
C	%	69.6	66.8	86
H	%	5.3	5.0	2.0
O	%	10	7.3	2.3
N	%	1.3	1.7	1.0
S	%	3.9	4.2	8.3
Ash	%	10	15	0.5

Table 2. Comparison of the simulation results and reference data.

	Illinois No. 6		Australia		Fluid coke	
	R (mol%)	S (mol%)	R (mol%)	S (mol%)	R (mol%)	S (mol%)
CO	41.0	41.0	35.2	35.4	47.1	47.2
H_2	29.80	30.1	29.9	29.6	24.3	23.7
CO_2	10.2	10.0	12.8	12.8	13.2	13.3
H_2O	17.1	16.8	20.3	20.0	12.7	13.0
CH_4	0.3	0.15	0.02	0.22	0.09	0.33
N_2	0.80	0.9	0.63	0.63	0.4	0.3
H_2S	1.1	1.01	1.14	1.10	2.2	2.07
COS		0.04		0.25		0.10
Error		0.26		0.14		0.21

Table 3. Model input.

Input	Unit	Data
Slurry rate	t/d	650
Slurry concentration	kg coal/kg slurry	0.66
Oxygen purity	Vol%	98
Oxygen/coal	kg O_2/kg dry coal (no ash)	0.96
Pressure	MPa	4.0
Temperature	°C	1350
Heat loss	H.H.V.%	2
Ultimate analysis	%	
C	%	71.5
H	%	4.97
O	%	11.15
N	%	1.07
S	%	2.16
Ash	%	9.15

Table 4. Comparison of dry syngas output content and industry data.

	CO	H_2	CO_2	CH_4 + Sulphide + N_2 + Ar
Industry	48.82	36.58	14.41	0.19
Simulation	49.54	35.69	12.79	1.98

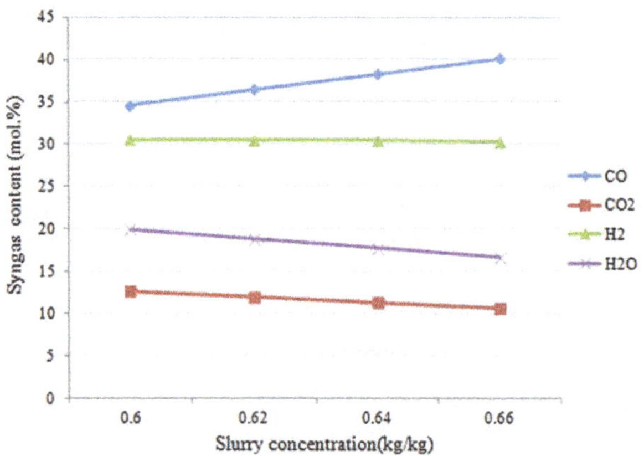

Figure 4. Syngas content change with coal slurry concentration unit (kg/kg).

where *p, v* are the pressure and volume value of different states, and *n* is the coefficient in the polytrophic process. The value of *n* in this study is set as 0.997 to stabilize the air temperature, which is required for isothermal condition.

In the isobaric heat exchanger, the gas phase air temperature is reduced to the target, and the difference in enthalpy between the input and output states is the transferred heat.

The throttle valve module is based on the Joule–Thompson effect (Perry & Green, 1984), and liquefaction of the air which has been cooled enough consists of an isenthalpic expansion. The pressure loss over the valve is calculated by the Thermolib valve block. This isenthalpic process will cause full liquefaction of the air. The valve block is based on the following equations:

$$\dot{H}_{\text{out}} = \dot{H}_{\text{in}}, \tag{44}$$

$$\dot{m}_{\text{out}} = \dot{m}_{\text{in}}, \tag{45}$$

$$p_{\text{out}} = p_{\text{in}} - k(\text{pos})\dot{m}, \tag{46}$$

where $k(\text{pos})$ is the function of valve position. The value is calculated based on a well-validated lookup table in the Thermolib toolbox.

In this study, the liquefied air temperature is 70 K (Karmarkar, 2005). The liquid phase air is than fractionally distilled to generate pure nitrogen and oxygen. The media flow in the model is set as a vector carrying the state parameters, which include flow rate, temperature, pressure, enthalpy, entropy, contents concentration and gas phase fraction.

8. Results and discussion

The whole process model implemented in Simulink and Thermolib is shown in Figure 3. Three types of coal are applied to test the working process of the model. The model input variables and model parameters are listed in Table 1. The simulation results of the final steady state and their associated reference data (Watkinson et al., 1991) are given in Table 2. For error analysis, the mean absolute error between the simulation results and reference data is given in Table 2.

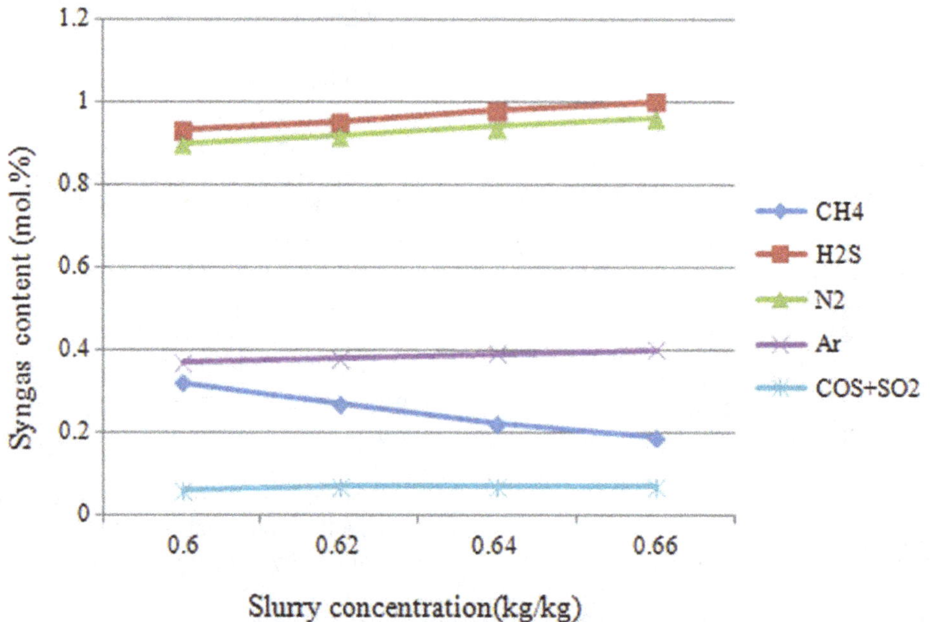

Figure 5. Syngas content change with coal slurry concentration unit (kg/kg).

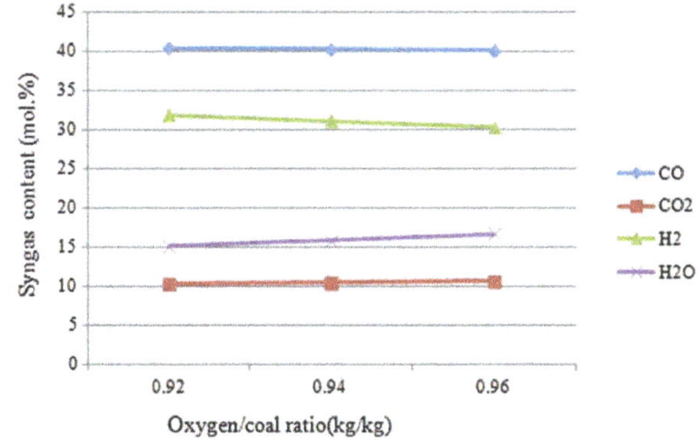

Figure 6. Syngas content change with oxygen/coal ratio unit (kg/kg).

Column R denotes the reference data of syngas contents concentration in mole percentage, the column S presents the simulation results of three coal types. The comparison in Table 2 shows that the simulation results match well with the reference data (Watkinson et al., 1991).

The gasifier simulation results are then compared with the reference data from the Lu-nan fertilizer factory. The model input data are given in Table 3. Comparison of the predicted output dry syngas content results and industry data is given in Table 4.

Table 4 shows that the simulation results can match well with the industry data. This means that the assumptions and the mathematic model used in the modelling work are reasonable.

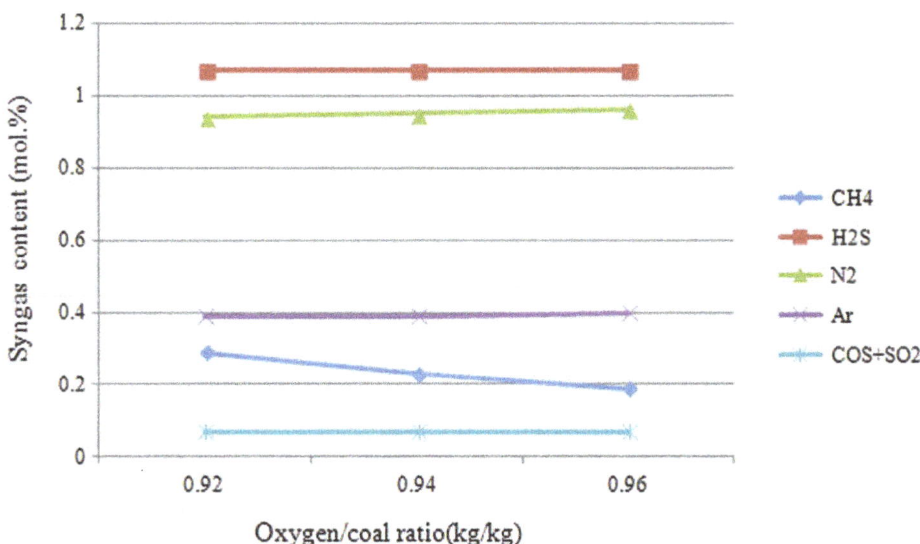

Figure 7. Syngas content change with oxygen/coal ratio unit (kg/kg).

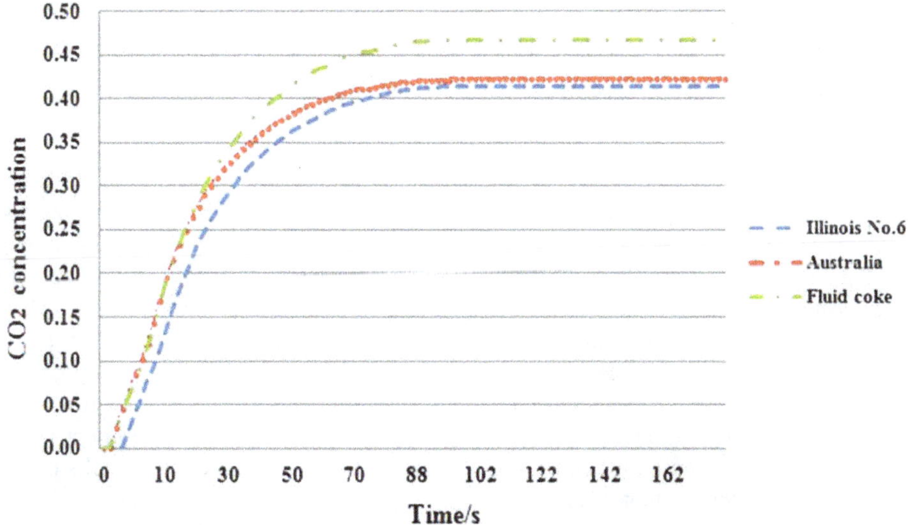

Figure 8. Shifted syngas CO_2 concentration dynamic change.

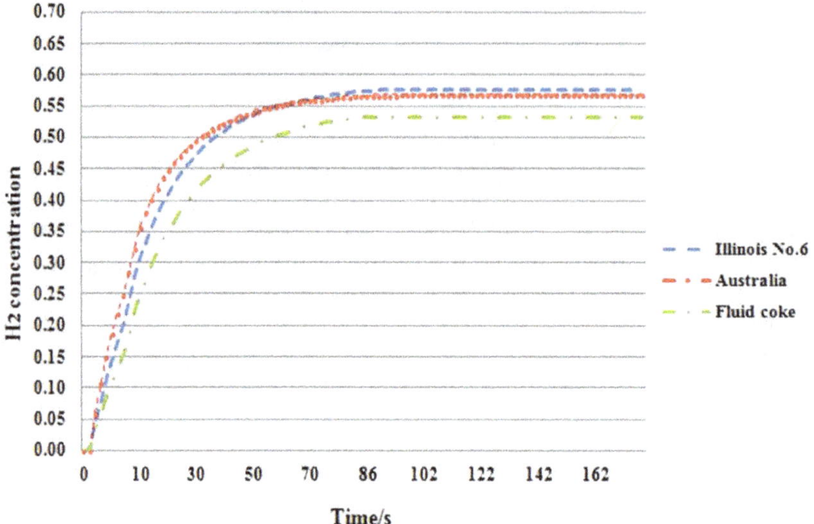

Figure 9. Shifted syngas H_2 concentration dynamic change.

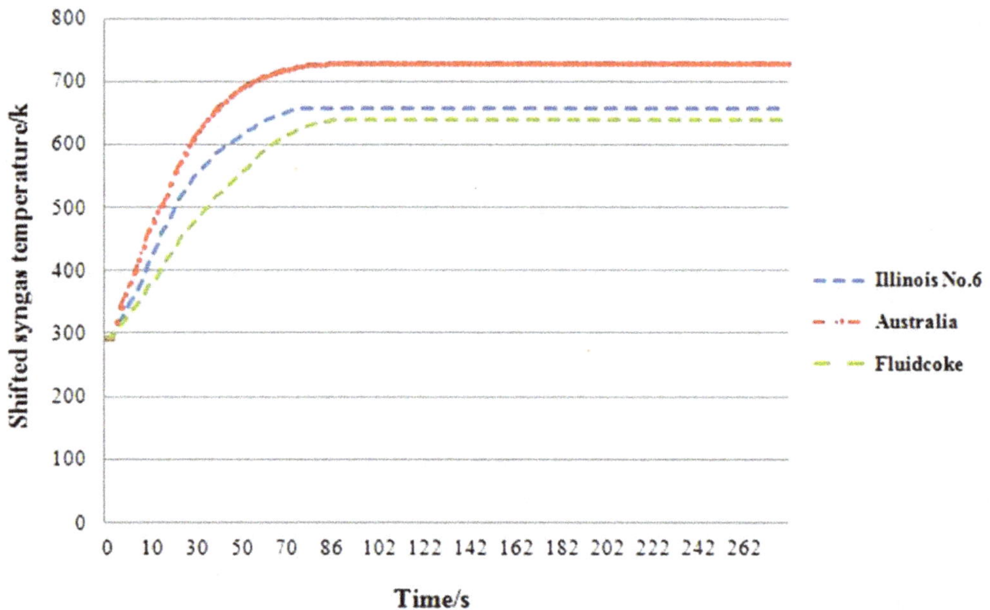

Figure 10. Shifted syngas temperature dynamic change.

To further test the model, the change in oxygen/coal ratio and slurry concentration's effects on the syngas content are studied. Figures 4 and 5 show the slurry concentration's effect on syngas contents. When the slurry concentration increases from 60% to 66%, the CO content increases; while H_2 remains stable, CO_2 and CH_4 decrease. When slurry concentration increases, more coal feed enters the gasifier; but when the oxygen/coal ratio remains stable, it means the oxygen input will rise as well, thus enhancing the gasification process and inducing content change. The results well match with Azuhata's experiment (Azuhata, Hedman, & Smoot, 1986).

Figures 6 and 7 show the syngas content change with oxygen/coal ratio. The rise in this ratio means increase in the oxygen supply, which will enhance the combustion and raise the gasifier temperature, thus enhancing the gasification process. But it will also consume more CO and H_2 released from the volatile, resulting in the decrease in CO and H_2 content and the increase in CO_2 content. The results match with Azuhata's experiment (Azuhata et al., 1986) and Vamvuka's simulation results (Vamvuka, Woodburn, & Senior, 1995a, 1995b).

Three different types of coal are applied to the simulation to test the process dynamic response. The coal feeding

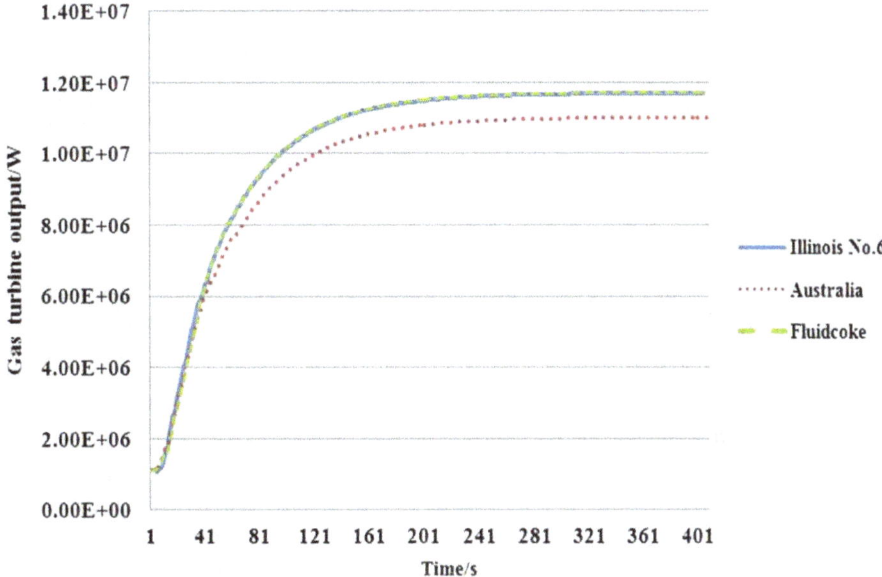

Figure 11. Gas turbine power output dynamic change.

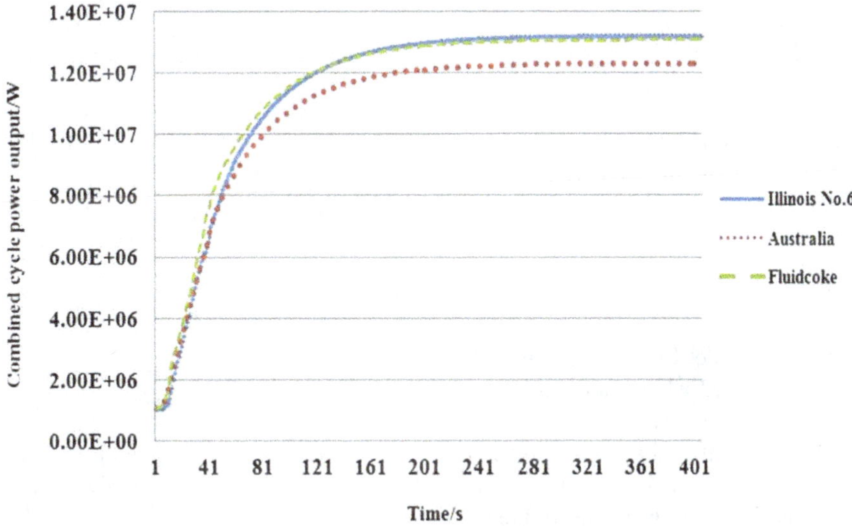

Figure 12. Combined cycle power output dynamic change.

is the input of the whole system. In this test, the coal flow rate will be inputted to the gasifier model and simulated to generate syngas output. The syngas flow rises from 0.1 to 100 mol/s within 100 s and it will first enter the WGS module for shift reaction.

The concentration of CO_2 and H_2 and the temperature of the shifted syngas are shown in Figures 8–10.

For the three types of coal, a similar trend is observed for the dynamic change in CO_2, H_2 concentration and syngas temperature. The simulation results prove that the WGS module can enhance the H_2 and CO_2 extraction; the carbon sequestration and hydrogen combustion process will thus benefit from it. Meanwhile, the shift reaction is exothermic, and the temperature of the shifted syngas rises in the WGS. In a real power plant, the heat can be used to raise HP steam to HRSG, but in this study, the heat recovery process is not considered.

The dynamic behaviours of the gas turbine power output are shown in Figure 11, while the net power output of the combined cycle is shown in Figure 12. The power consumptions of the compressors coupled with turbines are considered.

The power outputs of the gas turbine and combined cycle show similar trends for the three types of coal. The thermal efficiency of the gas turbine cycle is 30–33%. Integrated with HRSG, the combined cycle thermal efficiency is 51–55%. Hence, the HRSG is important in improving the power generation of the IGCC process.

For the ASU module, the liquefied air flow rate is set as 150 mol/s, which can satisfy the demand for the Texaco

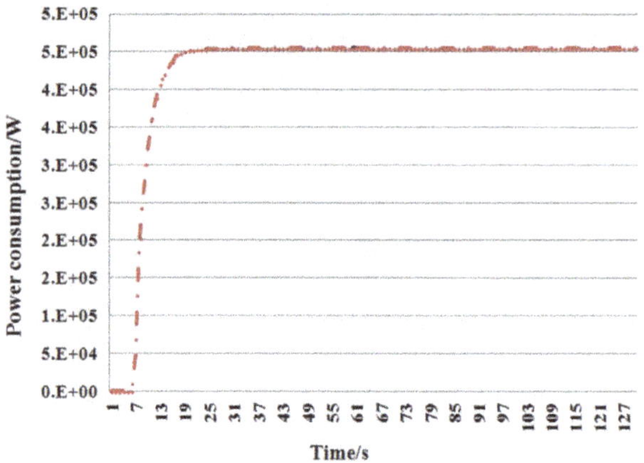

Figure 13. ASU compressor power consumption dynamic change.

gasifier use based on the model input oxygen/dry coal ratio. The compressor consumption dynamic change is shown in Figure 13. In this paper, the power consumption of air liquefaction process is mainly caused by compressor usage.

In this model, pure oxygen and wet air are used as the oxidant in the gasifier and the gas turbine, respectively. In the real industry, the ASU will not only provide oxygen to the gasifier but also integrate with the gas turbine; but in this paper, since NO_x is not considered as the combustion product, integration of the ASU and the gas turbine is not included.

9. Conclusion

In this paper, a mathematical model for an IGCC process using Taxaco gasifier is developed which can predict the syngas contents. The model is derived by applying the engineering principle, the chemical equilibrium, mass balance and energy balance. The simulation results of dry syngas match well with the industry data. The influence of coal slurry concentration and oxygen/coal ratio on the syngas contents is studied as CO, CO_2 and H_2 contents can reveal the syngas quality. To enhance the gasification process and generate raw syngas with high quality, higher slurry concentration and oxygen supply is shown to be one of the suitable solutions. The dynamic behaviour of WGS shows its importance in improving syngas quality potential to heat steam. To improve the syngas quality and prepare for further carbon capture, WGS plays a key role. HRSG improves the thermal efficiency of the combined cycle and the overall power-generation process. ASU can lead to efficiency loss of the whole power plant since the liquefaction and transportation processes will both consume a large amount of energy.

Disclosure statement

No potential conflict of interest was reported by the authors.

Funding

This work was supported by Engineering and Physical Sciences Research Council (EPSRC, EP/I010955/1). The validation data used in the gasification model are provided by Tsinghua University. The author Yue Wang is partially supported by Chinese Scholarship Council (CSC).

References

Azuhata, S., Hedman, P. O., & Smoot, L. D. (1986). Carbon conversion in an atmospheric-pressure entrained coal gasifier. *Fuel, 65*, 212–217.

Beér, J. M. (2000). Combustion technology developments in power generation in response to environmental challenges. *Progress in Energy and Combustion Science, 26*, 301–327.

Brown, B. W., Smoot, L. D., & Hedman, P. O. (1986). Effect of coal type on entrained gasification. *Fuel, 65*, 673–678.

Buskies, U. (1996). The efficiency of coal-fired combined-cycle powerplants. *Applied Thermal Engineering, 16*, 959–974.

Casella, F., & Colonna, P. (2012). Dynamic modeling of IGCC power plants. *Applied Thermal Engineering, 35*, 91–111.

Chen, X., He, M. Y., Spitsberg, I., Fleck, N. A., Hutchinson, J. W., & Evans, A. G. (2004). Mechanisms governing the high temperature erosion of thermal barrier coatings. *Wear, 256*, 735–746.

Colonna, P., & van Putten, H. (2007). Dynamic modeling of steam power cycles. Part I – modeling paradigm and validation. *Applied Thermal Engineering, 27*, 467–480.

Govind, R., Shah, J. (1984). Modeling and simulation of an entrained flow coal gasifier. *AIChE, 30*, 79–91.

Incropera, F. P., & DeWitt, D. P. (1985). *Fundamentals of heat and mass transfer*. New York, NY: Wiley.

Karmarkar, M. (2005). *Jacob Consultancy UK, University of Nottingham, E.ON UK, Power Asset Modelling, Watergrid, Mitsui Babcock and EPRI.* Impact of CO_2 removal on coal gasificaiton based fuel plants final Report, 57–58.

Lichty, L. C. (1967). *Combustion engine processes*. New York, NY: McGraw-Hill.

Lu, S., & Hogg, B. W. (2000). Dynamic nonlinear modelling of power plant by physical principles and neural networks. *International Journal of Electrical Power & Energy Systems, 22*, 67–78.

Maurstad, O. (2005). *An overview of coal based integrated gasification combined cycle (IGCC) technology*. Cambridge, MA: Massachusetts Institute of Technology, 8 Publication no. LFEE 2005–002 WP.

Ni, Q., & Williams, A. (1995). A simulation study on the performance of an entrained-flow coal gasifier. *Fuel, 74*, 102–110.

Peng, D. Y., & Robinson, D. B. (1976). A new two-constant equation of state. *Industrial & Engineering Chemistry Fundamentals, 15*, 59–64, 1976/02/01.

Perry, R. H.Green, D. W., et al. (1984). *Perry's chemical engineers' handbook*. New York, NY: McGraw-Hill.

van Putten, H., & Colonna, P. (2007). Dynamic modeling of steam power cycles: Part II – simulation of a small simple Rankine cycle system. *Applied Thermal Engineering, 27*, 2566–2582.

Smoot, L. D., & Smith, P. J. (1979). *Pulverized-coal combustion and gasification*. New York, NY: Plenum Press. Ch.13.

Sun, B., Liu, Y., Chen, X., Zhou, Q., & Su, M. (2011). Dynamic modeling and simulation of shell gasifier in IGCC. *Fuel Processing Technology, 92*, 1418–1425.

Timmerhaus, K. D., & R. P. Reed. (2007). *Cryogenic engineering: Fifty years of progress*. New York, NY: Springer.

Ubhayakar, S. K., Stickler, D. B., & Gannon, R. E. (1977). Modelling of entrained-bed pulverized coal gasifiers. *Fuel, 56*, 281–291.

Vamvuka, D., Woodburn, E. T., & Senior, P. R. (1995a). Modelling of an entrained flow coal gasifier. 1. Development of the model and general predictions. *Fuel, 74*, 1452–1460.

Vamvuka, D., Woodburn, E. T., & Senior, P. R. (1995b). Modelling of an entrained flow coal gasifier. 1. Effect of operating conditions on reactor performance. *Fuel, 74*, 1461–1465.

Wang, Y., Wang, J., Guo, S., Lv, J., & Gao, Q. (13–14 September, 2013). *Dynamic modelling and simulation study of Texaco gasifier in an IGCC process*. 19th International Conference on Automation and Computing, London, UK. pp. 1, 6.

Watkinson, A. P., Lucas, J. P., & Lim, C. J. (1991). A prediction of performance of commercial coal gasifiers. *Fuel, 70*, 519–527.

Wen, C. Y., & Chaung, T. Z. (1979). Entrainment coal gasification modeling. *Industrial Engineering, Chemistry, Process Design and Development, 18*, 684–694.

Wong, K. (2012). *Thermodynamics for engineers*. Boca Raton, FL: CRC Press.

Yang, Z., Wang, Z., Wu, Y., Wang, J., Lu, J., Li, Z., & Ni, W. (2011). Dynamic model for an oxygen-staged slagging entrained flow gasifier. *Energy and Fuels, 25*, 3646–3656.

18

On improving Popov's criterion for nonlinear feedback system stability

Y.V. Venkatesh[a,b*]

[a] *(formerly) Electrical Sciences Division, Indian Institute of Science, Bangalore, India;;* [b] *Department of ECE, National University of Singapore, Singapore*

For the L_2-stability of a nonlinear single-input–single-output (SISO) feedback system, described by an integral equation and with the forward block transfer function $G(j\omega)$ and a first- and third-quadrant non-monotone nonlinearity $\varphi(\cdot) \in \mathcal{N}$ in the feedback path, we derive an interesting generalization of the celebrated criterion of Popov [(1962). Absolute stability of nonlinear systems of automatic control. *Automation and Remote Control*, *22*(8), 857–875]: $\Re(1 + j\alpha\omega)G(j\omega) > 0, 0 \leq \omega < \infty$, where $\alpha > 0$ is a constant. The generalization entails the addition of a general causal + anticausal O'Shea–Zames–Falb multiplier function whose time-domain L_1-norm is constrained by certain characteristic parameters (CPs) of the nonlinearity obtained from certain novel algebraic inequalities. If the nonlinearity is monotone or belongs to any prescribed subclass of $\mathcal{N}$, its CPs are reduced, thereby relaxing the time-domain constraint on the multiplier. An important special feature of the new stability results is a partial bridging of the significant gap between the Popov criterion and the stability results that appeared post-Popov in the form of considering monotone and other subclasses of nonlinearities in exchange for weakening the restrictions on the phase angle behaviour of $G(j\omega)$. Extensions to time-varying nonlinearities more general than those in the literature are also presented. Numerical examples are given to illustrate the theorems and to demonstrate their superiority over the existing literature.

Keywords: K–Y–P lemma; L_2-stability; Lur'e problem; Popov criterion; time-varying feedback systems

1. Introduction

In the analysis of problems arising in diverse areas of dynamical system design (such as satellite control, communication networks, and chemical plants), stability theory of nonlinear time-varying systems is an invaluable tool. Nonlinear differential and integral equations, which are typically used as mathematical models to describe such systems, are linearized (or perturbed) around, for instance, a periodic solution. The perturbed behaviour of a dynamical system is found to be more accurately modelled by nonlinear time-varying differential and integral equations, of which the latter are more general than the former, since they can describe infinite-dimensional systems and include the former as a special case. In this context, the feedback system of Figure 1(a) plays an important role in the analysis and synthesis of dynamical systems which are in practice an interconnection of subsystems subject to switching operations. The system of Figure 1(a) consists of a linear time-invariant part in the forward path and a nonlinear time-varying gain in the feedback path. A special case of Figure 1(a) is Figure 1(b) in which the feedback block is a time-invariant nonlinear gain. Primarily inspired by the classic papers of Nyquist and Bode on the frequency-domain analysis (and synthesis) of systems modelled by a special case of Figure 1(b), namely, a system with a constant linear gain in the feedback path, research workers have considered extension of similar ideas to the analysis of stability (and instability) of the system of Figure 1(a), the main subject of this paper, described by

$$v(t) = f(t) - k(t)\varphi(\sigma(t));$$

$$\sigma(t) = \sum_{m=0}^{\infty} g_m v(t - \tau_m) + \int_0^{\infty} g(\tau)v(t - \tau)\,d\tau, \quad (1)$$

where $\delta(t - \tau_m)$ is the Dirac delta function at instant $t = \tau_m$; $\sum_{m=0}^{\infty} g_m \delta(t - \tau_m) + g(t)$ is the impulse response of the time-invariant forward block with constant real sequences $\{g_m\}, \{\tau_m\}$, in which $\tau_m \geq 0, \forall m$; $g(\cdot)$ is a real-valued function in $[0, \infty)$; and $\sum_{m=0}^{\infty} |g_m| + \int_0^{\infty} |g(t)|\,dt < \infty$, i.e. $\{g_m\} \in \ell_1$, and $g(\cdot) \in L_1$; real-valued gain $k(\cdot)$ assumes values in $[0, \infty)$; $f(\cdot), v(\cdot), \sigma(\cdot)$ are, respectively, the input, 'error' signal and output of the system; and $\varphi(\cdot)$, a real-valued function on $(-\infty, \infty)$, is a memoryless, first- and third-quadrant nonlinearity having the following basic properties: $\varphi(0) = 0$; there exist positive constants q_1 and q_2 with $q_1 < q_2$ such that $q_1\sigma^2 \leq \varphi(\sigma)\sigma \leq q_2\sigma^2$, $\sigma \neq 0$. We call the class of such nonlinearities $\mathcal{N}$, with its subclass of monotone nonlinearities denoted by $\mathcal{M}$. If the monotone nonlinearity also has the

*Email: yv.venkatesh@gmail.com

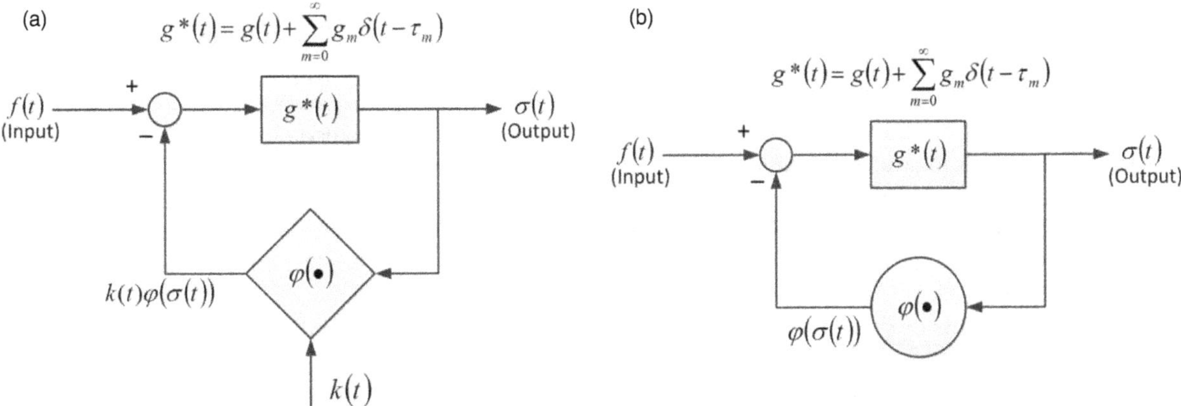

Figure 1. Block-schematics of systems under consideration. See text for details: (a) time-varying feedback system and (b) time-invariant feedback system.

'odd' property, i.e. $\varphi(-\sigma) = -\varphi(\sigma), \sigma \neq 0$, then its class is denoted by $\mathcal{M}_o$. The time-varying gain $k(\cdot)$ and the nonlinearity $\varphi(\cdot)$ in combination assume values in $[0, \infty)$. The transfer function of the forward block is given by $G(\mathrm{j}\omega)$. For convenience in some manipulations, we denote the operator representing the time-invariant forward block by $\mathcal{G}(\cdot)$. Note that the class $\mathcal{N}$ of nonlinearities was originally proposed by Lur'e and Postnikov (1944) in their pioneering Lyapunov method-based work on a special case of Equation (1), namely, a differential equation with $k(t)$ replaced by a constant gain $K \in [0, \infty)$. In this case, the system is governed by the following time-invariant nonlinear integral equation:

$$v(t) = f(t) - K\varphi(\sigma(t));$$

$$\sigma(t) = \sum_{m=0}^{\infty} g_m v(t - \tau_m) + \int_0^{\infty} g(\tau) v(t - \tau)\, \mathrm{d}\tau. \quad (2)$$

We consider the problems of L_2-stability of the systems governed by Equation (1) and by Equation (2). To this end, let $L_2[0, \infty)$ be the linear space of real-valued functions $x(\cdot)$ on $[0, \infty)$ with the property that $\int_0^{\infty} |x(t)|^2\, \mathrm{d}t < \infty$, and equipped with the norm, $\|x(\cdot)\| = (\int_0^{\infty} |x(t)|^2\, \mathrm{d}t)^{1/2}$. The nonlinear system described by Equation (1) is L_2-stable if $v \in L_2[0, \infty)$ for $f \in L_2[0, \infty)$, and an inequality of the type $\|v\| \leq C\|f\|$ holds where C is a constant.

1.1. Brief survey of literature

The class of (closed-loop) control systems *initiated* by Lur'e (1951) is governed, in the notation used by Kalman (1963), by the following equations:

$$\frac{\mathrm{d}\underline{x}}{\mathrm{d}t} = F\underline{x} - \underline{g}\varphi(\sigma); \quad \frac{\mathrm{d}\xi}{\mathrm{d}t} = -\underline{g}\varphi(\sigma); \quad \sigma = \underline{h}'\underline{x} + \rho\xi, \quad (3)$$

where the prime $'$ denotes transpose; σ, ξ, ρ are real scalars; $\underline{x}, \underline{g}, \underline{h}$ are real n-vectors; and F is real $n \times n$ matrix. The nonlinearity $\varphi(\sigma)$ is a real-valued continuous function which belongs to the class A_κ: $\varphi(0) = 0, 0 < \varphi(\sigma)\sigma < \kappa\sigma^2$. The problem then is to establish conditions for the global asymptotic stability – also called 'absolute stability' – of Equation (3) for any $\varphi \in A_\kappa$. It is well known that V. M. Popov established the following frequency-domain inequality for the absolute stability of Equation (3) in his celebrated paper (Popov, 1962):

$$\Re(2\alpha\rho + j\omega\beta)\left[\underline{h}'(j\omega I - F)^{-1}\underline{g} + \frac{\rho}{j\omega}\right] \geq 0$$
$$\text{for all real } \omega \quad (4)$$

holds for $2\alpha\rho = 1$ and some $\beta \geq 0$.

Earlier Lur'e (1951) had formulated the *algebraic* problem of finding necessary and sufficient conditions on ρ, g, h and F for the existence of a Lyapunov function comprising a quadratic form in $(\underline{x}, \sigma)$ and an integral of $\varphi(\sigma)$ to guarantee absolute stability of Equation (3) for class A_κ, where $\kappa = \infty$. It was left to Kalman (1963) to provide the complete answer in the form of a lemma, now known as the Kalman–Yakubovich–Popov (K–M–Y) lemma, containing certain matrix equations (and inequations), to the question of the existence of a Lyapunov function which guarantees absolute stability whenever Popov's criterion (4) is satisfied. See, in this context, Aizerman and Gantmacher (1964), LaSalle (1962), Lefschetz (1965) and Yakubovich (1962). For a generalization of this result (and of the circle criterion (Narendra and Goldwyn, 1964; Sandberg, 1964)) to input–output stability in an abstract setting, see the pioneering work of Zames (1966).

In the first 10 years or so after the publication of Popov's paper, a very large number of papers concentrated on the ramifications of Popov's criterion, as applied mainly to systems described by differential equations. In the course of analysing the challenge posed by the conjecture of Aizerman (1948) in light of Popov's criterion, Brockett and Willems (1965) relaxed the assumptions made on the phase angle behaviour of $G(\mathrm{j}\omega)$ in exchange for (i) the choice of nonlinearity classes which are subsets of $\mathcal{N}$ and (ii) the use of multiplier functions more general than $(1 + \mathrm{j}\alpha\omega)$ so that the 'real-part' conditions (in the frequency domain) are satisfied.

Brockett and Willems (1965) presented a list of third- and fourth-order $G(s)$ for which the null solutions of the linearized versions of the (time-invariant) system are asymptotically stable in the large for all positive gains K, but the Popov criterion cannot predict stability for $\varphi(\cdot) \in \mathcal{N}$, 'regardless of any additional assumptions placed on the coefficients of the system.' To deal with such problems, Brockett and Willems (1965) assumed monotone and odd monotone nonlinearities, and used RC-RL impedance functions. On the other hand, Lakshmi Thathachar, Srinath, and Ramapriyan (1966) explored the use of more general impedance (and hence positive real) functions with complex poles and zeros, and invoked the K–Y–P lemma. In this context, see also Narendra and Neuman (1966), Thathachar (1970) and Dewey and Jury (1966) for related results. However, the most significant result is due to O'Shea (1967) who proposed causal + anticausal functions with a time-domain L_1-norm constraint on the multiplier. See Zames and Falb (1968) for the infinite dimensional system counterpart based on the theory of positive operators. The multiplier used by O'Shea and Zames and Falb (hereafter called the OZF multiplier) is the most general multiplier function known in stability literature.

Remark 1 The system we consider is governed by the (infinite-dimensional) integral equation (1) which is more general than the (finite-dimensional) differential equation (3). When the following replacements are made, Equation (1) *effectively* reduces to the special case of Equation (3): (a) $k(t)$ is replaced by a constant gain $K \in [0, \infty)$; (b) $g(t)$ by the inverse Fourier transform $\mathfrak{g}(t) \doteq [\underline{h}'(\mathrm{j}\omega I - F)^{-1}\underline{g} + \rho/\mathrm{j}\omega]$; (c) $\sum_{m=0}^{\infty} g_m v(t - \tau_m)$ on the right-hand side of the second equation in Equation (1) by $\sigma(0)\mathfrak{g}(t)$ to represent the response to the initial condition $\sigma(0)$ of the finite dimensional system; (d) $f(t) \equiv 0, t \in [0, \infty)$. However, we consider the L_2-stability of Equation (1), for all $\varphi(\cdot) \in \mathcal{N}$, as against the global asymptotic stability of Equation (3). Furthermore, note that the stability results obtained can be made applicable to the case of finite feedback gain by appropriate transformations: if the feedback gain is confined to the sector $[0, \bar{K})$, then the transfer function $G(s)$ is replaced by $(G(s) + 1/\bar{K})$; and the time-varying gain $k(t)$ by $(k(t)/(1 + k(t)/\bar{K}))$. See Thathachar, Srinath, and Krishna (1966) and Huang, Venkatesh, Xiang, and Lee (2014a).

Remark 2 The impact of the K–Y–P lemma has been felt across the whole spectrum of system analysis and synthesis. Its applications to stability analysis are well known and too numerous to be listed here. In fact, elegant and extensive use has been made of the lemma in deriving stability conditions using multipliers having complex poles and zeros in the Lyapunov framework in Lakshmi Thathachar et al. (1966). For others, see the book Boyd, El Ghaoui, Feron, and Balakrishnan (1994) (pp. 123–128, 131–135). However, the K–Y–P lemma as such is not used in this paper because (i) we consider infinite-dimensional systems for which no counterparts of the K–Y–P lemma seem to exist; and, more importantly, (ii) for time-varying systems and for exploring anti-causal multiplier functions, it is not clear how to generate candidate Lyapunov functions, using the K–Y–P lemma or otherwise.

1.2. Motivation

A motivation to generalize the Popov criterion is the multiplier form of the Nyquist criterion which in the most general form reads: the system (2) with $\varphi(\sigma) \equiv \sigma$ is asymptotically stable for all constant gains $K \in [0, \bar{K})$, if there exists a frequency function, $Z(\mathrm{j}\omega)$ such that $-\pi/2 \le \arg\{(Z(\mathrm{j}\omega)\} \le \pi/2$ and $-\pi/2 < \arg\{Z(\mathrm{j}\omega)(G(\mathrm{j}\omega) + 1/\bar{K})\} < \pi/2$, where 'arg' denotes 'the phase angle of.' Alternatively, $\Re Z(\mathrm{j}\omega) \ge 0$, and $\Re Z(\mathrm{j}\omega)(G(\mathrm{j}\omega) + 1/\bar{K}) > 0$, $\omega \in (-\infty, \infty)$, where '$\Re$' denotes 'the real part of.' The last two frequency-domain inequalities are called below (for convenience) as merely *real-part conditions*. See Brockett and Willems (1965) and Thathachar (1970). Note that neither a restriction of causality nor a time-domain L_1-norm (or otherwise) constraint has been imposed on $Z(\mathrm{j}\omega)$. In particular, the chasm that exists between what we recognize as a very general frequency function $Z(\mathrm{j}\omega)$ of the Nyquist criterion for linear time-invariant system (obtained from Equation (2) by setting $\varphi(\sigma) \equiv \sigma$) and the Popov multiplier function $(1 + \mathrm{j}\alpha\omega)$ for the (absolute stability of the) nonlinear time-invariant system (2) is too big. In fact, a possible way of resolving Aizerman's conjecture Aizerman (1948) is to isolate classes of $G(\mathrm{j}\omega)$ functions which obey Nyquist's criterion and for which $(1 + \mathrm{j}\alpha\omega)$ for some $\alpha > 0$ is also a multiplier function. In this context, it is to be noted that the *transition* from the class $\mathcal{N}$ of nonlinearities to the class of linear gains is too abrupt.

In the present paper, we adopt a converse approach to deal with Aizerman's conjecture. More, specifically, in order to derive L_2-stability criteria for the system (2) with $\varphi(\cdot) \in \mathcal{N}$ we explore the use of the OZF multiplier itself. The results can specialized to be applicable to systems governed by differential equations of the type considered by Popov and others. It is known that under certain general conditions, L_2-stability of systems governed by differential equations can be shown to be equivalent to their absolute stability.

Another motivation (to generalize the Popov criterion) arose from some recent stability results for multi-input–muliti-output (MIMO) time-varying continuous- and discrete-time nonlinear systems, which constitute a generalizations of the system (1). See Huang, Venkatesh, Xiang, and Lee (2014b) and Venkatesh (2014). It was found that for continuous-time MIMO L_2-stability as also for discrete-time MIMO ℓ_2-stability, certain algebraic inequalities involving multi-variable nonlinearities in generalized two-variable polynomial forms (extending quadratic and bi-quadratic forms) can be exploited to modify (and partially dispense with) the existing constraints on multiplier functions commonly used in the literature. Characteristic parameters (CPs) of the nonlinearity obtained from the extremal values of the ratios of polynomial forms in two variables govern the time-domain constraints on the multiplier.

1.3. *Main contributions*

We establish two stability theorems: Theorem 1 for the time-invariant system (2) with $\varphi(\cdot) \in \mathcal{N}$ and Theorem 2 for the time-varying system (1) for the same class $\mathcal{N}$ of nonlinearities. In Theorem 1, the stability conditions are expressed in terms of a causal + anticausal multiplier function $Z(\mathrm{j}\omega)$ subject to the following requirements: [C1-1] - $\Re Z(\mathrm{j}\omega)G(\mathrm{j}\omega) > 0,\ \omega \in (-\infty, \infty)$; and [C1-2] – a time-domain constraint on the inverse Fourier transform of $Z(\mathrm{j}\omega)$, depending on certain CPs of the nonlinearity to be defined later below. On the other hand, in Theorem 2, the stability requirements are [C2-1] – same as [C1-1] of Theorem 1; [C2-2] – a global bound on positive and negative lobes of the normalized rate of variation $\theta(t) \doteq \mathrm{d}k(t)/\mathrm{d}t/k(t)$; and [C2-3] – this is a modified version of [C1-2] of Theorem 1 that depends on the choice of the global bounds in [C2-2].

1.4. *Organization of the paper*

The next section (Section 2) is concerned with the main assumptions, problem formulation, and mathematical preliminaries, including the definitions of the CPs of two classes $\mathcal{N}$ and $\mathcal{M}$ of nonlinearities needed in the proof of the stability theorems. Section 3 presents the first main stability result (Theorem 1) of the paper (meant for the L_2-stability of the time-invariant nonlinear system (1)) with $\varphi(\cdot) \in \mathcal{N}$. In Section 4, the second main result of the paper (Theorem 2) deals with the L_2-stability of the nonlinear time-varying system (1) with $\varphi(\cdot) \in \mathcal{N}$. In a subsection, the L_2-stability of the same system is analysed when there is a periodic switching of nonlinearities. Examples are given in Section 5 with a view to illustrate the application of the theorems and to exhibit the superiority of the stability conditions over those in the literature. In Section 6, the new stability results are critically examined for a *theoretical* comparison with the literature, and some open problems are presented. Section 7 concludes the paper, followed by appendices which contain proofs of the various lemmas used in the proofs of the theorems.

2. Assumptions, problem formulation, and preliminaries

For the system (1), the impulse response of the linear block is assumed to be in $L_1 \cap L_2$, and when the nonlinearity $\varphi(\sigma)$ is replaced by $q_2\sigma$ (with constant $q_2 > 0$) and time-varying gain $k(t)$ by the constant gain $K \in [0, \infty)$, its solutions are in $L_1 \cap L_2$, which implies that the zeros of $(1 + KG(s))$ for $K \in [0, \infty)$ lie strictly in the left-half ($\Re s < -\delta \le 0$) of the complex plane.

Problem formulation: Given $G(s)$ satisfying the above assumptions, find conditions (i) on $\varphi(\cdot) \in \mathcal{N}$ for the L_2-stability of the system (2); and (ii) on $\varphi(\cdot) \in \mathcal{N}$ and $k(t)$ for the L_2-stability of the system (1).

Preliminaries: For any real-valued function $x(\cdot)$ on $[0, \infty)$ and any $T \ge 0$, we define the *truncated function* $x_T(\cdot)$ by: $x_T(t) = x(t)$ for $0 \le t \le T$; and $x_T(t) = 0$ for $t < 0$ and $t > T$. Let L_{2e} be the space of those real-valued functions $x(\cdot)$ on $[0, \infty)$ whose truncations $x_T(\cdot)$ belong to $L_2[0, \infty)$ for all $T \ge 0$. In order to establish stability of the system under consideration, we first assume infinite *escape time* for the solution of the system with $f \in L_2$ and the solution belongs to L_{2e}. We then show that, under certain conditions on $\varphi(\cdot)$, $k(t)$ and $G(\mathrm{j}\omega)$, the solution actually belongs to $L_2[0, \infty)$.

Consider the class of operators $\mathcal{Z} \in L_{2e} \to L_{2e}$, satisfying an equation of the type

$$\mathcal{Z}\{\sigma(t)\} = \sigma(t) + \alpha\frac{\mathrm{d}\sigma}{\mathrm{d}t} + \sum_{m=1}^{\infty}\{z_m\sigma(t-\tau_m) + z'_m\sigma(t+\tau'_m)\} + \int_{-\infty}^{\infty} z(\tau)\sigma(t-\tau)\,\mathrm{d}\tau, \tag{5}$$

where (real) constant $\alpha > 0$; the real sequences $\{z_m\}$ and $\{z'_m\}$ are in ℓ_1, i.e. $\sum_{m=1}^{\infty}(|z_m| + |z'_m|) < \infty$; sequences $\{\tau_m\}$ and $\{\tau'_m\}$ are in $[0, \infty)$; $z(\cdot)$ is a real-valued function on $(-\infty, \infty)$, and is in $L_1(-\infty, \infty)$, i.e. $\int_{-\infty}^{\infty} |z(t)|\,\mathrm{d}t < \infty$. Its Fourier transform is given by

$$Z(j\omega) = 1 + j\alpha\omega + \sum_{m=1}^{\infty}\{z_m(\mathrm{e}^{-j\tau_m\omega}) + z'_m(\mathrm{e}^{j\tau'_m\omega})\} + \int_{-\infty}^{\infty} z(t)\mathrm{e}^{-j\omega t}\,\mathrm{d}t. \tag{6}$$

The multiplier function used by O'Shea (1967) and Zames and Falb (1968) for dealing with the class $\mathcal{M}$ of monotone nonlinearities is Equation (6), and called earlier (above) as the OZF multiplier. For convenience in later manipulations, let

(1) z_m^+ and $-z_m^-$ denote, respectively, the positive and negative coefficients from the set $\{z_m\}$, i.e. $z_m = z_m^+ - z_m^-, m = 1, 2, \ldots$; $z_m'^+$ and $-z_m'^-$ denote, respectively, the positive and negative coefficients from the set $\{z'_m\}$. i.e. $z'_m = z_m'^+ - z_m'^-, m = 1, 2, \ldots$;
(2) $z_c(t) = z(t)$ for $t \ge 0$; and $z_a(t) = z(t)$ for $t < 0$, so that $z(t) = z_c(t) + z_a(t)$ for $t \in (-\infty, \infty)$;
(3) $z_c^+(t) = z_c(t)$ if $z_c(t) > 0, t \in [0, \infty)$; else $z_c^+(t) = 0, t \in [0, \infty)$; $-z_c^-(t) = z_c(t)$ if $z_c(t) < 0, t \in [0, \infty)$;

else $z_c^-(t) = 0, t \in [0, \infty)$, i.e. $z_c(t) = z_c^+(t) - z_c^-(t)$, $t \in [0, \infty)$; and
(4) $z_a^+(t) = z_a(t)$ if $z_a(t) > 0, t \in (-\infty, 0]$; else $z_a^+(t) = 0, t \in (-\infty, 0]$. Similarly, $-z_a^-(t) = z_a(t)$ if $z_a(t) < 0, t \in (-\infty, 0]$; else $z_a^-(t) = 0, t \in (-\infty, 0]$, i.e. $z_a(t) = z_a^+(t) - z_a^-(t), t \in (-\infty, 0]$.

Theorems 1 and 2 (and their proofs) are based on new algebraic inequalities, involving the nonlinearity $\varphi(\cdot)$, which are motivated by the well-known algebraic inequality $-\frac{1}{2}(a^2+b^2) \leq ab \leq \frac{1}{2}(a^2+b^2)$ for arbitrary, real scalars a and b. Based on these inequalities, the characteristic parameters (CPs) of $\varphi(\cdot) \in \mathcal{N}, \mathcal{M}$ are defined as follows. Let $\Phi(\sigma) \doteq \int_0^\sigma \varphi(\xi)\, d\xi > 0$ for $\sigma \neq 0$; and for (real) $x, y \in (-\infty, \infty)$, let $\Psi(x,y) \doteq \{\varphi(x)x + \varphi(y)y\}$. Then (i) for $\varphi(.) \in \mathcal{N}$, $\nu_i\varphi(\sigma)\sigma \leq \Phi(\sigma) \leq \nu_s\varphi(\sigma)\sigma$ and $-\mu_i\Psi(x,y) \leq \varphi(x)y \leq \mu_s\Psi(x,y)$, where the CPs $\nu_i \geq 0, \nu_s > 0, \mu_i > 0$, and $\mu_s > 0$ are defined by

$$\nu_i \doteq \inf_{\substack{\varphi(\cdot)\in\mathcal{N} \\ x\neq 0}} \frac{\Phi(x)}{\varphi(x)x}; \quad \nu_s \doteq \sup_{\substack{\varphi(\cdot)\in\mathcal{N} \\ x\neq 0}} \frac{\Phi(x)}{\varphi(x)x};$$

$$-\mu_i \doteq \inf_{\substack{\varphi(\cdot)\in\mathcal{N} \\ x,y\neq 0}} \frac{\varphi(x)y}{\Psi(x,y)}; \quad \mu_s \doteq \sup_{\substack{\varphi(\cdot)\in\mathcal{N} \\ x,y\neq 0}} \frac{\varphi(x)y}{\Psi(x,y)}; \quad \text{and} \tag{7}$$

(ii) for $\varphi(.) \in \mathcal{M}$, $\eta_i\varphi(\sigma)\sigma \leq \Phi(\sigma) \leq \eta_s\varphi(\sigma)\sigma$; and $-\gamma_i\Psi(x,y) \leq \varphi(x)y \leq \gamma_s\Psi(x,y)$, where the CPs $\eta_i \geq 0, \eta_s > 0, \gamma_i > 0$, and $\gamma_s > 0$ are defined by

$$\eta_i \doteq \inf_{\substack{\varphi(\cdot)\in\mathcal{M} \\ x\neq 0}} \frac{\Phi(x)}{\varphi(x)x}; \quad \eta_s \doteq \sup_{\substack{\varphi(\cdot)\in\mathcal{M} \\ x\neq 0}} \frac{\Phi(x)}{\varphi(x)x};$$

$$-\gamma_i \doteq \inf_{\substack{\varphi(\cdot)\in\mathcal{M} \\ x,y\neq 0}} \frac{\varphi(x)y}{\Psi(x,y)}; \quad \gamma_s \doteq \sup_{\substack{\varphi(.)\in\mathcal{M} \\ x,y\neq 0}} \frac{\varphi(x)y}{\Psi(x,y)}. \tag{8}$$

Note that $\eta_s \leq 1$ for $\varphi(\cdot) \in \mathcal{M}$. See Table 1 for typical values of μ_i and μ_s for $\varphi(\cdot) \in \mathcal{N}$ and $\mathcal{M}$. In the last item of Table 1, the slope of $\varphi(\sigma)$ is negative for $x \in [1.214, 2.1275]$. It is conjectured that, for an arbitrary $\varphi(\cdot) \in \mathcal{N}$, the upper limit of $\mu_i = \mu_s = 1$.

3. Main results-1

With the preliminaries settled, we now present the first main result of the paper.

Table 1. Typical values of μ_i and μ_s for $\varphi(\cdot) \in \mathcal{N}$ and $\mathcal{M}$.

No.	Nonlinearity $\varphi(\cdot)$	Class of $\varphi(\cdot)$	μ_i	μ_s
1	$(x - 0.3x^3 + 0.03x^5)$	$\mathcal{N}$	0.7892	0.7892
2	$(x + 0.3x^3 + 0.03x^5)$	$\mathcal{M}$	0.636	0.636
3	$(x - 0.3x^3 + 0.1x^5)$ for $x < 0$ $(x - 0.3x^3 + 0.03x^5)$ for $x \geq 0$	$\mathcal{N}$	0.9389	0.7892

THEOREM 1 *The nonlinear system (1) with $\varphi(\cdot) \in \mathcal{N}$ and $k(t)$ replaced by a constant gain $K \in [0, \infty)$ is L_2-stable, if there exists a multiplier function $Z(\mathrm{j}\omega)$ of the form Equation (6) such that [H-1] for some positive constant δ, $\Re Z(\mathrm{j}\omega)G(\mathrm{j}\omega) \geq \delta > 0$, $\omega \in (-\infty, \infty)$; and [H-2] $\sum_{m=1}^{\infty}\{(\mu_i z_m^+ + \mu_s z_m^-) + (\mu_i z_m'^+ + \mu_s z_m'^-)\} + \int_0^\infty (\mu_i z_c^+(\tau) + \mu_s z_c^-(\tau))\, d\tau + \int_{-\infty}^0 (\mu_i z_a^+(\tau) + \mu_s z_a^-(\tau))\, d\tau) < \frac{1}{2}$, where μ_i and μ_s are defined by Equation (7).*

COROLLARY T1-1 *If $\varphi(\cdot) \in \mathcal{M}$, Theorem 1 holds if, in condition [H-2] μ_i and μ_s, are replaced respectively by γ_i and γ_s as defined by Equation (8).*

The proof of Theorem 1, which follows the strategy developed in Venkatesh (1978), requires the following lemma which is proved in Appendix 2.

LEMMA 1 *With the multiplier function $Z(j\,\omega)$ defined by Equation (6), the integral*

$$\lambda_1(T) \doteq \int_0^T \varphi(\sigma_T(t))\mathcal{Z}\{\sigma_T(t)\}\, dt \tag{9}$$

satisfies the inequality $\lambda_1(T) \geq \alpha(\Phi(\sigma_T(T)) - \Phi(\sigma_T(0)))$, where $\Phi(\sigma) = \int_0^\sigma \varphi(\xi)\, d\xi$, for all $\sigma_T(t)$ in the domain of $\mathcal{Z}$ and for all $T \geq 0$, if condition [H-2] of Theorem 1 is satisfied. Note that $\Phi(\sigma) > 0$ for $\sigma \neq 0$.

Proof of Theorem 1. Consider the integral, for any $T > 0$,

$$\rho_1(T) \doteq \int_0^T f_T(t)\mathcal{Z}\{\mathcal{G}\{v_T(t)\}\}\, dt, \tag{10}$$

where $\mathcal{G}\{v_T(t)\} \doteq \int_0^t g(\tau)v_T(t-\tau)\, d\tau$. It follows from $f_T(t) = v_T(t) + K\varphi(\sigma_T(t))$ in Equation (1) that

$$\rho_1(T) = \int_0^T v_T(t)\mathcal{Z}\{\mathcal{G}\{v_T(t)\}\}\, dt + \int_0^T K\varphi(\sigma_T(t))\mathcal{Z}\{\sigma_T(t)\}\, dt. \tag{11}$$

Let $V_T(\mathrm{j}\omega)$ denote the Fourier transform of $v_T(t)$. (Note that the subscript T in $V_T(\mathrm{j}\omega)$ refers to the fact that the original time-function is truncated. There is no truncation of the Fourier transform.) Applying the Parseval theorem (see Appendix 1) to the first integral on the right-hand side of Equation (11), we obtain

$$\int_0^T v_T(t)\mathcal{Z}\{\mathcal{G}\{v_T(t)\}\}\, dt = \frac{1}{2\pi}\int_{-\infty}^{\infty} V_T(-\mathrm{j}\omega)Z(\mathrm{j}\omega)G(\mathrm{j}\omega)V_T(\mathrm{j}\omega)\, d\omega. \tag{12}$$

Invoking the condition [H-1] of Theorem 1, $\Re Z(\mathrm{j}\omega) G(\mathrm{j}\omega) \geq \delta > 0$ for some $\delta > 0$, we obtain the following inequality:

$$\int_0^T v_T(t)\mathcal{Z}\{\mathcal{G}\{v_T(t)\}\}\,\mathrm{d}t = \frac{1}{2\pi}\int_{-\infty}^{\infty} Z(\mathrm{j}\omega)G(\mathrm{j}\omega)|V_T(\mathrm{j}\omega)|^2\,\mathrm{d}\omega \geq \frac{\delta}{2\pi}\|v_T\|^2. \tag{13}$$

By virtue of Lemma 1, the second integral right-hand side of Equation (11),

$$\int_0^T K\varphi(\sigma_T(t))\mathcal{Z}\{\sigma_T(t)\}\,\mathrm{d}t \geq \alpha(\Phi(\sigma_T(T)) - \Phi(\sigma_T(0))), \tag{14}$$

where, to recall, $\Phi(\sigma) = \int_0^\sigma \varphi(\xi)\,\mathrm{d}\xi$. By applying the Parseval theorem to Equation (10), combining the result with Equations (13) and (14), and noting that $\Phi(\sigma) \geq 0, \forall\sigma$, we obtain

$$\delta\|v_T\|^2 \leq \alpha\Phi(\sigma_T(0)) + 2\pi\int_0^T f_T(t)\mathcal{Z}\{\mathcal{G}\{v_T(t)\}\}\,\mathrm{d}t = \int_{-\infty}^{\infty} F_T(-\mathrm{j}\omega)Z(\mathrm{j}\omega)G(\mathrm{j}\omega)V_T(\mathrm{j}\omega)\,\mathrm{d}\omega. \tag{15}$$

Now using the Cauchy–Schwarz inequality in Equation (15) leads to

$$\int_{-\infty}^{\infty} F_T(-\mathrm{j}\omega)Z(\mathrm{j}\omega)G(\mathrm{j}\omega)V_T(\mathrm{j}\omega)\,\mathrm{d}\omega \leq \sup_{-\infty<\omega<\infty} |Z(\mathrm{j}\omega)G(\mathrm{j}\omega)|\|f_T\|\|v_T\|. \tag{16}$$

Note that $\sup_{-\infty<\omega<\infty}|Z(\mathrm{j}\omega)G(\mathrm{j}\omega)|$ is finite by virtue of the assumptions on $Z(\cdot)$ and $G(\cdot)$. Let $C = \sup_{-\infty<\omega<\infty} | Z(\mathrm{j}\omega)G(\mathrm{j}\omega) |$ and $\alpha_0 \doteq \alpha\Phi(\sigma_T(0))$, which is independent of T and can be assumed bounded. Then, from Equations (15) and (16), we obtain the inequality $\delta\|v_T\|^2 \leq \alpha_0 + C\|f_T\|\|v_T\|$ which is valid for all $T > 0$. Since δ, C and α_0 are independent of T, we conclude that $\|v\| \leq C'\|f\| + \sqrt{\alpha_0' + C'^2/4\|f\|^2}$, where $\alpha_0' = \alpha_0/\delta$ and $C' = C/\delta$. The theorem is proved. ■

4. Main results-2

We now derive L_2-stability conditions for the system (1). To this end, let $h(t)$ be a nonnegative, integrable and bounded function on $t \in [0,\infty)$, and let $\varpi(t) \doteq \mathrm{e}^{-\int_{t_0}^t h(\tau)\,\mathrm{d}\tau}$. Assume that the integral $\int_{t_0}^t h(\tau)\,\mathrm{d}\tau \leq M < \infty, t \in [0,\infty)$ and $0 < \epsilon \leq \lim_{t\in\infty}\int_{t_0}^t h(\tau)\,\mathrm{d}\tau \leq M < \infty$. Then, $\varpi(t)$ is a bounded positive function. Note that $(\mathrm{d}\varpi/\mathrm{d}t)/\varpi(t) = -h(t)$, which is non-positive. We assume that (i) $k(t) \in [\epsilon,\infty)$ for $t \geq 0$, where (constant) $\epsilon > 0$, is a piecewise-continuous function of bounded variation with first-order (i.e. jump-)discontinuities in ℓ_1; and (ii) $k(t)$ is made up of the continuous part, $k_c(t)$, and the discontinuous part $k_d(t)$: discontinuities at instants t_{m+} correspond to positive jumps, α_m^+; and instants t_{m-} corresponding to negative jumps, α_{m-}^-.

The derivative of $k(t)$ is then given by

$$\frac{\mathrm{d}k}{\mathrm{d}t} = \frac{\mathrm{d}k_c}{\mathrm{d}t} + \sum_m\{\alpha_m^+\delta(t-t_{m+}) + \alpha_m^-\delta(t-t_{m-})\}. \tag{17}$$

Furthermore, let $\theta(t) = \mathrm{d}k/\mathrm{d}t/k(t)$. At the positive discontinuities, t_{m+}, of $k(t)$, the value of $k(t)$ is, by convention, taken as $k(t_{m+}^-)$ where t_{m+}^- is the instant just to the left of t_{m+}. Similarly, at the negative discontinuities, t_{m-}, of $k(t)$, the value of $k(t)$ is taken as $k(t_{m-}^-)$ where t_{m-}^- is the instant just to the left of t_{m-}. Note that, based on the assumptions on $k(t)$, $k(t_{m-}^-) \neq 0$, and $k(t_{m+}^-) \neq 0$, $t \geq 0$. Furthermore, let $\theta^+(t) = \theta(t)$, for $\theta(t) > 0$; $\theta^+(t) = 0$, for $\theta(t) \leq 0$; and $\theta^-(t) = \theta(t)$, for $\theta(t) < 0$; $\theta^-(t) = 0$, for $\theta(t) \geq 0$. Also, let $\theta_c(t) = \mathrm{d}k_c/\mathrm{d}t/k_c(t)$. Note that $\theta(t) = \theta^+(t) + \theta^-(t)$, where

$$\theta^+(t) = \theta_c^+(t) + \sum_m \psi_{m+}\delta(t-t_{m+}); \quad \text{and}$$

$$\theta^-(t) = \theta_c^-(t) + \sum_m \psi_{m-}\,\delta(t-t_{m-}). \tag{18}$$

where $\psi_{m+} \doteq \alpha_m^+/k(t_{m+}^-)$ and $\psi_{m-} \doteq \alpha_m^-/k(t_{m-}^-)$. We need some additional preliminaries. By the statement that $\varpi(t)\mathrm{e}^{-\xi t}k(t)$ for $t \in [0,\infty)$ is nonincreasing, we mean that

$$\frac{\mathrm{d}(\varpi(t)k(t)\mathrm{e}^{-\xi t})}{\mathrm{d}t} = \mathrm{e}^{-\xi t}\left\{\frac{\mathrm{d}\varpi}{\mathrm{d}t} - \xi\varpi(t)\right\}k(t) + \mathrm{e}^{-\xi t}\varpi(t)\frac{\mathrm{d}k(t)}{\mathrm{d}t} \leq 0, \quad t \in [0,\infty). \tag{19}$$

And, similarly, by the statement that $\varpi(t)\mathrm{e}^{\zeta t}k(t)$ for $t \in [0,\infty)$ is nondecreasing, we mean that

$$\frac{\mathrm{d}(\varpi(t)k(t)\mathrm{e}^{\zeta t})}{\mathrm{d}t} = \mathrm{e}^{\zeta t}\left\{\frac{\mathrm{d}\varpi}{\mathrm{d}t} + \zeta\varpi(t)\right\}k(t) + \mathrm{e}^{\zeta t}\varpi(t)\frac{\mathrm{d}k(t)}{\mathrm{d}t} \geq 0, \quad t \in [0,\infty). \tag{20}$$

In view of the assumptions on $\varpi(t)$ and $k(t)$, it can be shown that (19) and (20) together reduce to the following inequality:

$$-\zeta \leq \left(\frac{\mathrm{d}\varpi/\mathrm{d}t}{\varpi(t)}\right) + \theta(t) \leq \xi. \tag{21}$$

THEOREM 2 *The system (1) with $\varphi(\cdot) \in \mathcal{N}$; $k(t) \in [\epsilon,\infty)$ for $t \geq 0$, where (constant) $\epsilon > 0$; and $\|v_\varepsilon\|^2 \doteq \int_0^\infty \mathrm{e}^{2\varepsilon t}(v(t))^2\,\mathrm{d}t$, where $\varepsilon > 0$ is an arbitrarily small number, is exponentially L_2-stable in the sense that $\|v_\varepsilon\| \leq C_1\|f\| + \sqrt{C_0 + (C_1^2/4)\|f\|^2}$, where C_1, C_2 are constants, if there exist a multiplier function $Z(\mathrm{j}\omega)$ defined* *by Equation* (6); *a bounded positive function* $\varpi(\cdot)$ *as defined above; and nonnegative constants,* ξ, ζ, *such that [H-1]: (for the* $\ddot{I}_{\xi}$ *defined above)* $\sup_{-\infty<\omega<\infty} \|Z(j\omega-\varepsilon)G(j\omega-\varepsilon)\| < \infty$; *and* $\Re Z((j\omega-\varepsilon)G(j\omega-\varepsilon)) > 0, \omega \in (-\infty, \infty)$; $[H-2]$: $\varpi(t)e^{-\xi t}k(t)$ *is nonincreasing and* $\varpi(t)e^{\zeta t}k(t)$ *is nondecreasing for all* $t \in [0, \infty)$; $[H-3]$: $\alpha\nu_s\xi + \sum_{m=1}^{\infty}\{(1+e^{\xi\tau_m})(\mu_i z_m^+ + \mu_s z_m^-) + (1+e^{\zeta\tau'_m})(\mu_i z_m'^+ + \mu_s z_m'^-)\} + \int_0^{\infty}\{(1+e^{\xi\tau})(\mu_i z_c^+(\tau) + \mu_s z_c^-(\tau)\} d\tau + \int_{-\infty}^{0}\{(1+e^{-\zeta\tau})(\mu_i z_a^+(\tau) + \mu_s z_a^-(\tau)\} d\tau < 1$, *where* ν_s, μ_i, *and* μ_s *are defined in Equation* (7).

The proof of Theorem 2 depends on Lemma 2 given below. The proofs of both are given in Appendix 3. (Also see in this context Venkatesh, 1978, pp. 572–573.)

LEMMA 2 *With the operator* $\mathcal{Z}$ *defined by Equation* (5), *the integral*

$$\lambda_2(T) \doteq \int_0^T \varpi(t)k(t)\varphi(\sigma_T(t))\mathcal{Z}\sigma_T(t)\, dt > -\alpha\varpi(0)k(0)\Phi(\sigma_T(0))), \tag{22}$$

for all σ_T *in the domain of* $\mathcal{Z}$ *and for all* $T \geq 0$, *if conditions [H-2] and [H-3] of Theorem 2 are satisfied.*

Condition [H-2] of Theorem 2 is equivalent to an appropriate bound on the *normalized* rate of variation of $k(t)$, as made explicit by the following corollary.

COROLLARY L2-1 *With* $\theta^+(t)$ *and* $\theta^-(t)$ *as defined above, if* $\varpi(t)e^{-\xi t}k(t)$ *is nonincreasing and* $\varpi(t)e^{\zeta t}k(t)$ *is nondecreasing for* $t \in [0, \infty)$, *then, for some positive constants* N_1 *and* N_2, *for all finite* $T > 0$ *and for all* $t_0 \geq 0$,

$$\frac{1}{T}\int_{t_0}^{T+t_0} \theta^+(t)\, dt \leq N_1, \quad -N_2 \leq \frac{1}{T}\int_{t_0}^{T+t_0} \theta^-(t)\, dt; \tag{23}$$

$$\text{for} \quad \xi \neq \zeta, \quad \lim_{T\to\infty} \frac{1}{T}\int_{t_0}^{T+t_0} \theta^+(t)\, dt \leq \xi,$$

$$-\zeta \leq \lim_{T\to\infty} \frac{1}{T}\int_{t_0}^{T+t_0} \theta^-(t)\, dt, \tag{24}$$

but, for $\xi = \zeta$, $\theta^+(t)$ *and* $\theta^-(t)$ *are unrestricted.*

If $k(t)$ is piecewise constant, then $k_c(t)$ is identically zero. Let M denote the number of discontinuities in a finite interval $T > 0$. Then, by invoking Equations (18), (24) becomes,

$$\frac{1}{T}\sum_{m_0}^{m_0+M} \psi_{m+} \leq N_1, -N_2 \leq \frac{1}{T}\sum_{m_0}^{m_0+M} \psi_{m-} \quad \text{and}$$

$$\lim_{T\to\infty}\frac{1}{T}\sum_m \psi_{m+} \leq \xi, -\zeta \leq \lim_{T\to\infty}\frac{1}{T}\sum_m \psi_{m-}, \tag{25}$$

but, for $\xi = \zeta$, $\sum_m \psi_{m+}$ and $\sum_m \psi_{m-}$ are unrestricted. The proof of Lemma 2 is similar to that of Lemma 2 of Venkatesh (1978) (pp. 572, 577–579), and is hence omitted. When $k(t)$ is periodic with period $\mathfrak{p}$, then Equations (23)–(25) reduce respectively to

$$\frac{1}{\mathfrak{p}}\int_0^{\mathfrak{p}} \theta^-(t)\, dt \geq -\zeta, \quad \frac{1}{\mathfrak{p}}\int_0^{\mathfrak{p}} \theta^+(t)\, dt \leq \xi; \tag{26}$$

$$\left(\frac{1}{\mathfrak{p}}\right)\sum_{m,t_{m-}\in\mathfrak{D}} \psi_m(t_{m-}) \geq -\zeta, \quad \left(\frac{1}{\mathfrak{p}}\right)\sum_{m,t_{m+}\in\mathfrak{D}} \psi_m(t_{m+}) \leq \xi, \tag{27}$$

where $\mathfrak{D}$ denotes the semi-closed interval $(0, \mathfrak{p}]$.

4.1. Switching nonlinearities

We now consider the case of switching between nonlinearities in Equation (1). To this end, let the nonlinear gain consists of two nonlinearities $\varphi_1(\cdot)$ and $\varphi_2(\cdot)$, both belonging to class $\mathcal{N}$, which switch from one to the other periodically, with the *apparent* dwell times of $\varphi_1(\cdot)$ and $\varphi_2(\cdot)$ given, respectively, by $\mathfrak{d}_1$ and $\mathfrak{d}_2$. The fundamental period of switching is $\mathfrak{p} \doteq \mathfrak{d}_1 + \mathfrak{d}_2$, Let $k_1(t), k_2(t) \in [\epsilon, \infty)$ for $t \geq 0$ and for some $\epsilon > 0$, be periodic with the same period $\mathfrak{p}$ and active, respectively, during the regimes of $\varphi_1(\cdot)$, and $\varphi_2(\cdot)$. In Equation (1), with $u(t)$ denoting the step function, let

$$k(t)\varphi(\sigma) = k_1(t)\varphi_1(\sigma)\sum_{m=0}^{\infty}(u(t-m\mathfrak{p}) - u(t-m\mathfrak{p}-\mathfrak{d}_1)) + k_2(t)\varphi_2(\sigma)\sum_{m=0}^{\infty} \times (u(t-m\mathfrak{p}-\mathfrak{d}_1) - u(t-(m+1)\mathfrak{p})). \tag{28}$$

For simplicity in later manipulations, let $U_1(t) \doteq \sum_{m=0}^{\infty}(u(t-m\mathfrak{p}) - u(t-m\mathfrak{p}-\mathfrak{d}_1))$ and $U_2(t) \doteq \sum_{m=0}^{\infty}(u(t-m\mathfrak{p}-\mathfrak{d}_1) - u(t-(m+1)\mathfrak{p}))$, so that Equation (28) becomes $k(t)\varphi(\sigma) = k_1(t)U_1(t)\varphi_1(\sigma) + k_2(t)U_2(t)\varphi_2(\sigma)$. Note that $U_1(t)$ and $U_2(t)$ are periodic with period $\mathfrak{p}$. We now establish a corollary to Theorem 2 using a special case of the multiplier operator $\mathcal{Z}$ defined by Equation (5). Let

$$\mathcal{Z}_p\{\sigma(t)\} \doteq \sigma(t) + \sum_{m=1}^{\infty}\{z_m\sigma(t-m\mathfrak{p}) + z'_m\sigma(t+m\mathfrak{p})\}, \tag{29}$$

where, as before, the real sequences $\{z_m\}$ and $\{z'_m\}$ are in ℓ_1, i.e. $\sum_{m=1}^{\infty}(|z_m| + |z'_m|) < \infty$; sequences $\{\tau_m\}$ and $\{\tau'_m\}$ are in $[0, \infty)$. Its Fourier transform is given by

$$Z_p(j\omega) = 1 + \sum_{m=1}^{\infty}\{z_m e^{-jm\mathfrak{p}\omega} + z'_m e^{jm\mathfrak{p}\omega}\}. \tag{30}$$

For (real) $x, y \in (-\infty, \infty)$ and with $r = 1, 2$, let $\Psi_r(x, y) \doteq \{\varphi_r(x)x + \varphi_r(y)y\}$. Then (i) for $\varphi_r(\cdot) \in \mathcal{N}$, $-\mu_{i,r}\Psi_r(x, y) \leq$

$\varphi_r(x)y \leq \mu_{s,r}\Psi_r(x,y)$, where the CPs $\mu_{i,r} > 0$, and $\mu_{s,r} > 0$ are defined by

$$-\mu_{i,r} \doteq \inf_{\substack{\varphi_r(\cdot)\in\mathcal{N}\\ x,y\neq 0}} \frac{\varphi_r(x)y}{\Psi_r(x,y)}, \quad \mu_{s,r} \doteq \sup_{\substack{\varphi_r(\cdot)\in\mathcal{N}\\ x,y\neq 0}} \frac{\varphi_r(x)y}{\Psi_r(x,y)},$$

$$r = 1,2, \quad \mu_{i,*} \doteq \max(\mu_{i,1}, \mu_{i,2}),;$$

$$\mu_{s,*} \doteq \max(\mu_{s,1}, \mu_{s,2}); \tag{31}$$

(ii) for $\varphi_r(\cdot) \in \mathcal{M}$, $-\gamma_{i,r}\Psi_r(x,y) \leq \varphi_r(x)y \leq \gamma_{s,r}\Psi_r(x,y)$, where the CPs $\gamma_{i,r} > 0$, and $\gamma_{s,r} > 0$ are defined by

$$-\gamma_{i,r} \doteq \inf_{\substack{\varphi_r(\cdot)\in\mathcal{M}\\ x,y\neq 0}} \frac{\varphi_r(x)y}{\Psi_r(x,y)}, \quad \gamma_{s,r} \doteq \sup_{\substack{\varphi_r(\cdot)\in\mathcal{M}\\ x,y\neq 0}} \frac{\varphi_r(x)y}{\Psi_r(x,y)},$$

$$r = 1,2, \quad \gamma_{i,*} \doteq \max(\gamma_{i,1}, \gamma_{i,2}),$$

$$\gamma_{s,*} \doteq \max(\gamma_{s,1}, \gamma_{s,2}). \tag{32}$$

COROLLARY T2-1 *The system* (1) *with* $\varphi_1(\cdot), \varphi_2(\cdot) \in \mathcal{N}$, *and* $k_1(t), k_2(t)$ *as defined above, is* L_2*-stable, if there exist a multiplier operator* $\mathcal{Z}_p$ *defined by Equation* (29) *such that [H-1]:* $\sup_{-\infty<\omega<\infty} \|Z_p(j\omega)G(j\omega)\| < \infty$; *and* $\Re Z_p((j\omega)G(j\omega)) > 0, \omega \in (-\infty,\infty)$; *and [H-2]:* $\sum_{m=1}^{\infty}\{\mu_{i,*}(z_m^+ + z_m'^+) + \mu_{s,*}(z_m^- + z_m'^-)\} < \frac{1}{2}$, *where* $\mu_{i,*}$ *and* $\mu_{s,*}$ *are as defined in Equation* (31).

Its proof is similar to the proof of Theorem 1 and depends on the following corollary which is proved in Appendix 4.

COROLLARY L2-2 *With the operator multiplier* $\mathcal{Z}_p$ *defined by* (29), *the integral*

$$\lambda_{2p}(T) \doteq \int_0^T k(t)\varphi(\sigma_T(t))\mathcal{Z}_p\{\sigma_T(t)\}\,dt \tag{33}$$

satisfies the inequality $\lambda_{2p}(T) \geq 0$, *for all* $\sigma_T(t)$ *in the domain of* $\mathcal{Z}_p$ *and for all* $T \geq 0$ *if condition [H-2] of Corollary T2-1 is satisfied.*

5. Examples

The difference (in form) between the Popov criterion and the circle criterion is that the former uses the (frequency domain) multiplier function $(1 + jq\omega)$, where the real constant $q > 0$, but the latter (i.e. the circle criterion) uses none. The starting point, then, for the illustrations to follow is that the phase angle behaviour of $G(j\omega)$ is such that L_2-stability *cannot* be established by either the circle criterion as applied to the nonlinear time-varying systems with $\varphi(\cdot) \in \mathcal{N}$, i.e. $\Re G(j\omega) \not> 0$, $\omega \in (-\infty,\infty)$, or the Popov criterion as applied to the nonlinear time-invariant system, i.e. there does not exist a (real) constant $q > 0$ such that $\Re(1 + jq\omega)G(j\omega) \not> 0$, $\omega \in (-\infty,\infty)$. For an application of the new theorems to the examples, we recall that the CPs ν_s, μ_i and μ_s of the nonlinearity $\varphi(\cdot) \in \mathcal{N}$ are defined by Equation (7); the CPs γ_i and γ_s for $\varphi(\cdot) \in \mathcal{M}$ are defined by Equation (8); and $\bar{K}$ denotes the upper limit of the time-varying gain $k(t)$ in the case of Equation (1), and of the constant gain K in Equation (2). Here, we implicitly assume, without loss of generality, that $0 < \varphi(\sigma)\sigma \leq \sigma^2$ for all $\sigma \neq 0$.

Example 1 In the sixth-order system of O'Shea (1967) (pp. 726–727), the k_2 (in O'Shea's notation) corresponds to the sector limit of the nonlinearity. Since we assume gain-transformed system in our theorems, the function

$$G_1(s) + \left(\frac{1}{k_2}\right) = \left(\frac{(s+0.005)(s+0.1)(s+1000)}{(s+0.0001)(s+2)(s+50)}\right)^2 \tag{34}$$

(where we have used subscript 1 for the transfer function of the forward block to distinguish it from our use of the notation) corresponds to our $G(s)$. We choose a multiplier function of the form

$$Z(j\omega) = 1 + j\alpha_0\omega + \frac{\alpha_1}{(j\omega + \beta_1)} + \frac{\alpha_2}{(j\omega + \beta_2)}, \tag{35}$$

where the parameters $\alpha_0, \alpha_1, \alpha_2, \beta_1$ and β_2 are to be computed subject to the constraint that the real part condition [H-1] of Theorems 1 and 2 is to be satisfied. O'Shea's multiplier function corresponds to $\alpha_0 = 0.01927; \alpha_1 = 1.8992; \beta_1 = -2.0; \alpha_2 = -0.000247; \beta_2 = 0.005$. This set can be slightly improved – by way of (subsequently) relaxing the constraints on the CPs of $\varphi(\cdot)$ or/and on the normalized rate of variation of $k(t)$ when applied to system (1), but without attempting an explicit cancellation of poles and zeros – to $\alpha_0 = 0.01927; \alpha_1 = 1.684; \beta_1 = -2.0; \alpha_2 = -0.00025; \beta_2 = 0.005$. It is found that an entirely different and improved set of parameters is $\alpha_0 = 0.0049; \alpha_1 = 1.576; \beta_1 = -1.9; \alpha_2 = -0.0; \beta_2 = 0.0$.

Application of Theorem 1: If we use O'Shea's multiplier function, we can conclude that the system (2) with $\varphi(\cdot) \in \mathcal{N}$ is L_2-stable if, from condition [H-2], the inequality $\mu_s < 0.5005$ is satisfied. The conclusions of using the slightly improved O'Shea's multiplier are almost the same. In contrast, if we use the new parameters, condition [H-2] leads to the inequality $\mu_s < 0.6028$ which is an improvement over the earlier inequality. A contribution of Theorem 1 is that the nonlinear time-invariant system (2), while being absolutely stable, according to OShea, for $\varphi(\cdot) \in \mathcal{M}$, is, *in fact*, stable for $\varphi(\cdot) \in \mathcal{N}$, if the CP μ_s of the nonlinearity obeys the inequality given above.

Application of Theorem 2: From the same O'Shea's multiplier function, we can conclude that the system (1) with $\varphi(\cdot) \in \mathcal{N}$ is (exponentially) L_2-stable if there exist nonnegative constants ξ and ζ such that condition [H-2] is satisfied; and with CPs ν_s and μ_s defined by Equation (7) and $0 \leq \xi < 0.005$ and $0 \leq \zeta < 2$, the inequality $\{0.01927\nu_s\xi + 0.999\mu_s + \mu_s(0.0002473\int_0^\infty e^{(\xi-0.005)t}\,dt + 1.899\int_{-\infty}^0 e^{(2-\zeta)t}\,dt) < 1\}$, obtained from condition [H-3], is satisfied. Evidently, we can strike a trade-off among ξ, ζ, ν_s and μ_s to satisfy the last inequality. For instance, suppose we set $\xi = 0.003$ and $\zeta = 1.5$, then the inequality becomes $(0.00006\nu_s + 4.87943\mu_s) < 1$, In other words, the system (1) is (exponentially) L_2-stable if (i) the nonlinearity with $\varphi(\cdot) \in \mathcal{N}$ has CPs ν_s and μ_s which together satisfy the last inequality; and, with $\xi = 0.003$ and $\zeta = 1.5$, (ii) the positive lobes $\theta^+(t)$ and negative lobes $\theta^-(t)$ of the normalized rate of variation of the time-varying gain $k(t)$ satisfy the inequalities (23), (24), and/or (25), depending on the nature of $k(t)$. The optimized parameters of O'Shea's multiplier function improve the above conclusions marginally.

In contrast, the consequences of using the new set of parameters are as follows. With $\xi \geq 0$ and $0 \leq \zeta < 1.9$, the inequality obtained from condition [H-3] is $\{0.0049\nu_s\xi + \mu_s(0.8295 + 1.576\int_{-\infty}^{0} e^{(1.9-\zeta)t}\,dt) < 1\}$, Suppose we set $\xi = 0.003$ and $\zeta = 1.5$ as before, then the last inequality becomes $(0.000015\nu_s + 4.7695\mu_s) < 1$ which allows a larger sub-class of nonlinearities in $\mathcal{N}$ than the inequality obtained above for O'Shea's multiplier function. Note that the constraints on $\theta^+(t)$ and $\theta^-(t)$ of the normalized rate of variation of $k(t)$ are the same as with the use of the O'Shea multiplier function (because the values of ξ and ζ have been retained).

Note that Corollary T2-1 (meant for the case of a periodic switching of nonlinearities) of Theorem 2 cannot be applied to the present problem because the multiplier function does not have the periodic structure of Equation (30).

Example 2 O'Shea (1967) also considers the example of Dewey and Jury (1966) with the transfer function $G_1(s) = 40/(s(s+1)(s^2+0.8s+16))$. The time-invariant system (2) with $\varphi(\sigma) \equiv \sigma$ with the above $G_1(s)$ is asymptotically stable for $k \in [\epsilon, 1.76)$ where constant $\epsilon > 0$. For an application of our theorems, we deal with the gain-transformed

$$G(s) \doteq \left(G_1(s) + \left(\frac{1}{1.76}\right)\right). \tag{36}$$

According to O'Shea (1967), the multiplier function

$$Z(j\omega) = 1 + j10^{-14}\omega + j0.999\sin(1.1118\omega) \tag{37}$$

satisfies the real-part condition as well as the time-domain constraint on it as required by his Theorem 2 (on p. 725 of the quoted reference) for the absolute stability of the system (2) with odd-monotone (i.e. class $\mathcal{M}_o$ of) nonlinearities having slopes in the sector $(\epsilon, 1.76)$, where $\epsilon > 0$.

In contrast, from condition [H-2] of our Theorem 1, the system is L_2-stable, in fact, for $\varphi(\cdot) \in \mathcal{N}$, if its CPs μ_i and μ_s satisfy the inequality $(\mu_i + \mu_s) < 1.001$. When we use this multiplier in our Theorem 2, meant for the time-varying system (1), we need to satisfy the following inequality: $10^{-14}\nu_s\xi + 0.4995(\mu_i e^{\xi} + \mu_s e^{\zeta}) < 0.001$, on the basis of which any desired trade-off can be struck among (i) the CPs ν_s, μ_i, μ_s of $\varphi(\cdot) \in \mathcal{N}$; and (ii) the (global) upper bound ξ on the positive lobes $\theta^+(t)$ and (global) lower bound ζ on the negative lobes $\theta^-(t)$ of the normalized rate of variation of $k(t)$. In the application of our Theorems 1 and 2, note that there is no explicit bound on the slope of the nonlinearity. Note further that Corollary T2-1 of Theorem 2 cannot be used because the term $j10^{-14}\omega$ in the multiplier function affects the periodic structure of the multiplier function as required by Equation (30). Hence, nonlinear switching cannot be handled by this multiplier.

Interestingly, the following multiplier functions without the $(j\omega)$ term also satisfy not only the real-part condition [H-1] of Theorem 2, but also the periodicity requirement of its Corollary T2-1:

$$Z_1(j\omega) = 1 + j1.109\sin(1.1\omega); \tag{38}$$

$$Z_2(j\omega) = 1 + j1.62\sin(1.082\omega); \quad \text{and} \tag{39}$$

$$Z_3(j\omega) = 1 + j2.6\sin(1.067\omega). \tag{40}$$

We can use these three multipliers separately in applying Theorem 2 to the system (1) with $\varphi(\cdot) \in \mathcal{N}$ and $k(t)$ periodic with the fundamental period $\mathfrak{p}$. Note that such an application corresponds to a special case of Corollary T2-1 (meant for the case of a periodic switching of nonlinearities), since we consider only one of each $k(t)$ and $\varphi(\cdot)$. As a consequence, we find that the system (1) is L_2-stable if (i) $\mathfrak{p} = 1.1$, and $(\mu_i + \mu_s) < 0.9091$; or (ii) $\mathfrak{p} = 1.082$, and $(\mu_i + \mu_s) < 0.6713$; or (iii) $\mathfrak{p} = 1.067$, and $(\mu_i + \mu_s) < 0.3846$. A by-product of the results is that smaller the period of $k(t)$, the more severe is the constraint on the CPs of the nonlinearity. There are *no* restrictions on the rate of variation of $k(t)$. To apply Corollary T2-1 in full, let us now assume that, in the system (1), two nonlinearities $\varphi_1(\cdot)$ and $\varphi_2(\cdot)$, both belonging to class $\mathcal{N}$, switch from one to the other periodically with period $\mathfrak{p}$, and the associated time-varying gains $k_1(t)$ and $k_2(t)$ also have the same period, then the system (1) is L_2-stable, if, in the last three inequalities (above in this paragraph), μ_i and μ_s are replaced, respectively, by $\mu_{i,*}$ and $\mu_{s,*}$, defined by Equation (31). It may be further noted that, in the applications of Theorem 2 and Corollary T2-1, no constraints are imposed on the dwell-time characteristics of $k(t)$ or of switching nonlinearities.

Example 3 In the course of illustrating integral quadratic constraint-based system analysis, which includes absolute stability, Megretski and Rantzer (1997) (pp. 822–824) consider a third-order system with

$$G(s) = \frac{s^2}{(s^3 + 2s^2 + 2s + 1)}, \tag{41}$$

a saturation nonlinearity and a unit-gain element characterized by delay parameter θ_0, as a result of which the transfer function of the forward block becomes $G(s)\mathrm{e}^{-\theta_0 s}$. The authors analyse the absolute stability of the system without and with delay.

The linear system obtained from Equation (2) by setting $\varphi(\sigma) \equiv \sigma$ is asymptotically stable for $k \in [0, \infty)$. For nonlinear time-varying systems, the circle criterion gives the upper bound ($\bar{K}$) on the gain as approximately 8.13; and for a successful application of the Popov criterion to the system (2), this gain is approximately 8.9.

In this example of Megretski and Rantzer (1997), the aspect relevant to the application of our theorems is the reference to an odd monotone (i.e. class $\mathcal{M}_o$ of) nonlinearities (of which saturation nonlinearity is an example). In this case, $\bar{K}$ is allowed to be arbitrarily large. To establish the absolute stability of the system *without delay*, they suggest the use of an OZF multiplier function of the form

$$Z(\mathrm{j}\omega) = 1 + \frac{\alpha}{(\beta - j\omega)}. \tag{42}$$

It is found (from our experimental work) that $\alpha = -1.3704$ and $\beta = 1.4$. The assumption of an odd monotone nonlinearity is crucial to an application of either Theorem 2 of O'Shea (1967) (p. 725) (or, equivalently, the corresponding theorem in Zames and Falb (1968)). Furthermore, when there is non-zero delay, there is a need to arrive at the Routh–Hurwitz limit for the feedback gain of the system (2) after setting $\varphi(\sigma) \equiv \sigma$.

Our stability Theorems 1 and 2 do not restrict the sign of the impulse response function of the multiplier; and the same multiplier is applicable to the nonlinearity class $\mathcal{N}$ for both the systems (2) and (1). In this particular case, the consequences of applying our theorems are as follows. From Theorem 1, the system (2) with $\varphi(\cdot) \in \mathcal{N}$ is L_2-stable, if $\mu_s < 0.5108$. From Theorem 2, the system (1) with $\varphi(\cdot) \in \mathcal{N}$ is L_2-stable, if, for $0 \le \zeta < 1.4$, the inequality $(0.7143 + 1/(1.4 - \zeta))\mu_s < 1$ is satisfied. In the latter case, we can trade-off between μ_s and ζ. Suppose we set $\zeta = 1$, then the last inequality reduces to $\mu_s < 0.3111$. Since ξ can be allowed to be arbitrarily large, there is no upper bound on the positive lobes $\theta^+(t)$ of the normalized rate of variation of the time-varying gain $k(t)$. On the other hand, since $\zeta = 1$, the (global) lower bound on the negative lobes $\theta^-(t)$ of the normalized rate of variation of the time-varying gain $k(t)$ must satisfy the inequalities (23), (24), and/or (25), depending on the nature of $k(t)$.

For the same problem, typically and without any attempt to optimize the parameters of the multiplier function, we find the following: (i) by setting $\bar{K} = 11.4$, the multiplier function

$$Z_1(\mathrm{j}\omega) = (1 - \mathrm{j}0.199 \sin 3.57\omega); \quad \text{and} \tag{43}$$

(ii) by setting $\bar{K} = 16.0$, the multiplier function

$$Z_2(\mathrm{j}\omega) = (1 - \mathrm{j}0.46 \sin 2.7\omega), \tag{44}$$

satisfy separately the condition [H-1] of our Theorem 2 (as also of Theorem 1) with $G(\mathrm{j}\omega)$ replaced by $(G(\mathrm{j}\omega) + 1/\bar{K})$. The corresponding stability results are as follows. With $\varphi(\cdot) \in \mathcal{N}$ and $k(t)$ periodic with the fundamental period $\mathfrak{p}$, the system (1) is L_2-stable if i) for $\bar{K} = 11.4, \mathfrak{p} = 3.57$, and $(\mu_i + \mu_s) < 5.0251$; and (ii) for $\bar{K} = 16.0$, $\mathfrak{p} = 2.7$, and $(\mu_i + \mu_s) < 2.1739$. From these results, it is evident that, for the L_2-stability of Equation (1), the larger the upper bound $\bar{K}$ is, the smaller will be the period of $k(t)$ and the more severe the constraint on the CPs of the nonlinearity.

It is found that, when there is delay in the system, $\bar{K}$ is to be restricted. Even with such a restriction, designing a suitable multiplier is quite complicated. A typical result is as follows. With $\theta_0 = 0.2$ and $\bar{K} = 8.8$, the Popov-multiplier function $Z(\mathrm{j}\omega) = 1 + \mathrm{j}0.0375\omega$ satisfies condition [H-1] of Theorem 2, in which we replace $G(\mathrm{j}\omega)$ by $\mathrm{e}^{-\theta_0\omega}G(\mathrm{j}\omega)$. Therefore, the system (1) with $G(s)$ replaced by $\mathrm{e}^{-\theta_0 s}G(s)$ and with $\varphi(\cdot) \in \mathcal{N}$ is (exponentially) L_2-stable if (from condition H-3]) the inequality $0.0375(\nu_s\xi) < 1$, or $\nu_s\xi < 26.6667$, is satisfied, in which we can strike a trade-off between ξ and ν_s. (Note that ζ can assume arbitrarily large values.) Suppose we set $\xi = 50$, then the CP $\nu_s < 0.5333$. The (global) upper bound on the positive lobes $\theta^+(t)$ of the normalized rate of variation of the time-varying gain $k(t)$ must satisfy the inequalities (23), (24), and/or (25), depending on the nature of $k(t)$, with $\xi = 50$. Note that Corollary T2-1 of Theorem 2 cannot be considered for the system (1) with switching nonlinearities, because the multiplier does not have the structure of Equation (29).

Example 4 The fourth-order system considered by Brockett and Willems (1965) (Part 2, p. 410) has the transfer function

$$G(s) = \frac{(10\,s + 1)(2\,s + 1)}{(s^2 + 20\,s + 400)(2\,s^2 + 5\,s + 4)}, \tag{45}$$

(which is Equation (22) in the same reference). Brockett and Willems (1965) conclude that the time-invariant nonlinear feedback system (2) with $\varphi(\cdot) \in \mathcal{M}_o$ is absolutely stable. However, it is found that there does exist the Popov-multiplier function $(1 + 0.0401s)$ using which the Popov criterion $\Re((1 + \mathrm{j}0.0401\omega)G(\mathrm{j}\omega)) > 0$, $-\infty < \omega < \infty$ is verified. Therefore, the time-invariant nonlinear feedback system is absolutely stable for $\varphi(\cdot) \in \mathcal{N}$. In other words, there is *no need* to use a multiplier function meant for monotone nonlinearities as found in Brockett and Willems (1965) (Part 2).

With the above choice of the Popov-multiplier function, we apply our Theorem 2 to the time-varying nonlinear feedback system with $\varphi(\cdot) \in \mathcal{N}$. Note that (i) the condition [H-1] of Theorem 2 is satisfied for an arbitrarily small $\varepsilon > 0$; (ii) and the multiplier function $Z(\mathrm{j}\omega)$ defined by Equation (6) now has only one term, namely $\mathrm{j}\alpha\omega$. As a

Table 2. Application of Theorem 1. See text for details.

Example No.	G (s)	Multiplier function	Multiplier parameters	CPs of $\varphi(\cdot)$	Constraints on CPs
1	(34)	(35)	$(\alpha_0, \alpha_1, \alpha_2, \beta_1, \beta_2)$	μ_s	–
			(0.01927, 1.8992, −2.0, −0.000247, 0.005) (O'Shea)	μ_s	$\mu_s < 0.5005$
			(0.01927, 1.684, −2.0, −0.00025, 0.005) (Improved)	μ_s	$\mu_s < 0.5006$
			(0.0049, 1.576, −1.9, −0.0, 0.0) (New)	μ_s	$\mu_s < 0.6028$
2	(36)	(37)	See (37)	μ_i, μ_s	$(\mu_i + \mu_s) < 1.001$
3	(41)	(42)	(α, β) (−1.3704, 1.4)	μ_s	$\mu_s < 0.5108$

Table 3. Application of Theorem 2. See Table 2 for the $G(s)$ and corresponding multiplier functions, and the text for details.

Example no.	A typical choice of (ξ, ζ)	Constraint on CPs of $\varphi(\cdot)$	Constraints[a] on $(\theta^+(t), \theta^-(t))$
1	$\xi = 0.003, \zeta = 1.5$	$(0.00006\nu_s + 4.87943\mu_s) < 1$	(23) and (24), or (25), or (26) and (27)
	Same as above	$(0.000052\nu_s + 4.69\mu_s) < 1$	Same as above
	Same as above	$(0.000015\nu_s + 4.7695\mu_s) < 1$	Same as above
2	(ξ, ζ)	$10^{-14}\nu_s\xi + 0.4995(\mu_i e^{\xi} + \mu_s e^{\zeta}) < 1.001$	(23) and (24), or (25)
3	$\zeta = 1; 0 \leq \xi < \infty$	$\mu_s < 0.3111$	$\theta^+(t) < \infty$; $\theta^-(t)$ obeys (23) and (24), or (25) with $\zeta = 1$.

[a] See text for details.

consequence, in the condition [H-3] of Theorem 2, the summation involving $z_c(\tau)$ and $z_a(\tau)$ vanishes, and ζ can be allowed to be arbitrarily large. Therefore, the system (1) is (exponentially) L_2-stable if (from condition H-3]) the inequality $0.0401(\nu_s\xi) < 1$, or $\nu_s\xi < 24.9376$, is satisfied. Now invoking Corollary A17 of Theorem 2, we can strike a trade-off between the (global) bound ξ on the positive lobes $\theta^+(t)$ of the normalized rate of variation of $k(t)$ and the CP ν_s of the nonlinearity $\varphi(\cdot) \in \mathcal{N}$ for the (exponential) L_2-stability of the system (1). Suppose we set $\xi = 25$, then the CP $\nu_s < 0.9975$. Note that Corollary T2-1 of Theorem 2 cannot be considered for the system (1) with switching nonlinearities, because the multiplier does not have the structure of Equation (29).

Tables 2–4 contain a summary of the computational results obtained from the above Examples 5.1–5.3 for the L_2-stability of:

(i) system (2), using Theorem 1 and with $\varphi(\cdot) \in \mathcal{N}$ whose CPs are defined by Equation (7); and
(ii) system (1), using Theorem 2 with the same $\varphi(\cdot)$ as for system (2), and with $k(t)$ aperiodic if the multiplier function is aperiodic (in the frequency domain), and $k(t)$ periodic if the multiplier function is periodic (in the frequency domain).

For Example 5.4 and also for Example 5.3 with delay, the Popov multiplier can be used to satisfy the real-part condition of Theorems 1 and 2, i.e. there is no need for extra terms in the multiplier function. Therefore, these two cases are not included in Tables 2 and 3. Table 4 is meant for an application of Theorem 2 to the special case of a periodic $k(t)$ with period $\mathfrak{p}$. In this table, the following details are omitted to avoid repetition of the contents of Table 2: $G(s)$ and the multiplier function parameters corresponding to Examples 5.1–5.3. Note that, with a multiplier function which is periodic in the frequency domain, Corollary T2-1 contains no restrictions on the rate of variation of the time-varying gain(s).

Table 4. Application of Theorem 2 to Examples 5.2 and 5.3 for periodic $k(t)$ with period $\mathfrak{p}$. See text for details.

Example no.	Multiplier function	$\mathfrak{p}$	Constraint on CPs of $\varphi(\cdot)$
2	(38)	1.1	$(\mu_i + \mu_s) < 0.9091$
	(39)	1.082	$(\mu_i + \mu_s) < 0.6713$
	(40)	1.067	$(\mu_i + \mu_s) < 0.3846$
$3, \bar{K} = 11.4$	(43)	3.57	$(\mu_i + \mu_s) < 5.0251$
$3, \bar{K} = 16.0$	(44)	2.7	$(\mu_i + \mu_s) < 2.1739$

6. Comparisons and critique

(1) There seem to be *no* results in the literature on the stability of nonlinear time-invariant/time-varying feedback systems with $\varphi(\cdot) \in \mathcal{N}$ in which a multiplier distinct from the Popov multiplier $(1 + j\alpha\omega)$ has been used.

(2) The L_2-stability conditions of Venkatesh (1978) (Theorem 1, p. 571) are a special case of Theorem 2. In fact, as applied to the class $\mathcal{M}$ of monotone nonlinearities, Theorem 2 of the present paper can be considered as an alternative form of Theorem 1 of Venkatesh (1978), in the

sense that the CPs γ_i and γ_s obtained from Equation (8) are distinct from the CPs δ_i and δ_s that appear in Theorem 1 of Venkatesh (1978). The problem of establishing which CPs of the class $\mathcal{M}$ of nonlinearities lead to more relaxed constraints on the (normalized) rate of variation of $k(t)$ seems to be open.

(3) In Huang et al. (2014b), the only theorem that has some tangible relationship with Theorem 2 of the present paper is Theorem 4B on page 25. However, the multiplier function in Theorem 4B of that reference is the counterpart of the Popov multiplier, i.e. it is a special case of Equation (6). A generalization of that theorem, namely, Theorem 4B, to correspond to Theorem 2 of the present paper is an open problem. The other theorems of Huang et al. (2014b) meant for nonlinear MIMO systems apply to nonlinearities which are monotone (and its variations). Even in these cases, the multiplier function is a special case of (the matrix counterpart of) Equation (6). More specifically, the former does not contain the matrix counterparts of the causal and anti-causal functions $z_c(t)$ and $z_a(t)$ of Equation (6).

(4) As far as Example 5.2 above is concerned, if we were to use the matrix counterpart of the O'Shea multiplier in Theorem 4B of Huang et al. (2014b), then the stability theorems are applicable only to those systems in which the vector nonlinearities have the property of path independence for line integrals involving them, and also possess the property of monotonicity (or its variations). However, if we were to use the (matrix) counterparts of the new multipliers, Equations (38)–(40), of Example 5.2 in the theorems of Huang et al. (2014a), then the vector nonlinearities in Huang et al. (2014b) need to belong to the monotone class (or its variations). In other words, the stability conditions of Huang et al. (2014a) cannot be applied to the case when the vector nonlinearities belong to (the vector counterpart of) class $\mathcal{N}$. A generalization of the theorems of Huang et al. (2014a) to hold for the case of vector nonlinearities belonging to class $\mathcal{N}$ is an open problem.

(5) While we can include any type of restriction on nonlinearities in the definition of their CPs, it would be highly desirable to improve Theorems 1 and 2 by invoking more effective CPs of the class $\mathcal{N}$ of nonlinearities such that, when we specialize the theorems to apply to a linear time-invariant system, the bound on the time-domain L_1-norm of the multiplier tends to an arbitrarily large value. In effect, the goal is to arrive at the Nyquist criterion as a limiting case of a stability result for nonlinear systems having a nonlinearity belonging to class $\mathcal{N}$. In this context, it is interesting to compare this with the introduction of the class of power-law (monotone) nonlinearities in Brockett and Willems (1965) and its exploitation in Thathachar (1970). Note that the problem posed above is of a different nature.

(6) In the framework used in the paper, we cannot estimate the domain of attraction of the origin. In fact, the relationship between L_2-stability and the domain of attraction as studied in the literature (using Lyapunov functions) does not seem to be known. See Hu, Huang, and Lin (2004) for an application of linear matrix inequalities (in a Lyapunov framework) to compute domains of attraction. From another point of view, the implications of Theorem 2 for robust absolute stability need further study. In this context, see, for instance, Liu and Molchanov (2002) in which a Lyapunov framework has been used to derive some robust absolute stability criteria for certain types of time-varying uncertainties and multiple time-varying nonlinearities. Similarly, it would be interesting to study the relationship between Theorems 1 and 2 and the results in, for instance, Wada, Ikeda, Ohta, and Siljak (1998) on parametric absolute stability to deal with parametric uncertainties and input reference values.

(7) To recall, Section 2 and its subsection on switching nonlinearities deal with the L_2-stability of the system (1) under the general assumption that the system is asymptotically stable when $\varphi(\sigma) \equiv \sigma$ and $k(t) \equiv K \in [0, \infty)$. It is not known how to modify the framework adopted here to deal with the possibility of stabilizing an unstable time-invariant system by switching operations on either the time-varying gain or the nonlinearity (or both).

(8) As mentioned in the introduction, the KYP-lemma proves that Popov's stability criterion for the system (2) with $\varphi(\cdot) \in \mathcal{N}$ is equivalent to the existence of a Lyapunov function comprising a quadratic form and an integral of the nonlinearity. There exist various generalization of this lemma. See, for instance, Iwasaki and Hara (2000). An interesting open problem is to find possible Lyapunov function candidates for Theorem 1, as also their generalization for Theorem 2. Such candidates, if they exist, facilitate computation of finite domains of attraction on the basis of Theorems 1 and 2.

(9) One of the goals of researchers is to find a graphical interpretation (in the frequency domain) of the real-part condition of Theorems 1 and 2. In contrast with the simple, elegant and completely graphical version of Popov's theorem for time-invariant nonlinear systems, the variations on that theorem (using more general multiplier functions) not only lack simplicity in the frequency domain, but also require satisfaction of time-domain integral inequalities involving the inverse Fourier transforms of the multiplier functions. See Venkatesh (1978) (pp. 573–575) for one of the few attempts in this genre for time-varying nonlinear feedback systems with $\varphi(\cdot) \in \mathcal{M}$. A similar procedure can be adopted for $\varphi(\cdot) \in \mathcal{N}$, but is more complex. It is possible, however, to convert the real-part condition and the time-domain constraints on the multiplier function of Theorems 1 and 2 to a non-convex optimization problem. Details are omitted due to lack of space.

(10) Since we deal with periodically switching nonlinearities (with periodic gains having the same period) as one of the applications of the main results, the very interesting survey paper of Shorten, Wirth, Mason, Wulff, and King (2007) on the stability of switching and hybrid systems is relevant here. Note, however, that Shorten et al. (2007) do not consider generalization of the Popov theorem to systems with time-varying nonlinearities with $\varphi(\cdot) \in \mathcal{N}$, and described by integral equations. Zevin and Pinsky (2005) present frequency-domain absolute (asymptotic) stability (and instability) conditions for a system, which is described by a Volterra equation and controlled by a nonlinear sector-restricted feedback having a time-varying delay, to be absolutely (asymptotically) stable (and unstable). The stability conditions are independent of the delay. Interestingly, these authors provide examples of systems satisfying the Aizerman conjecture. However, they do not consider time-varying feedback systems of the type (1). On the other hand, Dehghan and Ong (2012) introduce the concepts of *dwell-time invariance* and *maximal constraint admissible dwell-time-invariant set* for discrete-time switching systems under dwell-time switching, and derive a necessary and sufficient condition for asymptotic stability of the origin of the switching systems under dwell-time switching. In contrast, we consider continuous-time systems and dispense with dwell-time considerations for the L_2-stability of time-varying nonlinear systems described by Equation (1). See Venkatesh (2014) for more general results (dispensing with dwell-time considerations) for the ℓ_2-stability of discrete-time MIMO systems.

(11) Based on the computational experiments that gave typical values of μ_i and μ_s for specific nonlinearities, as listed in Table 1, it is conjectured that, when nothing is known about the precise structure of $\varphi(\cdot) \in \mathcal{N}$, the upper limit of both μ_i and μ_s is 1. A consequence of this is the following conjecture:

Generalized Popov Theorem: The nonlinear system (1) with $\varphi(\cdot) \in \mathcal{N}$ and $k(t)$ replaced by a constant gain $K \in [0, \infty)$ is L_2-stable, if there exists a multiplier function $Z(\mathrm{j}\omega)$ of the form Equation (6) such that [H-1] for some positive constant δ, $\Re Z(\mathrm{j}\omega)G(\mathrm{j}\omega) \geq \delta > 0$, $\omega \in (-\infty, \infty)$; and [H-2] $\sum_{m=1}^{\infty}(|z_m| + |z'_m|) + \int_0^{\infty} |z_c(\tau)| d\tau + \int_{-\infty}^{0} |z_a(\tau)| d\tau < \frac{1}{2}$.

Conjectured Generalization of Theorem 1 of Venkatesh (1978) (p. 571): The system (1) with $\varphi(\cdot) \in \mathcal{N}$; $k(t) \in [\epsilon, \infty)$ for $t \geq 0$, where (constant) $\epsilon > 0$; and $\|v_\varepsilon\|^2 \doteq \int_0^{\infty} \mathrm{e}^{2\varepsilon t}(v(t))^2 \, \mathrm{d}t$, where $\varepsilon > 0$ is an arbitrarily small number, is *exponentially* L_2-stable in the sense that $\|v_\varepsilon\| \leq C_1\|f\| + \sqrt{C_0 + (C_1^2/4)\|f\|^2}$, where C_1, C_2 are constants, if there exist a multiplier function $Z(\mathrm{j}\omega)$ defined by Equation (6); a bounded positive function $\varpi(\cdot)$ as defined above; and nonnegative constants, ξ, ζ, such that [H-1]: (for the $\ddot{I}ţ$ defined above) $\sup_{-\infty<\omega<\infty} \|Z(\mathrm{j}\omega - \varepsilon)G(\mathrm{j}\omega - \varepsilon)\| < \infty$; and $\Re Z((\mathrm{j}\omega - \varepsilon)G(\mathrm{j}\omega - \varepsilon)) > 0, \omega \in (-\infty, \infty)$; [H-2]: $\varpi(t)\mathrm{e}^{-\xi t}k(t)$ is nonincreasing and $\varpi(t)\mathrm{e}^{\zeta t}k(t)$ is nondecreasing for all $t \in [0, \infty)$; and [H-3]: $\alpha \nu_s \xi + \sum_{m=1}^{\infty}\{(1 + \mathrm{e}^{\xi \tau_m})|z_m| + (1 + \mathrm{e}^{\zeta \tau'_m})|z'_m|\} + \int_0^{\infty}(1 + \mathrm{e}^{\xi\tau})|z_c(\tau)| \, \mathrm{d}\tau + \int_{-\infty}^{0}(1 + \mathrm{e}^{-\zeta\tau}) |z_a(\tau)| \, \mathrm{d}\tau < 1$, where ν_s is defined by (7).

7. Conclusions

For the L_2-stability of time-invariant and time-varying single-input–single-output feedback systems with *non-monotone* nonlinearities, we have derived new frequency-domain criteria in terms of the transfer function of the linear time-invariant part and a general multiplier function originally employed for *monotone nonlinearities*. The results provide a preliminary bridge between the Popov criterion (for first- and third-quadrant nonlinearities) and the results of the literature on monotone and other nonlinearities for time-invariant nonlinear systems. This bridge is established via certain CP of the nonlinearities obtained from new algebraic inequalities. In some sense, the results can be treated as a quantified (and somewhat *baroque*) improvement of Popov's criterion whose necessity or otherwise *cannot* be established from the presented results. Without doubt, Popov's criterion has an everlasting and apparently impregnable beauty. Examples are given not only to illustrate the theorems, but also to demonstrate their superiority over the existing stability conditions of the literature.

In common with many of the results in the literature, a limitation of the framework used in the paper is that it appears to be impossible to find L_2-stabilization conditions in the frequency domain for an unstable transfer function of the forward block or for a feedback system in which the gain is in the unstable Routh–Hurwitz sector. On the other hand, the same framework in an extended form has been used to derive new frequency-domain L_2-stability conditions for continuous-time MIMO systems in Huang et al. (2014b), and ℓ_2-stability conditions for discrete-time MIMO systems in Venkatesh (2014) for (in both the cases) aperiodic/periodic time-varying and nonlinear feedback gains.

Acknowledgments

The author wishes to express grateful thanks to the expert reviewers for their critical comments and valuable suggestions which have led to the present, significantly improved version of the paper.

Disclosure statement

No potential conflict of interest was reported by the authors.

References

Aizerman, M. A. (1948). On one problem concerning the stability in the 'large' of dynamic systems. *Uspexi Matemacheskii Nauk*, *4*(4), 186–188.

Aizerman, M. A., & Gantmacher, F. R. (1964). *Absolute stability of regulator systems*. San Francisco, CA: Holden-Day.

Boyd, S., El Ghaoui, L., Feron, E., & Balakrishnan, V. (1994). *Linear matrix inequalities in systems and control theory*. SIAM studies in applied mathematics. Philadelphia: SIAM.

Brockett, R. W., & Willems, J. L. (1965). Frequency-domain stability criteria – Parts 1 and 2. *IEEE Transaction on Automatic Control*, *10*(3, 4), 407–413, 255–261.

Dehghan, M., & Ong, C. J. (2012). Discrete-time switching linear system with constraints: Characterization and computation of invariant sets under dwell-time consideration. *Automatica*, *48*(5), 964–969.

Dewey, A., & Jury, E. (1966). A stability inequality for a class of nonlinear feedback systems. *IEEE Transaction on Automatic Control*, *11*(1), 54–62.

Hu, T., Huang, B., & Lin, Z. (2004). Absolute stability with a generalized sector condition. *IEEE Transaction on Automatic Control*, *49*(4), 535–548.

Huang, Z. H., Venkatesh, Y. V., Xiang, C., & Lee, T. H. (2014). Frequency-domain L_2-stability conditions for switched linear and nonlinear SISO systems. *International Journal of Systems Science*, *45*(3), 682–703.

Huang, Z., Venkatesh, Y. V., Xiang, C., & Lee, T. H. (2014). Frequency-Domain L_2-stability conditions for time-varying linear and nonlinear MIMO systems. *Control Theory and Technology*, *12*(1), 1–22.

Iwasaki, T., & Hara, S. (2000). Generalized KYP lemma: Unified frequency domain inequalities with design applications. *IEEE Trans. on Automatic Control*, *50*(1), 41–59.

Kalman, R. E. (1963). Lyapunov functions for the problem of Lure in automatic control. *Proceedings of the National Academy of Sciences*, *49*(2), 201–205.

Lakshmi Thathachar, M. A., Srinath, M. D., & Ramapriyan, H. K. (1966). On a modified Lur'e problem. *IEEE Transaction on Automatic Control*, *11*(1), 62–68.

LaSalle, J. P. (1962). Complete stability of a nonlinear control system. *Proceedings of the National Academy of Sciences*, *48*(3), 600–603.

Lefschetz, S. (1965). Liapunov stability and controls. *SIAM Journal on Control*, *3*(1), 1–6.

Liu, D., & Molchanov, A. (2002). Criteria for robust absolute stability of time-varying nonlinear continuous-time systems. *Automatica*, *38*(4), 627–637.

Lur'e, A. I. (1951). *Neikotorlye Nelineinye Zadachi Teorii Avtom'aticheskogo Regulirovaniya*. Gostekhizdat: Moscow. Book in Russian.

Lur'e, A. I., & Postnikov, V. N. (1944). On the theory of stability of control systems. *Prikladnaya Matematica i Mekhanika*, *8*(3), 246–248.

Megretski, A., & Rantzer, A. (1997). System analysis via integral quadratic constraints. *IEEE Transaction on Automatic Control*, *42*(5), 819–830.

Narendra, K., & Goldwyn, R. (1964). A geometrical criterion for the stability of certain nonlinear non-autonomous systems. *IEEE Transaction on Circuit Theory*, *11*(3), 406 407.

Narendra, K. S., & Neuman, C. P. (1966). Stability of a class of differential equations with a single monotone nonlinearity. *SIAM Journal on Control*, *4*(2), 295–308.

O'Shea, R. P. (1967). An improved frequency time domain stability criterion for autonomous continuous systems. *IEEE Transaction on Automatic Control*, *12*(6), 719–724.

Popov, V. M. (1962). Absolute stability of nonlinear systems of automatic control. *Automation and Remote Control*, *22*(8), 857–875.

Sandberg, I. W. (1964). A frequency domain condition for the stability of feedback systems containing a single time varying non-linear element. *Bell System Technical Journal*, *43*, 1601–1638.

Shorten, R., Wirth, F., Mason, O., Wulff, K., & King, C. (2007). Stability criteria for switched and hybrid systems. *SIAM Review*, *49*(4), 545–592.

Thathachar, M. A. L. (1970). Stability of systems with power-law nonlinearities. *Automatica*, *6*, 721–730.

Thathachar, M. A. L., Srinath, M. D., & Krishna, G. (1966). Stability with a nonlinearity in a sector. *IEEE Transaction on Automatic Control*, *11*(4), 311–312.

Venkatesh, Y. V. (1978). Global variation criteria for the L_2-stability of nonlinear time varying systems. *SIAM Journal on Mathematical Analysis*, *19*(3), 568–581.

Venkatesh, Y. V. (2014). On the ℓ_2-stability of time-varying linear and nonlinear discrete-time MIMO systems. *Control Theory and Technology*, *12*(3), 250–274.

Wada, T., Ikeda, M., Ohta, Y., & Siljak, D. D. (1998). Parametric absolute stability of Lur'e systems. *IEEE Transaction on Automatic Control*, *43*(11), 1649–1653.

Yakubovich, V. A. (1962). The solution of certain matrix inequalities in automatic control theory. *Doklady Akademii Nauk SSSR*, *143*, 1304–1307.

Zames, G. (1966). On the input–output stability of time-varying nonlinear feedback systems, Parts 1 and 2. *IEEE Transaction on Automatic Control*, *11*(2, 3), 465–476, 228–238.

Zames, G., & Falb, P. L. (1968). Stability conditions for systems with monotone and slope restricted nonlinearities. *SIAM Journal on Control*, *6*, 89–108.

Zevin, A. A., & Pinsky, M. A. (2005). Absolute stability criteria for a generalized Lur'e problem with delay in the feedback. *SIAM Journal on Control and Optimization*, *43*(6), 2000–2008.

Appendix 1. Parseval's theorem

Suppose $f_1(\cdot)$ and $f_2(\cdot)$ are real-valued functions defined on $[0, \infty)$, and belong to the class of $L_1 \cap L_2$ functions. Then

$$\int_0^\infty f_1(t) f_2(t)\,\mathrm{d}t = \frac{1}{2\pi}\int_{-\infty}^{\infty} F_1(\mathrm{j}\omega)F_2(-\mathrm{j}\omega)\,\mathrm{d}\omega,$$

where F_1, F_2 are Fourier transforms of $f_1(t)$ and $f_2(t)$.

Appendix 2. Proof of Lemma 1

The integral of Equation (9) can be rewritten as

$$\lambda_1(T) = \int_0^T \varpi(\sigma_T(t)) \left\{ \sigma_T(t) + \alpha\frac{\mathrm{d}\sigma_T}{\mathrm{d}t} + \sum_{m=1}^{\infty} z_m \sigma_T(t-\tau_m) + z'_m \sigma_T(t+\tau'_m) + \int_{-\infty}^{\infty} z(\tau)\sigma_T(t-\tau)\,\mathrm{d}t \right\} \mathrm{d}t. \quad \text{(A1)}$$

We split Equation (A1) into its components and simplify wherever required as follows. Let

$$\lambda_{1-1}(T) \doteq \alpha \int_0^T \left\{\frac{\mathrm{d}\sigma_T}{\mathrm{d}t}\right\} \varphi(\sigma_T(t))\,\mathrm{d}t = \alpha\int_{\sigma_T(0)}^{\sigma_T(T)} \varphi(\sigma_T)\,\mathrm{d}\sigma_T = \alpha(\Phi(\sigma_T(T)) - \Phi(\sigma_T(0))), \quad \text{(A2)}$$

where $\Phi(\sigma) = \int_0^\sigma \varphi(\xi)\,\mathrm{d}\xi > 0$ for $\sigma \neq 0$.

We assume that interchanges of the operations of summation and integration and of two integrals (one with respect to τ and the other with respect to t) in Equation (A1) are valid. Let

$$\lambda_{1-2}(T) \doteq \int_0^T \sum_{m=1}^{\infty} z_m \sigma_T(t-\tau_m)\varphi(\sigma_T(t))\,dt$$

$$= \sum_{m=1}^{\infty}(z_m^+ - z_m^-)\int_0^T \sigma_T(t-\tau_m)\varphi(\sigma_T(t))\,dt, \quad \text{(A3)}$$

$$\lambda_{1-3}(T) \doteq \int_0^T \sum_{m=1}^{\infty} z'_m \sigma_T(t+\tau'_m)\varphi(\sigma_T(t))\,dt$$

$$= \sum_{m=1}^{\infty}(z_m'^+ - z_m'^-)\int_0^T \sigma_T(t+\tau'_m)\varphi(\sigma_T(t))\,dt, \quad \text{(A4)}$$

$$\lambda_{1-4}(T) \doteq \int_0^T \varphi(\sigma_T(t))\left\{\int_0^{\infty} z_c(\tau)\sigma_T(t-\tau)\,d\tau\right\}dt$$

$$= \int_0^{\infty}(z_c^+(\tau) - z_c^-(\tau))\left\{\int_0^T \varphi(\sigma_T(t))\sigma_T(t-\tau)\,dt\right\}d\tau, \quad \text{(A5)}$$

$$\lambda_{1-5}(T) \doteq \int_0^T \varphi(\sigma_T(t))\left\{\int_{-\infty}^{0} z_a(\tau)\sigma_T(t-\tau)\,d\tau\right\}dt$$

$$= \int_{-\infty}^{0}(z_a^+(\tau) - z_a^-(\tau))\left\{\int_0^T \varphi(\sigma_T(t))\sigma_T(t-\tau)\,dt\right\}d\tau. \quad \text{(A6)}$$

We invoke Equation (7) defining the CPs of $\varphi(\cdot)$ to reduce Equation (A3) to the following inequality:

$$\lambda_{1-2}(T) \geq -\sum_{m=1}^{\infty}(\mu_i z_m^+ + \mu_s z_m^-) \times \int_0^T \{\varphi(\sigma_T(t))\sigma_T(t) + \varphi(\sigma_T(t-\tau_m))\sigma_T(t-\tau_m)\}\,dt, \quad \text{(A7)}$$

in which the integral involving the integrand $\varphi(\sigma_T(t-\tau_m))\sigma_T(t-\tau_m)$ can be simplified as follows by changing the variable of integration to $\eta \doteq (t-\tau_m)$:

$$\int_0^T \varphi(\sigma_T(t-\tau_m))\sigma_T(t-\tau_m)\,dt = \int_{-\tau_m}^{T-\tau_m} \varphi(\sigma_T(\eta))\sigma_T(\eta)\,d\eta, \quad \text{(A8)}$$

which, on noting that $\sigma_T(\eta) = 0$ for $\eta < 0$ and changing the (dummy) variable of integration back to t, becomes

$$\int_0^T \varphi(\sigma_T(t-\tau_m))\sigma_T(t-\tau_m)\,dt = \int_0^{T-\tau_m} \varphi(\sigma_T(t))\sigma_T(t)\,dt. \quad \text{(A9)}$$

We now recall the property of the nonlinearity that $\varphi(\sigma)\sigma > 0, \forall\sigma \neq 0$ to reduce Equation (A9) to the inequality

$$\int_0^T \varphi(\sigma_T(t-\tau_m))\sigma_T(t-\tau_m)\,dt \leq \int_0^T \varphi(\sigma_T(t))\sigma_T(t)\,dt, \quad \text{(A10)}$$

which in combination with Equation (A7) gives

$$\lambda_{1-2}(T) \geq -2\sum_{m=1}^{\infty}(\mu_i z_m^+ + \mu_s z_m^-)\int_0^T \varphi(\sigma_T(t))\sigma_T(t)\,dt. \quad \text{(A11)}$$

Along similar lines, we can show that

$$\lambda_{1-3}(T) \geq -2\sum_{m=1}^{\infty}(\mu_i z_m'^+ + \mu_s z_m'^-)\int_0^T \varphi(\sigma_T(t))\sigma_T(t)\,dt. \quad \text{(A12)}$$

We now consider $\lambda_{1-4}(T)$ as defined in Equation (A5), and adopt (with minor obvious modifications) the procedure employed above for $\lambda_{1-2}(T)$. With respect to the integral with the integrand $\varphi(\sigma_T(t-\tau))\sigma_T(t-\tau)\,dt$, we change the variable of integration to $\eta \doteq (t-\tau)$, and simplify as before (by invoking the property of truncated functions) to arrive at the following inequality:

$$\lambda_{1-4}(T) \geq -2\left(\int_0^{\infty}(\mu_i z_c^+(\tau) + \mu_s z_c^-(\tau))\,d\tau\right) \times \int_0^T \varphi(\sigma_T(t))\sigma_T(t)\,dt. \quad \text{(A13)}$$

Similarly, we can show that

$$\lambda_{1-5}(T) \geq -2\left(\int_{-\infty}^{0}(\mu_i z_a^+(\tau) + \mu_s z_a^-(\tau))\,d\tau\right) \times \int_0^T \varphi(\sigma_T(t))\sigma_T(t)\,dt. \quad \text{(A14)}$$

Using the inequalities (A11)–(A14) along with Equation (A2) in Equation (A1), we obtain

$$\lambda_1(T) \geq \int_0^T \left\{1 - \left(2\sum_{m=1}^{\infty}(\mu_i z_m^+ + \mu_s z_m^-) + (\mu_i z_m'^+ + \mu_s z_m'^-) + \int_0^{\infty}(\mu_i z_c^+(\tau) + \mu_s z_c^-(\tau))\,d\tau + \int_{-\infty}^{0}(\mu_i z_a^+(\tau) + \mu_s z_a^-(\tau))\,d\tau\right)\varphi(\sigma_T(t))\sigma_T(t)\,dt\right\} + \alpha(\Phi(\sigma_T(T)) - \Phi(\sigma_T(0))), \quad \text{(A15)}$$

from which we conclude that $\lambda_1(T) \geq \alpha(\Phi(\sigma_T(T)) - \Phi(\sigma_T(0)))$ if condition [H-2] of Theorem 1 is satisfied. The lemma is proved.

Appendix 3. Proof of Lemma 2

The integral of Equation (22) can be rewritten as

$$\lambda_2(T) = \int_0^T \varphi(t)k(t)\varphi(\sigma_T(t)) \times \left\{\sigma_T(t) + \alpha\frac{d\sigma_T}{dt} + \sum_{m=1}^{\infty}(z_m\sigma_T(t-\tau_m) + z'_m\sigma_T(t+\tau'_m) + \int_{-\infty}^{\infty} z(\tau)\sigma_T(t-\tau)\,d\tau\right\}dt. \quad \text{(A16)}$$

Let $\lambda_2(T) \doteq \lambda_{2-1}(T) + \lambda_{2-2}(T)$, where

$$\lambda_{2-1}(T) = \int_0^T \varpi(t)k(t)\varphi(\sigma_T(t)) \left\{\beta_1\sigma_T(t) + \alpha\frac{d\sigma_T}{dt} + \sum_{m=1}^{\infty}(z_m\sigma_T(t-\tau_m) \times + \int_0^{\infty} z_c(\tau)\sigma_T(t-\tau)\,d\tau\right\}dt; \quad \text{and} \quad \text{(A17)}$$

$$\lambda_{2-2}(T) = \int_0^T \varpi(t)k(t)\varphi(\sigma_T(t)) \times \left\{\beta_2\sigma_T(t) + \sum_{m=1}^{\infty}(z'_m\sigma_T(t+\tau'_m) + \int_{-\infty}^{0} z_a(\tau)\sigma_T(t-\tau)\,d\tau\right\} dt, \quad \text{(A18)}$$

where positive constants β_1 and β_2 are chosen such that $(\beta_1 + \beta_2) = 1$ a As in the proof of Lemma 1, we assume that interchanges of summation and integration operations; and of two integrals (one with respect to τ and the other with respect to t) are valid. We split Equation (A17) into its components and simplify wherever required as follows:

$$\lambda_{2-1a}(T) \doteq \beta_1 \int_0^T \varpi(t)k(t)\varphi(\sigma_T(t))e^{-\xi t}\{e^{\xi t}\sigma_T(t)\}\,dt; \lambda_{2-2a}(T) \doteq \alpha \int_0^T \varpi(t)k(t)\varphi(\sigma_T(t))e^{-\xi t}\left\{e^{\xi t}\frac{d\sigma_T}{dt}\right\}dt; \quad \text{(A19)}$$

$$\lambda_{2-3a}(T) \doteq \sum_{m=1}^{\infty} z_m \left\{\int_0^T \varpi(t)k(t)\varphi(\sigma_T(t))e^{-\xi t}(e^{\xi t}\sigma_T(t-\tau_m))\,dt\right\}; \quad \text{(A20)}$$

$$\lambda_{2-4a}(T) \doteq \int_0^{\infty} z_c(\tau) \times \left\{\int_0^T \varpi(t)k(t)\varphi(\sigma_T(t))e^{-\xi t}(e^{\xi t}\sigma_T(t-\tau))\,dt\right\}d\tau; \quad \text{(A21)}$$

$$\lambda_{2-1b}(T) \doteq \beta_2 \int_0^T \varpi(t)k(t)\varphi(\sigma_T(t))e^{\zeta t}\{-e^{\zeta t}\sigma_T(t)\}\,dt; \quad \text{(A22)}$$

$$\lambda_{2-3b}(T) \doteq \sum_{m=1}^{\infty} z'_m \left\{\int_0^T \varpi(t)k(t)\varphi(\sigma_T(t))e^{\zeta t}(e^{-\zeta t}\sigma_T(t+\tau'_m))\,dt\right\};$$

and (A23)

$$\lambda_{2-4b}(T) \doteq \int_{-\infty}^{0} z_a(\tau) \times \left\{\int_0^T \varpi(t)k(t)\varphi(\sigma_T(t))e^{\zeta t}(e^{-\zeta t}\sigma_T(t-\tau))\,dt\right\}d\tau. \quad \text{(A24)}$$

Since $\varpi(t)k(t)e^{-\xi t}$ is nonincreasing, by the second mean value theorem, there is a point T' in [0, T] for which the integrals of Equations(A19)–(A21) become

$$\lambda_{2-1a}(T) = \varpi(0)k(0)\beta_1 \int_0^{T'} e^{\xi t}\varphi(\sigma_T(t))\sigma_T(t)\,dt; \lambda_{2-2a}(T) = \varpi(0)k(0)\alpha \int_0^{T'} e^{\xi t}\varphi(\sigma_T(t))\frac{d\sigma_T}{dt}\,dt; \quad \text{(A25)}$$

$$\lambda_{2-3a}(T) = \varpi(0)k(0)\sum_{m=1}^{\infty} z_m \times \left\{\int_0^{T'} e^{\xi t}\varphi(\sigma_T(t))\sigma_T(t-\tau_m)\,dt\right\}; \quad \text{and (A26)}$$

$$\lambda_{2-4a}(T) = \varpi(0)k(0)\int_0^{\infty} z_c(\tau) \times \left\{\int_0^{T'} e^{\xi t}\varphi(\sigma_T(t))\sigma_T(t-\tau)\,dt\right\}d\tau. \quad \text{(A27)}$$

The second integral in Equation (A25) can be simplified by integrating by parts and by invoking the definition of the CP ν_s in Equation (7) to obtain

$$\int_0^{T'} e^{\xi t}\varphi(\sigma_T(t))\frac{d\sigma_T}{dt}\,dt = e^{\xi T'}\Phi(\sigma_T(T')) - \Phi(\sigma_T(0)) - \xi\int_0^{T'} e^{\xi t}\Phi(\sigma_T(t))\,dt \geq e^{\xi T'}\Phi(\sigma_T(T')) - \Phi(\sigma_T(0)) - \nu_s\xi\int_0^{T'} e^{\xi t}\varphi(\sigma_T(t))\sigma_T(t)\,dt, \quad \text{(A28)}$$

where $\Phi(\sigma) = \int_0^{\sigma}\varphi(\xi)\,d\xi > 0$ for $\sigma \neq 0$. Inequality (A28) in combination with the definition of $\lambda_{2-2a}(T)$ in Equation (A25), leads to the inequality

$$\lambda_{2-2a}(T) \geq \varpi(0)k(0)\left\{e^{\xi T'}\Phi(\sigma_T(T')) - \Phi(\sigma_T(0)) - \nu_s\xi\int_0^{T'} e^{\xi t}\varphi(\sigma_T(t))\sigma_T(t)\,dt\right\}, \quad \text{(A29)}$$

As far as integrals (A26) and (A27) are concerned, by changing the variables of integration in them respectively to $u \doteq (t-\tau_m)$ and $u \doteq (t-\tau)$, exploiting the property of (time-)truncated functions, invoking the definition of CPs μ_i and μ_s in Equation (7), and recalling the nonnegativity of $\varphi(\sigma)\sigma$, as was done in the proof of Lemma 1, we get the following inequalities:

$$\lambda_{2-3a}(T) \geq -\varpi(0)k(0)\left\{\sum_{m=1}^{\infty}((1+e^{\xi\tau_m})(\mu_i z_m^+ + \mu_s z_m^-))\right\} \times \left\{\int_0^{T'} e^{\xi t}\varphi(\sigma_T(t))\sigma_T(t)\,dt\right\}, \quad \text{(A30)}$$

$$\lambda_{2-4a}(T) \geq -\varpi(0)k(0)\left\{\int_0^{\infty}(1+e^{\xi\tau})(\mu_i z_c^+(\tau) + \mu_s z_c^-(\tau))\,d\tau\right\} \times \left\{\int_0^{T'} e^{\xi t}\varphi(\sigma_T(t))\sigma_T(t)\,dt\right\}, \quad \text{(A31)}$$

We use Equations (A25) and (A29)–(A31) in Equation (A17) to obtain the inequality

$$\lambda_{2-1}(T) \geq \alpha\varpi(0)k(0)e^{\xi T'}(\Phi(\sigma_T(T')) - \Phi(\sigma_T(0))) + \varpi(0)k(0) \times \left\{\beta_1 - \alpha\nu_s\xi - \sum_{m=1}^{\infty}((1+e^{\xi\tau_m})(\mu_i z_m^+ + \mu_s z_m^-)) - \int_0^{\infty}(1+e^{\xi\tau})(\mu_i z_c^+(\tau) + \mu_s z_c^-(\tau))\,d\tau\right\} \times \int_0^{T'} e^{\xi t}\varphi(\sigma_T(t))\sigma_T(t)\,dt, \quad \text{(A32)}$$

from which we conclude that $\lambda_{2-1}(T) > -\alpha\varpi(0)k(0)\Phi(\sigma_T(0))$ if

$$\alpha v_s\xi + \sum_{m=1}^{\infty}((1+e^{\xi\tau_m})(\mu_i z_m^+ + \mu_s z_m^-)) + \int_0^{\infty}(1+e^{\xi\tau})(\mu_i z_c^+(\tau) + \mu_s z_c^-(\tau))\,d\tau < \beta_1. \quad (A33)$$

We now consider $\lambda_{2-2}(T)$ defined by Equation (A18). Since $\varpi(t)k(t)e^{\zeta t}$ is nondecreasing, by the (extended) second mean value theorem, there is a point T'' in [0, T] for which the integrals of Equations (A1)–(A3) respectively become

$$\lambda_{2-1b}(T) = \beta_2\varpi(T)k(T)e^{\zeta T}\int_{T''}^{T} e^{-\zeta t}\varphi(\sigma_T(t))\sigma_T(t)\,dt; \quad (A34)$$

$$\lambda_{2-3b}(T) = \varpi(T)k(T)e^{\zeta T}\sum_{m=1}^{\infty} z'_m \times \left\{\int_{T''}^{T} e^{-\zeta t}\varphi(\sigma_T(t))\sigma_T(t+\tau_m)\,dt\right\}; \quad \text{and (A35)}$$

$$\lambda_{2-4b}(T) = \varpi(T)k(T)e^{\zeta T}\int_{-\infty}^{0} z_a(\tau) \times \left\{\int_{T''}^{T} e^{-\zeta t}\varphi(\sigma_T(t))\sigma_T(t-\tau)\,dt\right\}d\tau. \quad (A36)$$

In the integral of Equation (A35), let $u \doteq (t+\tau'_m)$, invoke the property of (time-)truncated functions, recall the definitions of the CPs μ_i and μ_s in Equation (7), and follow the line of simplification adopted above for Equation (A26) to get

$$\lambda_{2-3b}(T) \geq -\varpi(T)k(T)e^{\zeta T}\left\{\sum_{m=1}^{\infty}(1+e^{\zeta\tau'_m})(\mu_i z_m'^{+} + \mu_s z_m'^{-})\right\} \times \int_{T''}^{T} e^{-\zeta t}\varphi(\sigma_T(t))\sigma_T(t)\,dt. \quad (A37)$$

Similarly, from Equation (A36), by changing the variable of integration in its integral to $u \doteq (t-\tau)$, and following a procedure similar to the above, we obtain

$$\lambda_{2-4b}(T) \geq -\varpi(T)k(T)e^{\zeta T} \times \left\{\int_{-\infty}^{0}(1+e^{\zeta\tau})(\mu_i z_a^+(\tau) + \mu_s z_a^-(\tau))\,d\tau\right\} \times \int_{T''}^{T} e^{-\zeta t}\varphi(\sigma_T(t))\sigma_T(t)\,dt. \quad (A38)$$

We combine Equations (A34), (A37) and (A38) with Equation (A18) to get the inequality

$$\lambda_{2-2}(T) \geq \varpi(T)k(T)e^{\zeta T}\left\{\beta_2 - \sum_{m=1}^{\infty}((1+e^{\zeta\tau'_m})(\mu_i z_m'^{+} + \mu_s z_m'^{-})) - \int_{-\infty}^{0}(1+e^{-\zeta\tau})(\mu_i z_a^+(\tau) + \mu_s z_a^-(\tau))\,d\tau\right\} \times \int_{T''}^{T} e^{-\zeta t}\varphi(\sigma_T(t))\sigma_T(t)\,dt, \quad (A39)$$

from which we conclude that $\lambda_{2-2}(T) > 0$ if

$$\sum_{m=1}^{\infty}((1+e^{\zeta\tau'_m})(\mu_i z_m'^{+} + \mu_s z_m'^{-})) + \int_{-\infty}^{0}(1+e^{-\zeta\tau})(\mu_i z_a^+(\tau) + \mu_s z_a^-(\tau))\,d\tau \times \int_{T''}^{T} e^{-\zeta t}\varphi(\sigma_T(t))\sigma_T(t)\,dt < \beta_2. \quad (A40)$$

Since $\lambda_2(T) = \lambda_{2-1}(T) + \lambda_{2-2}(T)$, from Equations (A33) and (A40) we conclude that

$$\lambda_2(T) > -\alpha\varpi(0)k(0)\Phi(\sigma_T(0)), \quad (A41)$$

if conditions [H-2] and [H-3] of Theorem 2 are satisfied. Lemma 2 is proved.

Proof Consider the integral, for any $T > 0$,

$$\rho_2(T) \doteq \int_0^T \varpi(t)f_T(t)\mathcal{Z}\mathcal{G}\{v_T(t)\}\,dt, \quad (A42)$$

where $\mathcal{G}\{v_T(t)\} = \int_0^t g(\tau)v_T(t-\tau)\,d\tau$. It follows from $f_T(t) = v_T(t) + k(t)\varphi(\sigma_T(t))$ in Equation (1) that

$$\rho_2(T) = \int_0^T \varpi(t)v_T(t)\mathcal{Z}\{\mathcal{G}\{v_T(t)\}\}\,dt + \int_0^T \varpi(t)k(t)\varphi(\sigma_T(t))\mathcal{Z}\{\sigma_T(t)\}\,dt. \quad (A43)$$

From condition [H-1], there exists an $\varepsilon > 0$, however small, such that $\Re Z(j\omega-\varepsilon)G(j\omega-\varepsilon) > \delta > 0$ for $\omega \in (-\infty,\infty)$. There exists a $\varepsilon > 0$ from condition [H-1] of the theorem on the basis of which the first integral on the right-hand side of Equation (A43) is rewritten as follows:

$$\int_0^T \varpi(t)v_T(t)\mathcal{Z}\{\mathcal{G}\{v_T(t)\}\}\,dt = \int_0^T \varpi(t)e^{-2\varepsilon t}(v_T(t)e^{\varepsilon t})(e^{\varepsilon t}\mathcal{Z}\mathcal{G}\{v_T(t)\})\,dt. \quad (A44)$$

Now, from the assumptions made on $\varpi(t)$, $e^{-2\varepsilon t}\varpi(t)$ is non-increasing. From the second mean value theorem as applied to Equation (A44), there exists a $T' \in [0, T]$ such that

$$\int_0^T \varpi(t)v_T(t)\mathcal{Z}\{\mathcal{G}\{v_T(t)\}\}\,dt = \varpi(0)\int_0^{T'}(v_T(t)e^{-\varepsilon t})(e^{-\varepsilon t}\mathcal{Z}\mathcal{G}\{v_T(t)\}\,dt. \quad (A45)$$

We let $V_T(j\omega)$ denote, as before, the Fourier transform of (the time-truncated function) $v_T(t)$, and apply the Parseval theorem to the integral on the right-hand side of Equation (A45). For this process, we note that there is no loss of generality in assuming that the upper limit T' of the integral can be replaced by T itself, and $v_T(t)$ can be set to zero in the interval $(T', T]$. We get

$$\int_0^T \varpi(t)v_T(t)\mathcal{Z}\{\mathcal{G}\{v_T(t)\}\}\,dt = \frac{\varpi(0)}{2\pi}\int_{-\infty}^{\infty} V_T(-(j\omega-\varepsilon))Z(j\omega-\varepsilon) \times G(j\omega-\varepsilon)V_T(j\omega-\varepsilon)\,d\omega. \quad (A46)$$

Invoking the condition [H-1] of the theorem, namely, for some $\delta > 0$, $\Re Z(j\omega-\varepsilon)G(j\omega-\varepsilon) \geq \delta > 0, \omega \in (-\infty,\infty)$, the following inequality holds:

$$\int_0^T \varpi(t) v_T(t) \mathcal{Z}\{\mathcal{G}\{v_T(t)\}\} \, dt$$

$$= \frac{\varpi(0)}{2\pi} \int_{-\infty}^{\infty} Z(j\omega - \varepsilon) G(j\omega - \varepsilon) |V_T(j\omega - \varepsilon)|^2 d\omega$$

$$\geq \frac{\varpi(0)\delta}{2\pi} \int_0^T e^{2\varepsilon t} (v_T(t))^2 \, dt. \qquad \text{(A47)}$$

By virtue of Lemma 2, the second integral right-hand side of Equation (A43),

$$\int_0^T \varpi(t) k(t) \varphi(\sigma_T(t)) \mathcal{Z}\{\sigma_T(t)\} \, dt \geq -\alpha \varpi(0) k(0) \Phi(\sigma_T(0))), \qquad \text{(A48)}$$

where, to recall, $\Phi(\sigma) = \int_0^\sigma \varphi(\xi) \, d\xi$. With (an arbitrarily small) $\varepsilon > 0$, we rewrite Equation (A42) as follows:

$$\rho_2(T) = \int_0^T \varpi(t) e^{-\varepsilon t} f_T(t) (e^{\varepsilon t} \mathcal{Z}\mathcal{G}\{v_T(t)\}) \, dt. \qquad \text{(A49)}$$

Based on the assumed property of $\varpi(t)$, we infer that $e^{-\varepsilon t}\varpi(t)$ is nonincreasing. Invoking the second mean value theorem in Equation (A48), there exists a $T' \in [0, T]$ such that

$$\rho_2(T) = \varpi(0) \int_0^{T'} f_T(t) (e^{\varepsilon t} \mathcal{Z}\mathcal{G}\{v_T(t)\}) \, dt. \qquad \text{(A50)}$$

Noting that there is no loss of generality, as before, in assuming that $f_T(t) = 0$ and $v_T(t) = 0$ for $t \in [T', T]$, we apply the Parseval theorem to Equation (A49) to obtain

$$\rho_2(T) = \frac{\varpi(0)}{2\pi} \int_{-\infty}^{\infty} F_T(-j\omega) Z(j\omega - \varepsilon) G(j\omega - \varepsilon) V_T(j\omega - \varepsilon) \, d\omega, \qquad \text{(A51)}$$

which can be reduced to an inequality (in steps) as follows:

$$\rho_2(T) \leq \frac{\varpi(0)}{2\pi} \sup_{-\infty<\omega<\infty} Z(j\omega - \varepsilon) G(j\omega - \varepsilon) \times \int_0^\infty f_T(t) (e^{\varepsilon t} v_T(t)) \, dt$$

$$\leq \frac{\varpi(0)}{2\pi} \sup_{-\infty<\omega<\infty} Z(j\omega - \varepsilon) G(j\omega - \varepsilon) \|f_T\| \times \sqrt{\int_0^\infty e^{2\varepsilon t} (v_T(t))^2 \, dt} \qquad \text{(A52)}$$

the last step being based on an application of the Cauchy–Schwartz inequality. With the relationship (A43) in mind, we now combine the inequalities (A50), (A47) and (A46) to get

$$\frac{\varpi(0)\delta}{2\pi} \int_0^T e^{2\varepsilon t} (v_T(t))^2 \, dt$$

$$\leq \alpha \varpi(0) k(0) \Phi(\sigma_T(0))) + \frac{\varpi(0)}{2\pi} \sup_{-\infty<\omega<\infty} \times Z(j\omega - \varepsilon) G(j\omega - \varepsilon) \|f_T\| \sqrt{\int_0^T e^{2\varepsilon t} (v_T(t))^2 \, dt}. \qquad \text{(A53)}$$

Noting that $\sup_{-\infty<\omega<\infty} |Z(j\omega - \varepsilon) G(j\omega - \varepsilon)| \doteq C$ is finite by virtue of the assumptions on $Z(\cdot)$ and $G(\cdot)$, Equation (A51) can be simplified to give

$$\int_0^T e^{2\varepsilon t} (v_T(t))^2 \, dt$$

$$\leq \frac{2\pi\alpha}{\delta} k(0) \Phi(\sigma_T(0)) + \frac{C}{\delta} \|f_T\| \left(\int_0^T e^{2\varepsilon t} (v_T(t)) \, dt \right)^{1/2}$$

$$\leq C_0 + C_1 \|f_T\| \left(\int_0^T e^{2\varepsilon t} (v_T(t))^2 \, dt \right)^{1/2}, \qquad \text{(A54)}$$

where $C_0 \doteq (2\pi\alpha/\delta) k(0) \Phi(\sigma_T(0))$ and $C_1 \doteq (C/\delta)$ are finite. From Equation (A52), we get the following inequality:

$$\int_0^T e^{2\varepsilon t} (v_T(t))^2 \, dt \leq C_0 + C_1 \|f_T\| \left(\int_0^T e^{2\varepsilon t} (v_T(t))^2 \, dt \right)^{1/2}, \qquad \text{(A55)}$$

which is valid for all $T > 0$. Since C_0 and C_1 are independent of T, we conclude that $\|v_\varepsilon\| \leq C_1 \|f\| + \sqrt{C_0 + C_1^2/4\|f\|^2}$, where $\|v_\varepsilon\|^2 \doteq \int_0^\infty e^{2\varepsilon t} (v(t))^2 \, dt$. The theorem is proved. ■

Appendix 4. Proof of Corollary L2-1

The integral of Equation (33) can be rewritten as

$$\lambda_{2p}(T) = \int_0^T k(t) \varphi(\sigma_T(t)) \times \left\{ \sigma_T(t) + \sum_{m=1}^{\infty} (z_m \sigma_T(t - m\mathfrak{p})) + z'_m \sigma_T(t + m\mathfrak{p}) \right\} dt. \qquad \text{(A56)}$$

As before, we assume that an interchange of the operations of summation and integration in Equation (A55) is valid. In Equation (101), we replace $k(t)\varphi(\sigma)$ by the right-hand side of Equation (28), and split $\lambda_{2p}(T)$ in terms of its components and simplify as follows.

Let

$$\lambda_{2p-1}(T) \doteq \int_0^T \sum_{m=1}^{\infty} z_m \sigma_T(t - m\mathfrak{p}) k(t) \varphi(\sigma_T(t)) \, dt$$

$$= \sum_{m=1}^{\infty} (z_m^+ - z_m^-) \int_0^T \sigma_T(t - m\mathfrak{p}) k(t) \varphi(\sigma_T(t)) \, dt$$

$$= \sum_{m=1}^{\infty} (z_m^+ - z_m^-) \int_0^T \sigma_T(t - m\mathfrak{p}) k_1(t) U_1(t) \varphi_1(\sigma_T(t)) + k_2(t) U_2(t) \varphi_2(\sigma_T(t)) \, dt. \qquad \text{(A57)}$$

In the last integral of the right-hand side of Equation (A56), we change the variable of integration from t to $\xi \doteq (t - m\mathfrak{p})$, invoke Equation (31) defining the CPs of $\varphi_1(\cdot)$ and $\varphi_2(\cdot)$, use the periodicity property of $U_1(t)$ and $U_2(t)$ along with the properties of the time-truncated function – the last step being similar to what was done in the proof of Lemma 1 – and recall that $k(t)\varphi(\sigma_T(t)) = \{k_1(t)\varphi_1(\sigma_T(t)) + k_2(t)\varphi_2(\sigma_T(t))\}$ to reduce Equation (A56) to the following inequality:

$$\lambda_{2p-1}(T) \geq -\sum_{m=1}^{\infty}(\mu_{i,1}z_m^+ + \mu_{s,1}z_m^-) \times \int_0^T [k_1(t)U_1(t)\{\varphi_1(\sigma_T(t))\sigma_T(t) + \varphi_1(\sigma_T(t-m\mathfrak{p}))\sigma_T(t-m\mathfrak{p})\}]\,\mathrm{d}t + (\mu_{i,2}z_m^+ + \mu_{s,2}z_m^-) \times \int_0^T [k_2(t)U_2(t)\{\varphi_2(\sigma_T(t))\sigma_T(t) + \varphi_2(\sigma_T(t-m\mathfrak{p})) \times \sigma_T(t-m\mathfrak{p})\}]\,\mathrm{d}t$$

$$\geq -2\sum_{m=1}^{\infty}(\mu_{i,*}z_m^+ + \mu_{s,*}z_m^-)\int_0^T k(t)\varphi(\sigma_T(t))\sigma_T(t)\,\mathrm{d}t. \tag{A58}$$

Similarly, let

$$\lambda_{2p-2}(T) \doteq \int_0^T \sum_{m=1}^{\infty} z'_m\sigma_T(t+\tau'_m)k(t)\varphi(\sigma_T(t))\,\mathrm{d}t$$

$$= \sum_{m=1}^{\infty}(z_m'^+ - z_m'^-)\int_0^T \sigma_T(t+\tau'_m)k(t)\varphi(\sigma_T(t))\,\mathrm{d}t$$

$$= \sum_{m=1}^{\infty}(z_m'^+ - z_m'^-)\int_0^T \sigma_T(t+m\mathfrak{p})k_1(t)U_1(t)\varphi_1(\sigma_T(t)) + k_2(t)U_2(t)\varphi_2(\sigma_T(t))\,\mathrm{d}t. \tag{A59}$$

Following the line of simplification and reduction of $\lambda_{2p-1}(T)$ above, we can show that $\lambda_{2p-2}(T)$ obeys the following inequality:

$$\lambda_{2p-2}(T) \geq -2\sum_{m=1}^{\infty}(\mu_{i,*}z_m'^+ + \mu_{s,*}z_m'^-) \times \int_0^T k(t)\varphi(\sigma_T(t))\sigma_T(t)\,\mathrm{d}t. \tag{A60}$$

Using the inequalities (A58) and (A59) in (A55), we obtain

$$\lambda_{2p}(T) \geq \left\{1 - 2\sum_{m=1}^{\infty}(\mu_{i,*}(z_m^+ + z_m'^+) + \mu_{s,*}(z_m^- + z_m'^-))\right\} \times \int_0^T k(t)\varphi(\sigma_T(t))\sigma_T(t)\,\mathrm{d}t \tag{A61}$$

from which we conclude that $\lambda_{2p}(T) \geq 0$, if condition [H-2] of Corollary T2-1 is satisfied. The lemma is proved.

19

A novel artificial intelligent control system to suppress the vibration of a FGM Plate

Jalal Javadi Moghaddam* and Ahmad Bagheri

Department of Mechanical Engineering, Faculty of Engineering, University of Guilan, PO Box 3756, Rasht, Iran

In this paper, an adaptive neuro-fuzzy sliding-mode-based genetic algorithm (ANFSGA) control system is proposed to control functionally graded material (FGM) plates. The model of the FGM plate is considered by the finite element method based on the classical laminated plate theory. Moreover, to show the performance of the proposed ANFSGA intelligent control system, a traditional sliding-mode control (SMC) system and an adaptive neuro-fuzzy (ANF) SMC system are designed to suppress the vibrations of the FGM plate as a comparison. The proposed genetic algorithm control system uses the ANF SMC system in the crossover and mutation operation. In this way, the online learning ability can be used by adjusting the control parameters to deal with external disturbance. The control objective is to drive the system state to the original equilibrium point and thus, the asymptotically stability of the proposed control system can be achieved.

Keywords: neuro fuzzy; FGM plates; sliding mode; on line; genetic algorithm

1. Introduction

FGMs offer good possibilities for optimizing engineering structures to achieve high performance and material efficiency. It is well known that functionally gradient materials are used in high-temperature conditions. That is why they have successful applications as electronic devices, optical films, anti-wear and anticorrosion coatings and biomaterials.

In the solid mechanical field, many researchers have investigated the application of piezoelectric materials as the sensors and actuators for the purpose of monitoring and controlling which is used in active structural systems.

Advanced reinforced composite structures incorporating piezoelectric sensors and actuators are increasingly becoming important due to the development of adaptive structures.

These structures offer potential benefits in a wide range of engineering applications such as vibration and noise suppression, shape control and precision positioning. Parashkevova, Ivanova, and Bontcheva (2004) developed an optimal design and defined two cost functions of functionally graded plates. Turteltaub (2002) introduced a numerical procedure to determine an optimal material layout of a functionally graded material (FGM) within the context of a transient phenomenon. Numerous approaches have been introduced into the analysis of plates with bonded piezoelectric sensors and actuators. Piezoelectric materials are more promising to use in structural mechanics in a way to develop adaptive structures. This is due to their coupled mechanical and electrical properties.

Reddy (2004) and Robbins and Reddy (1991) investigated the transformed piezoelectric and dielectric coefficients, and presented the finite element model of a piezoelectrically actuated beam by using four different displacement-based one-dimensional beam theories. He, Ng, Sivashanker, and Liew (2001) presented a finite element formulation based on the classical laminated plate theory for the shape and vibration control of the FGM plates with integrated piezoelectric sensors and actuators. Moita, Correia, Martins, Mota Soares, and Mota Soares (2006) presented a finite element formulation based on the classical laminated plate theory for laminated structures with integrated piezoelectric layers or patches, acting as sensors and actuators. Based on the first-order shear deformation theory, Liew, He, and Kitipornchai (2004) developed a generic finite element formulation to account for the coupled mechanical and electrical responses of FGM shells with piezoelectric sensors and actuator layers. The large-scale shell structures with distributed piezoelectric components of complicated geometrical configurations are approximated by the hybrid strain or mixed formulation based on lower order triangular shell finite elements investigated by To and Chen (2007).

In the control field, variable structure control with sliding mode, which is commonly known as sliding-mode control (SMC), is a nonlinear control strategy that is well

*Corresponding author. Email: jalaljavadimoghaddam@gmail.com

known for its robustness characteristics. Many methods based on sliding mode have been developed to control the dynamic systems; in particular, Bagheri and Moghaddam (2009) developed decoupled adaptive neuro-fuzzy SMC system methods for the chaos control problem in a system without precise system model information. Wai and Lee (2004) investigated a double-inductance double-capacitance resonant driving circuit and a sliding-mode fuzzy-neural-network control system for the motion control of an linear piezoelectric ceramic motor. Bagheri and Javadi Moghaddam (2010) developed artificial intelligence control system for underwater vehicle. Lin and Wai (2003) presented adaptive and fuzzy-neural-network sliding-mode controllers for the motor-quick-return servomechanism.

A total sliding-mode-based genetic algorithm control system for a linear piezoelectric ceramic motor driven by a newly designed hybrid resonant inverter is discussed by Wai and Tu (2007).

In this paper, the traditional SMC and the adaptive neuro-fuzzy (ANF) SMC and also the adaptive neuro-fuzzy sliding-mode-based genetic algorithm (ANFSGA) control system are presented to control the FGM plate in a vibration problem. It can be understood that the proposed control system can be easily used to other mechanical and electrical systems.

2. Model description

A cantilevered (CFFF) FGM plate with the integrated sensors and actuators is shown in Figure 1. The piezoelectric actuator layer and the piezoelectric sensor layer are distributed uniformly on the top and bottom layers of the laminated plate, respectively. The region between the two surfaces is made of the combined aluminum oxide and Ti-6A1-4 V materials. The material properties of the FGM plate are graded through the thickness direction according to a volume fraction power law distribution. The material properties can be easily found in the literature (Liew et al., 2004; Touloukian, 1967).

2.1. *Mathematical model using the classical laminated plate theory (CLPT)*

In the CLPT theory, the displacement field is presented by the following form:

$$\{u\} = \begin{Bmatrix} u_1 \\ u_2 \\ u_3 \end{Bmatrix} = \begin{Bmatrix} u_0 \\ v_0 \\ w_0 \end{Bmatrix} - \begin{Bmatrix} z\dfrac{\partial w_0}{\partial x} \\ z\dfrac{\partial x_0}{\partial y} \\ 0 \end{Bmatrix} = [H]\{\bar{u}\}, \quad (1)$$

$$\{\bar{u}\} = \left\{ u_0,\ v_0, w_0, \frac{\partial w_0}{\partial x}, \frac{\partial w_0}{\partial y} \right\}^{\mathrm{T}}, \quad (2)$$

$$[H] = \begin{bmatrix} 1 & 0 & 0 & -z & 0 \\ 0 & 1 & 0 & 0 & -z \\ 0 & 0 & 1 & 0 & 0 \end{bmatrix}, \quad (3)$$

where $\{\bar{u}\}$ is the midplane displacement.

u_0, v_0, w_0 are displacements in the x, y and z directions, and $\partial w_0/\partial x, \partial w_0/\partial y$ are rotations of the yz and xz planes due to bending.

The strains according to the displacement field in Equation (1) are given by

$$\begin{Bmatrix} \varepsilon_1 \\ \varepsilon_2 \\ \varepsilon_6 \end{Bmatrix} = \begin{Bmatrix} \dfrac{\partial u_0}{\partial x} \\ \dfrac{\partial v_0}{\partial y} \\ \dfrac{\partial u_0}{\partial y} + \dfrac{\partial v_0}{\partial x} \end{Bmatrix} - z \begin{Bmatrix} \dfrac{\partial^2 w_0}{\partial x^2} \\ \dfrac{\partial^2 w_0}{\partial y^2} \\ 2\dfrac{\partial^2 w_0}{\partial x \partial y} \end{Bmatrix}. \quad (4)$$

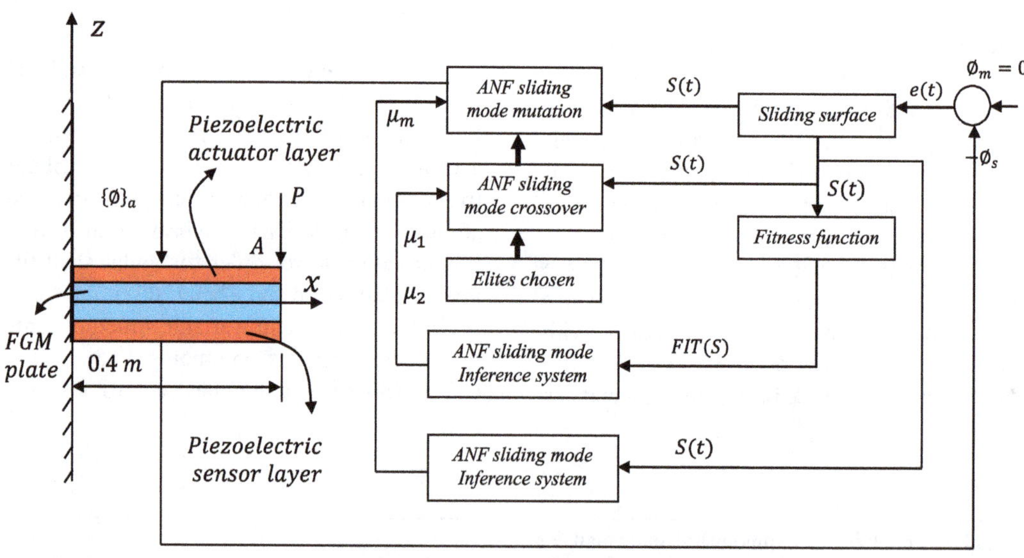

Figure 1. The block diagram of an ANFSGA control system and 2D plot of the FGM plate with distributed piezoelectric layer as an actuator on top and a sensor on bottom.

The equations of equilibrium and electrostatics are given as follows:

$$\sigma_{ij,j} + f_{bi} = \rho \ddot{u}_i, \tag{5}$$

$$D_{i,i} = 0. \tag{6}$$

In the quasi-static and plane stress formulation analysis, the constitutive relationship for the FGM lamina in the principal material coordinates of the lamina can be considered by the following form:

$$\sigma_{ij} = c_{ijkl}\varepsilon_{kl} - e_{ijk}E_k, \tag{7}$$

$$D_k = e_{ijk}\varepsilon_{ij} + k_{kl}E_l, \tag{8}$$

where $E_i = -\Phi_{,i}$ and Φ is the electric potential, σ_{ij} denotes stress, ε_{ij}, E_i and D_i are the strain, electric field and the electric displacements respectively. c_{ijkl} is the elastic coefficients, $[e]$ and $[k]$ are accordingly, the piezoelectric stress constants and the dielectric permittivity coefficients for a constant elastic strain. The symbol ρ is the density of the plate which varies according to the following form:

$$\rho(z) = (\rho_{\mathrm{T}} - \rho_{\mathrm{A}})\left(\frac{(2z+h)}{2h}\right)^n + \rho_{\mathrm{A}}. \tag{9}$$

The relationship between piezoelectric stress constants and the piezoelectric strain can be obtained by the following form:

$$e_{31} = d_{31}c_{11} + d_{32}c_{12}, \tag{10}$$

$$e_{32} = d_{31}c_{12} + d_{32}c_{22}. \tag{11}$$

In the present paper, the effective mechanical properties' definitions of the plate are assumed to vary through the thickness of the uniform plate and can be written as

$$c_{ij}(z) = \left(c_{ij}^{\mathrm{T}} - c_{ij}^{\mathrm{A}}\right)\left(\frac{(2z+h)}{2h}\right)^n + c_{ij}^{\mathrm{A}}. \tag{12}$$

Here, the simple power law distribution method is used, where c_{ij}^{T} and c_{ij}^{A} are the corresponding elastic properties of the Ti-6A1-4 V and aluminum oxide, n and h are the power law index and thickness of the plate, respectively.

According to Hamilton's principle and by using the above equations, the variational form of the equations of motion for the FGM plate can be written as

$$\int_{t_0}^{t_1}\int_{\mathrm{V}} (-\rho\ddot{u}_i\delta u_i - \sigma_{ij}\delta\varepsilon_{ij} + D_i\delta E_i)\,\mathrm{d}v\mathrm{d}t + \int_{t_0}^{t_1}\int_{\mathrm{V}} f_{\mathrm{b}i}\delta u_i \mathrm{d}v\,\mathrm{d}t + \int_{t_0}^{t_1} f_{\mathrm{c}i}\delta u_i \mathrm{d}t + \int_{t_0}^{t_1}\int_{\mathrm{S}} (f_{\mathrm{s}i}\delta u_i + q\delta\varnothing)\,\mathrm{d}s\mathrm{d}t = 0 \tag{13}$$

here q is the surface charge, t_0 and t_1 are arbitrary time intervals, the symbols 'v' and 's' represent the volume and surface of the solid, respectively. $f_{\mathrm{b}i}$, f_{ci} and f_{si} denote the body force, concentrated load and specified traction, respectively.

2.2. *Finite element model*

In this section, a finite element model of the FGM plate as a plant is introduced. The displacements and electric potential at the element level can be defined in terms of nodal variables by the following form:

$$\{u\} = [H][N_u]\{u^e\}, \tag{14}$$

$$\{\varnothing\} = [N_\varnothing]\{\varnothing^e\}, \tag{15}$$

where $[N_u]$ and $[N_\varnothing]$ are the shape functions, which include linear interpolation functions and non-conforming Hermite cubic interpolation functions. These shape functions can be found in the literature (He et al., 2001; Reddy, 2004). $\{u^e\}$ is the generalized nodal displacements and $\{\varnothing^e\}$ is the nodal electric potentials.

The infinitesimal engineering strains that are associated with the displacements are given by

$$\{\varepsilon\} = [B_u]\{u^e\}, \tag{16}$$

where the strain matrices $[B_u] = [[B_{u1}][B_{u2}][B_{u3}][B_{u4}]] = [A_u] - z[C_u]$ and $[B_{ui}] = [A_{ui}] - z[C_{ui}]$ are for $i = 1, 2, 3$ and 4.

$[A_{ui}]$ and $[C_{ui}]$ are derivative matrixes of linear and non-conforming Hermite cubic interpolation functions, respectively (Reddy, 2004).

The electric field vector $\{E\}$ can be expressed in terms of nodal variables as

$$\{E\} = -\nabla\varnothing = -[B_\varnothing][\varnothing^e], \tag{17}$$

where $[B_\varnothing] = \nabla[N_\varnothing]$. Substituting Equations (7), (8), (14), (16) and (17) into Equation (13), and assembling the element equations yield

$$[M_{uu}]\{\ddot{u}\} + [K_{uu}]\{u\} + [K_{u\varnothing}]\{\varnothing\} = \{F_m\}, \tag{18}$$

$$[K_{\varnothing u}]\{u\} - [K_{\varnothing\varnothing}]\{\varnothing\} = \{F_q\}. \tag{19}$$

The matrices and vectors are given by

$$[M_{uu}] = \sum_{\text{elem}}\sum_{K=1}^{NL}\int_{-1}^{1}\int_{-1}^{1}[N_u]^T[I][N_u]|J|\,\mathrm{d}\xi\mathrm{d}\eta, \tag{20}$$

$$[K_{uu}] = \int_{\mathrm{V}}[B_u]^{\mathrm{T}}[C][B_u]\,\mathrm{d}v = \sum_{\text{elem}}\sum_{K=1}^{NL}\int_{-1}^{1}\int_{-1}^{1}([A_u]^{\mathrm{T}}[A][A_u] - [A_u]^{\mathrm{T}}[B][C_u] - [C_u]^{\mathrm{T}}[B][A_u] + [C_u]^{\mathrm{T}}[Q][C_u])|J|\mathrm{d}\xi\mathrm{d}\eta, \tag{21}$$

$$[K_{u\varnothing}] = \int_{\mathrm{V}}[B_u]^{\mathrm{T}}[e]^{\mathrm{T}}[B_\varnothing]\,\mathrm{d}v = \sum_{\text{elem}}\sum_{K=1}^{NL}(z_{K+1} - z_K)\int_{-1}^{1}\int_{-1}^{1}([A_u][e]^{\mathrm{T}}[B_\varnothing] - \frac{1}{2}(z_{K+1} + z_K)[C_u]^{\mathrm{T}}[e]^{\mathrm{T}}[B_\varnothing])|J|\,\mathrm{d}\xi\mathrm{d}\eta, \tag{22}$$

$$[K_{\varnothing u}] = \int_{\mathrm{v}} [B_{\varnothing}]^{\mathrm{T}}[e][B_u]\,\mathrm{d}v$$

$$= \sum_{\text{elem}} \sum_{K=1}^{NL} (z_{K+1} - z_K) \int_{-1}^{1}\int_{-1}^{1} ([B_{\varnothing}]^{\mathrm{T}}[e][A_u] - \frac{1}{2}(z_{K+1} + z_K)[B_{\varnothing}]^{\mathrm{T}}[e][C_u])|J|\,\mathrm{d}\xi\,\mathrm{d}\eta, \quad (23)$$

$$[K_{\varnothing\varnothing}] = \int_{\mathrm{v}} [B_{\varnothing}]^{\mathrm{T}}[k][B_{\varnothing}]\,\mathrm{d}v = \sum_{\text{elem}} \sum_{K=1}^{NL} (z_{K+1} - z_K) \times \int_{-1}^{1}\int_{-1}^{1} [B_{\varnothing}]^{\mathrm{T}}[k][B_{\varnothing}]|J|\,\mathrm{d}\xi\,\mathrm{d}\eta, \quad (24)$$

$$\{F_m\} = \int_{\mathrm{v}} [N]^{\mathrm{T}}[H]^{\mathrm{T}}[f_{\mathrm{b}}]\,\mathrm{d}v + \int_{\mathrm{sf}} [N]^{\mathrm{T}}[H]^{\mathrm{T}}[f_{\mathrm{s}}]\,\mathrm{d}s + [N]^{\mathrm{T}}[H]^{\mathrm{T}}[f_{\mathrm{c}}] \quad (25)$$

$$\{F_q^e\} = \int_{S_q} [N_{\varnothing}]^{\mathrm{T}}\{q\}\,\mathrm{d}s, \quad (26)$$

where

$$[I] = \left(\int_{z_K}^{Z_{K+1}} (\rho_{\mathrm{T}} - \rho_{\mathrm{A}}) \left(\frac{(2z+h)}{2h} \right)^n [H]^{\mathrm{T}}[H] + \rho_{\mathrm{A}}[H]^{\mathrm{T}}[H] \right) \mathrm{d}z, \quad (27)$$

$$[A,\, B,\, Q] = \int_{z_K}^{Z_{K+1}} \left(([C]^{\mathrm{T}} - [C]^{\mathrm{A}}) \left(\frac{(2z+h)}{2h} \right)^n (1,\, z,\, z^2) + [C]^{\mathrm{A}}(1,\, z,\, z^2)\mathrm{d}z \right). \quad (28)$$

Substituting Equation (19) into Equation (18), one can obtain

$$[M_{uu}]\{\ddot{u}\} + ([K_{uu}] + [K_{u\varnothing}][K_{\varnothing\varnothing}]^{-1}[K_{\varnothing u}])\{u\} = \{F_m\} + [K_{u\varnothing}][K_{\varnothing\varnothing}]^{-1}\{F_q\}. \quad (29)$$

where $\{F_q\}$ for the sensor and actuator layer can be written as

$$\{F_q\} = \begin{Bmatrix} \{F_q\}_{\mathrm{s}} \\ \{F_q\}_{\mathrm{a}} \end{Bmatrix} = \begin{bmatrix} [K_{\varnothing u}]_{\mathrm{s}} & 0 \\ 0 & [K_{\varnothing u}]_{\mathrm{a}} \end{bmatrix} \begin{Bmatrix} \{u\}_{\mathrm{s}} \\ \{u\}_{\mathrm{a}} \end{Bmatrix} - \begin{bmatrix} [K_{\varnothing\varnothing}]_{\mathrm{s}} & 0 \\ 0 & [K_{\varnothing\varnothing}]_{\mathrm{a}} \end{bmatrix} \begin{Bmatrix} \{\varnothing\}_{\mathrm{s}} \\ \{\varnothing\}_{\mathrm{a}} \end{Bmatrix}. \quad (30)$$

Here, the subscript 's' denotes the sensors and subscript 'a' represents the actuators.

For the sensor layer, the applied charge $\{F_q\}$ is zero and the converse piezoelectric effect is assumed negligible. Using Equation (19), the sensor output is

$$\{\varnothing\}_{\mathrm{s}} = [K_{\varnothing\varnothing}]_{\mathrm{s}}^{-1}[K_{\varnothing u}]_{\mathrm{s}}\{u\}_{\mathrm{s}} \quad (31)$$

and the sensor charge due to deformation from Equation (19) is

$$\{F_q\}_{\mathrm{s}} = [K_{\varnothing u}]_{\mathrm{s}}\{u\}_{\mathrm{s}}. \quad (32)$$

For the actuator layer, from Equation (19), $\{F_q\}_{\mathrm{a}}$ can be written in the following form:

$$\{F_q\}_{\mathrm{a}} = [K_{\varnothing u}]_{\mathrm{a}}\{u\}_{\mathrm{a}} - [K_{\varnothing\varnothing}]_{\mathrm{a}}\{\varnothing\}_{\mathrm{a}}. \quad (33)$$

As mentioned above and substituting Equations (32) and (33) into Equation (30), thus Equation (30) can be expressed as

$$\{F_q\} = [K_{\varnothing u}]\{u\} - [K_{\varnothing\varnothing}]_{\mathrm{a}}\{\varnothing\}_{\mathrm{a}} \quad (34)$$

substituting Equation (34) into Equation (29) and by using some mathematics operations one can obtain

$$[M_{uu}]\{\ddot{u}\} + [C_{\mathrm{s}}]\{\dot{u}\} + [K_{uu}]\{u\} = \{F_m\} - [K_{u\varnothing}]_{\mathrm{a}}\{\varnothing\}_{\mathrm{a}}, \quad (35)$$

where $[C_{\mathrm{s}}] = a[M_{uu}] + b[K_{uu}]$ is the damping matrix, a and b are Rayleigh's coefficients.

3. Control system

In this section the control objective is to find a control law $\{\varnothing\}_{\mathrm{a}}$ so that the desired sensor output $\{\varnothing\}_m(t)$ as a state can be tracked by the sensor output $\{\varnothing\}_{\mathrm{s}}(t)$. Here, to suppress the vibration, the fact that the state (mode shape) of the plate goes to equilibrium point should be considered, therefore $\{\varnothing\}_m(t) = 0$ and the corresponding sensor output $\{\varnothing\}_{\mathrm{s}}(t) \to 0$. It is noted that the proposed control systems can be used in forced vibrations problem by selecting a proper $\{\varnothing\}_m(t)$. Based on this strategy, the mode shapes of the plate can be held on arbitrary trajectory or desired mode shape.

3.1. Traditional sliding mode

In this section a traditional sliding-mode control (TSMC) system is designed and fabricated to suppress vibrations of the FGM plate. To achieve the control objective, the following tracking error vector can be defined $e(t) = \{\varnothing\}_m(t) - \{\varnothing\}_{\mathrm{s}}(t)$, Moreover, the sliding surface can be expressed as

$$S(t) = \left(\frac{\mathrm{d}}{\mathrm{d}t} + \lambda \right)^2 \int_0^t e(\tau)\,\mathrm{d}\tau, \quad (36)$$

where λ is a positive constant. Note that since the function $S(t) = 0$ when $t = 0$, there is no reaching phase as in the traditional sliding-mode control (Lin & Hsu, 2004; Slotine

& Li, 1991). Differentiating $S(t)$ with respect to time and using Equation (31), one can obtain:

$$\dot{S} = \ddot{e}(t) + 2\lambda\dot{e}(t) + \lambda^2 e(t), \tag{37}$$

$$\dot{S} = \ddot{\varnothing}_m - [K_{\varnothing\varnothing}]_s^{-1}[K_{\varnothing u}]_s\{\ddot{u}\}_s + 2\lambda(\dot{\varnothing}_m - \dot{\varnothing}_s) + \lambda^2(\varnothing_m - \varnothing_s), \tag{38}$$

Now, $\{\ddot{u}\}$ can be expressed as

$$\{\ddot{u}\} = [M_{uu}]^{-1}(\{F_m\} - [K_{u\varnothing}]_a\{\varnothing\}_a - [C_s]\{\dot{u}\} - [K_{uu}]\{u\}) \tag{39}$$

substituting Equation (39) in Equation (38) one can obtain

$$\dot{S} = \ddot{\varnothing}_m - [K_{\varnothing\varnothing}]_s^{-1}[K_{\varnothing u}]_s[M_{uu}]^{-1}(\{F_m\} - [K_{u\varnothing}]_a\{\varnothing\}_a - [C_s]\{\dot{u}\} - [K_{uu}]\{u\}) + 2\lambda(\dot{\varnothing}_m - \dot{\varnothing}_s) + \lambda^2(\varnothing_m - \varnothing_s). \tag{40}$$

The aforementioned tracking problem is to design a control law $\{\varnothing\}_a$ so that the state remains on the surface $S(t) = 0$ for all times. In designing the sliding-mode control system, first of all the equivalent control law $\{\varnothing\}_{aeq}$, which will determine the dynamic of the system on the sliding surface, can be found. The equivalent control law is derived from recognizing

$$\dot{S}|_{\{\varnothing\}_a=\{\varnothing\}_{aeq}} = 0. \tag{41}$$

Substituting Equation (44) into Equation (43) and rearranging yield

$$\{\varnothing\}_{aeq} = -\left([K_{\varnothing\varnothing}]_s^{-1}[K_{\varnothing u}]_s[M_{uu}]^{-1}[K_{u\varnothing}]_a\right)^{-1}\left(\ddot{\varnothing}_m - [K_{\varnothing\varnothing}]_s^{-1}[K_{\varnothing u}]_s[M_{uu}]^{-1}\left(\{F_m\} - [C_s]\{\dot{u}\} - [K_{uu}]\{u\}\right) + 2\lambda\left(\dot{\varnothing}_m - \dot{\varnothing}_s\right) + \lambda^2\left(\varnothing_m - \varnothing_s\right)\right). \tag{42}$$

Thus, given $\dot{S}(t) = 0$, the dynamics of the system on the sliding surface for $t > 0$ is given by

$$\ddot{e}(t) + 2\lambda\dot{e}(t) + \lambda^2 e(t) = 0. \tag{43}$$

In this controller if the system parameters are perturbed or unknown, the equivalent control design cannot guarantee the performance specified by Equation (43). Moreover, the stability of the controlled system may be destroyed. To ensure the system performance designed by Equation (43) under the existence of the uncertainties, a robust controller $\{\varnothing\}_{aR}$ is designed by the following form:

$$\{\varnothing\}_{aR} = -k_k\ \mathrm{sign}(S), \tag{44}$$

where k_k is the gain control. Finally, the TSMC control system can be obtained as

$$\{\varnothing\}_a = \{\varnothing\}_{aeq} + \{\varnothing\}_{aR}. \tag{45}$$

3.2. *ANF sliding mode*

The architecture diagram of the neuro-fuzzy inference mechanism is depicted in Figure 2. The ANF sliding-mode controller is composed of a neuro-fuzzy network with the online learning algorithm.

Let ***input*** $= [S(t), \dot{S}(t)]$ and ***output*** $= \{\varnothing\}_a$ be the input and output variables to the ANF sliding-mode system, respectively.

3.2.1. *Description of ANF*

In the proposed controller, the four-layer NN is used (Figure 2). Layers I–IV represent the inputs to the network, the membership functions, the fuzzy rule base and the outputs of the network, respectively.

3.2.2. *Layer I: input layer*

Inputs and outputs of nodes in this layer are represented as

$$\mathrm{net}_1^{\mathrm{I}} = S(t),\ y_1^{\mathrm{I}} = f_1^{\mathrm{I}}(\mathrm{net}_1^{\mathrm{I}}) = \mathrm{net}_1^{\mathrm{I}} = S(t), \tag{46}$$

$$\mathrm{net}_1^{\mathrm{I}} = \dot{S}(t),\ y_2^{\mathrm{I}} = f_2^{\mathrm{I}}(\mathrm{net}_2^{\mathrm{I}}) = \mathrm{net}_2^{\mathrm{I}} = \dot{S}(t), \tag{47}$$

where y_1^{I} and y_2^{I} are outputs of the input layer. In this layer, the weights are unity and fixed.

3.2.3. *Layer II: membership layer*

In this layer, each node performs a fuzzy set and the Gaussian function is adopted as a membership function

$$\mathrm{net}_{1,j}^{\mathrm{II}} = -\frac{\left(x_{1,j}^{\mathrm{II}} - m_{1,j}^{\mathrm{II}}\right)^2}{\left(\sigma_{1,j}^{\mathrm{II}}\right)^2},\quad y_{1,j}^{\mathrm{II}} = f_{1,j}^{\mathrm{II}}\left(\mathrm{net}_{1,j}^{\mathrm{II}}\right) = \exp\left(\mathrm{net}_{1,j}^{\mathrm{II}}\right), \tag{48}$$

$$\mathrm{net}_{1,k}^{\mathrm{II}} = -\frac{\left(x_{2,k}^{\mathrm{II}} - m_{2,k}^{\mathrm{II}}\right)^2}{\left(\sigma_{2,k}^{\mathrm{II}}\right)^2},\quad y_{2,k}^{\mathrm{II}} = f_{2,k}^{\mathrm{II}}\left(\mathrm{net}_{2,k}^{\mathrm{II}}\right) = \exp\left(\mathrm{net}_{2,k}^{\mathrm{II}}\right), \tag{49}$$

where $m_{1,j}^{\mathrm{II}}$, $m_{2,k}^{\mathrm{II}}$ and $\sigma_{1,j}^{\mathrm{II}}$, $\sigma_{2,k}^{\mathrm{II}}$ are the mean and the standard deviation of the Gaussian function, respectively. The variables $x_{1,j}^{\mathrm{II}}$ and $x_{2,k}^{\mathrm{II}}$ are the outputs of layer I.

3.2.4. *Layer III: rule layer*

This layer includes the rule base used in the fuzzy logic control. Each node in this layer which multiplies the input signals and outputs can be expressed as follows:

$$\mathrm{net}_{jk}^{\mathrm{III}} = \left(x_{1,j}^{\mathrm{III}} \times x_{2,k}^{\mathrm{III}}\right),\quad y_{jk}^{\mathrm{III}} = f_{jk}^{\mathrm{III}}\left(\mathrm{net}_{jk}^{\mathrm{III}}\right) = \mathrm{net}_{jk}^{\mathrm{III}} \tag{50}$$

here $x_{1,j}^{\mathrm{III}}$ and $x_{2,k}^{\mathrm{III}}$ are the outputs of layer II. The values of link weights between the membership layer and rule base layer are unity.

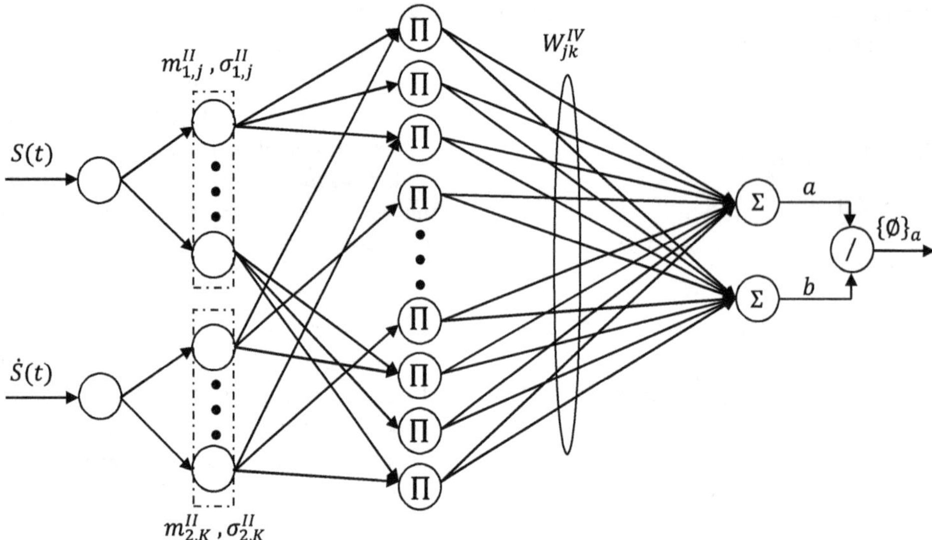

Figure 2. Schematic diagram of the neuro-fuzzy network.

3.2.5. Layer IV: output layer

This layer represents the inference and defuzzification which are used in the fuzzy logic system. For defuzzification, the center of area method is used. Therefore, the following form can be obtained:

$$a_i = \sum_j \sum_k W_{jk}^{\mathrm{IV}} y_{jk}^{\mathrm{III}}, \quad b_i = \sum_j \sum_k y_{jk}^{\mathrm{III}},$$
$$\mathrm{net}_{0\,i}^{\mathrm{IV}} = \frac{a_i}{b_i}, \quad y_{0\,i}^{\mathrm{IV}} = f_0^{\mathrm{IV}}(\mathrm{net}_{0\,i}^{\mathrm{IV}}) = \frac{a_i}{b_i}, \tag{51}$$

where y_{jk}^{III} is the output of the rule layer, a_i and b_i are the numerator and the denominator of the function used in the center of area method according to each degree and W_{jk}^{IV} is the center of the output membership functions used in the fuzzy logic system, respectively. The aim of the learning algorithm is to adjust the weights of W_{jk}^{IV}, $m_{1,j}^{\mathrm{II}}$, $m_{2,k}^{\mathrm{II}}$ and $\sigma_{1,j}^{\mathrm{II}}$, $\sigma_{2,k}^{\mathrm{II}}$. Finally, y_0^{IV} is the output of the proposed inference system.

The online learning algorithm is a gradient descent search algorithm in the space of network parameters. The Lyapunov function is chosen as $(1/2)S^2(t)$. The aim is to minimize the derivative of the Lyapunov function with respect to time or $S(t)\dot{S}(t)$.

3.2.6. Online learning algorithm

The error expression for the input of Layer IV can be expressed as follows:

$$\delta_{0\,i}^{\mathrm{IV}} = -\frac{\partial S(t)\dot{S}(t)}{\partial y_{0\,i}^{\mathrm{IV}}} \frac{\partial y_{0\,i}^{\mathrm{IV}}}{\partial\, \mathrm{net}_i^{\mathrm{IV}}} = \varsigma_1 S(t), \tag{52}$$

where ς_1 is the learning rate for W_{jk}^{IV}. Therefore, the changing of W_{jk}^{IV} is written as

$$\dot{W}_{jk}^{\mathrm{IV}} = -\frac{\partial S(t)\dot{S}(t)}{\partial\, \mathrm{net}_{0\,i}^{\mathrm{IV}}} \frac{\partial\, \mathrm{net}_{0\,i}^{\mathrm{IV}}}{\partial a_i} \frac{\partial a_i}{\partial W_{jk}^{\mathrm{IV}}} = \frac{1}{b_i} \delta_{0\,i}^{\mathrm{IV}} y_{jk}^{\mathrm{III}}. \tag{53}$$

Since the weights in the rule layer are unified, only the approximated error term needs to be calculated and propagated by the following equation:

$$\delta_{jk\,i}^{\mathrm{III}} = -\frac{\partial S(t)\dot{S}(t)}{\partial\, \mathrm{net}_{0\,i}^{\mathrm{IV}}} \frac{\partial\, \mathrm{net}_{0\,i}^{\mathrm{IV}}}{\partial y_{1,j}^{\mathrm{III}}} \frac{\partial y_{1,j}^{\mathrm{III}}}{\partial\, \mathrm{net}_{jk\,i}^{\mathrm{III}}} = \frac{1}{b_i} \delta_{0\,i}^{\mathrm{IV}} \left(W_{jk}^{\mathrm{IV}} - \partial y_{0\,i}^{\mathrm{IV}} \right). \tag{54}$$

The error received from Layer III is computed as

$$\delta_{1,j\,i}^{\mathrm{II}} = \sum_k \left[\left(-\frac{\partial S(t)\dot{S}(t)}{\partial\, \mathrm{net}_{jk\,i}^{\mathrm{III}}} \right) \frac{\partial\, \mathrm{net}_{jk\,i}^{\mathrm{III}}}{\partial y_{1,j}^{\mathrm{II}}} \frac{\partial y_{1,j\,i}^{\mathrm{II}}}{\partial\, \mathrm{net}_{1,j\,i}^{\mathrm{II}}} \right] = \sum_k \delta_{jk\,i}^{\mathrm{III}} y_{jk\,i}^{\mathrm{III}}, \tag{55}$$

$$\delta_{2,k\,i}^{\mathrm{II}} = \sum_j \left[\left(-\frac{\partial S(t)\dot{S}(t)}{\partial\, \mathrm{net}_{jk\,i}^{\mathrm{III}}} \right) \frac{\partial\, \mathrm{net}_{jk\,i}^{\mathrm{III}}}{\partial y_{2,k}^{\mathrm{II}}} \frac{\partial y_{2,k\,i}^{\mathrm{II}}}{\partial\, \mathrm{net}_{2,k\,i}^{\mathrm{II}}} \right] = \sum_j \delta_{jk\,i}^{\mathrm{III}} y_{jk\,i}^{\mathrm{III}}. \tag{56}$$

The update laws of $m_{1,j}^{\mathrm{II}}$, $m_{2,k}^{\mathrm{II}}$ and $\sigma_{1,j}^{\mathrm{II}}$, $\sigma_{2,k}^{\mathrm{II}}$ also can be obtained by the gradient decent search algorithm, it means:

$$\dot{m}_{1,j\,i}^{\mathrm{II}} = -\frac{\partial S(t)\dot{S}(t)}{\partial\, \mathrm{net}_{1,j\,i}^{\mathrm{II}}} \frac{\partial\, \mathrm{net}_{1,j\,i}^{\mathrm{II}}}{\partial m_{1,j\,i}^{\mathrm{II}}} = \varsigma_2 \delta_{1,j\,i}^{\mathrm{II}} \frac{2\left(x_{1,j}^{\mathrm{II}} - m_{1,j\,i}^{\mathrm{II}} \right)}{\left(\sigma_{1,j\,i}^{\mathrm{II}} \right)^2}, \tag{57}$$

$$\dot{m}^{\mathrm{II}}_{2,k\,i} = -\frac{\partial S(t)\,\dot{S}(t)}{\partial\,\mathrm{net}^{\mathrm{II}}_{2,k_i}}\frac{\partial\,\mathrm{net}^{\mathrm{II}}_{2,k_i}}{\partial m^{II}_{2,k_i}} = \varsigma_3 \delta^{\mathrm{II}}_{2,k\,i}\frac{2\left(x^{\mathrm{II}}_{2,k} - m^{\mathrm{II}}_{2,k_i}\right)}{\left(\sigma^{\mathrm{II}}_{2,k_i}\right)^2}, \tag{58}$$

$$\dot{\sigma}^{\mathrm{II}}_{1,j\,i} = -\frac{\partial S(t)\,\dot{S}(t)}{\partial\,\mathrm{net}^{\mathrm{II}}_{1,j_i}}\frac{\partial\,\mathrm{net}^{\mathrm{II}}_{1,j\,i}}{\partial \sigma^{\mathrm{II}}_{1,j\,i}} = \varsigma_4 \delta^{\mathrm{II}}_{1,j\,i}\frac{2\left(x^{\mathrm{II}}_{1,j} - m^{\mathrm{II}}_{1,j\,i}\right)^2}{\left(\sigma^{\mathrm{II}}_{1,j\,i}\right)^3}, \tag{59}$$

$$\dot{\sigma}^{\mathrm{II}}_{2,k\,i} = -\frac{\partial S(t)\,\dot{S}(t)}{\partial\,\mathrm{net}^{\mathrm{II}}_{2,k_i}}\frac{\partial\,\mathrm{net}^{\mathrm{II}}_{2,k_i}}{\partial m^{\mathrm{II}}_{2,k_i}} = \varsigma_5 \delta^{\mathrm{II}}_{2,k\,i}\frac{2\left(x^{\mathrm{II}}_{2,k} - m^{\mathrm{II}}_{2,k_i}\right)^2}{\left(\sigma^{\mathrm{II}}_{2,k_i}\right)^3}, \tag{60}$$

where ς_2, ς_3, ς_4, and ς_5 are the learning-rate parameters of the mean and the standard deviation of the Gaussian functions.

3.3. *ANFSGA control system*

In this section a control law-based genetic algorithm is designed to the FGM plate for tracking mode shapes, suppurating vibration and external disturbance rejection. The proposed control system is included in the SMC concept and the neuro-fuzzy sliding-mode-based evolutionary procedure. In order to achieve the control object, the evolutionary spirit of GA is embedded. The neuro-fuzzy approach is used to further ensure the correct evolutionary direction and decide the appropriate evolutionary step. In this section, the control law is made as the chromosome in GA with floating point coding can be considered. This process, which is a real one, is to be replaced by the ANF sliding-mode crossover method. An ANF sliding-mode mutation is used just after selecting the chromosomes. The first step after creating a generation is to calculate the fitness function of each member in the population. If the evolutionary direction is correct, the fittest control action can be obtained. In order to achieve the correct evolutionary direction and to ensure the stable system dynamic, the concept of the SMC system is embedded in the genetic operators to form the direction-based operators with the ANF sliding-mode evolutionary procedure.

Now, a fitness function is defined as an exponential term by the following form (Wai & Tu, 2007):

$$\mathrm{FIT}(S) = \exp[-\zeta \times (S(t)^2 + \dot{S}(t)^2)] \in [0, 1], \tag{61}$$

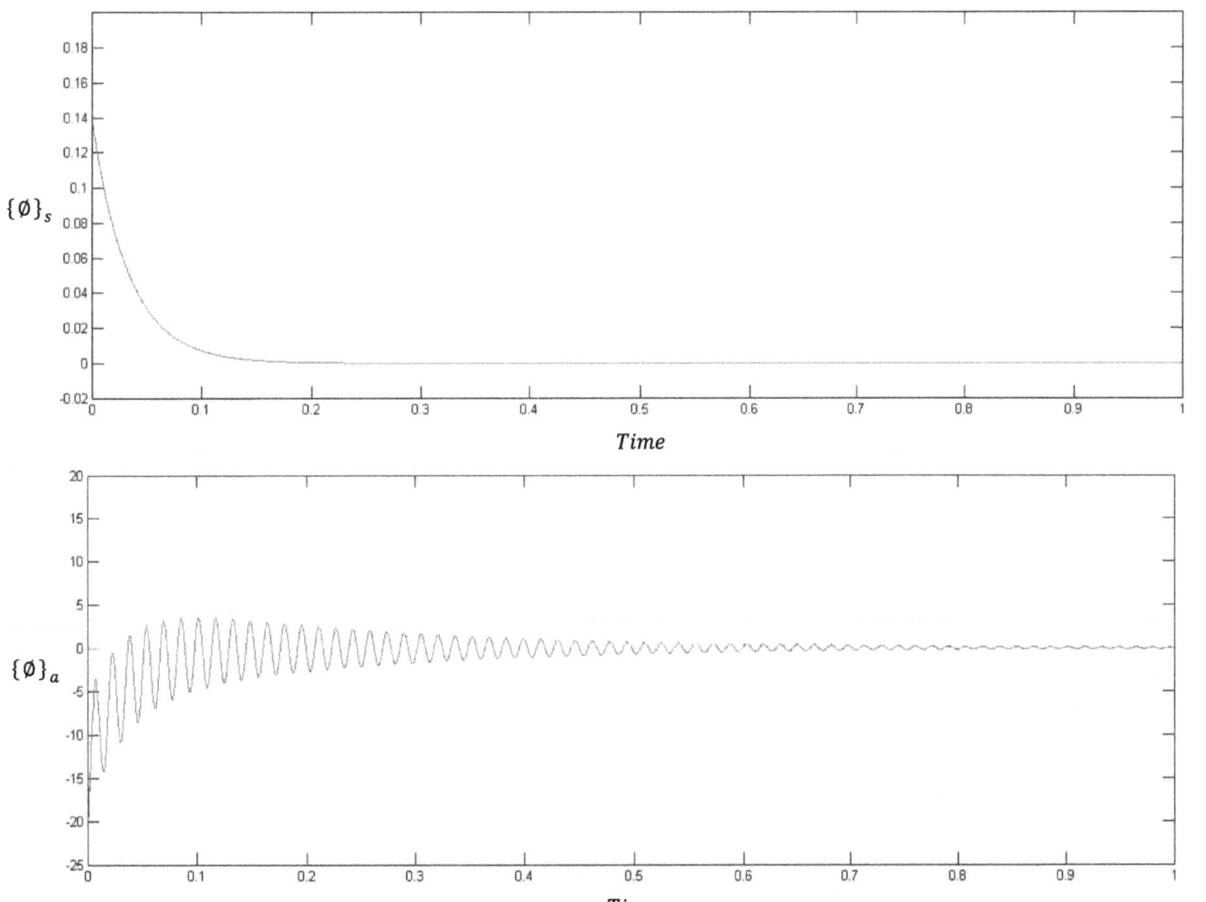

Figure 3. Simulation result of piezoelectric sensor and actuator due to the traditional sliding mode with robust controller.

where ζ is a positive constant, S is the sliding surface and $\dot{S}$ is the first derivative of S which is defined as Equation (36). The next step after evaluation is to create a new population from the current generation. The selection operation determines which chromosome participates in producing offspring for the next generation. Initially, the population is selected randomly, which means that several control actions are randomly selected from the operational region $[\{\varnothing\}_{a_{\min}}, \{\varnothing\}_{a_{\max}}]$. After comparing the fitness values of all the individuals, the best one is regarded as the elite. If the fitness value of the new control action is higher than all the previous ones, it will become the new elite.

Crossover operation is used to reshape the GA system, which can produce offspring by charging the features of the parent. In this study, the sliding surface is combined with the crossover operation by the following form:

$$\{\varnothing\}_{\mathrm{a_{GA,new}}} = \{\varnothing\}_{\mathrm{a_{GA,\,old}}} + \mu_1 \times S + \mu_2 \times \dot{S}, \tag{62}$$

where $\{\varnothing\}_{\mathrm{a_{GA,new}}}$ is the generated offspring, $\{\varnothing\}_{\mathrm{a_{GA,old}}}$ is the selected elitist chromosome of the last generation, μ_1 and μ_2 are the positive tuning parameters of S and $\dot{S}$, respectively. Here, the important problem is selecting the tuning parameters. The small tuning step may not satisfy the stability conditions. Therefore, an ANF sliding-mode system is used to produce the tuning coefficients. In this section the ANF sliding-mode mechanism of the previous section is considered to produce μ_1 and μ_2. Let ***input*** = FIT(S) for both ANF sliding-mode mechanisms and ***output***$_1 = \mu_1$ and ***output***$_2 = \mu_2$. For the two systems, different means and the standard deviations of the Gaussian function are used.

To avoid the problem of local optimization an ANF sliding-mode mechanism is used in mutation operation. Traditional mutation methods are not useful to produce better offspring in an online learning ability. Therefore, the stability of the system may be destroyed. If the control action cannot let the system dynamic stay on the sliding surface after fuzzy sliding-mode crossover, the mutation operation will further compel the system dynamic to close the sliding surface by using the fuzzy sliding-mode inference mechanism.

The offspring after mutation operation can be expressed as

$$\{\varnothing\}^{\Delta}_{\mathrm{a_{GA,new}}} = \{\varnothing\}_{\mathrm{a_{GA,new}}} + \mu_m, \tag{63}$$

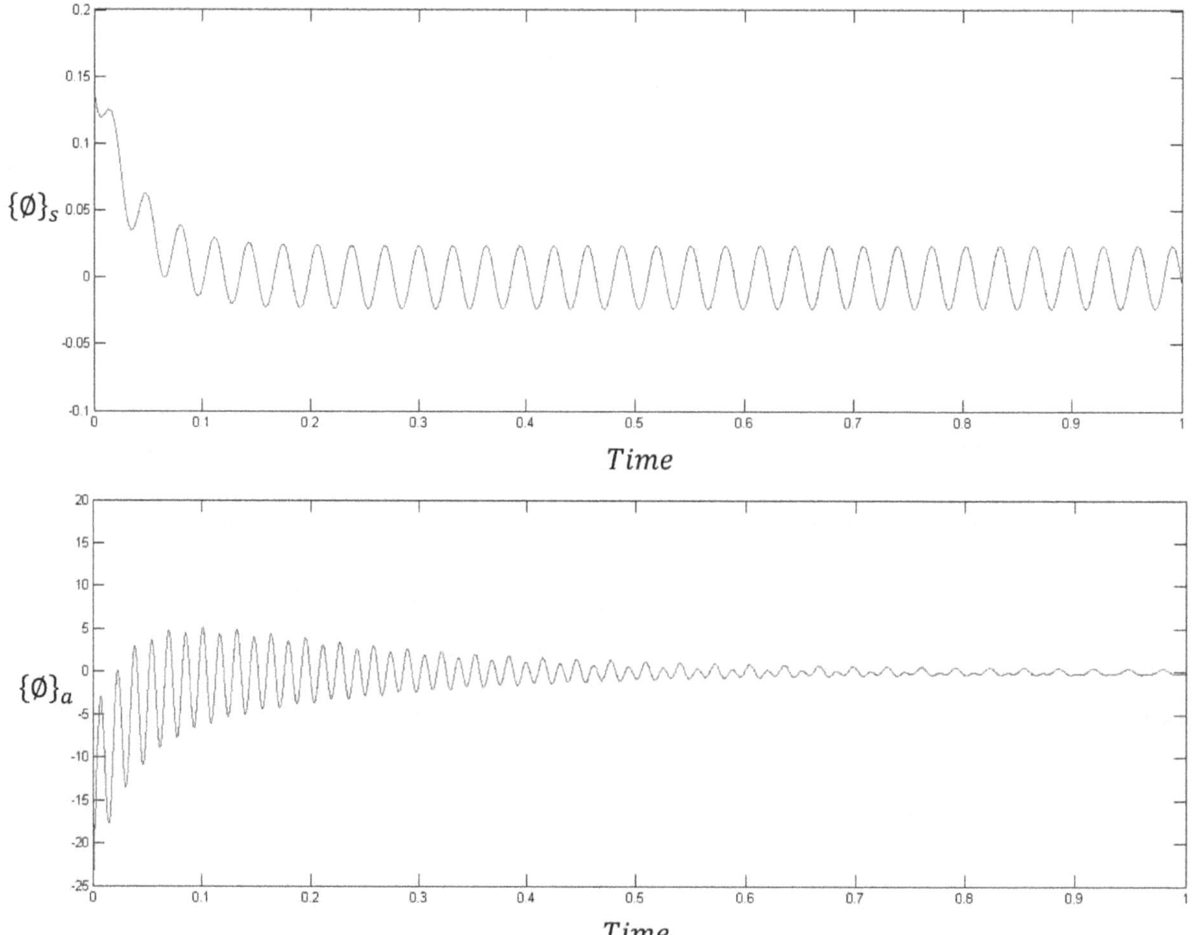

Figure 4. Simulation result of piezoelectric sensor and actuator due to the traditional sliding mode with robust controller in the disturbance condition.

where μ_m is the adjustment of mutation operation. $\{\varnothing\}^{\Delta}_{a_{GA,new}}$ is the offspring after mutation operation which is produced by the ANF sliding-mode inference mechanism. In this situation, the input to the ANF sliding-mode system is the sliding surface or ***input*** $= S(t)$. If the fitness value is lower than a specified value (FIT_B), mutation occurs. On the other hand, if the fitness value is higher than the specified value, the mutation idles.

The main process of the proposed GA-based controller is represented by the following pseudo-code:

Step 1 Select the size of population $[N]$ and the fitness function [FIT(S)].

Step 2 Generate the initial population.

Step 3 Evaluate the fitness value via (61) and sort the sequence to choose the elite $\{\varnothing\}_{a_{GA,old}}$.

Step 4 Do ANF sliding-mode crossover operation to generate $\{\varnothing\}_{a_{GA,new}}$ via (62).

Step 5 Compare the fitness value with the specified value (FIT_B), if it is not lower, then go to Step 7, otherwise follow the chart.

Step 6 Do ANF sliding-mode mutation operation to generate the $\{\varnothing\}^{\Delta}_{a_{GA,new}}$ via (63).

Step 7 Output control action.

Step 8 Program complete? If yes then it is the end, if not go to Step 3.

It is noted that in the proposed controller, for the crossover and mutation operations Equations (46)–(51) are used. The adaptive laws and the online learning algorithm are used in Equations (52)–(60).

The chattering phenomenon is a particular problem in the control algorithms. The chattering problem can reduce the control accuracy and destroy the stability of system. To find the smooth control action and reduce chattering phenomena, the following soft limit switching function f_{SL} is presented as

$$f_{SL}(S) = \frac{S(t)^2}{1 + S(t)^2}\tanh(S(t)). \tag{64}$$

4. Simulation results

The finite element model for the FGM plate is based on the general concept of solid mechanics and to ensure the accuracy, it is validated with different values of volume fraction

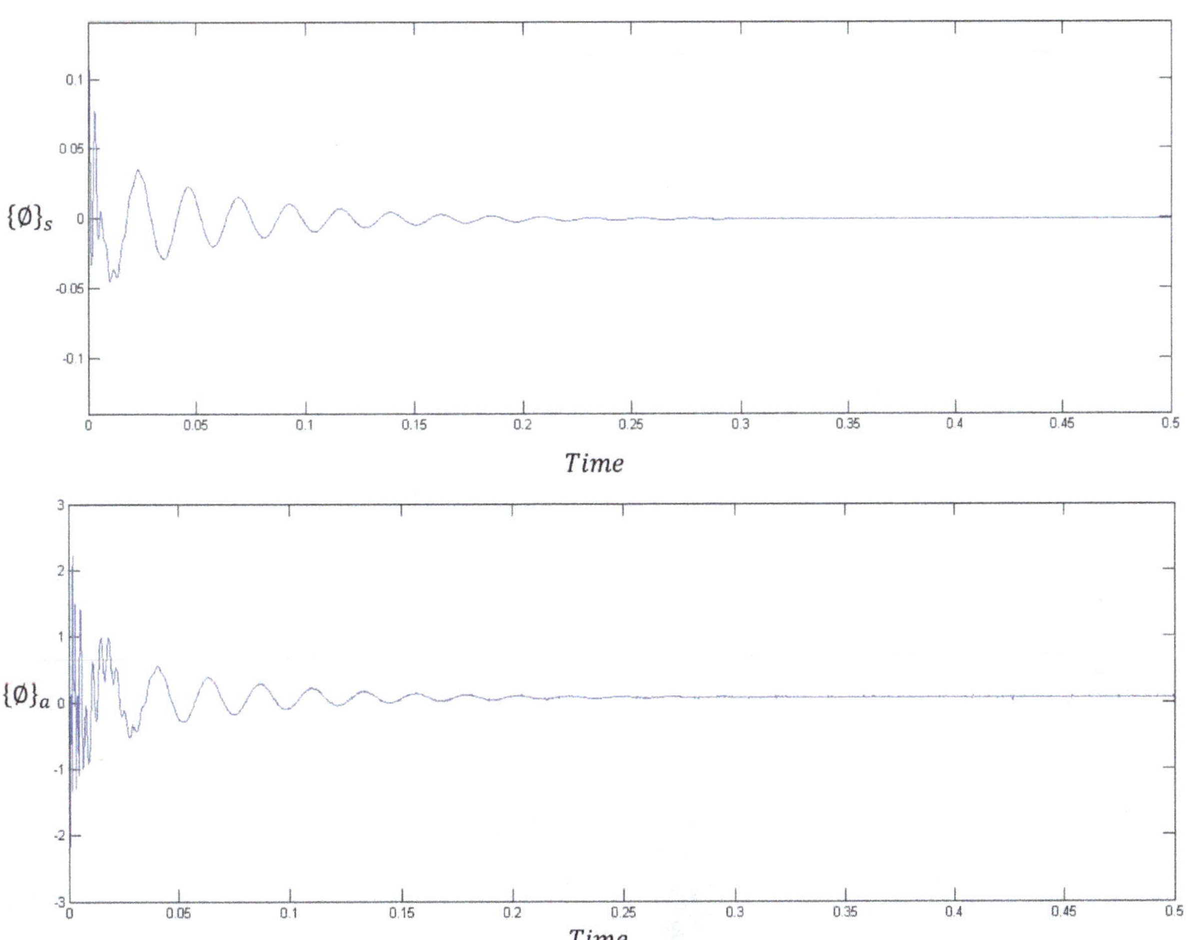

Figure 5. Simulation result of piezoelectric sensor and actuator due to the ANF SMC system.

power law exponent n and compared with the results of Bishop (1979) and Praveen and Reddy (1998).

The G-1195N piezoelectric films bond both the top and bottom surfaces of the FGM plate as shown in Figure 1. The plate is square with both length and width set as 0.4 m. The thickness of the plate is set as 5 mm, and each G-1195N piezoelectric layer has a thickness equal to 0.1 mm. The material properties of piezoelectric materials are elastic modulus $E = 63 \times 10^9\ \mathrm{N/m^2}$, Poison's ratio $\upsilon = 0.3$, density $\rho = 7600\ \mathrm{kg/m^3}$, piezoelectric constant $d_{31} = 254 \times 10^{-12}(\mathrm{m/V})$, piezoelectric constant $d_{32} = 254 \times 10^{-12}(\mathrm{m/V})$ and dielectric coefficients $k_{33} = 15 \times 10^{-9}(\mathrm{F/m})$. The material constants of the constituent of the FGM plate are listed as follows: for aluminum oxide, $E = 3.2024 \times 10^{11}\ \mathrm{N/m^2}$, $\upsilon = 0.2600$, density $\rho = 3750\ \mathrm{kg/m^3}$ and for Ti-6Al-4 V $E = 1.0570 \times 10^{11}\ \mathrm{N/m^2}$, $\upsilon = 0.2981$, density $\rho = 4429\ \mathrm{kg/m^3}$.

The cantilevered (CFFF) plate is considered as the boundary condition. For the vibration control analysis, 64(8 × 8) elements are used to model the FGM plate and to simplify the vibration analysis, the modal superposition algorithm is used and the first six modes are considered in this modal space analysis. An initial modal damping for each mode has been assumed to be 0.8%. A unit of force is imposed at point A of the FGM plate (Figure 1) in the vertical direction and is subsequently removed to generate motion from the initial displacement. Power law exponent for FGM plate is selected as $n = 5$.

In the design of proposed control systems, the effect of external disturbance are modeled as

$$\Delta\mathrm{Dis} = Am\,[\bar{K}_{uu}]\begin{bmatrix}\sin(\omega_d) & \cos(\omega_d) & \sin(\omega_d) & \cos(\omega_d) & \sin(\omega_d) & \cos(\omega_d)\end{bmatrix}^{\mathrm{T}}, \quad (65)$$

where $Am = 0.000001$ is the amplitude of disturbance, $[\bar{K}_{uu}]$ is the normalized matrix of $[K_{uu}]$ and $\omega_d = 200$ is the frequency of disturbance. Therefore, Equation (35) in the disturbance condition can be rewritten as

$$[\bar{M}_{uu}]\{\ddot{u}\} + [\bar{C}_{\mathrm{s}}]\{\dot{u}\} + [\bar{K}_{uu}]\{u\} = \{F_m\} - [\bar{K}_{u\varnothing}]_{\mathrm{a}}\{\varnothing\}_{\mathrm{a}} + \Delta\mathrm{Dis} \quad (66)$$

here $[\bar{M}_{uu}]$, $[\bar{C}_{\mathrm{s}}]$ and $[\bar{K}_{u\varnothing}]_{\mathrm{a}}$ are the normalized matrices of $[M_{uu}]$, $[C_{\mathrm{s}}]$ and $[K_{u\varnothing}]_{\mathrm{a}}$.

The simulation results are shown in Figures 3–8.

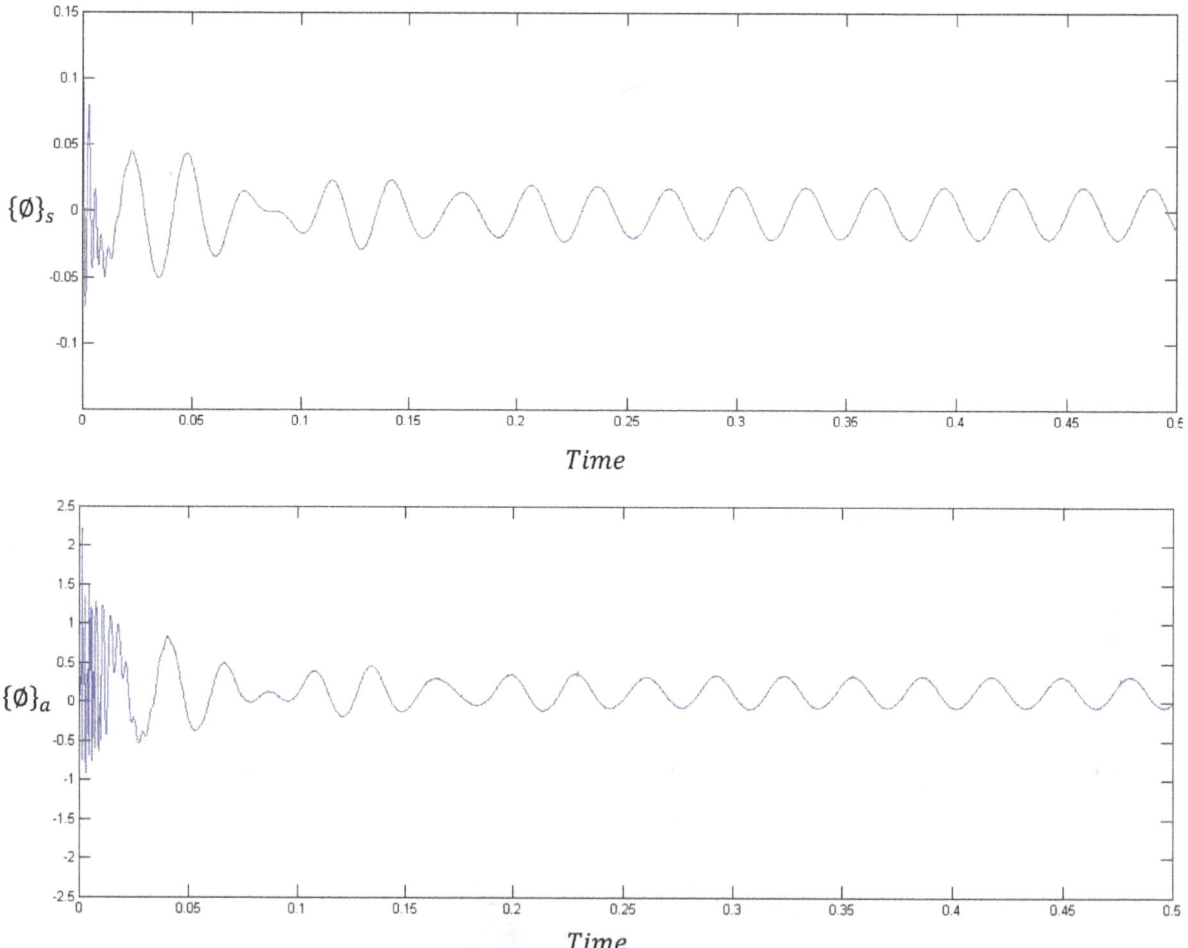

Figure 6. Simulation result of piezoelectric sensor and actuator due to the ANF SMC system in the disturbance condition.

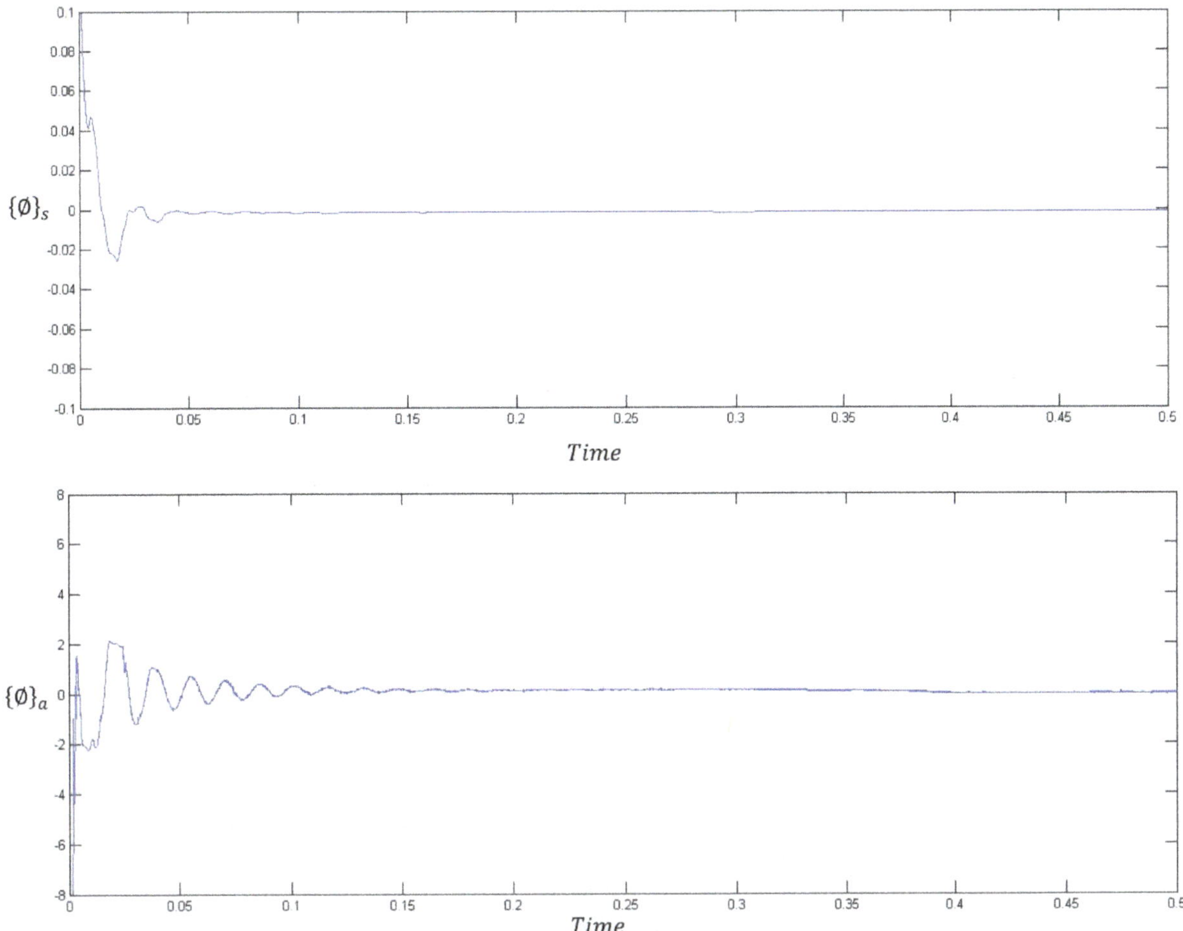

Figure 7. Simulation result of piezoelectric sensor and actuator due to the ANFSGA control system.

The effectiveness of the TSMC system is depicted in Figure 3 with $\lambda = 14.8$, Figure 4 shows that the ability of the TSMC system is improved by selecting $k_k = 1.5$ for the robust term in the disturbance condition.

The plots of the ANF SMC system (Figures 5 and 6 with control parameters $\varsigma_1 = 1.5$, $\varsigma_2 = \varsigma_3 = \varsigma_4 = \varsigma_5 = 0.05$ and $\lambda = 1.2$) show that the rate of the voltage which is applied on the actuator layer is smaller than the TSMC system.

In this study, 10 initial populations are randomly chosen from the reasonable region $[\{\varnothing\}_{a_{\min}} = -1, \{\varnothing\}_{a_{\max}} = 1]$ for the ANFSGA control system. The ability to suppress the vibration and reduce the external disturbance of the ANF-SGA control system rather than the ANF sliding mode and also the TSMC system are demonstrated in Figures 7 and 8. Figure 5 shows that the settling time in the response of the ANF SMC system is nearly 0.17, but Figure 7 shows that the response of the ANFSGA control system is nearly 0.05.

In the proposed control system, a small voltage can be used to drive the system states to the equilibrium point. Therefore, it is superior to the ANF sliding mode and TSMC system to suppress the vibrations. The control parameters of the ANFSGA control system in the crossover operation to produce μ_1 are $\varsigma_1 = 0.0001$, $\varsigma_2 = \varsigma_3 = \varsigma_4 = \varsigma_5 = 0.0001$ and also to produce μ_2 are $\varsigma_1 = 0.00005$, $\varsigma_2 = \varsigma_3 = \varsigma_4 = \varsigma_5 = 0.0001$. In the mutation operation, the control parameters of the ANFSGA control system are $\varsigma_1 = 0.0003$, $\varsigma_2 = \varsigma_3 = \varsigma_4 = \varsigma_5 = 0.0002$. The sliding surface parameter is selected as $\lambda = 1.2$. The threshold value to activate the mutation operation is applied as $\mathrm{FIT}_B = 0.1$ and the parameter fitness value $\zeta = -16.3$ is used. It can be regarded that the associated fuzzy sets with the Gaussian function for each input signal are divided into NE (negative), ZE (zero) and PO (positive). Moreover, the means of the Gaussian functions are set as -0.5, 0, 0.5 and the standard deviations of the Gaussian functions are set as 0.3 for the NE, ZE and PO neurons.

The reasonable region $[\{\varnothing\}_{\mathrm{a_{min}}} = -1, \{\varnothing\}_{\mathrm{a_{max}}} = 1]$ bounds the power fluctuation. The reasonable region only makes the system use $\{\varnothing\}_\mathrm{a}$ of interval $[\{\varnothing\}_{\mathrm{a_{min}}} = -1, \{\varnothing\}_{\mathrm{a_{max}}} = 1]$ and then the values close to 1 and -1 are produced by crossover and the mutation algorithm. Moreover, stability of the system is kept in this operation. Therefore, based on this method, the operator can set the reasonable region without tuning the control parameters to have a desired response.

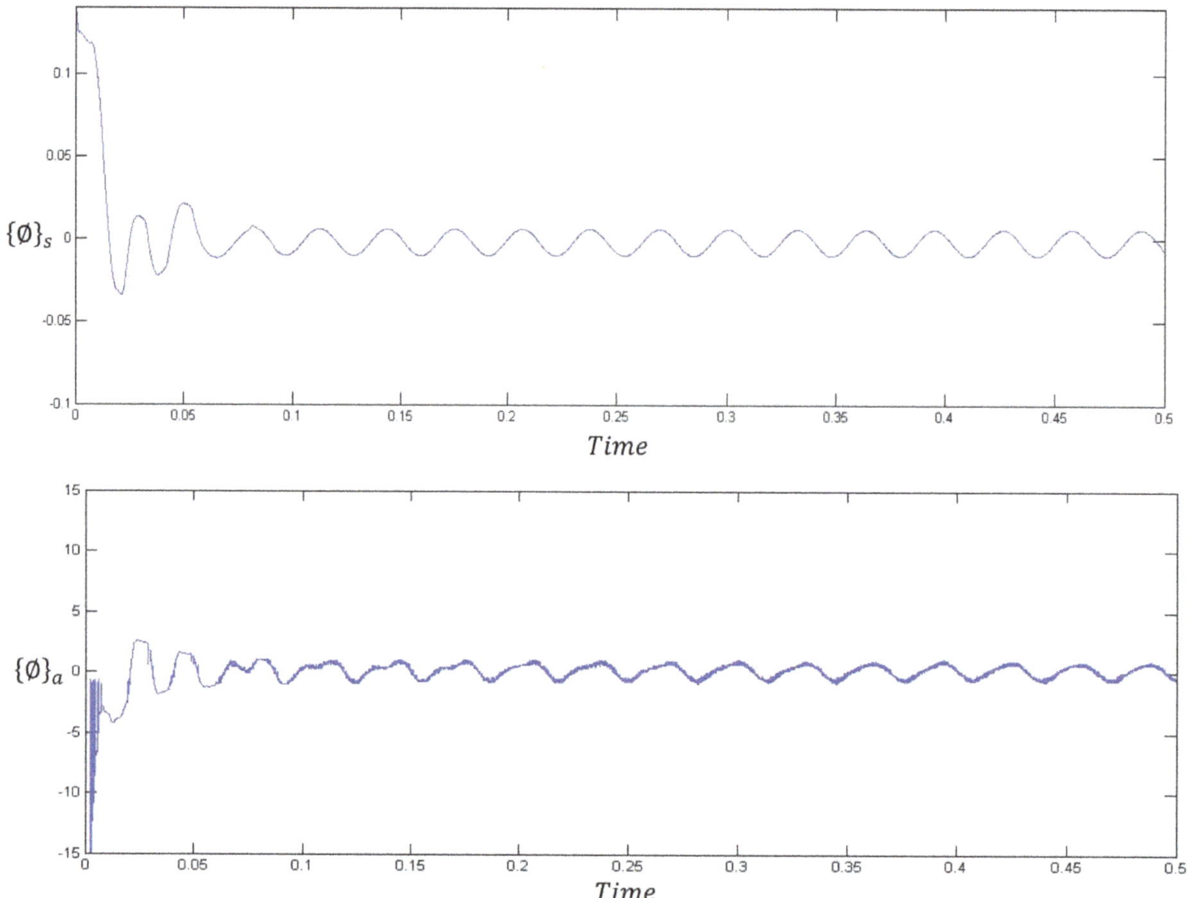

Figure 8. Simulation result of piezoelectric sensor and actuator due to the ANFSGA control system in the disturbance condition.

5. Conclusion

A general finite element model of the FGM plate has been introduced in this paper. A TSMC system has been designed to suppress the vibration of the FGM plate in the normal and disturbance conditions. The ANF sliding mode and the ANFSGA control system as the intelligent control methods have been successfully designed and effectively used to reduce the disturbance and eliminate the vibrations for the FGM plate. It is noted that, in the proposed controller, no constrained conditions and prior knowledge of the controlled plant have been used in the design process. Therefore, any information of the FGM plate is not utilized in the ANFSGA control system. The proposed controller is a flexible kind of control systems and it can be applied in another engineering applications.

Disclosure statement

No potential conflict of interest was reported by the authors.

References

Bagheri, A., & Javadi moghaddam, J. (2010). An adaptive neuro-fuzzy sliding mode based genetic algorithm control system for under water remotely operated vehicle. *Expert Systems with Applications, Expert Systems with Applications, 37*, 647–660.

Bagheri, A., & Moghaddam, J. J. (2009). Decoupled adaptive neuro-fuzzy (DANF) sliding mode control system for a Lorenz chaotic problem. *Expert Systems with Applications, 36*, 6062–6068.

Bishop, R. E. D. (1979). *The mechanics of vibration*. New York, NY: Cambridge University Press.

He, X. Q., Ng, T. Y., Sivashanker, S., & Liew, K. M. (2001). Active control of FGM plates with integrated piezoelectric sensors and actuators. *International Journal of Solids and Structures, 38*, 1641–1655.

Liew, K. M., He, X. Q., & Kitipornchai, S. (2004). Finite element method for the feedback control of FGM shells in the frequency domain via piezoelectric sensors and actuators. *Computer Methods in Applied Mechanics and Engineering, 193*, 257–273.

Lin, C. M., & Hsu, C. F. (2004). Adaptive fuzzy sliding-mode control for induction servomotor systems. *IEEE Transactions on Energy Conversion, 19*(2), 362–368.

Lin, F.-J., & Wai, R.-J. (2003). Adaptive and fuzzy neural network sliding-mode controllers for motor-quick-return servomechanism. *Mechatronics, 13*, 477–506.

Moita, J. M. S., Correia, V. M. F., Martins, P. G., Mota Soares, C. M. M., & Mota Soares, C. A. (2006). Optimal design in vibration control of adaptive structures using a simulated annealing algorithm. *Composite Structures, 75*, 79–87.

Parashkevova, L., Ivanova, J., & Bontcheva, N. (2004). Optimal design of functionally graded plates with thermo-elastic plastic behaviour. *Comptes Rendus Mécanique*, *332*, 493–498.

Praveen, G. N., & Reddy, J. N. (1998). Nonlinear transitent thermoelastic analysis of functionally graded ceramic-metal plates. *International Journal of Solids and Structures*, *35*, 4457–4476.

Reddy, J. N. (2004). *Mechanics of laminated composite plates and shells: Theory and analysis* (2nd ed.). Boca Raton: CRC Press.

Robbins, D. H., & Reddy, J. N. (1991). Analysis of piezoelectrically actuated beams using a layer wise displacement theory. *Computers and Structures*, *41*, 265–279.

Slotine, J. J. E., & Li, W. (1991). *Applied nonlinear control.* EnglewoodCli4s, NJ: Prentice-Hall.

To, C. W. S., & Chen, T. (2007). Optimal control of random vibration in plate and shell structures with distributedpiezoelectric components. *International Journal of Mechanical Sciences*, *49*, 1389–1398.

Touloukian, Y. S. (1967). *Thermophysical properties of high temperature solid materials*. New York, NY: Macmillian.

Turteltaub, S. (2002). Functionally graded materials for prescribed field evolution. *Computer Methods in Applied Mechanics and Engineering*, *191*, 2283–2296.

Wai, R.-J., & Lee, J.-D. (2004). Intelligent motion control for linear piezoelectric ceramic motor drive. *IEEE Transactions on Systems, Man, and Cybernetics – Part B: Cybernetics*, *34*(5), 2100–2111.

Wai, R.-J., & Tu, C.-H. (2007). Design of total sliding-mode-based genetic algorithm control for hybrid resonant-driven linear piezoelectric ceramic motor. *IEEE Transactions on Power Electrics*, *22*(2), 563–575.

20

The effect of torque feedback exerted to driver's hands on vehicle handling – a hardware-in-the-loop approach

S. Samiee[a,b]*, A. Nahvi[a], S. Azadi[a], R. Kazemi[a], A.R. Hatamian Haghighi[a] and M.R. Ashouri[a]

[a]*Faculty of Mechanical Engineering, K.N. Toosi University of Technology, Tehran, Iran;* [b]*Institute of Automotive Engineering, Graz University of Technology, Graz, Austria*

In this paper, road forces on tire are exerted on driver's hands via an equivalent torque applied to the steering wheel using a hardware-in-the-loop method to analyze the effect of steering torque feedback in vehicle handling. An electrical torque-feedback steering system is used for experimental validation. A 14-degree-of-freedom vehicle dynamic model, including engine, tire, and steering system mechanism are simulated. The required inputs such as throttle angle, brake demand, and steering wheel angular position are transmitted to the computer via an I/O interface card. Tire forces and steering gear torque are solved. This torque is then sent via an I/O interface card to a DC motor connected to the steering shaft. All equations are solved in real time. To investigate the influence of torque feedback on vehicle handling, several experiments are executed on 25 users. For this purpose, an experimental protocol is defined. In the experiments, the users had to drive along a specific path with constant speed using the designed electrical torque-feedback steering system. During the tests, the driving pattern of each user was recorded and the simulator's instantaneous position was compared with its desirable value. The results show that the torque feedback improved the driver's perception from the surrounding environment and enables her/him to handle the vehicle satisfactorily.

Keywords: vehicle handling; electrical steering; hardware-in-the-Loop

1. Introduction

Virtual reality systems expose humans to simulated environments and through interaction with the five senses they can simulate physical presence of the real world for the users (Burdea & Coiffet, 2003). Generally, driving simulators are divided into three categories including high-cost, medium-cost and low-cost (Gregersen, Falkmer, Dol, & Pardo, 2001). About 81% of driving simulators are used in research projects and the remaining 19% serve as training simulations for novice drivers (Straus, 2005). The importance of steering system with torque feedback in driving simulators makes it preferable even in low-cost simulations. Such a system increases the environmental interaction between the user and the simulator, and intensifies the user's sense of immersion and enhances the process of driver training. It is not only essential for training of novice drivers, but also can be utilized for evaluation and validation of the electrical steer control designs in brand-new vehicles in the car industry. Car manufacturers can use this system to evaluate and modify their conceptual designs without bearing the cost of initial modeling.

Replacement of mechanical and hydraulic systems by electronic ones has led to the by-wire technology. Enhancement of efficiency, increased safety and reliability, and reduction of manufacturing costs are the main advantages of electronic steering systems. If the mechanical steering system is fully replaced by a steer-by-wire system, the following benefits are achieved:

(1) Removal of steering shaft simplifies the internal design of the vehicle. Removal of various parts allows more space around the engine. Hence, there will be fewer difficulties for installation of the combustion engine and the steering system can be designed and installed as small units.
(2) There is no direct physical connection between tires and steering wheel and in the case of accidents, fewer impacts are transferred to the driver via steering wheel.
(3) The specifications of the steering system can be easily changed. Hence, desirable steer response and steering feel can be achieved based on the specified needs.
(4) Safety can be improved by providing computer-controlled intervention of vehicle controls with systems such as Electronic Stability Control, Adaptive Cruise Control, and Lane Keeping System.

*Corresponding author. Email: S.samiee@tugraz.at

A considerable number of research works have already been carried out on steer-by-wire systems. Kim and Song (2002) worked on the electronic power steering (EPS) control system. They focused on designing a control logic, which was able to reduce the torque applied to the driver, produce a different steering feel, and improve return-to-center performance. They implemented their proposed approach on a hardware-in-the-loop system to validate it. A robust control technique was used by Chen and Ulsoy (2002) to control the EPS system and provide the driver with desirable driving experiences. They conducted experiments in both time and frequency domains. The obtained results revealed that the robust control technique can reduce a vehicle's lateral position deviation and provide slight improvement to the marginal stability of the vehicle. This controller was implemented on the steering system of a human-in-the-loop driving simulator. Park, Han, and Lee (2005) studied steer-by-wire system control. They intended to improve the driving experience and enhance a vehicle stability and investigated the advantages of the removal of the steering shaft in the steer-by-wire system and creation of more space around the engine. A state feedback controller was developed by Yih and Gerdes (2004) with lateral slip as the state variable to improve handling of the vehicle. To measure lateral slip, an observer was designed using steering torque, steering angle, and yaw angle rate. The state feedback controller was implemented on a vehicle with the steer-by-wire system. They estimated the steering torque using the steer system's motor current and calculated the yaw angle rate based on a linear model of the vehicle and the steering system. Segawa, Nakano, Nishihara, and Kumamoto (2001). designed a controller to improve the stability of the vehicles with a steer-by-wire system and implemented it on a driving simulator. They demonstrated that the proposed controller is able to reduce the effect of the lateral winds on the vehicle's stability. Moreover, based on some experiments, they showed that the controller improves vehicle's stability in the case of driving on low-friction roads where behavior of the vehicle is different from the driver's input command. Yao (2006) transformed various driving scenarios, such as experiencing different steering feel, return-to-center performance, and fast and accurate wheel angle tracking based on driver's input into the control problems. Modern control techniques, system parameter identification, and state variable estimation were used to execute these scenarios. Finally, Yao implemented different control loops on seven vehicles with steer-by-wire systems and eight hardware-in-the-loop systems.

This paper focuses on the effect of torque feedback – exerted to driver's hands – on vehicle handling. An electrical torque-feedback steering system is designed and implemented in a driving simulator for experimental tests. The developed system is a segment of a by-wire control system and hence can be used for the construction and control of the steer-by-wire systems in domestic vehicles. Other applications of the proposed system include tele-operation of vehicles and driver behavior investigations such as works done in Friedrichs and Yang (2010) and Samiee, Azadi, Kazemi, Nahvi, and Eichberger (2014).

In this paper, forces applied to the driver's hand via the steering wheel will be identified and reproduced for implementation on the hardware. Vehicle's dynamics is modeled using equations of motion. Dynamic equations of tires are also solved instantaneously to obtain friction forces. Furthermore, geometry of the steering mechanism and tires are simulated to transform the generated forces on the tires to an equivalent torque on the steering wheel. Then, the equivalent torque is sent to the motor driver using an interface card to produce the required torque by providing the equivalent current. The generated torque is measured by a sensor, installed on the output of the gearbox, and is sent back to the computer using another interface card. Hence, the generated torque is compared with the reference (desirable) value and the error signal is fed into the controller.

All of the aforementioned calculations and the data transfers between computer and hardware must be carried out in real time. The used hardware and software are able to perform real-time calculations, generate graphical pictures, and communicate with interface cards simultaneously. The vehicle is modeled by a 14-degree-of-freedom (DOF) model. The dynamic model of tires is of a Fiala type. Other dynamics of the system are ignored and only kinematics is modeled for the generation of torques on the steering wheel based on the generated tire forces. A classical proportional-integrator-derivative (PID) control is used in this paper which demonstrates favorable performance according to the experiments results.

2. Dynamic modeling of vehicle in ADAMS

In this paper, the virtual vehicle's dynamic equations, tires dynamic equations, and geometry of the front wheels are validated by a full car model with 251-DOF, implemented in ADAMS/Car, which had been previously validated using experimental results of a real car. The characteristics of the subsystems of the vehicle modeled in ADAMS are as follows:

- Front wheel's suspension system is of McPherson type with linear spring.
- The compound suspension system with nonlinear spring is considered for rear wheels.
- The steering system is of rack-pinion type.
- Disk-type braking system is modeled for each of the four wheels.
- The power generation and transmission units are modeled by a series of mathematical equations, with throttle, brake, and gear as the inputs and the torque on the wheels as the output.

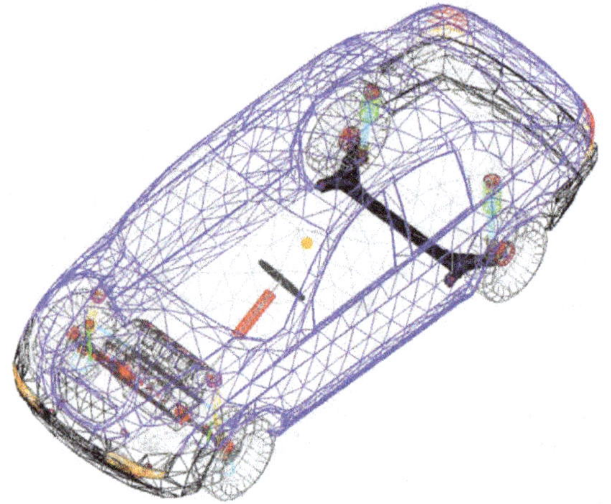

Figure 1. The vehicle's 251-DOF model implemented in ADAMS.

- The tires are represented by the Fiala model. The vehicle's body is modeled as a concentrated mass on the center of gravity of the vehicle.

By proper interconnection of the vehicle subsystems in ADAMS/Car, the full dynamic model of the vehicle is obtained. Figure 1 illustrates the vehicle model.

3. Simulation of vehicle's dynamics

A 14-DOF model is used to represent the vehicle's dynamics. The main advantage of the 14-DOF model against other simpler models (e.g. three- and seven-DOF models) is that it is composed of full six-DOF models for vehicle's body and suspension system dynamics. Figure 2 shows the suspension system of the 14-DOF model. It is assumed that each wheel has its own independent suspension system. For each wheel, there is a spring and damper, located between the body and suspension arm.

The parameters in Figure 2 are defined as follows:

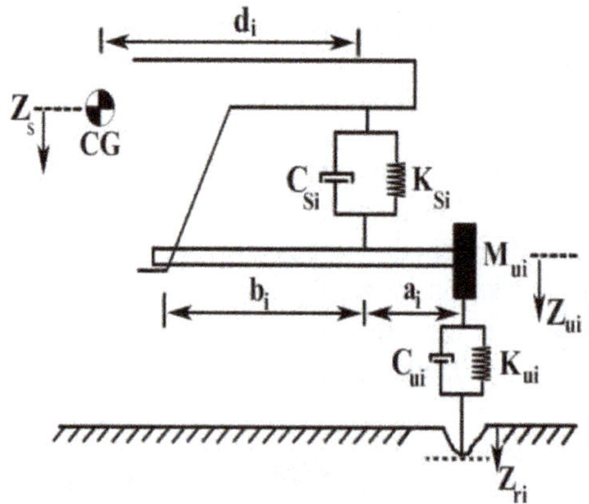

Figure 2. Suspension system of the 14-DOF model.

- The distance from wheel i to the connection point of spring and damper to the suspension arm is indicated by a_i.
- The distance between the connection point of the spring and damper to the suspension arm of wheel i and the connection point of the suspension arm to the vehicle's body is denoted by b_i.
- The distance from the spring and damper of wheel i to the longitudinal axis is shown by d_i.
- Parameters M_{ui} and Z_{ui} represent the ith unsprung mass and its height from the ground, respectively, and Z_{ri} indicates the height of the road beneath the ith wheel.
- Parameters K_{ui} and C_{ui} are the spring stiffness and equivalent damping coefficient of the ith tire, respectively.
- The spring stiffness and equivalent damping coefficient of the ith wheel are designated by K_{si} and C_{si}, respectively.
- And Z_s shows the distance from the center of gravity to the ground.

After deriving the equations of linear and angular motion for the vehicle and some algebraic manipulations, the 14-DOF equations of motion can be expressed as follows (Bastow, Howard, & Whitehead, 2004).

3.1. *Longitudinal equations of vehicle*

Using Newton's second law, the longitudinal dynamic equation of the vehicle can be presented as Equation (1),

$$M_t(\dot{u} + qw - rv) = X_1 + X_2 + X_3 + X_4 - F_{ax}, \quad (1)$$

where M_t is the total vehicle mass, $\dot{u}$ is the longitudinal acceleration of the vehicle's center of gravity, q is the pitch rate, w is the vertical velocity of the vehicle's center of gravity, r is the yaw rate, v is the lateral velocity of the vehicle's center of gravity, and F_{ax} is longitudinal air resistance force. In addition, X_i is found from Equation (2),

$$X_i = F_{xi}\cos(\delta_i) - F_{yi}\sin(\delta_i) \quad (2)$$

where $\delta_1 = \delta_{fr}$ is the steering angle of the front right wheel, $\delta_2 = \delta_{fl}$ is the steering angle of the front left wheel, and $\delta_3 = \delta_4 = 0$ are the steering angles of rear wheels, which are always equal to zero. Also, F_{xi} and F_{yi} represent the applied longitudinal and lateral forces on the ith tire, respectively.

3.2. *Lateral equations of vehicle*

Using Newton's second law, the lateral dynamic equation of the vehicle can be presented as Equation (3),

$$M_t(\dot{v} + ru - pw) = Y_1 + Y_2 + Y_3 + Y_4 - F_{ay}, \quad (3)$$

where p indicates the roll rate and F_{ay} shows the lateral air resistance force. Furthermore, Y_i can be stated as

Equation (4)

$$Y_i = F_{xi}\sin(\delta_i) + F_{yi}\cos(\delta_i). \tag{4}$$

3.3. *Vertical equations of vehicle*

The vertical dynamic equation can be shown by

$$M_s(\dot{w} + pv - qu) = \sum_{i=1}^{4} \frac{F_{si}}{R_{ri}}, \tag{5}$$

where M_s represents the sprung mass and F_{si} is the suspension force of wheel i, expressed by Equation (6),

$$\begin{aligned}
F_{s1} &= K_{s1}(Z_{u1} - Z_s + L_f\sin(\theta) + d_1\sin(\varphi)) + C_{s1}(w_{u1} - \\
&\quad w + L_f q\cos(\theta) + d_1 p\cos(\varphi)),\\
F_{s2} &= K_{s2}(Z_{u2} - Z_s + L_f\sin(\theta) - d_2\sin(\varphi)) + C_{s2}(w_{u2}\\
&\quad - w + L_f q\cos(\theta) - d_2 p\cos(\varphi)),\\
F_{s3} &= K_{s3}(Z_{u3} - Z_s - L_r\sin(\theta) + d_3\sin(\varphi)) + C_{s3}(w_{u3}\\
&\quad - w - L_r q\cos(\theta) + d_3 p\cos(\varphi)),\\
F_{s4} &= K_{s4}(Z_{u4} - Z_s - L_r\sin(\theta) - d_4\sin(\varphi)) + C_{s4}(w_{u4}\\
&\quad - w - L_r q\cos(\theta) - d_4 p\cos(\varphi)),
\end{aligned} \tag{6}$$

where θ and φ are pitch and roll angles of vehicle's body, respectively, and L_f and L_r indicate the distance of the vehicle's center of gravity from the front and rear axles, respectively. R_{ri} is a coefficient related to the geometry of the suspension system and is found from Equation (7),

$$R_{ri} = \frac{a_i + b_i}{b_i}. \tag{7}$$

3.4. *Roll angle*

Angular equations of vehicle around longitudinal axle represent Equation (8).

$$I_{xs}\dot{P} + (I_{zs} - I_{ys})\mathrm{qr} = \sum_{i=1}^{4} \overline{R_{ri}} F_{si} - h_{cg} \sum_{i=1}^{4} Y_i, \tag{8}$$

where h_{cg} is the distance of vehicle's center of gravity from ground in standstill and I_{xs}, I_{ys}, and I_{zs} indicate the moment of inertia of the sprung mass around longitudinal, lateral, and vertical axles, respectively. In addition, $\overline{R_{ri}}$ is a coefficient related to the geometry of the suspension and,

$$\begin{aligned}
\overline{R}_{r1} &= -d_1 + (d_1 - b_1)\frac{a_1}{a_1 + b_1},\\
\overline{R}_{r2} &= d_2 - (d_2 - b_2)\frac{a_2}{a_2 + b_2},\\
\overline{R_{r3}} &= -d_3 + (d_3 - b_3)\frac{a_3}{a_3 + b_3},\\
\overline{R_{r4}} &= d_4 - (d_4 - b_4)\frac{a_4}{a_4 + b_4}.
\end{aligned} \tag{9}$$

3.5. *Pitch angle*

Vehicle's equations around lateral axle represent Equation (10)

$$\begin{aligned}
I_{ys}\dot{q} + (I_{xs} - I_{zs})pr &= -L_f\left(\frac{F_{s1}}{R_{r1}} + \frac{F_{s2}}{R_{r2}}\right) + L_r\left(\frac{F_{s3}}{R_{r3}} + \frac{F_{s4}}{R_{r4}}\right)\\
&\quad + h_{cg}\sum_{i=1}^{4} X_i\sqrt{2}.
\end{aligned} \tag{10}$$

3.6. *Yaw angle*

Vehicle's equations around vertical axle result in Equation (11)

$$\begin{aligned}
I_z\dot{r} + (I_y - I_x)pq &= L_f(Y_1 + Y_2) - L_r(Y_3 + Y_4)\\
&\quad + \frac{T_f}{2}(X_1 - X_2) + \frac{T_r}{2}(X_3 - X_4) + \sum_{i=1}^{4} M_{zi},
\end{aligned} \tag{11}$$

where T_f is the distance of front wheels from each other, T_r is the distance between rear wheels, and I_x, I_y, and I_z represent the moment of inertia around longitudinal, lateral, and vertical axles, respectively.

3.7. *Unsprung masses (four equations)*

Equations of motion for unsprung masses lead to Equation (12).

$$M_{ui}\dot{w}_{ui} = K_{ui}(Z_{ri} - Z_{ui}) + C_{ui}(w_{ri} - w_{ui}) - \frac{F_{si}}{R_{ri}}, \tag{12}$$

where w_{ri} indicates the rate of change of the road input and w_{ui} is the rate of change of the motion, which is perpendicular to the unsprung mass.

3.8. *Wheels dynamics (four equations)*

Considering one DOF for each tire results in Equation (13)

$$I_{wi}\dot{\omega}_i = T_{wi} - R_w F_{xi} - T_{Ri} - T_b \tag{13}$$

where R_w is the effective radius of the tires, and T_b and T_{Ri} represent the braking and resistant torques, respectively. The dynamic parameters of the vehicle are presented in Table 1. The 14-DOF model is obtained after performing some algebraic manipulations on the linear and angular equations of motion. To validate the 14-DOF model, a full vehicle model in ADAMS/Car was used.

Two different maneuvers were applied to both the models. In the first maneuver, the vehicle was traveling with a constant velocity of 70 km/h and then a sinusoidal displacement with the frequency of 0.2 Hz and amplitude of 70 degrees are applied to the steering wheel. In the second maneuver, a step input of 70 degrees was applied to

Table 1. Dynamic parameters of the vehicle.

Distance of vehicle's center of gravity from front axle	$l_f = 1$ (m)
Distance of vehicle's center of gravity from rear axle	$l_r = 2.6$ (m)
Distance between front wheels	$T_f = 1.401$ (m)
Distance between rear wheels	$T_r = 1.408$ (m)
Distance of vehicle's center of gravity from ground in standstill	$h_{cg} = 0.497$ (m)
Mass of vehicle	$M_t = 924$ (kg)
Coefficient of front spring	$K_{s1} = 17,650$ (N/m)
Coefficient of rear spring	$K_{s2} = 22,600$ (N/m)
Coefficient of front damper	$C_{s1} = 1600$ (N s/m)
Coefficient of rear damper	$C_{s2} = 3730$ (N s/m)
Steer ratio	$N = 1 : 17.5$

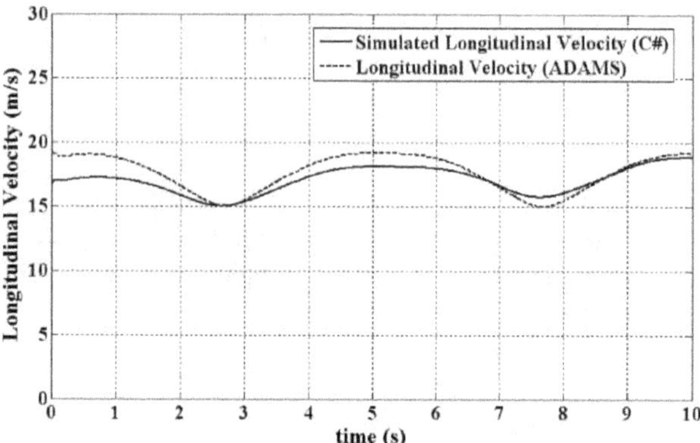

Figure 3. Longitudinal velocity of 14-DOF and ADAMS models (sinusoidal input of steering wheel).

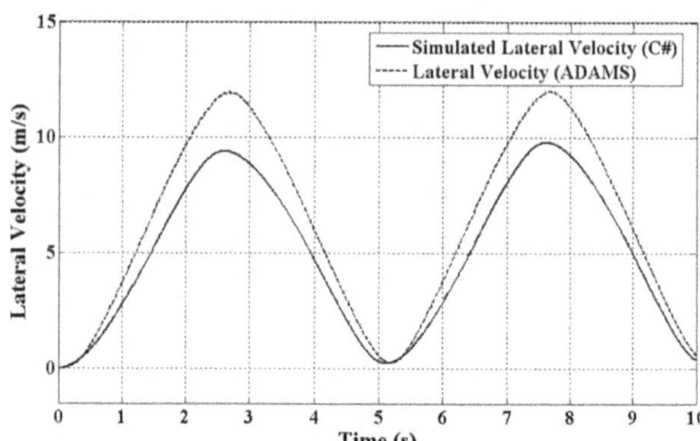

Figure 4. Lateral velocity of 14-DOF and ADAMS models (sinusoidal input of steering wheel).

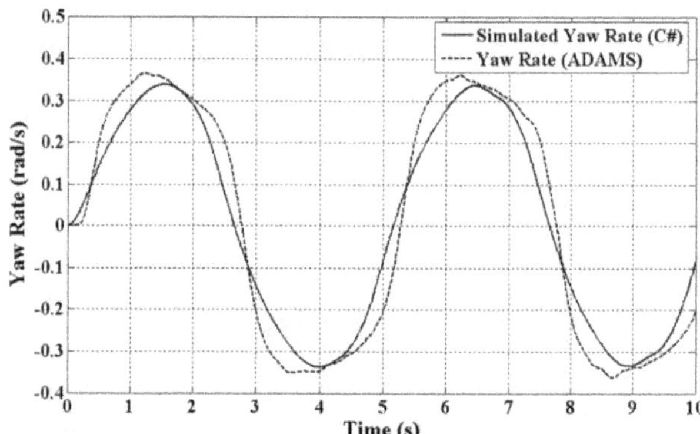

Figure 5. Yaw rate of 14-DOF and ADAMS models (sinusoidal input of steering wheel).

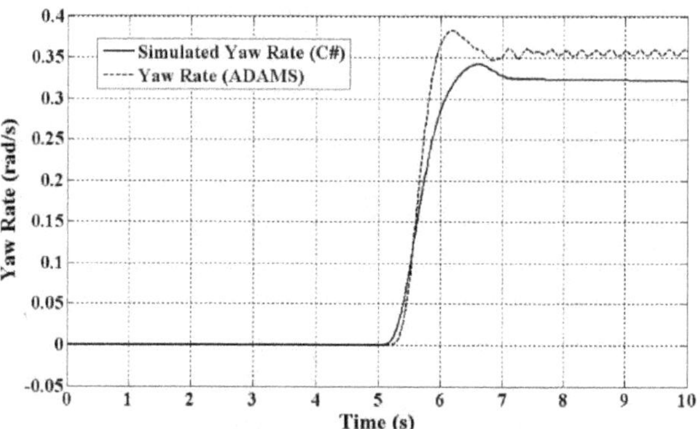

Figure 6. Yaw rate of 14-DOF and ADAMS models (step input of steering wheel).

the steering wheel of the vehicle, traveling with a constant velocity of 70 km/h.

Figures 3–6 show the results of the comparison between the dynamic parameters of both the models, including the longitudinal and lateral velocities and the rate of yaw angle. These figures demonstrate a close match between the outputs of both the models. The maximum longitudinal and lateral velocity errors for both the models are 3 m/s. Moreover, the 14-DOF model has accurately generated the rate of change of the yaw angle. The maximum error for the 14-DOF model in the case of step input maneuvering is 0.05 rad/s.

4. Dynamic simulation of tire

Correct modeling of tires significantly influences the process of the vehicle dynamic simulation. The applied forces are transferred to the vehicle through the tires. Therefore, the accurate calculation of the tire forces is the very first step toward the simulation of the vehicle. Different tire models have been proposed, which produce dependable estimations for tire forces and torques if the range of applied forces is not broad and tire deflection is slight. This paper uses the Fiala model to simulate the tire. In this model, it is assumed that tire contact to the road is rectangular and camber angle does not affect forces. The Fiala model is described as follows.

4.1. Calculation of longitudinal slip angle of tire

Longitudinal slip happens due to the difference between the velocity of the tire center and the contact patch center. According to Figure 7, theoretical calculation of longitudinal slip is as follows:

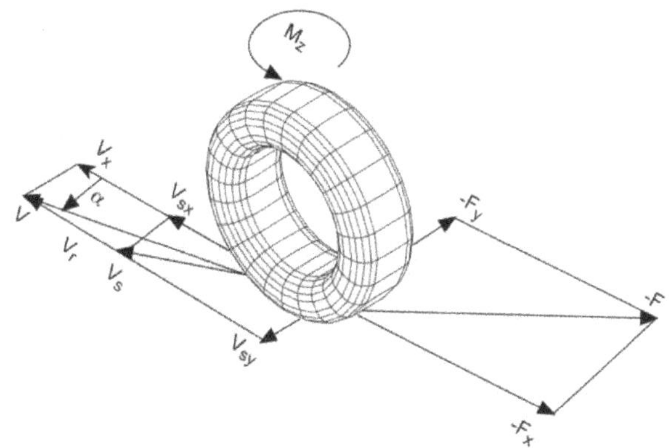

Figure 7. Velocity and force vectors directions (MSC Software Corporation, 2003).

The Fiala model uses the following equations for stable estimation of the slip.

4.1.1. Acceleration slip

$$S_s = \frac{V_r - V_x}{V_r} = \frac{(\omega_{\text{Actual}} - \omega_{\text{Free}})}{\omega_{\text{Actual}}}. \tag{14}$$

4.1.2. Braking slip

$$S_s = \frac{V_r - V_x}{V_x} = \frac{(\omega_{\text{Actual}} - \omega_{\text{Free}})}{\omega_{\text{Free}}}, \tag{15}$$

where V_r is the linear velocity of the contact point between tire and the road, obtained from Equation (16) and V_x indicates the image of the velocity of wheel center along the longitudinal axle of the wheel and is found from Equations (17) and (18)

$$V_r = \omega R_w, \tag{16}$$

$$\begin{aligned} V_{x\text{fr}} &= V_x - 0.5\text{r} \cdot T_f \cos\delta_{\text{fr}} + (V_y + r \cdot L_f) \sin \delta_{\text{fr}}, \\ V_{x\text{fl}} &= V_x + 0.5\text{r} \cdot T_f \cos\delta_{\text{fl}} + (V_y + r \cdot L_f) \sin \delta_{\text{fl}}, \end{aligned} \tag{17}$$

$$\begin{aligned} V_{x\text{rr}} &= (V_x - 0.5r \cdot T_r), \\ V_{x\text{rl}} &= (V_x + 0.5r \cdot T_r). \end{aligned} \tag{18}$$

In these equations, $V_{x\text{fr}}$, $V_{x\text{fl}}$, $V_{x\text{rr}}$, and $V_{x\text{lr}}$ represent the image of the velocity of wheel center along the longitudinal axle of the front right, front left, rear right, and rear left wheels, respectively, and R_w is the radius of the unloaded wheel.

4.2. Calculation of lateral slip angle of tire

The lateral slip angle of the tires is found from the difference between the longitudinal axle of tire and the axis of velocity vector on the tire plane (α in Figure 8).

The sign of the angle is positive in the upward direction and negative in the downward direction. The slip can be stated based on the state variables such as longitudinal,

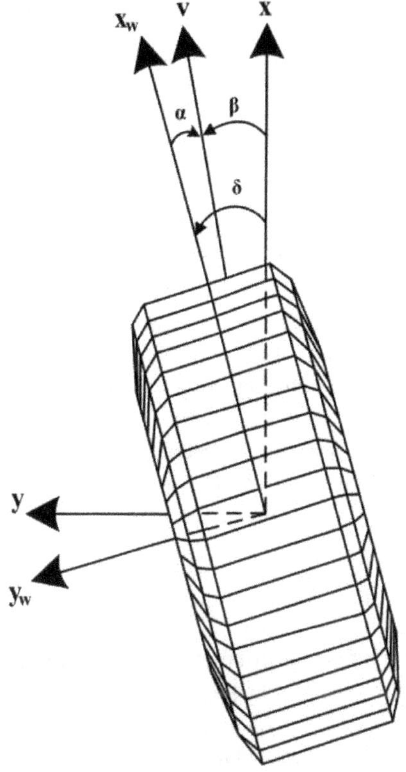

Figure 8. Slip angle of the front tire.

lateral, and yaw velocities. The lateral slip is expressed by Equation (19)

$$\begin{aligned} \alpha_{\text{fr}} &= \delta_{\text{fr}} - a\tan\left(\frac{(V_y + r \cdot L_f)}{(V_x - 0.5r \cdot T_f)}\right), \\ \alpha_{\text{fl}} &= \delta_{\text{fl}} - a\tan\left(\frac{(V_y + r \cdot L_f)}{(V_x + 0.5r \cdot T_f)}\right), \\ \alpha_{\text{rr}} &= -a\tan\left(\frac{(V_y - r \cdot L_r)}{(V_x - 0.5r \cdot T_r)}\right), \\ \alpha_{\text{rl}} &= -a\tan\left(\frac{(V_y - r \cdot L_f)}{(V_x + 0.5r \cdot T_r)}\right). \end{aligned} \tag{19}$$

4.3. Calculation of vertical force

In the Society of Automotive Engineers coordinate system, vector $+Z$ faces downwards (towards the road). Hence, the vertical force of the road on the tire in the contact point is always negative. The vertical force is expressed by Equation (20),

$$F_z = \min(0.0, (F_{ZK} + F_{ZC})), \tag{20}$$

where F_{ZK} is the vertical force due to the vertical stiffness of the tire and is found from Equation (21):

$$F_{ZK} = -\text{VS} \times \text{TD}, \tag{21}$$

where VS and TD indicate vertical stiffness and tire deflection, respectively. Moreover, F_{ZC} is the vertical force due to the tire damping, expressed by Equation (22):

$$F_{zc} = -\mathrm{VD} \times \frac{\mathrm{d}}{\mathrm{d}t}(\mathrm{TD}), \quad (22)$$

where VD represents vertical damping of the tire.

4.4. *Calculation of longitudinal force*

The longitudinal force on the tires in any instance depends on the vertical force, F_Z, instantaneous friction coefficient, U, the ratio of the longitudinal slip, S_s, and the lateral slip angle α in that instance. Also, total slip ($S_{s\alpha}$), instantaneous friction coefficient (U), and critical longitudinal slip (S_{Critical}) must be calculated prior to calculation of the longitudinal force. The total slip is expressed by Equation (23),

$$S_{s\alpha} = \sqrt{S_s^2 + \tan^2(\alpha)}. \quad (23)$$

The instantaneous friction coefficient depends on the static friction coefficient ($U_{\max}$), dynamic friction coefficient ($U_{\min}$), and the total slip. The $U_{\max}$ represents the friction coefficient between the tire and the road at zero slip, which does not happen in practice as there always exists non-zero slip. On the other hand, $U_{\min}$ indicates friction coefficient between the tire and the road at 100% slip

$$U = U_{\min} + (U_{\max} - U_{\min}) \times S_{s\alpha}. \quad (24)$$

The critical slip (S_{Cr}) can be expressed by Equation (25)

$$S_{\mathrm{Cr}} = \left| \frac{U \times F_Z}{2 \times C_{\mathrm{Sl}}} \right|, \quad (25)$$

where C_{Sl} is the longitudinal slip constant found by the limit of changes in F_x with respect to S_s while slip tends to go to zero. For the critical mode, that is, when the slip is below its critical value ($|S_S| < S_{\mathrm{Cr}}$), the longitudinal force is calculated by Equation (26):

$$F_X = -C_{\mathrm{Sl}} \times S_S. \quad (26)$$

For the full slip mode when slip is higher than its critical value ($|S_S| > S_{\mathrm{Cr}}$), the longitudinal force is given by

$$F_X = -(F_{X1} - F_{X2}) \times \mathrm{sign}(S_S), \quad (27)$$

where $F_{X1}$and F_{X2} are stated as,

$$F_{X1} = U \times F_Z, \quad (28)$$

$$F_{X2} = \left| \frac{(U \times F_Z)^2}{4 \times |S_S| \times C_{\mathrm{Slip}}} \right|. \quad (29)$$

4.5. *Calculation of lateral force*

The lateral force on a tire is a function of the vertical force and the friction coefficient. Similar to the calculation of the longitudinal force, the Fiala model uses a critical lateral slip (α_{Cr}), to obtain the lateral force,

$$\alpha_{\mathrm{Cr}} = a\tan\left(\frac{3 \times U \times |F_Z|}{C_\alpha}\right), \quad (30)$$

where C_α is the lateral stiffness, found by the limit of changes in F_y with respect to α when α is close to zero. When the slip angle α reaches its critical value (α_{Cr}), the lateral force takes its maximum value ($U \times |F_Z|$). Furthermore, if $|\alpha| \leq \alpha_{\mathrm{Critical}}$ then F_y is given by

$$F_y = -U \times |F_z| \times (1 - H^3) \times \mathrm{sign}(\alpha), \quad (31)$$

where H is expressed by

$$H = 1 - \frac{C_\alpha \times |\tan(\alpha)|}{3 \times U \times |F_Z|}. \quad (32)$$

For the full slip mode, that is, $|\alpha| > \alpha_{\mathrm{Cl}}$, F_y is given by

$$F_y = -U \times |F_Z| \times \mathrm{sign}(\alpha). \quad (33)$$

Table 2 summarizes some dynamic parameters of the tire.

Dynamic equations of the tire were coded in C# and then two aforementioned maneuvers were used for validation. The forces applied to the tire from the road, derived from dynamic equations, were compared with those obtained through the full car model in ADAMS.

Table 2. Dynamic parameters of tire.

Unload tire radius	$R_\omega = 0.283$ (m)
Mass of tire	$M_{\mathrm{tire}} = 14.9$ (kg)
Friction coefficient at full slip	$U_{\min} = 0.7$
Friction coefficient at zero slip	$U_{\max} = 0.9$
Vertical damping factor of tire	VD = 153,000 (N/m)
Vertical stiffness factor of tire	VS = 100 (N s/m)

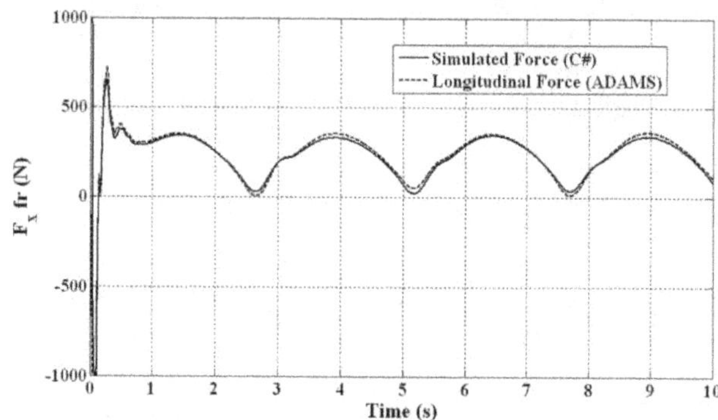

Figure 9. Longitudinal force on the tire of ADAMS and C# models (sinusoidal input of the steering wheel).

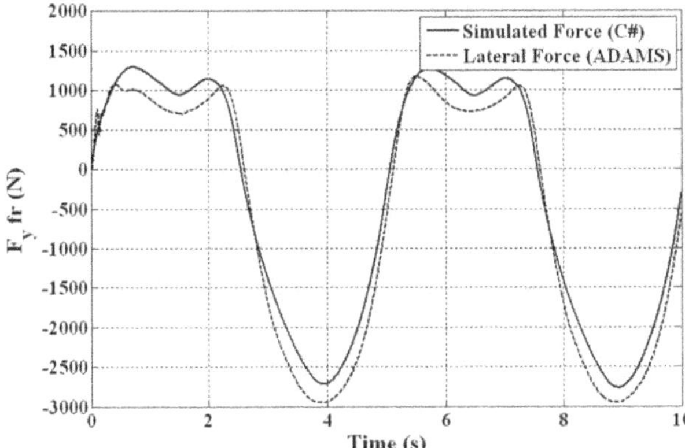

Figure 10. Lateral force on the tire of ADAMS and C# models (sinusoidal input of the steering wheel).

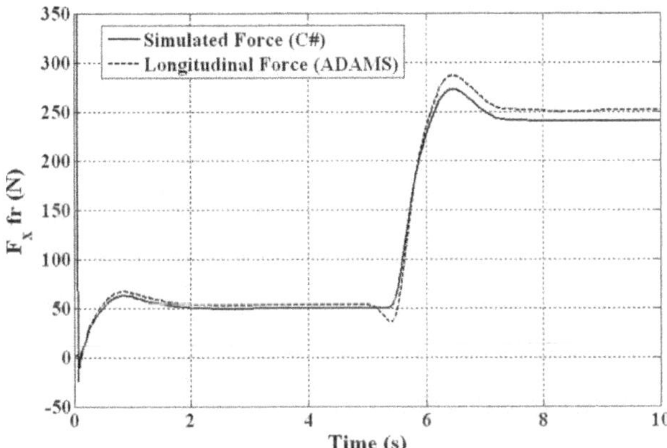

Figure 11. Longitudinal force on the tire of ADAMS and C# models (step input of the steering wheel).

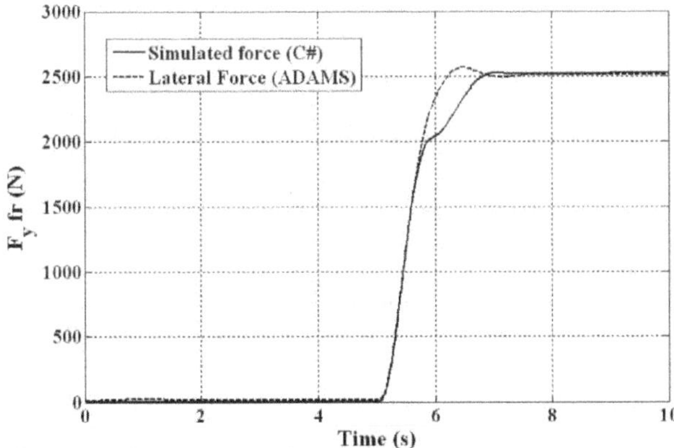

Figure 12. Lateral force on the tire of ADAMS and C# models (step input of the steering wheel).

Figures 9–12 illustrate the results of comparison of the tire longitudinal and lateral forces.

It is clear that the simulated model has accurately produced longitudinal and lateral forces. The maximum error for the lateral force is 200 N, obtained in the sinusoidal input maneuvering. The average error over one cycle is less than 100 N, indicating an error level below 5%. This value for longitudinal force is below 50 N.

5. Simulation of front wheels kinematics

The effect of tire forces and torques on the steering wheel must be calculated. In real vehicles, these forces and torques are transferred to the steering wheel through linkage of the front wheels. In the simulated steering system, the linkage connecting the steering box to the steering wheel is incorporated into the control system in a hardware-in-the-loop form and hence its dynamic modeling is not required. However, the linkage from the tire to the steering box does not exist and therefore must be simulated. To perform this simulation, some simplifications are considered to reduce computational burden and allow real-time implementation with insignificant computational error. These simplifications are as follows:

- Wheel angle δ is calculated by dividing the steering angle by the pinion ratio. Hence, this angle is identical for both wheels. For the real car, the average value of the left and the right wheel's angle is considered.
- To calculate the torques on the steering wheel, only the influence of the longitudinal and lateral forces, applied from the road to the tire, and the steering angle are taken into account and the effect of other forces and torques is ignored.
- The contact point of the kingpin axle to the ground is considered as the center of rotation.
- The tie-rod shaft is modeled as a bi-force element and its deflection during movement is ignored. Hence, the force applied on the tie-rod is equivalent to the rack force.
- The ground-tire contact point is always constant.

After simplifications, the diagram of the applied forces on the tire is developed as illustrated in Figure 13.

Based on this figure, rack forces are obtained by vector summation of the forces from wheels on the tie-rod.

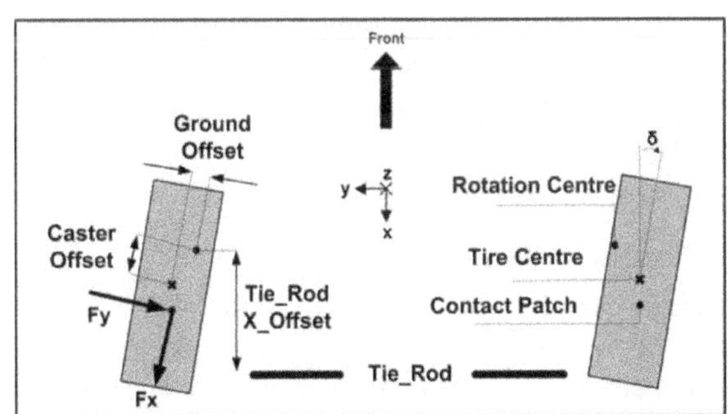

Figure 13. Diagram of the applied forces on the tire.

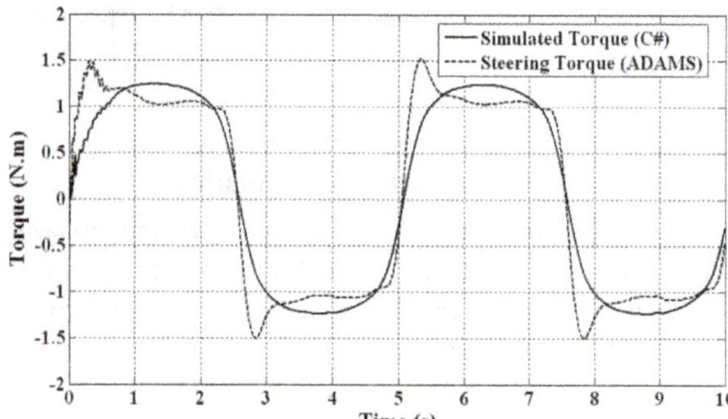

Figure 14. Torques on steering wheel of ADAMS and C# models (sinusoidal input of steering wheel).

The forces on the tie-rod are calculated by equations in Equation (34),

$$A = \left(\frac{1}{\mathrm{TO}_x + \mathrm{CO}}\right),$$
$$B = (F_y + (F_x \cdot \sin\delta))(\mathrm{CO_x} + \mathrm{Caster\_Offset}), \quad (34)$$
$$C = (F_x \cdot \cos\delta \cdot \mathrm{Ground\_Offset}).$$

where TO_x is the vertical distance of the tie-rod from the longitudinal rotation center of the tire and CO_x indicates the vertical distance of the tire–road contact center from the longitudinal rotation center of the tire.

5.1. *Simulation for the left wheel*

$$(F_{\mathrm{Tie\_Rod}})_{\mathrm{l}} = A(B + C). \quad (35)$$

5.2. *Simulation for the right wheel*

$$(F_{\mathrm{Tie\_Rod}})_{\mathrm{r}} = A(B - C). \quad (36)$$

The geometry of the front wheels was validated using the maneuvers described in Section 3. The developed torques on the pinion of the steering box were compared with those obtained by the ADAMS model during the maneuvers. Figures 14 and 15 illustrate this comparison.

It is seen from these figures that the maximum error for both the maneuvers is about 0.2 Nm. This error can be justified by the fact that only the effect of the longitudinal and lateral forces on development of torques on the steer is considered in the simulated model and the influence of other forces and torques, such as vertical force and rolling resistance torque, is ignored.

6. Hardware-in-the-loop implementation

In this paper, dynamic equations of the vehicle, friction force on tires and the resulting torques on the steering box are simulated in the computer, while elements of the steering system such as torque actuator, sensors, steering shaft, and steering wheel are physically realized in

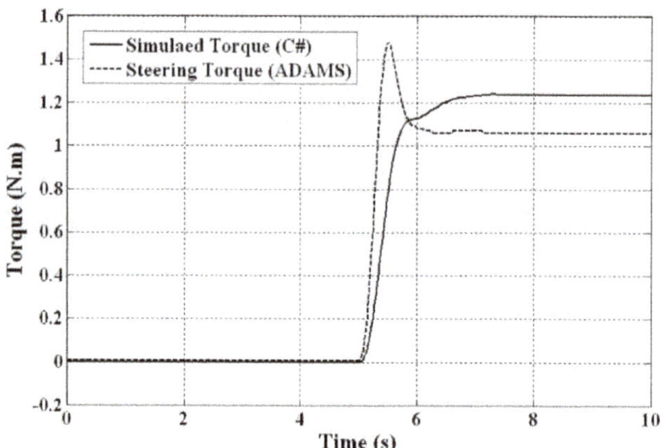

Figure 15. Torques on steering wheel of ADAMS and C# models (step input of steering wheel).

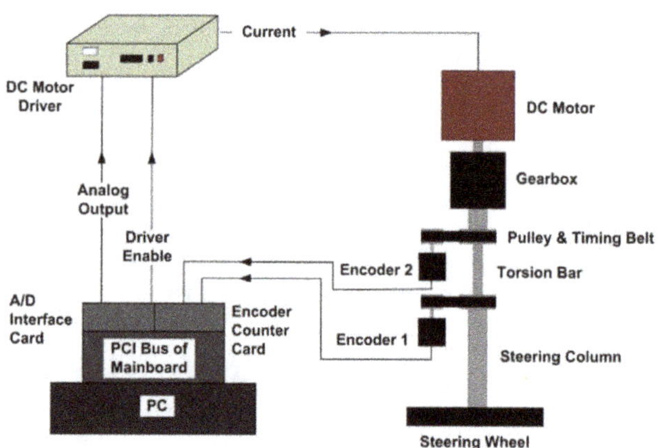

Figure 16. Hardware-in-the-loop system.

the hardware-in-the-loop system. An illustration of the hardware-in-the-loop system is shown in Figure 16.

This system uses a brushed DC motor, which is directly connected to the end of the steering shaft and produces the torques developed by the friction forces between the tire and the road. The used DC motor has the rated power of 400 W and produced 3.1 Nm torque at the maximum current of 16 A.

The simulated dynamic systems required the instantaneous angular position of the steering wheel to calculate the steering torques. Moreover, the control system needs the motor torque to properly control the applied torque. These requirements are fulfilled by the use of a torque sensor, which is shown in Figure 17.

In this scheme, the angular positions of both ends of the torsion bar are estimated by two 12-bit encoders. The changes in the angles of both ends of the torsion bar are transferred by pulleys and timing belts to the encoders. The pulleys have a 1:2 diameter ratio and are connected to high-resolution encoders (3600 pulse/revolution). The torque sensor resolution is 0.0375 Nm. The sensor works based on a simple principle that the applied torque on a rod is proportional to the difference in rotational angles of the

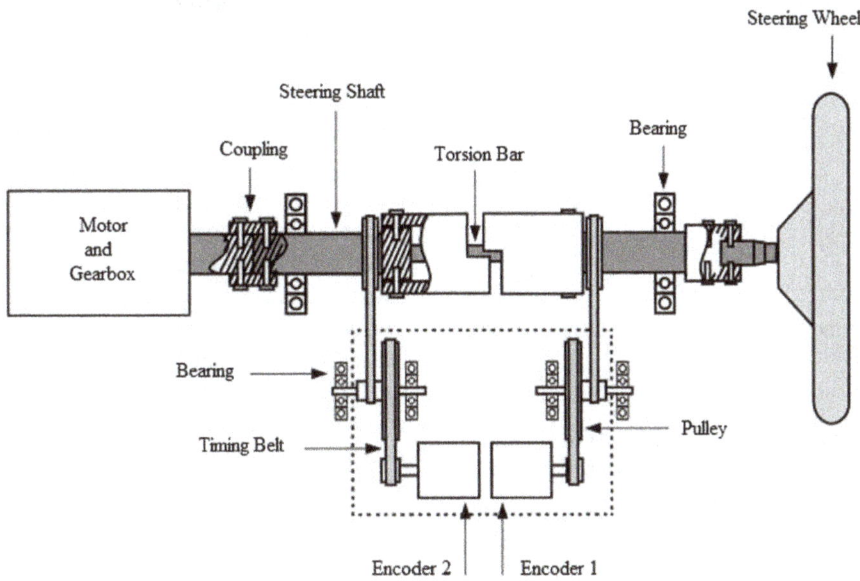

Figure 17. Torque sensor placement in the system.

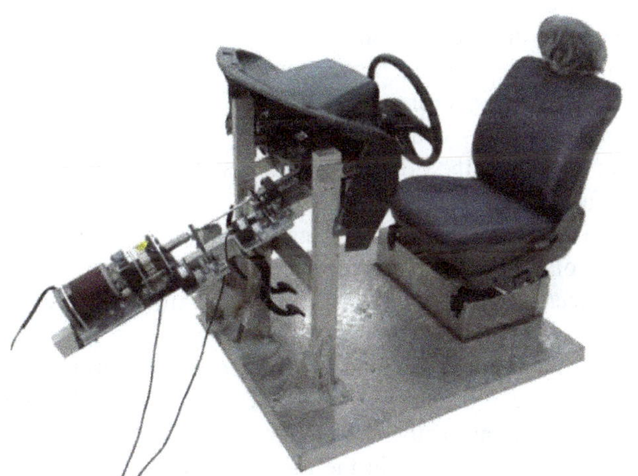

Figure 18. The hardware system made at the laboratory.

Figure 19. Implemented torque sensor and its components.

rod's two ends (Beer & Johnston, 2008),

$$T_{gb} = K_{tb} \cdot \Delta\theta = K_{tb} \cdot (\theta_{sw} - \theta_{gb}), \qquad (37)$$

where T_{gb} is the output torque of the gearbox, K_{tb} indicates stiffness coefficient of the torsion bar, which is about 186.6 N m./rad, and θ_{sw} and θ_{gb} are the rotation angles of the steering wheel and the gearbox, respectively. Hence, the motor torque is calculated and sent as a feedback signal to the computer. The picture of the implemented hardware system is shown in Figure 18 whereas a detailed overview of the torque sensor is illustrated in Figure 19.

7. Simulation of control system

Before hardware implementation of the control system, the controller was simulated in the computer to investigate its functionality and performance. Hence, all elements of the hardware system were modeled by equivalent dynamic equations as shown in Figure 20.

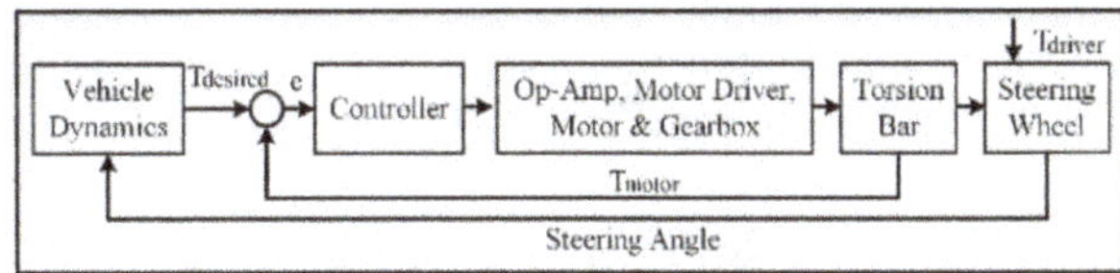

Figure 20. Control system block diagram.

In Figure 20, $T_{desired}$ is the desired torque which is calculated by the vehicle dynamic model and is compared with the motor torque (T_{motor}) and the generated error is fed to the controller. The controller's output signal is amplified by an operational amplifier and then sent to the driver

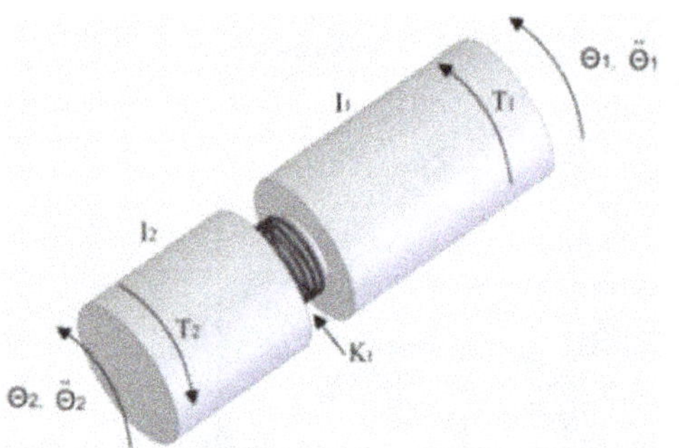

Figure 21. Simplified model of the steering system from the steering wheel to the pinion.

of the DC motor to produce the suitable current for generation of the desired torque. The motor torque is passed through the gearbox and then is measured by the torque sensor. The resultant of this torque and the driver torque, T_{driver}, are simultaneously applied to the steering wheel. The rotation angle of the steering wheel is the output of the system, which is re-used to solve the dynamic equations of the vehicle.

To derive the dynamic model of the steering system from the steering wheel to the connection point of motor and gearbox, the overall system is considered by two masses with rotational inertia. Since the mass and inertia of the torsion bar are negligible compared with other system's components, it is modeled by a torsional spring alone. Hence, the overall system is modeled by two rotational inertias interconnected by a torsion spring as shown in Figure 21.

In this figure, T_1 is equivalent to T_{driver}, applied by the driver to the steering wheel and T_2 is the motor torque after the gearbox and is expressed by

$$T_2 = T_m \cdot N_1 = T_{gb}. \tag{38}$$

Moreover, I_1 is the sum of the steering wheel inertia (I_{sw}) and the steering shaft inertia (I_{ss}). In addition, I_2 indicates the sum of the motor inertia (I_m), the gearbox inertia, I_{gb}, and the coupling inertia, $I_{coupling}$, between the gearbox shaft and the torsion bar (Craig, 2004),

$$I_1 = I_{sw} + I_{ss}, \tag{39}$$

$$I_2 = I_m \cdot N_1^2 + I_{gb} + I_{coupling}. \tag{40}$$

The equation governing DC motor is expressed by Equation (41) as stated in Dorf & Bishop (2004),

$$V_a = V_b + R_a I_a + L_a \frac{dI_a}{dt}, \tag{41}$$

where V_a, I_a, and R_a are armature voltage, current, and resistance, respectively, and L_a is the armature self-inductance. V_b indicates the induced voltage, which is proportional to motor speed. The relationship between the armature current and the output torque is expressed by

$$T_m = K_t I_a,$$

where K_t is the motor constant. The classical PID controller is used in this paper. To analyze the system response, a step input is applied. To find the controller coefficients, the Zeigler–Nichols approach is used. Then the coefficients are tuned for a more accurate response. Since the input of the actual system changes frequently, the transient response of the system is very important. To investigate the transient response, a sinusoidal input with 1 Hz frequency and unit amplitude is applied to the system. The system response reveals that the designed controller is able to desirably follow variable inputs.

8. Experimental results

After investigating the controller performance by simulation inputs, the control was implemented on hardware.

Before doing the main tests, and to evaluate systems performance in a virtual driving condition, a user was asked to use the system and follow the desired route, specified by graphical illustrations.

The desired steering shaft torque, obtained by dynamic equations, along with the real torque generated by the steering system of the driving simulator and measured by the torque sensor are shown in Figure 22.

The generated torque by the dynamic equations and controller is transferred to driver's hands through the motor connected to the steering shaft. The maximum delay for this case is below 0.05 s.

For more investigation of the steering system performance and the influence of torque feedback on driver's proper steering of the vehicle, various experiments were executed on 25 users in the driving simulator. To do so, first an experimental protocol and a driving scenario were defined. Then, the users had to drive on the route of Figure 23, using the steering system, as described in this paper.

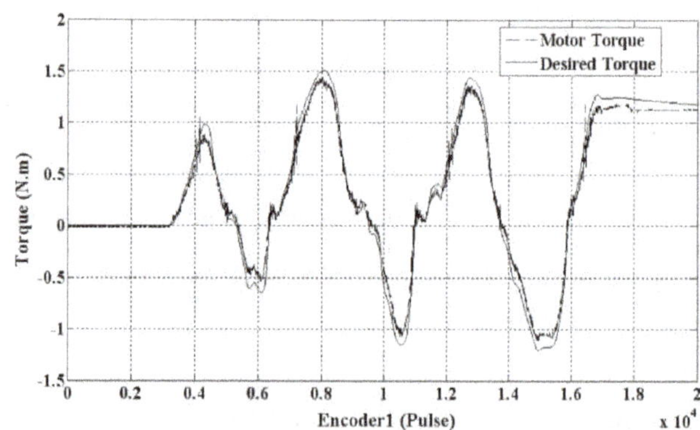

Figure 22. Desired torque and torque generated with the steering system's motor in a driving scenario.

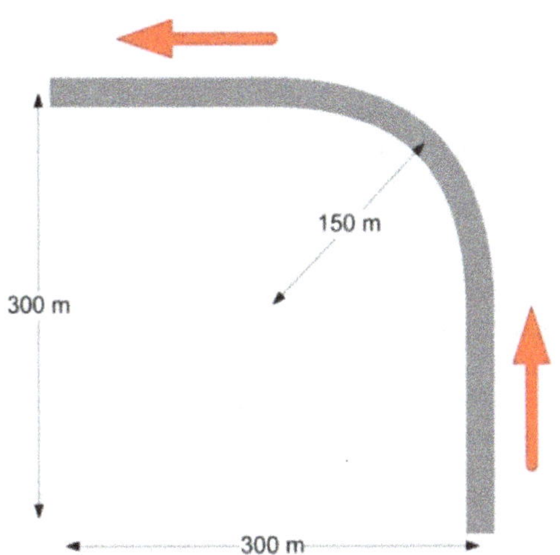

Figure 23. Schematic image of the path used in a driving scenario.

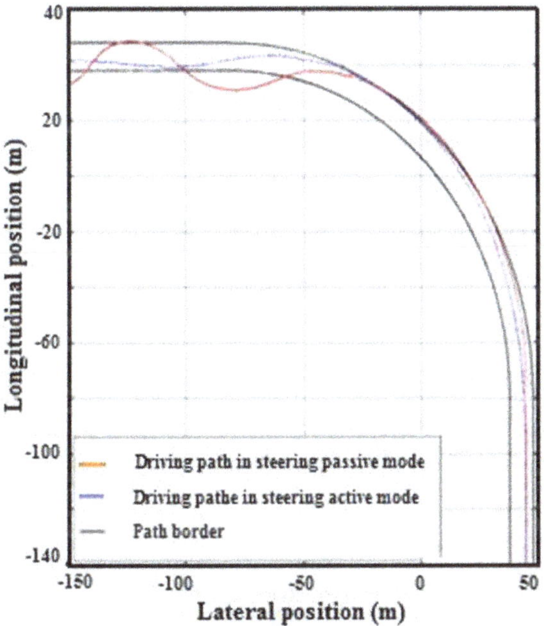

Figure 24. Different paths traveled by a chosen user while driving with two different steering modes.

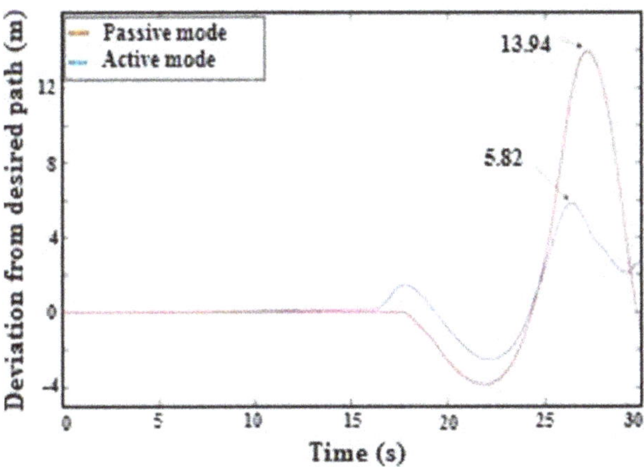

Figure 25. Deviation from desired position in two different steering modes (selected user result).

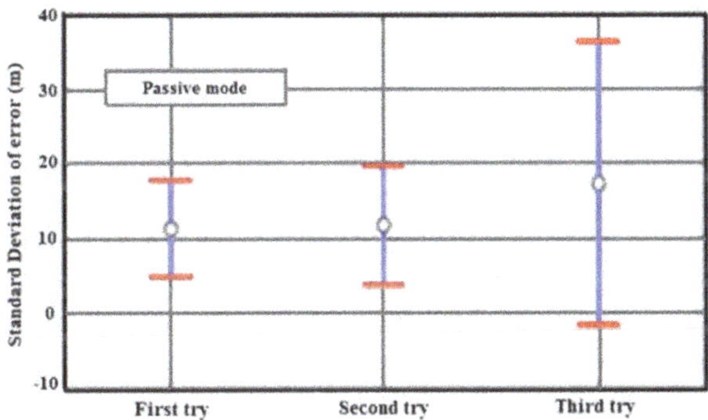

Figure 26. Maximum and standard deviation of error from the desired path (passive steering mode).

Before doing the main tests and to familiarize each participant with the path, each subject was asked to drive in the specified scenario once when there is torque feedback on the steering system (active mode) and another time when there is no torque feedback on the steering system (passive mode). For the main test, the users had to drive along the path with constant speed three times in the active steering mode and three times in the passive steering mode. During the experiment, the driving pattern of each user was recorded and the simulator's instantaneous position was compared with its desirable value (keeping within the road limits). Figure 24 shows the path passed by one of the users during the driving scenario for both active and inactive steering modes.

The deviation of the user from the desired position is shown in Figure 25. Obviously, it can be seen that there is a small deviation when the steering system with torque feedback is used.

To have a broader analysis, the results obtained for all users are considered. The average error of maximum deviations from the road for all 25 users and the standard deviation of this error for three experiments, in the passive mode, are shown in Figure 26.

In this figure, the white circles indicated the root mean square error of all the users in every experiment and the blue vertical lines show the standard deviation of the error.

It is seen that, on average, the maximum deviation of the users for the first, second, and third stage is 11.5, 12 and 17.2 m, respectively. Clearly, increasing the number of experiments has led to higher deviations due to the driver fatigue. The standard deviation of error has also increased with the number of experiments, which is not desirable.

The average of maximum errors during the three experiments in the active mode (with torque feedback) is depicted in Figure 27.

Comparing Figures 26 and 27 confirms the favorable influence of the torque feedback on error reduction.

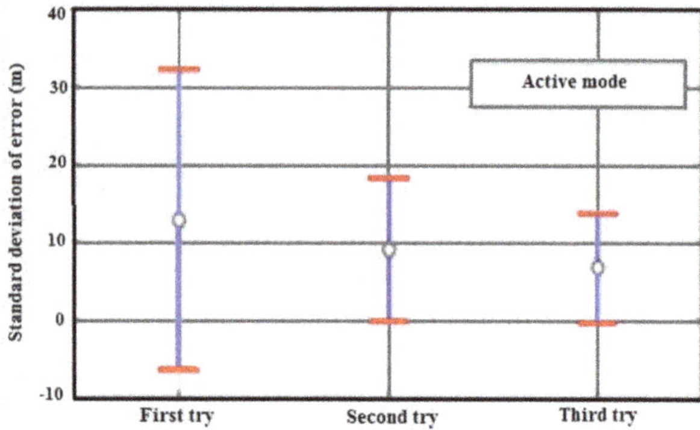

Figure 27. Maximum and standard deviation of error in comparison with the desired path (active steering mode).

Moreover, the torque feedback has improved users' performance during the experiments. For instance, the average of maximum errors in the first round of experiment is 13 m, whereas this value in the third round of experiments had been reduced to 7 m.

9. Conclusion

This paper presented the design, construction, and control of the electrical steering system for a driving simulator. The simulator can be used in driving simulation to have a deeper sense of immersion by the users as well as for construction of steer-by-wire systems. The system was then used to analyze the effect of torque feedback exerted to driver's hands on vehicle handling.

To produce the steering torques, the 14-DOF equations of the vehicle, the engine model, the dynamic equations of the wheels, and the steer geometry were simulated. The validity of all the models was investigated by comparison with a 251-DOF car model, developed and validated in ADAMS.

After construction of simulator, it was used by a user to travel along a specified path. The torque calculated by the dynamic equations was compared with those generated by the system actuator and it was seen that the system followed the generated torques. On the other hand, though the PID controller fulfilled the control requirements of the system, it was sensitive to variations of system parameters. Hence, using adaptive control schemes may improve system response.

Various experiments were conducted on several users to investigate the influence of the torque feedback on the steering wheel on proper steering of the vehicle. By comparing the results obtained with and without torque feedback during these experiments, it was shown that the torque feedback improved the driver's perception from the surrounding environment and enabled her/him to handle the vehicle satisfactorily.

Disclosure statement

No potential conflict of interest was reported by the authors.

References

Bastow, D., Howard, G., & Whitehead, J. P. (2004). *Car suspension and handling*. Warrendale, PA: SAE International.

Beer, F. P., & Johnston, E. R. (2008). *Mechanics of materials*. New York: McGraw Hill.

Burdea, G. C., & Coiffet, P. (2003). *Virtual reality technology*. Hoboken, NJ: John Wiley & Sons.

Chen, L. K., & Ulsoy, A. G. (2002). *Experimental validation of a robust steering assist controller on a driving simulator* (Vol. 3, pp. 2528–2533). Proceedings of the 2002 American Control Conference, Anchorage, AK, 2002.

Craig, J. J. (2004). *Introduction to robotics: Mechanics and control*. Upper Saddle River, NJ: Addison-Wesley.

Dorf, R. C., & Bishop, R. H. (2004). *Modern control systems*. Upper Saddle River, NJ: Prentice Hall.

Friedrichs, F., & Yang, B. (2010). *Drowsiness monitoring by steering and lane data based features under real driving conditions* (pp. 209–213). Paper presented at the 18th European Signal Processing Conference (EUSIPCO-2010), Denmark.

Gregersen, N. P., Falkmer, T., Dol, J., & Pardo, J. (2001). *Driving simulator scenarios and requirements*. EU; Competitive and Sustainable Growth Programme.

Kim, J. H., & Song, J. B. (2002). Control logic for an electric power steering system using assist motor. *Mechatronics*, *12*, 447–459.

MSC Software Corporation. (2003). MSC ADAMS 2003. *Help, Tire*.

Park, T., Han, C., & Lee, S. (2005). Development of the electronic control unit for the rack-actuating steer-by-wire using the hardware-in-the-loop simulation system. *Mechatronics*, *15*, 899–918.

Samiee, S., Azadi, S., Kazemi, R., Nahvi, A., & Eichberger, A. (2014). Data fusion to develop a driver drowsiness detection system with robustness to signal loss. *Sensors*, *14*, 17832–17847.

Segawa, M., Nakano, S., Nishihara, O., & Kumamoto, H. (2001). Vehicle stability control strategy for steer by wire system. *JSAE Review*, *22*, 383–388.

Straus, S. H. (2005). *New, improved, comprehensive, and automated driver's license test and vision screening system* (Report No. FHWA-AZ-04-559(1)). USA: U.S. Department of Transportation, Federal Highway Administration.

Yao, Y. (2006). *Vehicle steer-by-wire system control* (SAE Technical Paper). Detroit, MI: SAE.

Yih, P., & Gerdes, J. C. (2004). *Steer-by-wire for vehicle state estimation and control* (pp. 785–790). Paper presented at the Seventh International Symposium on Advanced Vehicle Control, The Netherlands.

21

Adaptive internal model-based suppression of torque ripple in brushless DC motor drives

Yoni Mandel[a*] and George Weiss[b]

[a]*Core Photonics, Tel Aviv 6971035, Israel;* [b]*School of EE, Tel Aviv University, Tel Aviv 6997801, Israel*

Permanent-magnet synchronous motors (PMSMs) are widely used as high-performance variable-speed drives. Ripple in the electric torque of such motors is often a source of vibration and tracking errors, especially at low speeds. We study the torque characteristics of PMSMs and propose a method to minimize the torque ripple. First, we establish a detailed model for the motor and present a Fourier analysis of the torque ripple, caused by the non-sinusoidal back electro-motive force (BEMF) and the cogging torque, where the main conclusion is that the frequencies present in the torque disturbance are integer multiples of six times the electric frequency. The resulting model is highly nonlinear. We propose an adaptive controller based on the internal model principle, where the resonant frequencies of the controller and the associated gains change according to the motor speed. This is achieved by replacing the time variable by the motor angle, which simplifies the nonlinear model. Our approach is passivity based and will work also for complex mechanical loads and several resonant frequencies. Simulation and experimental results are given to verify the new controller. We compare the performance of our adaptive algorithm with the well-known one from [Canudas de Wit, C., & Praly, L. (2000). Adaptive eccentricity compensation. IEEE Trans. Control Syst. Technol. 8, 757–766]. We find that it performs similarly in simple configurations, and it works also when the motor is part of a more complex system, for example, when the motor is connected to a load via a very flexible shaft.

Keywords: brushless DC; mutual torque; cogging torque; Park transformation; internal model; adaptive control

1. Introduction

A brushless DC motor (BLDCM) is a permanent-magnet synchronous machine (PMSM) connected to an inverter, usually operated by pulse width modulation and controlled by a processor that receives position measurements from a sensor (encoder) mounted on the motor shaft. BLDCMs offer high power density, reliability and efficiency. However, typical motors, driven by either rectangular or sine wave currents, exhibit torque pulsations or ripple.

At high speeds, torque ripple is mostly filtered out by the rotor inertia. However, at low speeds, the torque ripple produces undesirable speed variations, and inaccuracies in motion control. There are applications (e.g. computer tomography (Lee, Park, & Kwon, 2004), optics, machine tools (Jahns & Soong, 1996)) where very high precision is needed at low motor speeds.

In general, torque in a PMSM is developed by

- mutual torque,
- cogging torque and
- reluctance torque,

see Hanselman (1994) and Park, Park, Lee, and Harashima (2000). Imperfections of the motor geometry give rise to harmonics, which are usually multiples of six times the electric angular velocity, in the mutual torque (see also Degobert, Remy, Zeng, Barre, & Hautier, 2006; Hanselman, Hung, & Keshura, 1992). DC offset in the current sensors (and hence in the phase currents) leads to ripples in the mutual torque that are at the electric angular velocity, see the analysis in Gan and Qiu (2004).

There are two main control strategies for operating the inverter in a BLDCM: the "classical" one suits motors with trapezoidal back electro motive force (BEMF), and it produces rectangular current pulses lasting each for 60 electrical degrees. In this operation, only two phases are conducting at any time, but the torque will be close to constant, see Pillay and Krishnan (1989). Torque ripples will appear due to the imperfections of the current rectangles and of the BEMF – see Lu, Zhang, and Qu (2008) for methods to combat these ripples. The other strategy is by controlling the q component in order to control the electromagnetic torque. This leads to greater flexibility and it is better suited for PMSM with close to sinusoidal BEMF. In this paper, we consider only BLDCM operated under the latter (continuous) switching strategy.

Two general approaches have been proposed to reduce the torque ripple. One is to improve the motor's geometrical structure, see Hwang, Eom, Jung, Lee, and Kang (2001), Jahns and Soong (1996) and Ko and Kim (2004).

*Corresponding author. Email: mandel.yonatan@gmail.com

The second approach is to control the winding currents to overcome the disturbances, see for instance Ahn, Chen, and Dou (2005), Canudas de Wit and Praly (2000), Hung and Ding (1993), Le-Huy, Perret, and Feuillet (1986) and Malaize and Levine (2009). An interesting review of a wide range of design techniques for torque ripple minimization is in Jahns and Soong (1996). In Canudas de Wit and Praly (2000), an adaptive observer is designed to estimate the periodic torque disturbance. This estimate is then injected in order to cancel the disturbance. This algorithm needs a precise estimate of the friction in the system. Xu, Panda, Pan, Lee, and Lam (2004) use an iterative learning control (ILC) module in order to estimate the cyclic torque and record the reference current signals over one entire cycle. The ILC module then uses those signals to update the reference current for the next cycle. A gain-shaped sliding mode observer is used to estimate the torque ripple. In Gan and Qiu (2004), a gain scheduled robust two degree of freedom speed regulator, based on the internal model principle (IMP) and pole-zero placement, is developed to eliminate the torque ripples caused by DC offset (as discussed above). Their internal model changes its (single) resonant frequency according to the velocity reference. To prove local stability, the authors use linearization and the small gain theorem. Their proof uses the assumption that the disturbance frequency equals the reference electric velocity, which is debatable. In Hung and Ding (1993), a Fourier series decomposition is used to find a closed-form solution for the current harmonics that eliminate torque ripple and maximize efficiency simultaneously. In Park et al. (2000), the BEMF data, according to the rotor position, are measured and set up in a lookup table. Optimized reference current wave-forms are then obtained from the lookup table using the position and speed information from the shaft encoder. The motor currents are forced to track the reference currents. Petrović, Ortega, Stanković, and Tadmor (2000) design a passivity-based controller, relying on the principles of energy shaping and damping injection. The information about torque ripple harmonics used in the adaptation is extracted from the electrical subsystem, and a current controller is designed to achieve ripple minimization. Qian, Panda, and Xu (2004) implement an ILC scheme in the time domain to reduce periodic torque pulsations. A forgetting factor is introduced in this scheme to increase the robustness. However, this limits the extent to which torque pulsations can be suppressed. To eliminate this limitation, a modified ILC scheme is implemented in the frequency domain using Fourier series. Mattavelli, Tubiana, and Zigliotto (2005) propose the application of repetitive techniques to the current control in a field-oriented PMSM drive, where the q-axis current reference has been modified to achieve constant torque. Ferretti, Magnani, and Rocco (1998) present a compact model of the pulsating torque in a PMSM. An offline identification is proposed for the pulsating torque, to be used for suppressing the oscillations. In Ferretti, Magnani, and Rocco (1999), the technique of Ferretti et al. (1998) is extended to cope with variable motor speed using an adaptive compensator. Degobert et al. (2006), Ruderman, Ruderman, and Bertram (2013) and also Yepes et al. (2010) propose controllers that eliminate ripples of the mutual torque, using an IMP-based controller whose resonant frequencies are adjusted online according to the motor velocity.

In this paper, we propose a new type of controller to reduce the torque ripple caused by both the mutual torque and the cogging torque. *The idea is to use a resonant controller on the q-axis current reference but using the rotor angle in place of the time.* This controller may be called adaptive in the sense that the resonant frequencies (when regarded in the time domain) adjust themselves according to the motor speed.

In Section 2, we establish a model for the motor to cover all dynamics without any assumptions on the signals. In Section 3, we derive a formula for the electromagnetic torque and express the torque ripple caused by non-sinusoidal BEMF. In Section 4, the constant speed version of our controller is presented. The velocity loop, which uses the q-axis current, includes an internal model and a feedforward block. In Section 5, we introduce the adaptive version of the internal model, which works at variable speed. It is based on a transformation of linear differential equations, when time is replaced by motor angle. In Section 6, we give a short review of the Adaptive Eccentricity Algorithm from Canudas de Wit and Praly (2000) and Malaize and Levine (2009). In Section 7, we prove that, for the adaptive algorithm from Section 5, and under the restrictive assumption that the reference ω_{ref} is constant, the velocity error tends to zero. Simulation and experimental results are provided in Sections 8–10, with detailed comparisons with the controller of Canudas de Wit and Praly (2000). In particular, in Section 9 we consider a more complex system with a load connected to the motor via a flexible shaft.

2. Motor model – electrical part

We give the derivation of a mathematical model for a PMSM with one pair of poles per phase (similar to Zhong & Weiss, 2011).

Assume that the windings of a PMSM are connected in star, with each winding having the series resistance r. The basic voltage equations are

$$\tilde{e}_x = \frac{\mathrm{d}\Psi_x}{\mathrm{d}t} = -ri_x + v_x - v_{\mathrm{m}}, \tag{1}$$

where x can be one of a, b, c, $\tilde{e}_x$ are the induced voltages, Ψ_x are the phase winding flux linkages, i_x are the phase currents, and v_x are the phase voltages. The voltage v_m in the center of the star cannot be measured, but this is also not needed. Clearly,

$$i_a + i_b + i_c = 0. \tag{2}$$

We denote by θ the angle of the rotor such that $\theta = 0$ corresponds to the rotor, creating a flux parallel to the axis of winding a and in the direction of the flux created by $i_a > 0$.

Assuming a round (non-salient) rotor and a magnetically non-saturated machine, the flux linkage equations are

$$\begin{aligned}\Psi_a &= Li_a + Mi_b + Mi_c + F(\theta),\\ \Psi_b &= Mi_a + \mathrm{L}i_b + Mi_c + F\left(\frac{\theta - 2\pi}{3}\right),\\ \Psi_c &= Mi_a + Mi_b + Li_c + F\left(\frac{\theta + 2\pi}{3}\right),\end{aligned} \tag{3}$$

where L is the phase winding self-inductance, M is the mutual inductance between two windings (usually $M = -0.5\,\mathrm{L}$), $F(\theta)$ is the flux linkage through phase a due to the rotor, $F(\theta - 2\pi/3)$ is the flux linkage through phase b due to the rotor and similarly for phase c. Substituting Equation (2) in Equation (3), we get:

$$\begin{aligned}\Psi_a &= L_s i_a + F(\theta),\\ \Psi_b &= L_s i_b + F\left(\frac{\theta - 2\pi}{3}\right),\\ \Psi_c &= L_s i_c + F\left(\frac{\theta + 2\pi}{3}\right),\end{aligned} \tag{4}$$

where $L_s = L - M$ is the equivalent motor phase inductance. From electromagnetic field theory, we know that $\begin{bmatrix} LM \\ ML \end{bmatrix} > 0$. It follows that $L_s > 0$ (usually $L_s = 1.5\,\mathrm{L}$). In this model, we assume that L_s is independent of θ.

Rewriting Equation (4) in vector form, we get:

$$\underline{\Psi} = L_s \underline{i} + \underline{F}(\theta), \tag{5}$$

where $\underline{F}(\theta) = [F(\theta)F(\theta - 2\pi/3)F(\theta + 2\pi/3)]^{\mathrm{T}}$. Due to the motor structure, $F(\theta)$ is a periodic function with period 2π, so that we can write it as a Fourier series:

$$F(\theta) = \Psi_f \left[\sum_{n=0}^{\infty} \gamma_n \cos(n\theta) + \delta_n \sin(n\theta)\right],$$

where $\Psi_f > 0$ is a parameter that will be chosen later. If the rotor is symmetric, then $F(-\theta) = F(\theta)$ and hence the terms with $\sin(n\theta)$ disappear: $\delta_n = 0$.

We argue that $F(\theta)$ contains only odd harmonics. The symmetry of the rotor (the shape remains unchanged after a rotation of π) implies that $F(\theta + \pi) = -F(\theta)$. By subtracting $F(\theta + \pi)$ from $F(\theta)$, where

$$F(\theta) = \Psi_f \left[\sum_{k=0}^{\infty} \gamma_{2k} \cos(2k\theta) + \sum_{k=1}^{\infty} \gamma_{2k-1} \cos((2k-1)\theta)\right],$$

$$F(\theta + \pi) = \Psi_f \left[\sum_{k=0}^{\infty} \gamma_{2k} \cos(2k\theta) - \sum_{k=1}^{\infty} \gamma_{2k-1} \cos((2k-1)\theta)\right],$$

we get

$$\begin{aligned}2F(\theta) &= F(\theta) - F(\theta + \pi)\\ &= 2\Psi_f \sum_{k=1}^{\infty} \gamma_{2k-1} \cos((2k-1)\theta).\end{aligned}$$

Therefore,

$$F(\theta) = \Psi_f \sum_{n\mathrm{odd}} \gamma_n \cos(n\theta). \tag{6}$$

We normalize the coefficients such that $\sum_{n=1}^{\infty} \gamma_n = 1$. Then, we see that Ψ_f is the maximal flux linkage due to the rotor. We remark that for a "perfectly built" machine, we have no harmonics:

$$F(\theta) = \Psi_f \cos\theta. \tag{7}$$

By differentiating Equation (5), we get

$$\underline{\tilde{e}} = L_s \underline{\dot{i}} + \underline{e} \quad \text{where} \quad \underline{e} = \underline{\dot{F}}(\theta) = [e_a\, e_b\, e_c]^{\mathrm{T}}, \tag{8}$$

so that e_a, e_b, e_c are the BEMF on each phase due to the rotor movement. According to Equation (6), we have

$$e_a = -\omega \Psi_f \sum_{n\,\mathrm{odd}} \gamma_n n \sin(n\theta) \tag{9}$$

and similarly for e_b and e_c. Here, $\omega = \dot{\theta}$ is the mechanical angular speed. We remark that for a "perfectly built" motor,

$$e_a = -\omega\, \Psi_f \sin(\theta) \tag{10}$$

and similarly for e_b (with $\theta - 2\pi/3$ in place of θ) and for e_c. Comparing Equation (1) with Equation (8) and eliminating $\underline{\tilde{e}}$,we get:

$$L_s \underline{\dot{i}} = \underline{v} - r\underline{i} - \underline{e} - \underline{v}_m, \tag{11}$$

where $\underline{v} = [v_a\, v_b\, v_c]^{\mathrm{T}}$and $\underline{v}_m = [v_m\, v_m\, v_m]^{\mathrm{T}}$. Recall the Park transformation introduced in Park (1929), a unitary matrix U which transforms a vector from the a, b, c

coordinate system to the rotating d, q coordinate system:

$$\begin{bmatrix} f_d \\ f_q \\ f_0 \end{bmatrix} = U \cdot \begin{bmatrix} f_a \\ f_b \\ f_c \end{bmatrix},$$

where

$$U = \sqrt{\frac{2}{3}} \cdot \begin{bmatrix} \cos(\theta) & \cos\left(\theta - \frac{2\pi}{3}\right) & \cos\left(\theta + \frac{2\pi}{3}\right) \\ -\sin(\theta) & -\sin\left(\theta - \frac{2\pi}{3}\right) & -\sin\left(\theta + \frac{2\pi}{3}\right) \\ \frac{1}{\sqrt{2}} & \frac{1}{\sqrt{2}} & \frac{1}{\sqrt{2}} \end{bmatrix}. \tag{12}$$

Multiplying Equation (11) with the unitary matrix U from Equation (12), we get:

$$L_s U \cdot \begin{bmatrix} \dot{i}_a \\ \dot{i}_b \\ \dot{i}_c \end{bmatrix} = \begin{bmatrix} v_d \\ v_q \\ v_0 \end{bmatrix} - r \begin{bmatrix} i_d \\ i_q \\ i_0 \end{bmatrix} - \begin{bmatrix} e_d \\ e_q \\ e_0 \end{bmatrix} - \begin{bmatrix} 0 \\ 0 \\ \sqrt{3} v_m \end{bmatrix}.$$

Since, according to a short computation,

$$\begin{bmatrix} \dot{i}_d \\ \dot{i}_q \\ \dot{i}_o \end{bmatrix} = U \begin{bmatrix} \dot{i}_a \\ \dot{i}_b \\ \dot{i}_c \end{bmatrix} + \omega \begin{bmatrix} i_q \\ -i_d \\ 0 \end{bmatrix},$$

we get

$$L_s \cdot \begin{bmatrix} \dot{i}_d \\ \dot{i}_q \\ \dot{i}_o \end{bmatrix} - L_s \omega \begin{bmatrix} i_q \\ -i_d \\ 0 \end{bmatrix} = \begin{bmatrix} v_d \\ v_q \\ v_o \end{bmatrix} - r \begin{bmatrix} i_d \\ i_q \\ i_o \end{bmatrix} - \begin{bmatrix} e_d \\ e_q \\ e_o \end{bmatrix} - \begin{bmatrix} 0 \\ 0 \\ \sqrt{3} v_m \end{bmatrix}. \tag{13}$$

Notice that $i_o = 0$ (hence also $\dot{i}_o = 0$), due to Equation (2). Hence, according to the third equation (the third line) in (13), $v_o = e_o + \sqrt{3} v_m$.

Thus, the dynamic voltage equations in d, q coordinates are

$$L_s \dot{i}_d = v_d - r i_d - e_d + L_s \omega i_q,$$
$$L_s \dot{i}_q = v_q - r i_q - e_q - L_s \omega i_d.$$

For the case of a perfectly built motor as in Equation (7), using the Park transformation on Equation (10) we get, after some computation,

$$e_d = 0, \quad e_q = \sqrt{\frac{3}{2}} \cdot \omega \Psi_f. \tag{14}$$

3. The motor torque and BEMF

Torque production inside brushless PMSM is mainly due to the interaction between the permanent magnet field and the currents in the phase windings (mutual torque), and the interaction between the permanent magnets, located in the rotor, and the slotted iron structure of the stator (cogging torque). In this section, we derive formulas for these torques.

The total energy stored in a PMSM is

$$E_{\text{total}} = E_{\text{mag}} + E_{\text{kin}} = E_{\text{mag}} + \frac{1}{2} J \omega^2, \tag{15}$$

where E_{mag} is the energy stored in the magnetic field and E_{kin} is the kinetic energy of the rotating body with inertia J that contains the rotor. Differentiating Equation (15), we get:

$$\dot{E}_{\text{total}} = \dot{E}_{\text{mag}} + J \omega \dot{\omega}. \tag{16}$$

We denote by T_e the *electromagnetic torque* generated inside the motor, and by T_l the *mechanical load torque.* Substituting Newton's law, $J\dot{\omega} = T_e - T_l$, into Equation (16), we get:

$$\dot{E}_{\text{total}} = \dot{E}_{\text{mag}} + \omega (T_e - T_l). \tag{17}$$

On the other hand, $\dot{E}_{\text{total}}$ is the flow of power to the system.

This power is equal to the ingoing power minus the power lost in the resistors minus the power drawn by the load,

$$\dot{E}_{\text{total}} = \underline{v} \circ \underline{i} - r \underline{i} \circ \underline{i} - \omega T_l = (\underline{v} - r \underline{i}) \circ \underline{i} - \omega T_l, \tag{18}$$

where $\circ$ is the inner product in $\Re^3$. From Equations (17) and (18), we conclude:

$$\dot{E}_{\text{mag}} = (\underline{v} - r \underline{i}) \circ \underline{i} - \omega T_e.$$

Substituting Equation (11) into the above formula, we get:

$$\dot{E}_{\text{mag}} = (L_s \cdot \dot{\underline{i}} + \underline{e} + \underline{v}_m) \circ \underline{i} - \omega T_e. \tag{19}$$

Since $\underline{v}_m \circ \underline{i} = 0$, multiplying the above formula by $\mathrm{d}t$, we get:

$$\mathrm{d}E_{\text{mag}} = (L_s \cdot \dot{\underline{i}} + \underline{e}) \circ \underline{i} \cdot \mathrm{d}t - T_e \, \mathrm{d}\theta. \tag{20}$$

Integrating both sides of Equation (20), while assuming no motion of the rotor, to charge the magnetic field, we get (using that $\mathrm{d}\theta = 0, \underline{e} = 0$)

$$E_{\text{mag}} = E_0(\theta) + \int_0^T L_s \cdot \dot{\underline{i}} \circ \underline{i} \cdot \mathrm{d}t = E_0(\theta) + \int_0^T \frac{\mathrm{d}}{\mathrm{d}t} \left(\frac{1}{2} L_s \cdot \underline{i} \circ \underline{i} \right) \mathrm{d}t = E_0(\theta) + \frac{1}{2} (L_s \cdot \underline{i}) \circ \underline{i}, \tag{21}$$

where $E_0(\theta)$ is the energy in the magnetic field due to the rotor magnet (while there are no currents) at the angle θ.

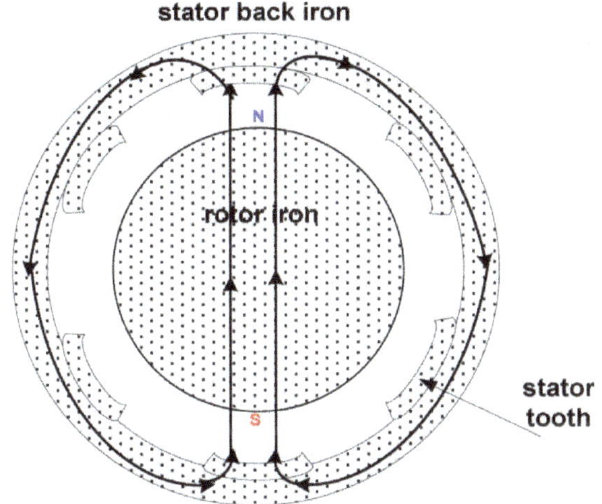

Figure 1. PMSM with stator slots.

In the sequel, we consider again the moving rotor. Differentiating Equation (21), we get

$$\dot{E}_{\text{mag}} = \frac{\mathrm{d}}{\mathrm{d}\theta} E_0(\theta) \cdot \omega + (L_s \cdot \underline{i}) \circ \dot{\underline{i}}.$$

From here and Equation (19), we conclude:

$$T_e = \frac{\underline{e} \circ \underline{i}}{\omega} - \frac{\mathrm{d}}{\mathrm{d}\theta} E_0(\theta). \quad (22)$$

The term $T_{\text{em}} = e \circ i / \omega$ is called the *mutual torque*, while $T_{\text{ec}} = -(\mathrm{d}/\mathrm{d}\theta)E_0(\theta)$ is called the *cogging torque*.

The mutual torque may be calculated using phase currents and back EMF instant values, both in (a, b, c) and in (d, q) coordinates (we use that $i_0 = 0$):

$$T_{\text{em}} = \frac{\underline{e} \circ \underline{i}}{\omega} = \frac{(U \cdot \bar{e}) \circ (U \cdot \bar{i})}{\omega} = \frac{i_d e_d + i_q e_q}{\omega}. \quad (23)$$

For a "perfectly built" motor, we have from Equation (14)

$$T_{\text{em}} = \frac{i_d \cdot 0 + i_q \cdot \sqrt{3/2}\Psi_f \omega}{\omega} = i_q \cdot \sqrt{3/2}\Psi_f. \quad (24)$$

As mentioned earlier, cogging torque is created by the interaction between the permanent magnets in the rotor and the slotted iron structure of the stator. Usually, the stator has slots to hold the windings, see Figure 1. Due to these slots, the rotor has preferred directions and cogging is produced.

The following expression (25) for the electromagnetic cogging torque is taken from Hwang et al. (2001):

$$T_{ec} = \sum_{n=0}^{\infty} K_n n \sin(n N_L \theta), \quad (25)$$

where K_n are constants determined by the machine's geometry and N_{L} is the least common multiple of the number of pairs rotor poles and the number of slots in the stator.

For the simple motor discussed earlier (one pair of stator poles per phase, one slot per stator pole and one pair of poles in the rotor), we have $N_{\text{L}} = 6$. More generally, N_{L} is a multiple of 6.

Recall that for a "non-perfectly built" motor, the BEMF due to the rotor movement can be described by Equation (9). Let us look at the third harmonic of the BEMF:

$$\begin{aligned}
e_{a_{\text{3rd harmonic}}} &= -\Psi_f \omega 3\gamma_3 \sin(3\theta), \\
e_{b_{\text{3rd harmonic}}} &= -\Psi_f \omega 3\gamma_3 \sin\left(3\left(\theta - \frac{2\pi}{3}\right)\right) \\
&= -\Psi_f \omega 3\gamma_3 \sin(3\theta), \\
e_{c_{\text{3rd harmonic}}} &= -\Psi_f \omega 3\gamma_3 \sin\left(3\left(\theta + \frac{2\pi}{3}\right)\right) \\
&= -\Psi_f \omega 3\gamma_3 \sin(3\theta).
\end{aligned}$$

Since $i_a + i_b + i_c = 0$, we have from Equation (23) that the contribution of the third harmonic to the torque is $(\underline{e}_{\text{3rdharmonic}} \circ \underline{i}/\omega) = 0$. A similar result can be derived for all the odd multiples of 3.

Therefore, we conclude that the harmonics which contribute to the mutual torque, T_{em}, are multiples of $\sin(n\theta)$, where $n = 6p + 1$ for $p = 0, 1, 2, \ldots$ and $n = 6p - 1$ for $p = 1, 2, 3, \ldots$.

Recall from Equation (23) that in d, q coordinates, the electromagnetic torque is $T_{\text{em}} = (i_d e_d + i_q e_q)/\omega$. Since i_d and i_q are inputs to be chosen by the user, it is useful to look at the expressions for e_{d} and e_{q}. If we compute e_{d} and e_{q} from (9) and (12), we obtain

$$\begin{aligned}
e_d &= -\sqrt{\frac{3}{2}}\omega \Psi_f \left[\sum_{p=1}^{\infty} [\gamma_{6p+1}(6p+1) \right. \\
&\quad \left. + \gamma_{6p-1}(6p-1)] \sin(6p\theta) \right], \\
e_q &= \sqrt{\frac{3}{2}}\omega \Psi_f \left[\gamma_1 + \sum_{p=1}^{\infty} [\gamma_{6p+1}(6p+1) \right. \\
&\quad \left. - \gamma_{6p-1}(6p-1)] \cos(6p\theta) \right].
\end{aligned} \quad (26)$$

Denoting for $p = 1, 2, 3, \ldots$

$$\eta_{dp} = -\sqrt{3/2}\Psi_f [\gamma_{6p+1}(6p+1) + \gamma_{6p-1}(6p-1)],$$

$$\eta_{qp} = \sqrt{3/2}\Psi_f [\gamma_{6p+1}(6p+1) - \gamma_{6p-1}(6p-1)],$$

and $\eta_{q0} = \sqrt{3/2}\Psi_{\text{f}}\gamma_1$, we get from Equation (26)

$$\begin{aligned}
e_d &= \omega \left[\sum_{p=1}^{\infty} \eta_{dp} \sin(6p\theta) \right], \\
e_q &= \omega \left[\eta_{q0} + \sum_{p=1}^{\infty} \eta_{qp} \cos(6p\theta) \right].
\end{aligned} \quad (27)$$

If the motor has Π pairs of poles per phase (instead of just one), then in Equations (25) and (27), we have to replace θ with $\Pi\theta$.

4. Velocity control (non-adaptive)

Figure 2 shows the current and velocity control loops of a BLDCM. Notice that $i_{d_{\text{ref}}} = 0$ and $i_{q_{\text{ref}}}$ is the sum of the velocity controller's output and a term from the feedforward path with transfer function $\alpha = (1/\eta_{q0})(\text{J}s + D)$, which has the role of making the response faster. Here, D is the viscous friction coefficient of the motor, so that $T_l = T_L + D\omega$, where T_{L} is the external load torque. It is recommended to include a saturation block in the feedforward path.

In the sequel, we allow the motor to have Π pairs of poles per phase (until now we had $\Pi = 1$), so that we rely on the modified version of Equation (27). We will assume complete and instantaneous control of the currents, hence $\varepsilon_d = 0$ and $\varepsilon_q = 0$ in Figure 2. If we operate in the linear range of the converter and the motor (no saturation), then this is due to the fact that the current loop is much faster than the velocity loop (typical bandwidths are 1 kHzfor the current loop and 100 Hz for the velocity loop, see also Gan and Qiu, (2004). Substituting Equation (27) in Equation (23), while assuming $i_{\text{d}_{\text{ref}}} = 0$ and complete current control, we get

$$T_{\text{em}} = i_{q_{\text{ref}}}\left[\eta_{q0} + \sum_{p=1}^{\infty} \eta_{qp} \cos(6p\Pi\theta)\right]. \tag{28}$$

Using Equation (28), the block diagram in Figure 2 simplifies to the one in Figure 3.

In the sequel, we consider a control loop as in Figure 3. The special structure of this feedback system and the fact that $i_{q_{\text{ref}}}$ is a function of the feedback result in a steady-state output ω whose spectral contents contain harmonics

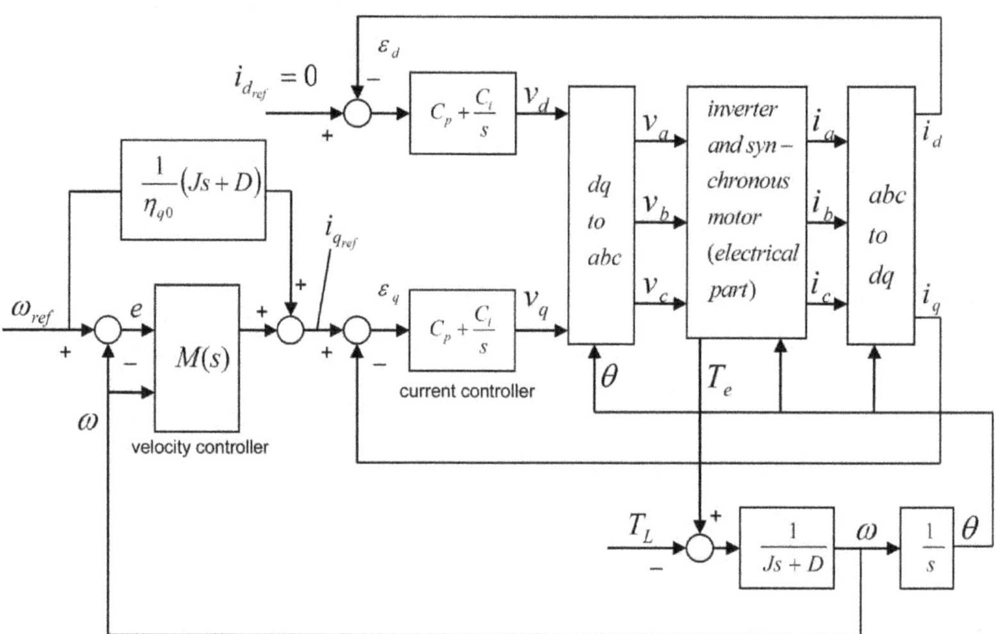

Figure 2. Full block diagram of the velocity control system. Here, T_L and T_e are the external load torque and the electromagnetic torque.

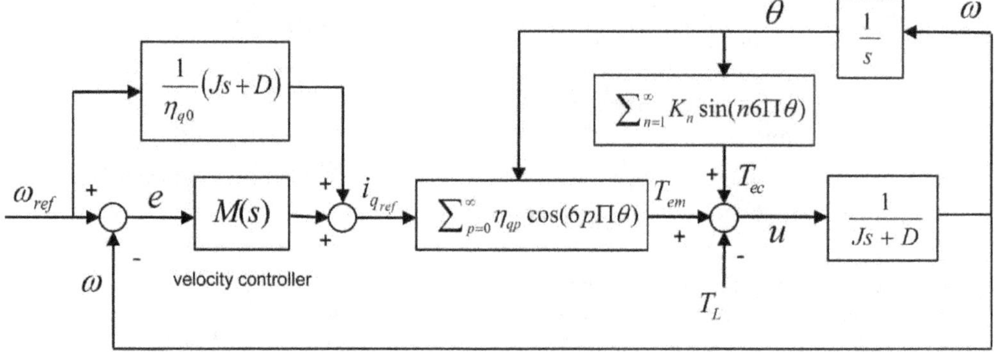

Figure 3. Block diagram of the simplified control system that corresponds to an ideal current control loop (i.e. no current tracking errors). The electromagnetic torque T_e from Figure 2 is $T_{\text{em}} + T_{\text{ec}}$, where T_{em} is the mutual torque and T_{ec} is the cogging torque.

whose frequency is a multiple of $6\Pi\omega_{\rm ref}$. Thus, in steady state, ω will be periodic, with the fundamental frequency $6\Pi\omega_{\rm ref}$. We will see this phenomenon in our simulations and experiments.

Based on the IMP, we propose for M a PI controller plus additional internal model terms:

$$M(s) = k_p + \frac{k_i}{s} + \sum_{m=1}^{N} \frac{k_m s}{s^2 + {\omega_m}^2}, \tag{29}$$

where $k_p, k_i, k_m > 0$ and $\omega_m = 6m\Pi\omega_{\rm ref}$ is a resonant frequency of M. Such internal models have been used for instance in Jayawardhana and Weiss (2009), Knobloch, Isidori, and Flockerzi (1993) or Spitsa, Kuperman, Weiss, and Rabinovici (2006). The idea of the IMP is that M has infinite gain at the frequencies ω_m, resulting in zero tracking errors at these frequencies, if the closed-loop system is stable. Since there are a possibly large number of resonant frequencies, and they are multiples of a fundamental frequency, another way to build M would be using delay lines, as in repetitive control, see, for instance, Weiss (1997), but we have not explored this further.

We mention that in the terminology of van der Schaft (2000), the controller $M$is strictly input passive.

5. An adaptive controller

In this section, we modify the controller in Equation (29) to accommodate a variable motor speed. Internal models with variable resonant frequencies have appeared in Canudas de Wit and Praly (2000), Esbrook, Tan, and Khalil (2011), Lu et al. (2008) and Spitsa et al. (2006).

Recall from Equations (22), (25) and (28) that the electromagnetic torque generated inside the motor is

$$\begin{aligned} T_e &= T_{\rm em} + T_{\rm ec} \\ &= i_{q_{\rm ref}} \left[\eta_{q0} + \sum_{p=1}^{\infty} \eta_{qp} \cos(6\Pi p\theta) \right] \\ &\quad + \sum_{n=0}^{\infty} K_n \, n \sin(nN_L\theta), \end{aligned}$$

where N_L is a multiple of 6Π.

From the above, it is obvious that the disturbance torque is periodic as a function of the motor positionθ, with frequencies which are usually multiples of 6Π. However, other integer frequencies may also appear, such as the frequency Π due to DC offsets (Gan & Qiu, 2004) (we mentioned this in the Introduction). Since the disturbance in the motor speed is a function of θ, we suggest using in the controller an internal model which is also a function of θ (instead of time). In order to do that, we visualize the input and the output of the controller as being functions of θ (instead of time). In this case, an internal model term to reduce the periodical disturbance, with frequency m, would be $k_m s/s^2 + m^2$. Overall, the controller would look as in Equation (29) and $\omega_{\rm ref} = 1$. However, this is now a transfer function in an unconventional frequency domain, corresponding to θ as the time variable. To express our controller in the conventional time domain, first we revert from the Laplace to the θ domain where the term corresponding to the frequency m is described by the differential equation

$$p_{\theta\theta} + p \cdot m^2 = k_m e_\theta. \tag{30}$$

Here, e is the input to the IMP controller and p is the output.

Converting Equation (30) to the time domain, while using

$$\begin{aligned} p_\theta &= \frac{{\rm d}p}{{\rm d}t} \cdot \frac{{\rm d}t}{{\rm d}\theta} = \dot{p} \cdot \frac{1}{\omega}, \\ p_{\theta\theta} &= \left(\dot{p} \cdot \frac{1}{\omega}\right)_\theta = \frac{\rm d}{{\rm d}t}\left(\dot{p} \cdot \frac{1}{\omega}\right)\frac{{\rm d}t}{{\rm d}\theta} \\ &= \left[\ddot{p} \cdot \frac{1}{\omega} - \dot{p}\frac{\dot{\omega}}{\omega^2}\right] \cdot \frac{1}{\omega} = \frac{\ddot{p}}{\omega^2} - \dot{p}\frac{\dot{\omega}}{\omega^3}, \end{aligned}$$

and similarly $e_\theta = ({\rm d}e/{\rm d}t) \cdot ({\rm d}t/{\rm d}\theta) = \dot{e} \cdot (1/\omega)$, we get:

$$\ddot{p} - \dot{p}\frac{\dot{\omega}}{\omega} + p \cdot m^2 \cdot \omega^2 = k_m \dot{e}\omega. \tag{31}$$

This is a linear differential equation with variable coefficients (that depend on $\omega, \dot{\omega}$). A digital controller will be able to implement a good approximation of such a subsystem. Since in Equation (31) we have a division by ω, it is recommended to impose a lower bound on ω in this division: For some small $\varepsilon > 0$, if $\omega < \varepsilon$, then replace $\dot{\omega}/\omega$ with $\dot{\omega}/\varepsilon$.

The term k_i/s in Equation (29) has to be expressed in the normal time domain as well. This is much easier and we get:

$$\dot{p} = k_i e\omega. \tag{32}$$

The block diagram of the control system with the adaptive internal model controller (AIMC) is the same as in Figure 2 except that M is replaced with the AIMC block that contains a realization of both Equations (31) and (32).

6. The adaptive eccentricity compensation algorithm

In Canudas de Wit and Praly (2000), an adaptive observer, called *adaptive eccentricity compensator* (AEC), is proposed for the velocity control of a BLDCM that estimates the angle-dependent periodic torque disturbances acting on an electric motor. A proof for stability is given for a proportional controller acting on a very simplified motor model (an integrator from torque to angular velocity) using a Lyapunov function. A friction predictor is added to cancel the Coulomb friction, which is feedforward, so that it cannot react to changes in the load torque. In our implementation of the controller from Canudas de Wit and Praly (2000),

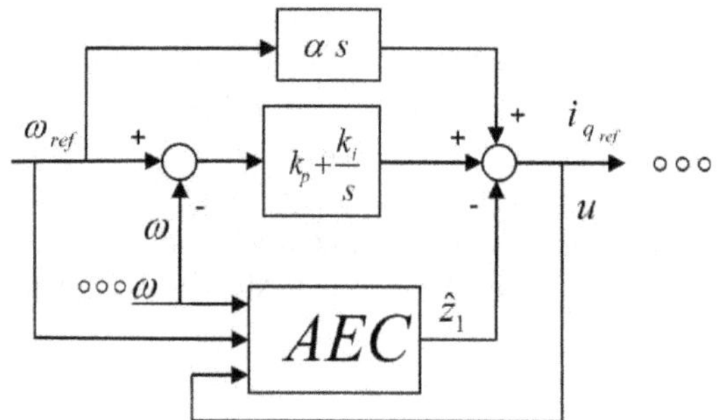

Figure 4. The velocity control loop with the AEC controller. The symbol ○○○ indicates a connection with the remaining part of the system, which is as in Figure 2.

used for comparison with our AIMC controller, we replace the friction predictor by an integral component in the controller, which can compensate any step load torque – this is in the PI block in the middle of Figure 4. Thus, we use a slightly better implementation of the AEC controller than the original one in Canudas de Wit and Praly (2000). The rest of the block diagram is as in Figure 2.

The controller of Canudas de Wit and Praly (2000) has been further developed in Malaize and Levine (2009) for the case of high precision positioning of a BLDCM (i.e. tracking of an angle reference signal). Thus, there is no need to filter out the noise originated in the discrete derivative of the encoder. On the other hand, tuning this new controller scheme is not that simple. The proof of stability in Malaize and Levine (2009) is more complex, and it too relies on a quadratic Lyapunov function. The adaptive observer structure, for the θ domain frequency m in Canudas de Wit and Praly (2000), is the following:

$$\begin{aligned} \alpha\dot{\hat{\omega}} &= u + \hat{z}_1 - h_0(\hat{\omega} - \omega), \\ \dot{\hat{z}}_1 &= \omega m^2 \hat{z}_2 - h_1(\hat{\omega} - \omega) - h_1(\omega_{\rm ref} - \omega), \\ \dot{\hat{z}}_2 &= -\omega \hat{z}_1. \end{aligned} \tag{33}$$

where h_0 and h_1 are positive constants to be chosen later. The control input, $u = i_{q_{\rm ref}}$, is given by the AEC block added to the output of a PI controller:

$$u = \alpha \cdot \dot{\omega}_{\rm ref} + k_p(\omega_{\rm ref} - \omega) + k_i \int_{t_0}^{t} (\omega_{\rm ref} - \omega)\, {\rm d}\sigma - \hat{z}_1 \tag{34}$$

7. A proof of convergence

We give a proof of convergence to zero of the speed tracking error e from Figure 3 under the restrictive assumption that the reference speed $\omega_{\rm ref}$ is constant. We shall also assume that the system remains throughout in the reasonable operating region where the angular velocity ω and the total torque u acting on the motor are positive. We emphasize that this is not the "standard" situation from internal model-based control, because the disturbance is a superposition of sinusoidal signals in the variable θ, and not in the time variable.

LEMMA. *Consider the system* **P** *with transfer function*

$$P(s) = \frac{1}{{\rm Js} + D}$$

with input u and output ω and let $\omega_{\rm ref} > 0$ be a constant. Introduce the speed error $e = \omega_{\rm ref} - \omega$, the reference torque $u_{\rm ref} = D\omega_{\rm ref}$, the torque error $v = u_{\rm ref} - u$ and the Hamiltonian $H = Je^2\omega_{\rm ref}/2$. If we regard **P** *with θ (the integral of ω) as the time variable, with v as input and with e as output, then it becomes a nonlinear time-invariant system $\mathbf{P}_{\rm NL}$. In the region where $u, \omega \geq 0$, $\mathbf{P}_{\rm NL}$ is passive, i.e. $H_\theta \leq ve$.*

Proof Since $\dot{e} = (-De + v)/J$, we have

$$\begin{aligned} H_\theta &= \frac{H}{\omega} = \frac{Je\dot{e}\omega_{\rm ref}}{\omega} = e(-De+v)\frac{\omega_{\rm ref}}{\omega} \\ &= \left[-De^2 + ev\left(1 - \frac{\omega}{\omega_{\rm ref}}\right)\right]\frac{\omega_{\rm ref}}{\omega} + ev \\ &= -\frac{[D\omega_{\rm ref} + v]e^2}{\omega} + ev = \frac{-ue^2}{\omega} + ev. \end{aligned}$$

We see that if $u, \omega \geq 0$ then $H_\theta \leq ev$. ■

PROPOSITION. *Consider the feedback system from* Figure 3 *but with a perfectly built motor, meaning that $\eta_{\rm qp} = 0$ for all $p > 0$, and with only finitely many terms in the sum defining the cogging torque $T_{\rm ec}$. Assume that the adaptive internal model M is built according to Equations* (31) *and* (32), *with the resonant frequencies (in the θ domain) matching the frequencies of the cogging torque. Assume that $\omega_{\rm ref} > 0$ and $T_{\rm L}$ are constant and the system remains in the region where $u, \omega \geq 0$. Then, $\lim_{t\to\infty} e(t) = 0$.*

Proof. Using simple block diagram manipulations (moving the summation point for the feedforward signal further ahead until the right end of the diagram), we can see that Figure 3 is equivalent to the one shown in Figure 5.

Now, we regard this feedback system in the θ domain (i.e. we use θ in place of the time). Then, the feedback loop in the right upper corner of Figure 5 disappears, because θ is no longer a state variable, and instead $T_{\rm ec} - T_{\rm L}$ is now a disturbance generated by a finite-dimensional linear exosystem that has simple eigenvalues on the imaginary axis ($T_{\rm L}$ is generated by an eigenvalue at zero). The linear system with parameters J and D at the right end of Figure 5 becomes the nonlinear passive system $\mathbf{P}_{\rm NL}$ described in the lemma, with the state e (passivity holds as long as $u, \omega \geq 0$). The block diagram becomes very simple, and we leave it to the reader to redraw it. This new block diagram is the same as Figure 2 in Jayawardhana and Weiss (2009), with $\eta_{\rm q0}M$ playing the role of the controller. It can be verified

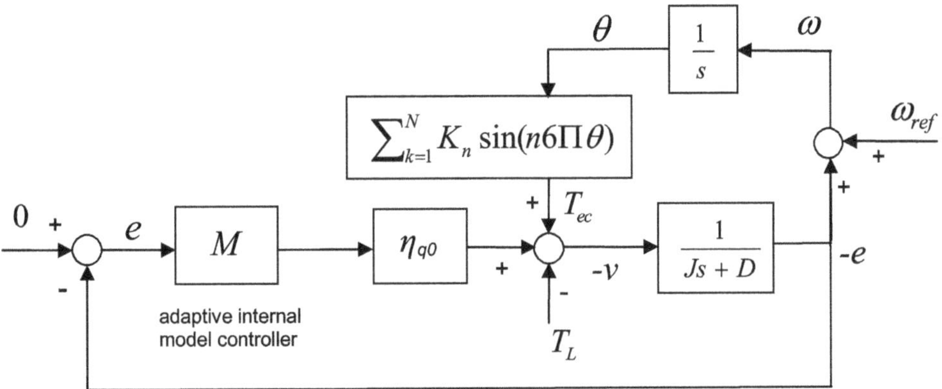

Figure 5. A block diagram equivalent to the one from Figure 3 when $\eta_{qp} = 0$ for all $p > 0$, and using an AIMC, as in the proposition.

that the assumptions in Theorem IV.3 from Jayawardhana and Weiss (2009) are satisfied, and this implies that indeed $e(t) \to 0$. ■

We remark that the conclusion of the proposition remains valid if we add an arbitrary L^2 component to the constant load torque T_L, and this follows from the same theorem in Jayawardhana and Weiss (2009).

8. Simulations without load

In this section (and also in the experiments), ω is measured by taking the discrete derivative from a high-resolution encoder with 20,000 counts per turn. Since the simulations and experiments are done at low motor speeds, a low-pass filter is used on these velocity measurements. We chose a second-order low-pass filter with natural frequency $\omega_n = 2\pi 200$ rad/sec and damping $\zeta = 0.7$. The angular acceleration (needed in Equation (31)) is computed by taking the discrete derivative of ω.

In order to facilitate the comparison of the AEC and AIMC controllers, the following assumptions are made:

- Since most of the ripple is due to the cogging, we assume $\eta_{q0} = \sqrt{3/2}\Psi_f$ and $\eta_{qp} = 0$ for $p \neq 0$.
- The motor has two pole pairs per phase, that is, $\Pi = 2$.
- D is assumed to be zero. Hence, comparing Figure 4 with Figure 2, we choose $\alpha = J/\eta_{q0}$.
- There is static friction on the rotor, 0.01 Nm, with sign depending on the direction of motion, of course.

We have obtained simulation results for the system shown in Figure 2, with: $\Psi_f = 0.0358$ V sec, $\eta_{q0} = 0.04384$ N m/A, $J = 332 \times 10^{-7}$ kg m^2, $\Pi = 2$, $K_1 = 0.006$ N m, $k_i = 10$ A, $k_p = 0.3125$ A sec, $L_s = 0.65 \times 10^{-3}$ H, $V_{bus} = 20$ V, $r = 0.5\Omega$. The coefficients of the PI part of the controller are chosen so that if we only use the PI part, we get a reasonable gain margin of 11 dB and a phase margin of 54°. For the first few simulations, the reference input is a step of size 2π rad/sec.

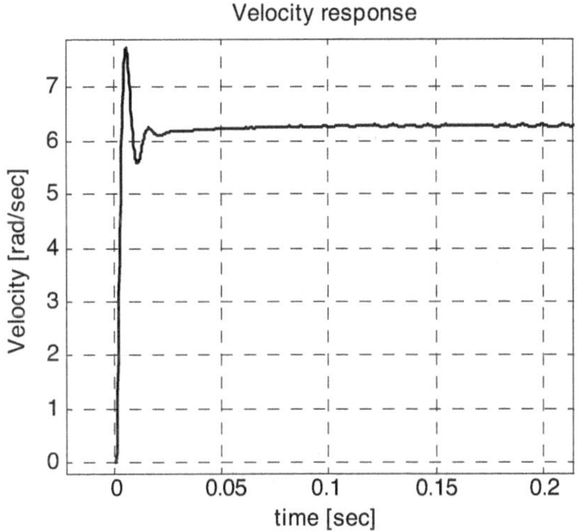

Figure 6. Velocity response for $T_{ec} = 0$, using a PI controller.

First, we simulate the velocity response for a "perfectly built" motor, that is, $T_{ec} = 0$. We see from Figure 6 that ω converges to ω_{ref}. The large overshoot is caused by the low-pass filter mentioned earlier. Next, we simulate the step response for a PMSM with cogging torque $T_{ec} = 0.006 \cdot \sin(12 \cdot \theta)$ (see Figure 7). We also show its steady-state frequency contents, that is, the absolute value of its fast Fourier transform (FFT) for the time interval [1,10] (sec). As we can see from Figure 7, the steady-state velocity is periodic, with the fundamental frequency $12\omega_{ref}$.

In the remaining simulations, we compare the two adaptive algorithms (AEC and AIMC). The PI part of the controllers is chosen as before. For the AEC controller, we choose $h_0, h_1 = 3$. For the AIMC controller, $k_1 = 1.8$. In both controllers, the gains are chosen such that if multiplied by 2 the system would become unstable.

In the next two figures, the reference input and the cogging torque are as in Figure 7. The velocity responses, while using the PI controller (alone) and the AIMC algorithm, are plotted in Figure 8. Figure 9 shows the

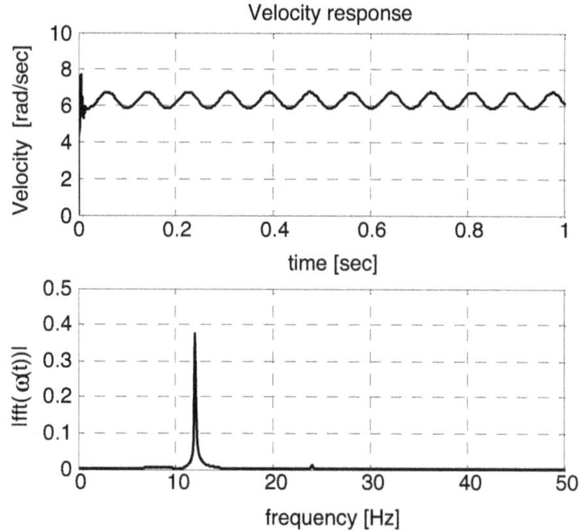

Figure 7. Velocity response and its steady-state frequency contents (FFT, absolute value) for a non-perfectly built motor with a PI controller.

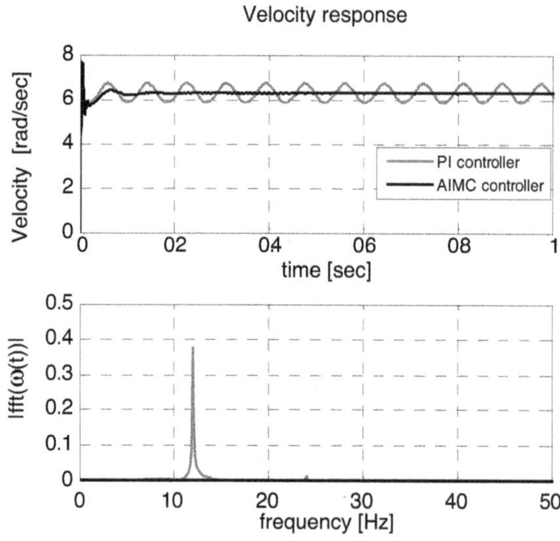

Figure 8. Velocity response and its steady-state frequency contents with a PI controller and with an AIMC controller, for the step velocity reference as in Figure 7.

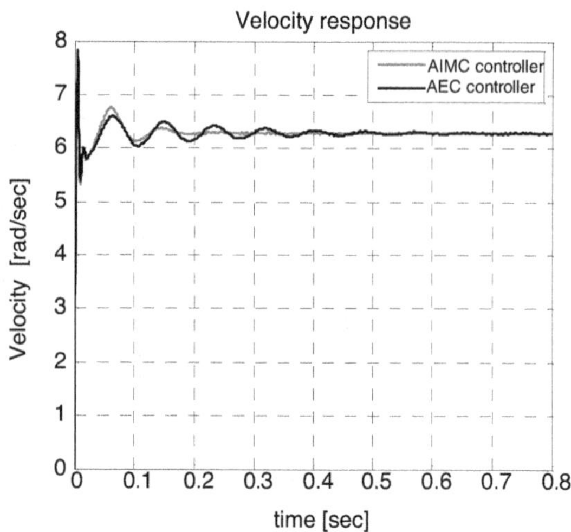

Figure 9. Comparison of the velocity response of both algorithms, for a step velocity reference, with cogging torque as in Figure 7.

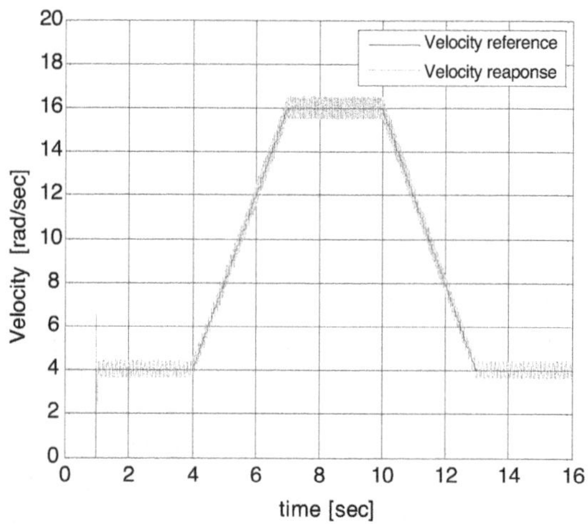

Figure 10. Velocity response for a variable reference velocity command using a simple PI controller, with cogging torque as in Figure 7.

velocity response with AEC algorithm compared with that of the AIMC algorithm.

To check the ability of both controllers to track a variable speed command, we changed the reference input. The new reference input and the velocity response using a simple PI controller are shown in Figure 10. The velocity responses, when using the adaptive algorithms, are shown in Figure 11. They are so close that we cannot distinguish them visually in Figure 11. A comparison between the velocity error signals of the two algorithms, after 3 s, is shown in Figure 12.

Next, we simulate a second harmonic for the cogging disturbance (though it does not exist in our experiment) and check both algorithms. We take $K_2 = 0.003\,\mathrm{N{\cdot}m}$ in Equation (25) (in addition to $K_1 = 0.006\,\mathrm{N\,m}$][· taken previously). To deal with the second harmonic as well, we add another AIMC block with $k_2 = 0.6$. The velocity response is shown in Figure 13. A comparison between the AIMC and the AEC for cogging with two harmonics is shown in Figure 14.

9. Simulation results with a flexible shaft and load

The AEC controller from Canudas de Wit and Praly (2000) was designed for a motor model that is just an integrator (from torque to angular velocity) and their proof of stability refers to this model. This model corresponds to a motor with a very simple mechanical load (just friction). The simulations in Section 8 use such a model and they show that in this case, the behavior of the control system

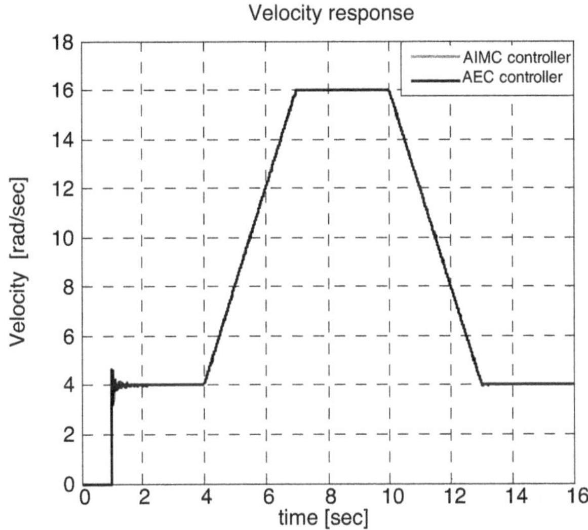

Figure 11. Velocity response for a variable speed command (as in Figure 10) for the two adaptive algorithms.

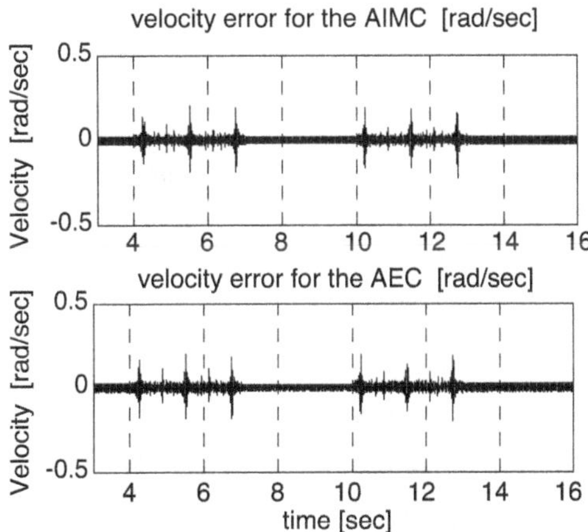

Figure 12. Velocity error for the same reference velocity and cogging torque as in Figure 10. The spikes are due to the resolution of the encoder.

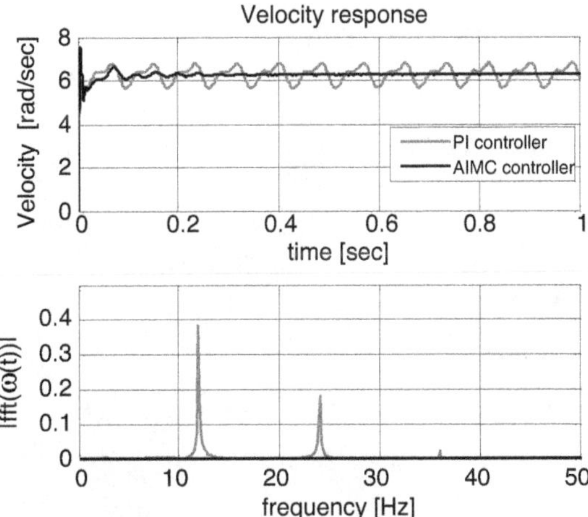

Figure 13. Velocity response and frequency contents in the presence of two cogging frequencies, using the PI controller and the AIMC algorithm, with $\omega_{\text{ref}} = 1$ Hz, leading to two resonant frequencies at 12 and 24 Hz.

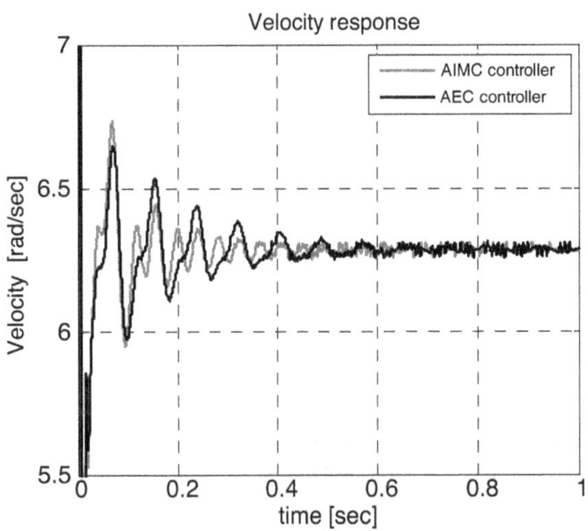

Figure 14. Velocity response in the presence of two cogging frequencies, using the AIMC and the AEC controllers, with a rotation speed of 1 Hz and two resonant frequencies at 12 and 24 Hz.

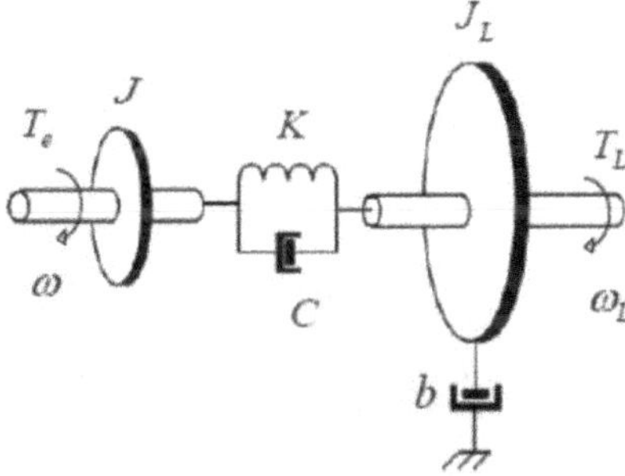

Figure 15. The motor connected to a load via a flexible shaft.

with the AEC controller and with our AIMC controller is very similar (assuming that the PI part of the controllers is the same).

In this section, we consider a motor that is connected to the load via a flexible shaft with internal viscous damping. The load has its own moment of inertia J_L and there is viscous friction between the load and the inertial reference system, with coefficient b. We denote by K and C the stiffness constant and the damping constant of the flexible shaft. The schematic representation of the mechanical system is shown in Figure 15.

We choose on purpose a very flexible (very soft) shaft, so that there will be a significant difference between the velocities at its two ends. The parameters are: $C = 0.005$ Nm sec/rad, $K = 0.1$ Nm/rad, $J_L = 332 \times 10^{-6}$ kg m^2, $b = 0.01$ Nm sec/rad.

For the simulations, we make the same assumptions as in Section 8 and we use the same motor parameters. The additions to the block diagram from Figure 2, in order to incorporate the flexible shaft and the load, are shown in Figure 16.

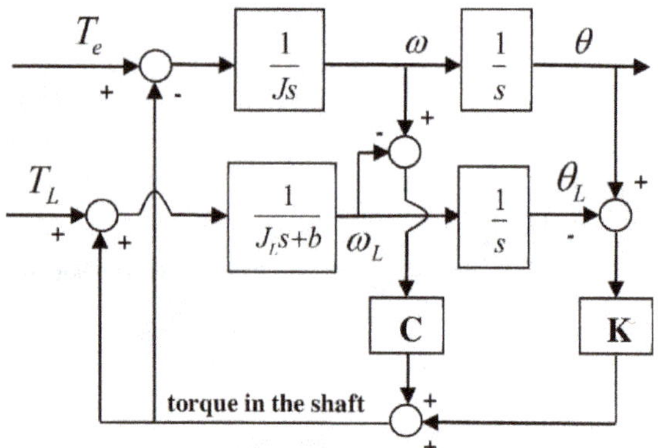

Figure 16. The block diagram of the mechanical system from Figure 15. This diagram should replace the two blocks in the right lower corner of Figure 2. The signal returning to the comparator on the left end of Figure 2 should be ω_L (if we want to control the velocity of the load). If the flexible shaft is sufficiently stiff, then, of course, $\omega_L \approx \omega$.

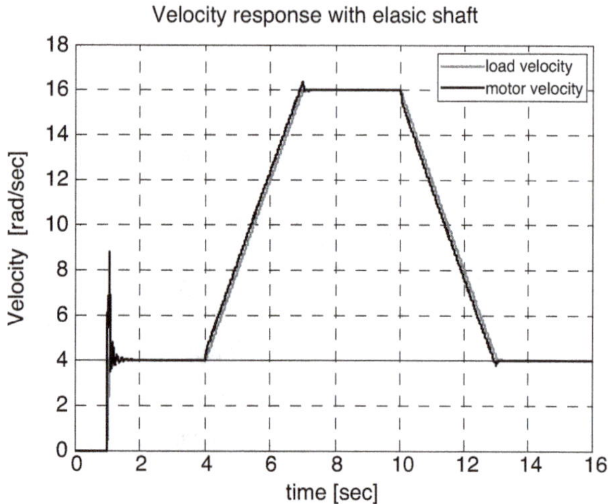

Figure 17. The motor and load velocities using a very flexible shaft and speed measurements from both the motor and the load. The adaptive controller is as described in Section 5, but it receives two different tracking errors e_L and e, as explained in the text.

One option is to close the velocity control feedback loop using the signalω, and to use exactly the same adaptive controller as in Section 5, with adjusted coefficients (we found that it performs well with $k_i$increased twice and k_1 decreased four times). When applying the reference signal from Figure 10, we observed that the cogging is not visible (like in Figure 11), but during the times when the reference signal increases or decreases between 4 and 16 rad/sec, there is an approximately constant velocity error of about 0.4 rad/sec (the load velocity ω_L lags behind ω). For an application where this velocity error is not acceptable, our solution is to mount another encoder on the load axis and use the tracking error $e_L = \omega_{ref} - \omega_L$ as the input of the PI part of the controller. The resonant part of the adaptive controller (described in (31)) should still receive the error signal $e = \omega_{ref} - \omega$, otherwise it becomes unstable.

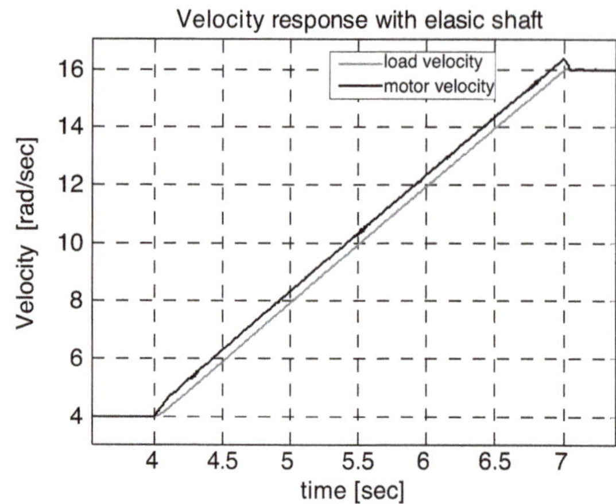

Figure 18. A zoom of Figure 17. Note that the motor velocity is ahead of the load velocity, which tracks the reference signal quite accurately.

With this control system, we obtain very good tracking for the reference signal from Figure 10, as shown in Figure 17. Figure 18 is a zoom of Figure 17.

10. Experimental results

For the experimental work, we used a RP23–54 24 V motor with an encoder of 20,000 counts/turn, a driver including the current sensors designed in the company Rafael and a dSpace setup, as shown in Figure 19. The sampling frequency for the current loop was 16 kHz and for the velocity loop, 4 kHz. The motor's nominal parameters are given in Section 8. The angular velocity was calculated inside the dSpace controller by taking the discrete derivative of the encoder output. We have closed the current loop in the dSpace controller.

10.1. A. Stability of the experimental system

The coefficients of the PI part of the controller have been chosen so that if we only use the PI part, we get a reasonable gain margin (8.5 dB) and phase margin (52°).

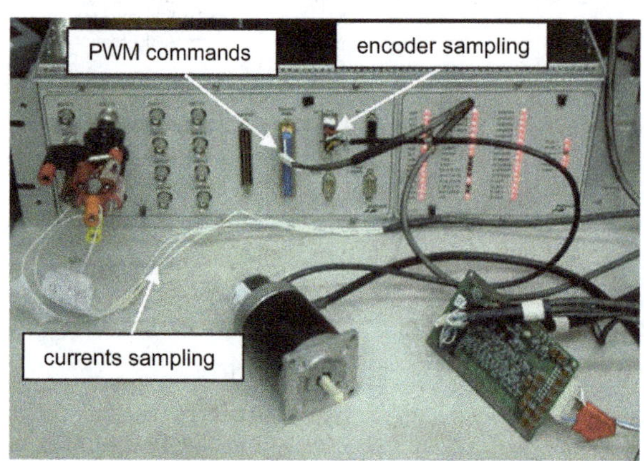

Figure 19. The equipment used for the experiments.

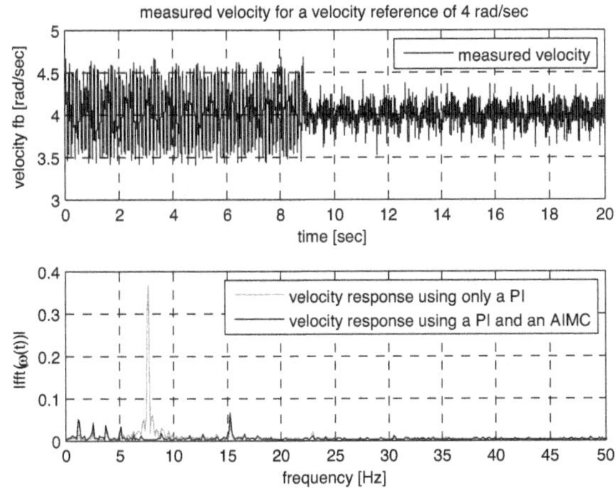

Figure 20. Velocity response and its steady-state frequency contents using a PI controller (0–9 s) and then an AIMC controller (from 9 to 20 s) for a velocity reference of 4 rad/sec. (The FFT was performed on the intervals 1–7 s for the PI, and 12–18 s for the AIMC).

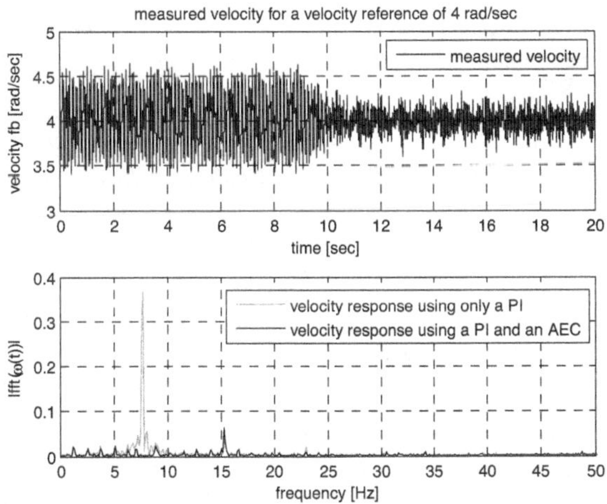

Figure 21. Velocity response and its steady-state frequency contents using a PI controller (0 up to 9 sec) and then an AEC controller (From 9 to 20 s) for a velocity reference of 4 rad/sec. (The FFT was performed on the intervals 1–7 s for the PI, and 12–18 s for the AEC).

10.2. B. Testing the adaptive algorithms

Figure 20 shows the motor velocity ω for $\omega_{\text{ref}} = 4$ rad/sec. We have used a PI controller up to 9 s, and then added an AIMC controller until 20 s. Figure 21 shows ω for the same ω_{ref} using a PI controller up to 9 s, and then switching to an AEC controller until 20 s. Since the AIMC and the AEC controllers give very similar results during the experiments (as can be seen by comparing Figures 20 and 21), in the next figures, we only give the results for the AIMC algorithm.

To check the adaptive controllers, the ω_{ref} command from Figure 10 was chosen. The velocity response using a PI controller versus an AIMC controller is shown in Figure 22. In order to see the reduction in the velocity error,

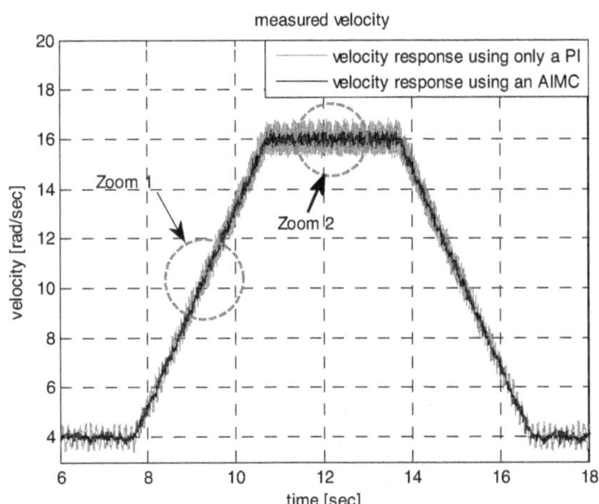

Figure 22. Velocity responses using a PI controller and an AIMC controller, for the reference command of Figure 10.

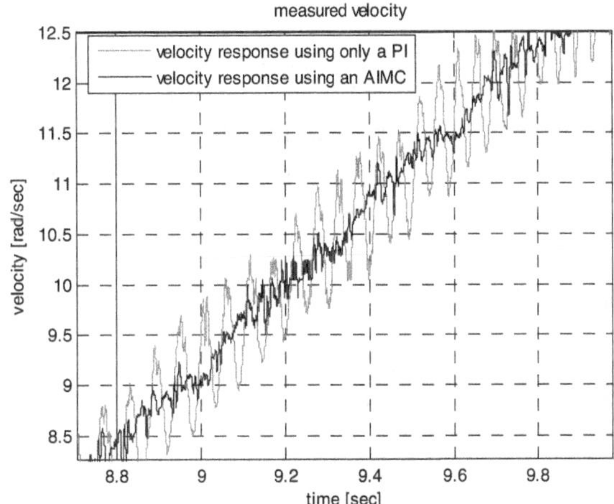

Figure 23. Velocity response using a PI versus an AIMC controller for a variable velocity reference, in the region marked "Zoom 1" in Figure 22.

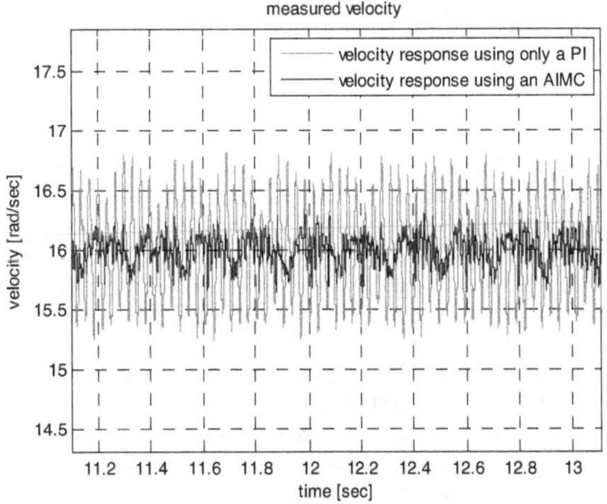

Figure 24. Velocity response using a PI versus an AIMC controller, for a constant reference command of 16 rad/sec, in the region marked as "Zoom 2" in Figure 22.

we view in Figures 23 and 24 two areas of the response, marked in Figure 22 as Zoom 1 and Zoom 2.

11. Conclusions

We have designed an internal model-based controller to suppress the torque ripple in a PMSM. The biggest challenge has been to get the controller to work at variable speed. To achieve this, we have used the rotor angle θ in place of the time variable in the formulation of the controller. We call this an adaptive internal model-based controller. Assuming a correct choice of resonant frequencies in the θ domain, we have proved convergence to zero of the tracking error under the restrictive assumption that ω_{ref} is constant. After giving a short review of the AEC from Canudas de Wit and Praly (2000), we have compared this controller with ours by simulations. The simulations show the ability of both controllers to track a variable ω_{ref} command while suppressing the effects of the cogging torque.

We have shown how to use our adaptive controller for a motor connected to a load via a flexible shaft and we have experimentally tested both controllers without load. The experiments confirm that both algorithms perform very well.

Acknowledgment

We are grateful to Dr Alex Ruderman from Nazarbayev University (Kazakhstan), formerly with the company Elmo Motion Control (Israel), for drawing our attention to this problem and providing background material.

Disclosure Statement

No potential conflict of interest was reported by the author(s).

Funding

This work was partly supported by the Israel Science Foundation grant 701/10.

References

Ahn, H. S., Chen, Y., & Dou, H. (2005). State periodic adaptive compensation of cogging and Coulomb friction in permanent-magnet linear motors. *IEEE Transaction on Magnetics*, *41*, 90–98.

Canudas de Wit, C., & Praly, L. (2000). Adaptive eccentricity compensation. *IEEE Transactions on Control Systems Technology*, *8*, 757–766.

Degobert, P., Remy, G., Zeng, J., Barre, P.-J., & Hautier, J.-P. (2006). *High performance control of the permanent magnet synchronous motor using self-tuning resonant controllers*. Proceedings of the South-eastern symposium on system theory, Cookeville, TN, USA.

Esbrook, A., Tan, X., & Khalil, H. K. (2011). *Tracking an unknown two-frequency reference using a frequency estimator-based servo compensator*. Proceedings of the 50th IEEE conference on decision and control, Orlando, FL, USA.

Ferretti, G., Magnani, G., & Rocco, P. (1998). Modeling, identification, and compensation of pulsating torque in permanent magnet AC motors. *IEEE Transaction on Industrial Electronics*, *45*, 912–920.

Ferretti, G., Magnani, G., & Rocco, P. (1999). *Torque ripple adaptive rejection in brushless motors*. Proceedings of the IEEE/ASME international conference on advanced intelligent mechatronics, Atlanta, USA.

Gan, W. C., & Qiu, L. (2004). Torque and velocity ripple elimination of AC permanent magnet motor control systems using the internal model principle. *IEEE Transaction on Mechatronics*, *9*, 436–447.

Hanselman, D. C. (1994). *Brushless permanent-magnet motor design*. New York: McGraw-Hill.

Hanselman, D., Hung, J. Y., & Keshura, M. (1992). *Torque ripple analysis in brushless permanent magnet motor drives*, Proceedings of the international conference on electrical machines, Manchester, UK, 823–827.

Hung, J. Y., & Ding, Z. (1993). Design of currents to reduce torque ripple in brushless permanent magnet motors. *IEE Proceedings B*, *140*, 260–266.

Hwang, S. M., Eom, J. B., Jung, Y. H., Lee, D. W., & Kang, B. S. (2001). Various design techniques to reduce cogging torque by controlling energy variation in permanent magnet motors. *IEEE Transaction on Magnetics*, *37*, 2806–2809.

Jahns, T. M., & Soong, W. L. (1996). Pulsating torque minimization techniques for permanent magnet AC motor drives – a review. *IEEE Transaction on Industrial Electronics*, *43*, 321–330.

Jayawardhana, B., & Weiss, G. (2009). State convergence of passive nonlinear systems with an L^2 input. *IEEE Transaction on Automatic Control*, *54*, 1723–1727.

Knobloch, H. W., Isidori, A., & Flockerzi, D. (1993). *Topics in control theory*. Basel: Birkhauser.

Ko, H. S., & Kim, K. J. (2004). Characterization of noise and vibration sources in interior permanent magnet brushless DC motors. *IEEE Transaction Magnetics*, *40*, 3482–3489.

Lee, K. W., Park, J. W., & Kwon, Y. A. (2004). *Research of speed ripple reduction method for computed tomography*. Proceedings of the annual conference on IEEE Industrial Electronics Society, Busan, Korea.

Le-Huy, H., Perret, R., & Feuillet, R. (1986). Minimization of torque ripple in brushless DC motor drives. *IEEE Transaction on Industrial Applications*, *22*, 748–755.

Lu, H., Zhang, L., & Qu, W. (2008). A new torque control method for torque ripple minimization of BLDC motors with unideal back EMF. *IEEE Transaction on Power Electronics*, *23*, 950–958.

Malaize, J., & Levine, J. (2009). *An observer based design for cogging forces cancellation in permanent magnet linear motors*. Proceeding of the 48th IEEE conference on. decision & control, Shanghai, China.

Mattavelli, P., Tubiana, L., & Zigliotto, M. (2005). Torque ripple reduction in PM synchronous motor drives using repetitive current control. *IEEE Transaction on Power Electronics*, *20*, 1423–1431.

Park, R. H. (1929). Two reactions theory of synchronous machines. *TAIEE*, *48*, 716–730. and Part II: *TAIEE*, 52, 338–350, 1933.

Park, S. J., Park, H. W., Lee, M. H., & Harashima, F. (2000). A new approach for minimum torque ripple maximum efficiency control of BLDC motor. *IEEE Transaction on Industrial Electronics*, *47*, 109–114.

Pillay, P., & Krishnan, R. (1989). Modeling, simulation and analysis of permanent-magnet motor drives, PART II: The brushless DC motor drive. *IEEE Transaction on Industry Applications*, *25*, 274–279.

Petrović, V., Ortega, R., Stanković, A., & Tadmor, G. (2000). Design and implementation of an adaptive controller for torque ripple minimization in PM synchronous motors. *IEEE Transaction on Power Electronics*, *15*, 871–880.

Qian, W., Panda, S. K., & Xu, J.-X. (2004). Torque ripple minimization in PM synchronous motors using iterative learning control. *IEEE Transaction on Power Electronics*, *19*, 272–279.

Ruderman, M., Ruderman, A., & Bertram, T. (2013). Observer-based compensation of additive periodic torque disturbances in permanent magnet motors. *IEEE Transaction on Industrial Informatics*, *9*, 1130–1138.

van der Schaft, A. (2000). *L_2-gain and passivity techniques in nonlinear control*. London: Springer-Verlag.

Spitsa, V., Kuperman, A., Weiss, G., & Rabinovici, R. (2006). *Design of a robust voltage controller for an induction generator in an autonomous power system using a genetic algorithm*. Proceeding of the ACC, Minneapolis, USA, 3475–3481.

Weiss, G. (1997). Repetitive control systems: Old and new ideas. In C. Byrneset al. (eds.), *System and control in the twenty-first century* (pp. 389–404). Basel: Birkhauser.

Xu, J. X., Panda, S. K., Pan, Y. J., Lee, T. H., & Lam, B. H. (2004). A modular control scheme for PMSM speed control with pulsating torque minimization. *IEEE Transaction on Industrial Electronics*, *51*, 526–536.

Yepes, A. G., Freijedo, F. D., Fernandez-Comesana, P., Malvar, J., Lopez, O., & Doval-Gandoy, J. (2010). *Torque ripple minimization in surface-mounted PM drives by means of PI + multi-resonant controller in synchronous reference frame*. Proceeding of the IECON, Glendale, AZ, USA.

Zhong, Q. C., & Weiss, G. (2011). Synchronverters: Inverters that mimic synchronous generators. *IEEE Transaction on Industrial Electronics*, *58*, 1259–1267.

22

Maneuver-based control of the 2-degrees of freedom underactuated manipulator in normal form coordinates

Carsten Knoll* and Klaus Röbenack

Faculty of Electrical and Computer Engineering, Institute of Control Theory, Technische Universität Dresden, Dresden, Germany

In this contribution, we provide a constructive way to transform a generic Lagrangian mechanical control system into the well-known Byrnes–Isidori normal form. Then, we restrict ourselves to the underactuated two-joint-manipulator in the horizontal plane. That system fails Brocketts condition and thus achieving point-to-point control is a challenging task. The system is analyzed in normal form coordinates, which allows to design a subordinate sliding mode controller and identify three discrete symmetries. Using these preliminary steps, finally a maneuver-based control scheme is proposed for equilibrium transition. Thereby each maneuver corresponds to a suitable sliding surface defined in normal form coordinates.

Keywords: underactuated manipulator; equilibrium transition; Lagrangian Byrnes–Isidori normalform; discrete symmetry; sliding mode control; maneuver-based control

1. Introduction

Underactuated mechanical systems are an interesting and illustrative field of control theory (Fantoni & Lozano, 2001; Spong, 2000). Many different control approaches have been applied to such systems, see, for example, Graichen, Treuer, and Zeitz (2007), Glück, Eder, and Kugi (2013), Riachy, Orlov, Floquet, Santiesteban, and Richard (2008), and Knoll and Röbenack (2013). Among them, underactuated manipulators (possessing some unactuated joints) are an important subclass, see De Luca, Iannitti, Mattone, and Oriolo (2002) and the references therein. Possible applications include areas where the reduction in weight and/or cost is desired while tasks can still be fulfilled with fewer actuators, like in space robotics (Yesiloglu & Temeltas, 2010; Yoshida, 1997) or when considering simple pick-and-place mechanisms (Oriolo & Nakamura, 1991).

Depending on how many joints lack an actuator, where they are located and which further assumptions are made, for example, regarding friction and potential forces, the control of underactuated manipulators is more or less difficult. In some cases, for example, the one discussed here, the system is not controllable, while in other cases even the property of linearizability via static feedback is fulfilled (Franch, Reyes, & Agrawal, 2013), allowing relatively easy controller design.

In this paper, we consider a 2-degrees of freedom (DOF) manipulator in the horizontal plane with an active joint at the base and a passive joint between the links. Motivated by pick-and-place tasks, our objective is the equilibrium transition of that system.

Due to Oriolo and Nakamura (1991) it is well-known that this manipulator does not fulfill the so-called Brockett condition (Brockett, 1983), which is necessary for the existence of a continuous differentiable feedback that stabilizes an isolated equilibrium point.

Because of the combination of an easy model with challenging control properties this system was subject to many studies before. Some of them, e.g. Arai and Tachi (1991) and Mareczek, Buss, and Schmidt (1999), rely on the presence of a holding brake which, however, might be seen as some kind of actuator. Other approaches assume the presence of considerable static friction, either explicitly (Mahindrakar, Rao, & Banavar, 2006), or implicitly (Scherm & Heimann, 2000) by means of the numerical stability of the algorithm. Like in the present contribution, the model investigated in Nakamura, Suzuki, and Koinuma (1997) and De Luca, Mattone, and Oriolo (2000) is assumed to be frictionless. The two approaches both use periodic input signals leading to high actuator activity, a longer duration and path length of the transition in comparison to the presence of a brake or static friction.

This paper contains two main contributions: Firstly, basing on the partial feedback linearization of a general underactuated system, we introduce its transformation to the "Lagrangian Byrnes–Isidori normal form (LBINF)" and give a simple proof of its global existence in terms of a closed-form formula. This representation allows for

*Corresponding author. Email: carsten.knoll@tu-dresden.de

any underactuated system to separate the effect of the input from system-inherent dynamics ("drift"), such that the corresponding vector fields are always orthogonal. In Olfati-Saber (2001, Section 3.7) a related but different approach has been proposed earlier, to transform an underactuated system to Byrnes–Isidori normal form, see also Choukchou-Braham, Cherki, Djemai, and Busawon (2014, Section 4.2). However, in that approach the existence of the transformation depends on an involutivity condition and the calculation depends on the solution of partial differential equations. Therefore, in general it is not possible to express that transformation in closed form, even if it exists. Furthermore, due to the simple structure of our transformation, some of the new state variables can be interpreted as quasi-velocities, see, for example, Cameron and Book (1997). This facilitates the controller design in the following.

The second contribution is, as an extension to Knoll and Röbenack (2010, 2011a), the presentation of a novel approach for the point-to-point control of the underactuated planar 2-DOF manipulator. In contrast to the existing approaches, it does neither rely on a brake or static friction nor on the application of periodic inputs.

This paper is organized as follows: In Section 2, we establish the transformation of a general underactuated mechanical system to the LBINF. Next, in Section 3, the model of the 2-DOF manipulator is stated and transformed into that representation. Section 4 is devoted to the analysis of the manipulator system in the new coordinates. Thereby, a generic subordinate sliding mode control is constructed and three discrete symmetry-mappings are identified, which substantially simplify the control design. Finally, in Section 5, the equilibrium transition is performed via successive execution of suitable maneuvers, each one corresponding to a special sliding surface defined in coordinates of the normal form.

The application of the control strategy proposed in Section 5 depends on a number of details not worthwhile to be included in the written text. Each single detail might be considered as a straightforward consequence of the outlined ideas. However, to facilitate the ability to reproduce and critically examine our results we provide the full source code (using the Python programming language) for the related simulations, see Knoll (2014a). This practice follows the argumentation of Ince, Hatton, and Graham-Cumming (2012).

2. Input–output linearization and Byrnes–Isidori normal form

2.1. *Input–output linearization of underactuated systems*

In this section, we consider a holonomic[1] mechanical system with n DOF. Its equations of motion, typically obtained from the Lagrangian formalism, read

$$\mathbf{M}(\mathbf{q})\ddot{\mathbf{q}} + \mathbf{C}(\mathbf{q}, \dot{\mathbf{q}}) + \mathbf{K}(\mathbf{q}, \dot{\mathbf{q}}) = \mathbf{B}(\mathbf{q})\tau. \tag{1}$$

Thereby, $\mathbf{q}$, $\dot{\mathbf{q}}$ and $\ddot{\mathbf{q}}$ denote the generalized coordinates, velocities and accelerations, respectively, $\mathbf{M}(\mathbf{q})$ is the (positive definite) mass matrix, $\mathbf{C}(\mathbf{q}, \dot{\mathbf{q}})$ describes the action of centrifugal and Coriolis forces and $\mathbf{K}$ collects the conservative and dissipative forces resulting from potential energy changes and friction. The m-dimensional vector τ contains the generalized forces from the actuators which are assigned to the scalar differential equations of (1) via the $n \times m$-matrix $\mathbf{B}(\mathbf{q})$. Obviously, the quantities $\mathbf{q}$, $\dot{\mathbf{q}}$ and $\ddot{\mathbf{q}}$ are time-dependent elements of $\mathbb{R}^n$. However for better readability the time argument is not explicitly written.

We assume that the system is underactuated, that is, $m < n$, and that the actuators are not redundant. Then, $\mathbf{B}(\mathbf{q})$ has full column rank for all $\mathbf{q} \in \mathbb{R}^m$ which implies the existence of $\mathbf{T}(\mathbf{q}) \in \mathbb{R}^{n\times n}$, such that $\mathbf{T}(\mathbf{q})\mathbf{B}(\mathbf{q}) = (\mathbf{I}_m, \mathbf{0})^{\mathrm{T}}$. By left-multiplying Equation (1) with $\mathbf{T}(\mathbf{q})$ and performing a suitable transformation of the coordinates, the system (1) can be rewritten as

$$\begin{pmatrix} \mathbf{M}_{11}(\mathbf{q}) & \mathbf{M}_{12}(\mathbf{q}) \\ \mathbf{M}_{12}^{\mathrm{T}}(\mathbf{q}) & \mathbf{M}_{22}(\mathbf{q}) \end{pmatrix} \begin{pmatrix} \ddot{\mathbf{q}}_1 \\ \ddot{\mathbf{q}}_2 \end{pmatrix} + \begin{pmatrix} \mathbf{C}_1(\mathbf{q}, \dot{\mathbf{q}}) \\ \mathbf{C}_2(\mathbf{q}, \dot{\mathbf{q}}) \end{pmatrix} + \begin{pmatrix} \mathbf{K}_1(\mathbf{q}, \dot{\mathbf{q}}) \\ \mathbf{K}_2(\mathbf{q}, \dot{\mathbf{q}}) \end{pmatrix} = \begin{pmatrix} \tau_1 \\ \mathbf{0} \end{pmatrix}. \tag{2}$$

For convenience, we also denote the new generalized coordinates $\mathbf{q} = (\mathbf{q}_1^{\mathrm{T}}, \mathbf{q}_2^{\mathrm{T}})^{\mathrm{T}}$. In this representation, the whole system splits up into an m-dimensional fully actuated subsystem with joint coordinates $\mathbf{q}_1$ and an $(n - m)$-dimensional subsystem without direct actuation.

For facilitation of the analysis and the controller design often an input–output linearization (also called "partial feedback linearization") of Equation (2) is performed (see, e.g. De Luca et al., 2002; Sastry, 1999) by applying the nonlinear static feedback

$$\begin{aligned} \tau_1 = {} & [\mathbf{M}_{11}(\mathbf{q}) - \mathbf{M}_{12}(\mathbf{q})\mathbf{M}_{22}^{-1}(\mathbf{q})\mathbf{M}_{12}^{\mathrm{T}}(\mathbf{q})]\mathbf{a} \\ & - \mathbf{M}_{12}(\mathbf{q})\mathbf{M}_{22}^{-1}(\mathbf{q})(\mathbf{C}_2(\mathbf{q}, \dot{\mathbf{q}}) + \mathbf{K}_2(\mathbf{q}, \dot{\mathbf{q}})) \\ & + \mathbf{C}_1(\mathbf{q}, \dot{\mathbf{q}}) + \mathbf{K}_1(\mathbf{q}, \dot{\mathbf{q}}). \end{aligned} \tag{3}$$

The resulting partial linearized system reads

$$\ddot{\mathbf{q}}_1 = \mathbf{a}, \tag{4a}$$

$$\ddot{\mathbf{q}}_2 = -\mathbf{M}_{22}^{-1}(\mathbf{q})(\mathbf{C}_2(\mathbf{q}, \dot{\mathbf{q}}) + \mathbf{K}_2(\mathbf{q}, \dot{\mathbf{q}}) + \mathbf{M}_{12}^{\mathrm{T}}(\mathbf{q})\mathbf{a}). \tag{4b}$$

According to Equation (4a), the new input $\boldsymbol{a}$ corresponds to the generalized accelerations of the actuated coordinates. In other words, the feedback (3) can be interpreted as inner control loop for the actuated joints.

In control theory, the input affine state-space representation of a system

$$\dot{\mathbf{x}} = \mathbf{f}(\mathbf{x}) + \mathbf{g}(\mathbf{x})\mathbf{u} \tag{5}$$

is very common. In this representation, $\mathbf{x}$ denotes the time-dependent state taking values in $\mathbb{R}^N$ (or an open subset thereof), $\mathbf{f}$ is the so-called drift vector field and $\mathbf{g}$ is a state-dependent $N \times m$ matrix assigning the components of the m-dimensional input $\mathbf{u}$ to the N differential equations in Equation (5). Clearly, by choosing the state components

$$\mathbf{x} = (\mathbf{q}_1^{\mathrm{T}}, \dot{\mathbf{q}}_1^{\mathrm{T}}, \mathbf{q}_2^{\mathrm{T}}, \dot{\mathbf{q}}_2^{\mathrm{T}})^{\mathrm{T}} =: (\mathbf{x}_1^{\mathrm{T}}, \mathbf{x}_2^{\mathrm{T}}, \mathbf{x}_3^{\mathrm{T}}, \mathbf{x}_4^{\mathrm{T}})^{\mathrm{T}} \tag{6}$$

and introducing n definitional equations of the form $\dot{x}_j = x_k$ the system of n second-order ODEs (4) can be rewritten in the form Equation (5) with $N = 2n$ and $\mathbf{u} = \boldsymbol{a}$:

$$\dot{\mathbf{x}}_1 = \mathbf{x}_2, \tag{7a}$$

$$\dot{\mathbf{x}}_2 = \mathbf{a}, \tag{7b}$$

$$\dot{\mathbf{x}}_3 = \mathbf{x}_4, \tag{7c}$$

$$\dot{\mathbf{x}}_4 = -\bar{\mathbf{M}}_{22}^{-1}(\mathbf{x}_1, \mathbf{x}_3)(\bar{\mathbf{C}}_2(\mathbf{x}) + \bar{\mathbf{K}}_2(\mathbf{x}) + \bar{\mathbf{M}}_{12}^{\mathrm{T}}(\mathbf{x}_1, \mathbf{x}_3)\mathbf{a}). \tag{7d}$$

The bar over the quantities indicates the reordering of the arguments to take the definition of the state (6) into account.

2.2. *Byrnes–Isidori normal form*

Equation (7d) shows that except in the special case where $\bar{\mathbf{M}}_{12}(\mathbf{x}_1, \mathbf{x}_3) \equiv 0$ the accelerations of the non-actuated joints are also influenced by $\boldsymbol{a}$. In terms of the input–output-normal form Equations (7c), (7d) represent the internal dynamics. As we will see in Sections 4.1 and 5 system analysis and control design are facilitated by the introduction of new coordinates, say $\mathbf{z} := (\mathbf{z}_1, \mathbf{z}_2, \mathbf{z}_3, \mathbf{z}_4)$, along with a state transformation which alters the internal dynamics such that it is not directly influenced by the input. This corresponds to a transformation to the Byrnes–Isidori normal form (Isidori, 1995), which in general refers to a decomposition of the system into mere integrator chains and an internal dynamics which does not depend on the input. In the scope of underactuated mechanical systems, we can restrict the generality and formulate an adapted definition:

DEFINITION 1 *A dynamical system of the form* (5) *is said to be in LBINF, if* $\mathbf{f}$ *and* $\mathbf{g}$ *are compatible with the representation*

$$\begin{pmatrix} \dot{\mathbf{z}}_1 \\ \dot{\mathbf{z}}_2 \\ \dot{\mathbf{z}}_3 \\ \dot{\mathbf{z}}_4 \end{pmatrix} = \begin{pmatrix} \mathbf{z}_2 \\ \mathbf{0} \\ \mathbf{f}_3(\mathbf{z}) \\ \mathbf{f}_4(\mathbf{z}) \end{pmatrix} + \begin{pmatrix} \mathbf{0} \\ \boldsymbol{I}_m \\ \mathbf{0} \\ \mathbf{0} \end{pmatrix} \mathbf{a}. \tag{8}$$

THEOREM 1 *Every mechanical system given in the form* (1) *can be transformed to the form* (8) *by a globally invertible change of state coordinates.*

Proof The equivalence between Equations (1) and (7) is straightforward.[2] Now, the LBINF coordinates are given by

$$\mathbf{z}_i := \mathbf{x}_i, \quad i = 1, 2, 3 \tag{9a}$$

and

$$\mathbf{z}_4 := \mathbf{x}_4 + \bar{\mathbf{M}}_{22}^{-1}(\mathbf{x}_1, \mathbf{x}_3)\bar{\mathbf{M}}_{12}^{\mathrm{T}}(\mathbf{x}_1, \mathbf{x}_3)\mathbf{x}_2. \tag{9b}$$

The first three components of the transformation are identical and the last one, given by Equation (9b) can be inverted globally as well as

$$\mathbf{x}_4 = \mathbf{z}_4 - \bar{\mathbf{M}}_{22}^{-1}(\mathbf{z}_1, \mathbf{z}_3)\bar{\mathbf{M}}_{12}^{\mathrm{T}}(\mathbf{z}_1, \mathbf{z}_3)\mathbf{z}_2. \tag{10}$$

To complete the proof, we notice that in the expression for the time derivative of $\mathbf{z}_4$ the coefficients of $\boldsymbol{a}$ nullify each other:

$$\begin{aligned} \dot{\mathbf{z}}_4 &= \underbrace{-\bar{\mathbf{M}}_{22}^{-1}(\bar{\mathbf{C}}_2 + \bar{\mathbf{K}}_2) - \bar{\mathbf{M}}_{22}^{-1}\bar{\mathbf{M}}_{12}^{\mathrm{T}}\mathbf{a}}_{\dot{\mathbf{x}}_4} \\ &\quad + \bar{\mathbf{M}}_{22}^{-1}\bar{\mathbf{M}}_{12}^{\mathrm{T}}\underbrace{\mathbf{a}}_{\dot{\mathbf{x}}_2} + \left(\frac{\mathrm{d}}{\mathrm{d}t}\left(\bar{\mathbf{M}}_{22}^{-1}\bar{\mathbf{M}}_{12}^{\mathrm{T}}\right)\right)\mathbf{z}_2 \\ &= -\bar{\mathbf{M}}_{22}^{-1}(\bar{\mathbf{C}}_2 + \bar{\mathbf{K}}_2) + \left(\frac{\mathrm{d}}{\mathrm{d}t}\left(\bar{\mathbf{M}}_{22}^{-1}\bar{\mathbf{M}}_{12}^{\mathrm{T}}\right)\right)\mathbf{z}_2. \end{aligned} \tag{11}$$

■

Remark 1 The vector $\mathbf{z}_4$ is a weighted sum of the actuated and non-actuated velocities and can be interpreted as a quasi-velocity (Cameron & Book, 1997). It obviously has the same physical dimension like $\mathbf{x}_4$ $(= \dot{\mathbf{q}}_2)$. In states with $\mathbf{x}_2 = \mathbf{0}$, we have $\mathbf{z}_4 = \mathbf{x}_4$, which is interesting for equilibrium transitions (see Section 4.1, Definition 2).

Remark 2 In contrast to the special Byrnes–Isidori normal form proposed earlier in Olfati-Saber (2001, Section 3.7), here no additional constraints regarding involutivity have to be fulfilled. The transformation to the normal form always exists and is given in closed form by Equation (9). Note that, for a general multi-input system (5) the Byrnes–Isidori normal form exists only if the distribution spanned by the input vector fields is involutive, cf. (Isidori, 1995, Proposition 5.1.2). However, for the mechanical system (7), the input vector fields depend only on $\mathbf{x}_1, \mathbf{x}_3$ but are zero in the first and third block (7a) and (7c), respectively. This structure implies that all Lie brackets between the input vector fields of Equation (7) are zero vector fields, that is, the associated distribution is involutive.

The LBINF representation has two main advantages: The system input $\boldsymbol{a}$ only occurs in the acceleration-equations of the actuated joints and can therefore be eliminated by interpreting the velocities as inputs themselves. This might be reasonable where a subordinate speed control is implemented in order to handle backslash and friction. Additionally, the systems motion through the state space is easier to understand and thus to control which will be illustrated in Section 4.1 and the following ones.

3. The underactuated manipulator

3.1. System description

From now on, we consider the underactuated two-degree-of-freedom horizontal manipulator with a friction-free non-actuated second joint, cf. Figure 1.

The configuration space for this system is considered to be $\mathbb{S}_1 \times \mathbb{S}_1$, that is, for the two angles the value of 2π is identified with the value 0. For the actual calculations this assumption does not matter, but it simplifies the transition scheme because any desired joint angle can be reached in both rotating directions.

By means of the Lagrangian formalism the equations of motions are obtained in the form of Equation (2). A subsequent partial linearization leads to the model

$$\dot{x}_1 = x_2, \tag{12a}$$

$$\dot{x}_2 = a, \tag{12b}$$

$$\dot{x}_3 = x_4, \tag{12c}$$

$$\dot{x}_4 = -\kappa x_2^2 \sin x_3 - (1 + \kappa \cos x_3)a, \tag{12d}$$

where the only remaining parameter κ is dimensionless and describes the distribution of mass on the second arm. Like many other mechanical systems the manipulator model (12) possesses time reversal symmetry (Knoll & Röbenack, 2011b) which will be used later for simplifying the trajectory planning, cf. Section 4.3.

The Jacobian linearization of this system is not controllable and, moreover, the system violates Brocketts necessary conditions (Brockett, 1983; Oriolo & Nakamura, 1991). In other words, there exists no continuous differentiable feedback of the state, that is, no continuous differentiable and static control algorithm, which stabilizes an isolated equilibrium point of the system.[3] Therefore, despite the simple model (12), the control of this system and especially equilibrium transitions are a challenging problem.

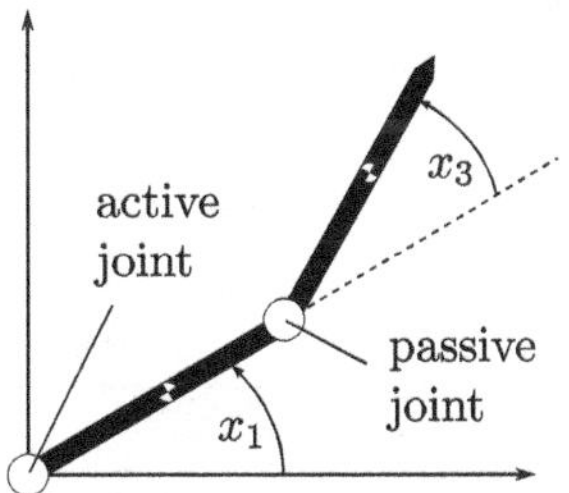

Figure 1. Underactuated manipulator model (top view).

As in the general case, discussed in the previous section, the input a affects both the double integrator subsystem for the active joint and the second subsystem for the passive joint. The transformation to the LBINF and its inverse are given by

$$z_i := x_i \quad \forall i \in \{1,2,3\}, \tag{13a}$$

$$z_4 := x_4 + (1 + \kappa \cos x_3)x_2 \tag{13b}$$

and

$$x_4 = z_4 - (1 + \kappa \cos z_3)z_2. \tag{13c}$$

This implies

$$\dot{z}_4 = -\kappa x_2 \sin x_3 (x_2 + x_4) \tag{14a}$$

$$= -\kappa z_2 \sin z_3 (z_4 - \kappa \cos z_3 z_2) \tag{14b}$$

and thus, the dynamics in LBINF read

$$\underbrace{\begin{pmatrix}\dot{z}_1\\ \dot{z}_2\\ \dot{z}_3\\ \dot{z}_4\end{pmatrix}}_{\dot{\mathbf{z}}} = \underbrace{\begin{pmatrix} z_2 \\ 0 \\ z_4 - (1+\kappa\cos z_3)z_2 \\ -\kappa z_2 \sin z_3 (z_4 - \kappa \cos z_3 z_2)\end{pmatrix}}_{\mathbf{f}(\mathbf{z})} + \underbrace{\begin{pmatrix}0\\1\\0\\0\end{pmatrix}}_{\mathbf{g}(\mathbf{z})} a. \tag{15}$$

4. System analysis in normal form

4.1. Projection to the z_2–z_4-plane

The model representation (15) has the advantage to separate the influence of the input a and the drift. In other words, the two vector fields $\mathbf{f}$ and $\mathbf{g}$ are always orthogonal to each other. Especially interesting is the projection into the two-dimensional subspace spanned by the second and fourth unit coordinate vector, that is, the z_2–z_4-plane:

$$\underbrace{\begin{pmatrix}\dot{z}_2\\ \dot{z}_4\end{pmatrix}}_{\dot{\bar{\mathbf{z}}}} = \underbrace{\begin{pmatrix} 0 \\ -\kappa z_2 \sin z_3 (z_4 - \kappa \cos z_3 z_2)\end{pmatrix}}_{\bar{\mathbf{f}}(\mathbf{z})} + \underbrace{\begin{pmatrix}1\\0\end{pmatrix}}_{\bar{\mathbf{g}}(\mathbf{z})} a, \tag{16}$$

where z_3 can be interpreted as a parameter rather than a component of the state. Obviously, $\bar{\mathbf{g}}$ and $\bar{\mathbf{f}}$ are parallel to the z_2 and z_4 axes, respectively. From a topological point of view, only considering the direction of the drift vector field $\bar{\mathbf{f}}$, the projected system is similar to a double integrator, see Knoll and Röbenack (2011a) for details. In other words, except for some singular z_3-values, in one of the four quadrants the drift always points towards the abscissa (dark gray) and in another one it points away from it (light gray), cf. Figure 2. The remaining two quadrants are crossed by the straight line

$$z_4 = \kappa \cos z_3 z_2 \tag{17}$$

which, together with the z_4-axis, indicates where the sign of the drift changes.

This situation makes the system dynamics much more predictable and hence allows the construction of a maneuver-based equilibrium transition.

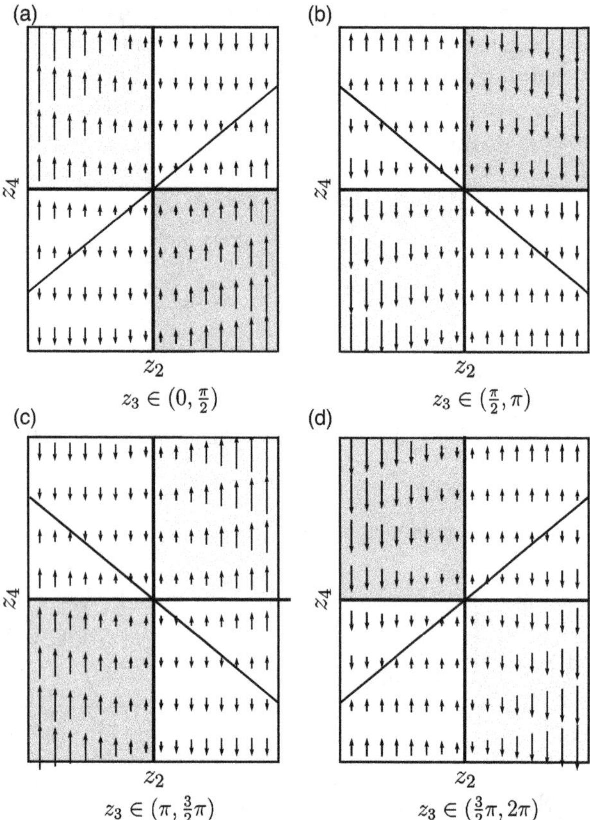

Figure 2. Drift vector field of the projected manipulator dynamics (16). For (almost) any value of z_3 in one quadrant the drift is topologically equivalent to the (negative) double integrator. In other words: the drift points towards (away from) the abcissa (dark and light shading, respectively). The diagonal line indicates where the drift changes its sign, see Equation (17).

4.2. Sliding mode control in the z_2–z_4-plane

These maneuvers will be performed by stabilizing the system to appropriate sliding surfaces which are constructed as follows. We consider the curve

$$z_2 = \varphi(z_4) \tag{18}$$

in the z_2–z_4-plane, where $\varphi(\cdot)$ is a differentiable function (up to a finite number of exception points where its derivative might be discontinuous). Associated to this "switching curve" is the "switching function"

$$\Phi(\mathbf{z}) := z_2 - \varphi(z_4). \tag{19}$$

Clearly, the condition $\Phi(\mathbf{z}) = 0$ constitutes a three-dimensional submanifold of the state space of system (15).

THEOREM 2 *The sliding surface given by $\Phi(\mathbf{z}) = 0$ is asymptotically stabilized by the feedback law*

$$v = \varphi'(z_4)f_4(\mathbf{z}) - \gamma_1 \operatorname{sign}(\Phi(\mathbf{z})) \mid \Phi(\mathbf{z})|^{\gamma_2} \tag{20}$$

with $\gamma_1 > 0$ and $\gamma_2 \in (0, 1)$.

Proof Attractiveness and invariance of the set $\{\mathbf{z} \in \mathbb{R}^4 \mid \Phi(\mathbf{z}) = 0\}$ is shown by the Lyapunov argument

$$\begin{aligned}\frac{1}{2}\frac{\mathrm{d}}{\mathrm{d}t}\Phi^2(\mathbf{z}) &= \Phi(\mathbf{z})\dot{\Phi}(\mathbf{z}) \\ &= \Phi(\mathbf{z})(\dot{z}_2 - \varphi'(z_4)\dot{z}_4) \\ &= \Phi(\mathbf{z})(v - \varphi'(z_4)f_4(\mathbf{z})) \\ &= -\gamma_1|\Phi(\mathbf{z})|^{1+\gamma_2} \le 0,\end{aligned} \tag{21}$$

cf. (Slotine & Li, 1991, Chapter 7). ■

Remark 3 If the system is in sliding regime, that is, if $\Phi(\mathbf{z}) = 0$, then it is governed by the so-called reduced dynamics

$$\dot{z}_3 = z_4 - (1 + \kappa \cos z_3)\varphi(z_4) =: \bar{f}_3(z_3, z_4) \tag{22a}$$

$$\dot{z}_4 = \kappa\varphi(z_4)\sin z_3(\kappa\varphi(z_4)\cos z_3 - z_4) =: \bar{f}_4(z_3, z_4), \tag{22b}$$

and therefore behaves like an autonomous system. On the other hand, the particular choice of $\varphi(\cdot)$ is a remaining degree of freedom to reach the control objective of the respective maneuver.

4.3. Symmetry properties

In this section, we identify some properties of the system which allow the direct construction of new solution trajectories from given solution trajectories. As mentioned above, the manipulator possesses time reversal symmetry because it is a conservative mechanical system (see Knoll & Röbenack, 2011b for further explanation). Using the reduced dynamics this is expressed by the observation

$$\bar{f}_3(-z_3, z_4) = \bar{f}_3(z_3, z_4), \tag{23a}$$

$$\bar{f}_4(-z_3, z_4) = -\bar{f}_4(z_3, z_4). \tag{23b}$$

Symmetry Property 1 (time reversal symmetry). Suppose $t \mapsto (z_3(t), z_4(t))$ is a solution of Equation (22) for $t \in [0, \tau]$. Then, another solution for this interval is given by

$$\hat{z}_3(t) := -z_3(\tau - t), \tag{24a}$$

$$\hat{z}_4(t) := z_4(\tau - t). \tag{24b}$$

Proof We differentiate $(\hat{z}_3, \hat{z}_4)$ w.r.t. time, apply the symmetry (23) to change the sign of the respective first arguments and then substitute using Equation (24):

$$\dot{\hat{z}}_3(t) = \bar{f}_3(\underbrace{-z_3(\tau - t)}_{\hat{z}_3(t)}, \underbrace{z_4(\tau - t)}_{\hat{z}_4(t)}), \tag{25a}$$

$$\dot{\hat{z}}_4(t) = \bar{f}_4(\underbrace{-z_3(\tau - t)}_{\hat{z}_3(t)}, \underbrace{z_4(\tau - t)}_{\hat{z}_4(t)}). \tag{25b}$$

The result is the system dynamics of Equation (22) but with the newly constructed solution. ■

This property allows to consider the submaneuvers of leaving and reaching an equilibrium point as essentially the same problem. The only difference is the sign of the time variable. Furthermore, there is a related property associated with Equation (23).

Symmetry Property 2 (recurrence property). Let $t \mapsto (z_3(t), z_4(t))$ be a solution of Equation (22) for $t \in [0, \tau]$ and suppose that $z_3(\tau) = k\pi$ with $k \in \mathbb{Z}$. Then, another solution for the interval $[\tau, 2\tau]$ is given by

$$\hat{z}_3(t) := 2z_3(\tau) - z_3(2\tau - t), \quad (26a)$$

$$\hat{z}_4(t) := z_4(2\tau - t). \quad (26b)$$

The proof of this property follows the same lines as the previous one and is therefore omitted. As a consequence of the recurrence property, in general there are periodic solutions during the sliding regime. The only reasons for non-periodic solutions are: (1) the system reaches an equilibrium, (2) the sliding regime is interrupted, for example, by a change of the switching curve, or (3) the solution ceases to exist due to final escape time. However, the latter case can be excluded if $|\varphi(\cdot)|$ is bounded.[4]

Now, we additionally assume point symmetry for the switching curve, that is, $\varphi(-z_4) = -\varphi(z_4)$, from which immediately follows

$$\bar{f}_3(-z_3, -z_4) = -\bar{f}_3(z_3, z_4), \quad (27a)$$

$$\bar{f}_4(-z_3, -z_4) = -\bar{f}_4(z_3, z_4). \quad (27b)$$

Symmetry Property 3 (quadrant equivalence). Let $t \mapsto (z_3(t), z_4(t))$ be a solution of Equation (22) for $t \in [0, \tau]$ and suppose $\varphi(\cdot)$ to be point symmetric. Then, another solution for $t \in [0, \tau]$ is given by

$$\hat{z}_3(t) := -z_3(t), \quad (28a)$$

$$\hat{z}_4(t) := -z_4(t). \quad (28b)$$

Again, the proof is similar to the argument used in the case of the time reversal symmetry and therefore omitted. The obvious consequence from the quadrant equivalence property is, that submaneuvers (or switching curves) only have to be planned for the first and second quadrant of the z_2–z_4-plane and then can be trivially adapted for the third and fourth quadrant, respectively.

4.4. *Parking regime*

Clearly, all equilibrium points of Equation (15) are located at the origin of the z_2–z_4-plane. The projected transition-trajectory must thus leave this point and reach this point somehow. Due to the loss of controllability near the equilibria, actually reaching such a state is quite hard. Considering the drift, this difficulty is reflected by the vanishing of the drift on the z_4-axis.

DEFINITION 2 *For system* (15) *all states with* $z_2 = 0$ *are called* parking regime.

This denomination results from the fact that whenever the active joint stands still (for finite time), the only non-vanishing time derivative in Equation (15) is $\dot{z}_3 = \text{const}$. In the z_2–z_4-projection the state then remains unchanged, that is, the projected system is "parking". However, the drift conditions in the plane of course do change with z_3. For the motion planning this property allows, simply to wait in the parking regime until suitable drift conditions for the subsequent maneuver are reached.

Another consequence of the investigation of the projected drift vector field is that the tangent of any solution trajectory of Equation (15) containing a parking regime must be parallel to the z_2-axis, where $z_2 = 0$ holds. Since equilibria also fulfill Definition 2, it is clear that trajectories leaving or reaching the origin must have the tangent line $z_4 = 0$, that is, the z_2-axis.

5. Equilibrium transition via maneuver-based control

5.1. *Maneuver overview*

We consider a desired transition between the initial state $\mathbf{z}^*$ and the final state $\mathbf{z}^\dagger$, both of which are assumed to be equilibrium points. The transition between these states can be split up into several maneuvers which are connected by time periods where the system is in the parking regime, that is, we have $z_2(t) = 0$ and thus $\dot{z}_3(t) = z_4(t) = \pm z_4^{\mathrm{p}}$. Therefore, any desired drift condition can be achieved simply by waiting. The maximal waiting time is given by $2\pi/z_4^{\mathrm{p}}$ and therefore $|z_4^{\mathrm{p}}|$ should be chosen sufficiently large. For the sake of simplicity, we consider $|z_4^{\mathrm{p}}|$ to be fixed for all occurring parking regimes.

To any of the maneuvers we can associate a curve in the z_2–z_4-plane, which can be stabilized by sliding mode control. Obviously, the first maneuver ("A") has to start in the initial equilibrium $\mathbf{z}^*$ and it ends on the z_4-axis in the parking regime. Conversely, the last maneuver ("D") has to reach the final equilibrium $\mathbf{z}^\dagger$ and it starts on the z_4-axis. As will be shown in Section 4.3, maneuvers A and D can be regarded as equivalent due to time reversal symmetry.

For the simplest combinations of initial and final states, these two maneuvers already suffice for equilibrium transition. In general, however, two more maneuvers are necessary. Maneuver B serves to change the sign of $\dot{z}_3$. In other words, it transfers the system from a parking regime with $z_4 = z_4^{\mathrm{p}}$ to another parking regime with $z_4 = -z_4^{\mathrm{p}}$, or vice versa. The objective of maneuver C is to adapt z_1 such that, after the final maneuver D, the active joint is in its desired position. Thereby we exploit the recurrence property, and,

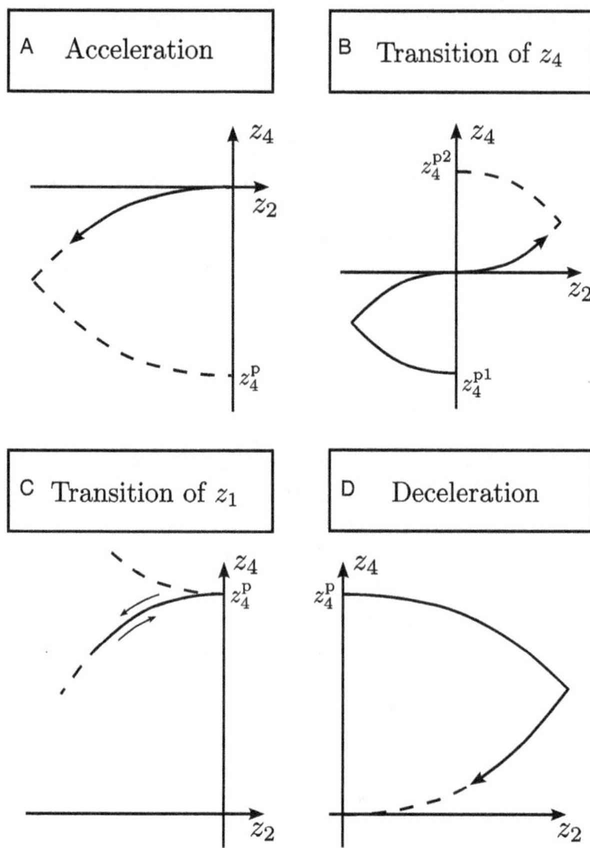

Figure 3. Principle succession of the maneuvers to perform an equilibrium point transition.

additionally, the fact that the coordinate z_1 is cyclic, that is, the dynamics of the system is independent of its value. Figure 3 shows how the switching curves of these four maneuvers might look like.

The further sections require addressing the components of the state at different stages of the transition. To this end, we introduce the following notation: The letters "A", "B", "C", and "D" used as head-index denominate the corresponding maneuver, the letter "p" indicates the parking regime and the symbols $*$ and $\dagger$ stand for the beginning and the end of a maneuver, respectively. For example, $z_3^{*\mathrm{A}}$ is the angle of the second joint in the start equilibrium, z_4^{pA} is the z_4 value in the parking regime associated with maneuver A and $z_1^{\dagger\mathrm{C}}$ is the angle of the first joint after the completion of maneuver C. The head index "eq" indicates the equilibrium value of a quantity – either at the start or end of the transition.

5.2. *Leaving and reaching equilibrium points*

In this section, we consider the transfer of the system from parking regime to an equilibrium (maneuver D), and its time reversal counterpart (maneuver A). Thereby, depending on the z_3 value of the concerned equilibrium, two cases can occur. Firstly, we discuss the case $z_3^{\mathrm{eq}} \in (\pi/2, \pi) \cup (\pi, \frac{3}{2}\pi)$, with the maneuvers D1 and A1. After that, we extend the results to $z_3^{\mathrm{eq}} \in (0, \pi/2) \cup (\frac{3}{2}\pi, 2\pi)$ for the construction of maneuvers A2 and D2.

From the investigation of the drift in Section 4.1, it becomes obvious that any projected trajectory containing a point with $z_2 = 0$ must have a tangent in this point which is parallel to the z_2-axis. Clearly, this also holds true for the switching curve (19). In other words: $\varphi(\cdot)$ must have infinite slope at its roots. This property is complied by power functions with exponents smaller then one. From Knoll and Röbenack (2011a), we know that for $z_2, z_4 > 0$ and $z_3 \in (\pi/2, \pi)$.

$$\varphi(z_4) := \begin{cases} (\mu\beta|z_4|)^{1/\beta} & \text{if } \dfrac{z_4^{\mathrm{p}} - z_4}{z_4^{\mathrm{p}}} > \dfrac{1}{2}, \\ (\mu\beta|z_4^{\mathrm{p}} - z_4|)^{1/\beta} & \text{else} \end{cases} \quad (29)$$

with $2 < \beta < 3$ is a suitable choice for maneuver D1 in the first quadrant.[5]

Of course, the state z_3 changes during the maneuver which causes two issues: Firstly, the drift conditions change during the execution and secondly, the final value of z_3 depends on the z_3-start value (of maneuver D1) and this dependency cannot be expressed in closed form. However, for the motion planning a closed-form expression is not necessary. Given the desired final value $z_3^{\dagger}$, the start value for maneuver D1 can be obtained by backward integration, that is, by stabilizing the system $\dot{\mathbf{z}} = -\mathbf{f}(\mathbf{z}) - \mathbf{g}(\mathbf{z})a$ on the same sliding surface.[6] The z_3-value where the system reaches the parking regime in backward running time, obviously is the start value z_3^* for maneuver D1 in forward time.

Regarding the first issue, it is clear that a mere *quantitative* change of the drift vector field is not a problem for the execution of maneuver D1. As long as the sign of the drift does not change, the reduced dynamics lead the system towards the equilibrium. If, however, z_3 leaves the interval $(\pi/2, \pi)$ the z_2–z_4-projection of the system dynamics are no longer topologically equivalent to the vector field depicted in Figure 2(b). In other words, if the drift changes its sign, the origin cannot be reached. From Equation (15) we extract

$$\dot{z}_3 = z_4 - (1 + \kappa \cos z_3) z_2 \quad (30)$$

and hence, all states with $\dot{z}_3 = 0$ must lay on the straight line

$$z_4 = (1 + \kappa \cos z_3) z_2. \quad (31)$$

For the sake of simplicity, we assume $\kappa < 1$. This so-called "strong inertial coupling" (Spong, 1998) can be reached for any combination of arm lengths by means of suitable constructional measures (Knoll, Leist, & Röbenack, 2011). Then, the straight line always passes from the third to the first quadrant of the z_2–z_4-plane. This means that, the switching curve (29) crosses this line somewhere in the first quadrant, and thus z_3 increases at the beginning of the

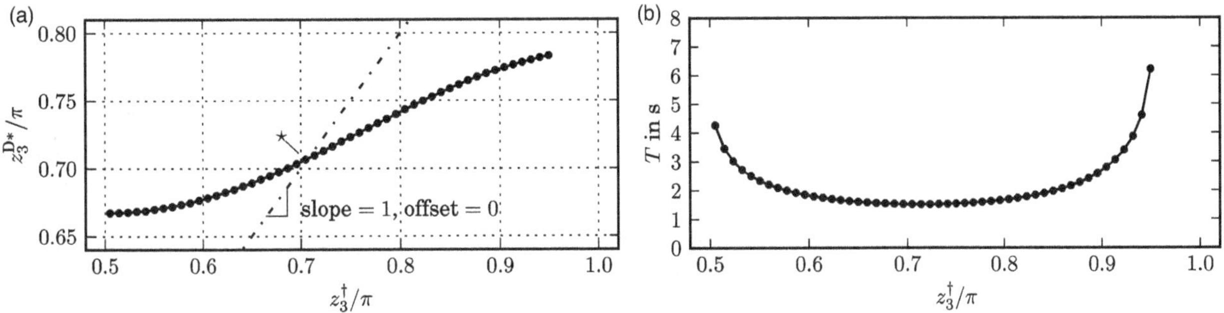

Figure 4. (a): The z_3-starting-value for maneuver D1 in dependency of the desired final passive joint angle $z_3^\dagger$. The bisecting line (dash-dotted) and the intersection point will be used for the construction of maneuver B (Section 5.3). (b): Duration of maneuver D1 for each final value. Each dot in both diagrams corresponds to a single simulation run.

maneuver and decreases at its end. From this qualitative consideration, we can suspect that the overall change of z_3 during maneuver D1 is "relatively small" and the construction (29) hence is a good choice for the sliding surface. This conjecture is confirmed by simulation experiments,[7] see Figure 4.

By applying the quadrant equivalence (Symmetry Property 3) and identifying $z_3 \in (-\pi, -\pi/2)$ with $z_3 \in (\pi, \frac{3}{2}\pi)$, this approach for maneuver D1 can be easily adapted for desired final conditions with $z_3^\dagger \in (\pi, \frac{3}{2}\pi)$. From Figure 2(c) follows that the switching line then lies in the third quadrant.

Due to time reversal symmetry it is clear, that (with slight adaption) the results for the deceleration maneuver also hold true for maneuver A1, that is, to leave the start equilibrium and reach a parking regime. In particular, the switching line through the first quadrant ($z_4^{\mathrm{P}} > 0$), which is used to reach equilibrium states with $z_3^\dagger \in (\pi/2, \pi)$, also serves to leave equilibria with $z_3^* \in (\pi, \frac{3}{2}\pi)$, while the switching line in the third quadrant (same curve mirrored along both coordinate axes, hence $z_4^{\mathrm{P}} < 0$) can either be used for breaking with $z_3^\dagger \in (\pi, \frac{3}{2}\pi)$ or accelerating with $z_3^* \in (\pi/2, \pi)$.

However, for the remaining two intervals $(0, \pi/2)$ and $(\frac{3}{2}\pi, 2\pi)$ the situation is different. In these cases, the double-integrator-like drift conditions lie in the second and fourth quadrant, see Figure 2(a) and 2(d). From the positive slope of the line (31) implicitly follows that in these quadrants the absolute value of $\dot{z}_3$ is relatively high which causes the drift to change its sign shortly after the maneuver start and therefore inhibits the same kind of maneuver as in the cases above.

The construction of a maneuver-pattern, which incorporates this sign change of the drift, is illustrated best by means of an acceleration-maneuver ("A2") for $z_3^* \in (\frac{3}{2}\pi, 2\pi)$ (fourth quadrant). The generalizations to the opposite time direction and the other quadrant can then be performed by the Symmetry properties 1 and 3.

The maneuver A2 is divided into two phases, each one associated with one value[8] of $\mathrm{sign}(\dot{z}_4)$. In each phase, the

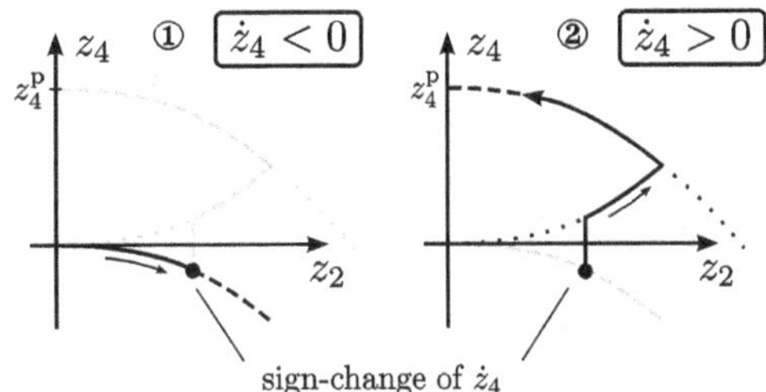

Figure 5. Schematic illustration of the two phases of maneuver A2. Phase 1: The drift points downwards until z_3 reaches a certain value. The sliding surface is defined by an adapted single branch version of Equation (29). Phase 2: Drift points upwards. A sliding surface similar to the maneuver A1 can be used, augmented with a lower bound on the value of $\varphi(\cdot)$ in the lower branch. Maneuver D2 uses the same switching curves. Then, due to time inversion, phase 1 and 2 switch its rolls and the direction of the arrows is reverted.

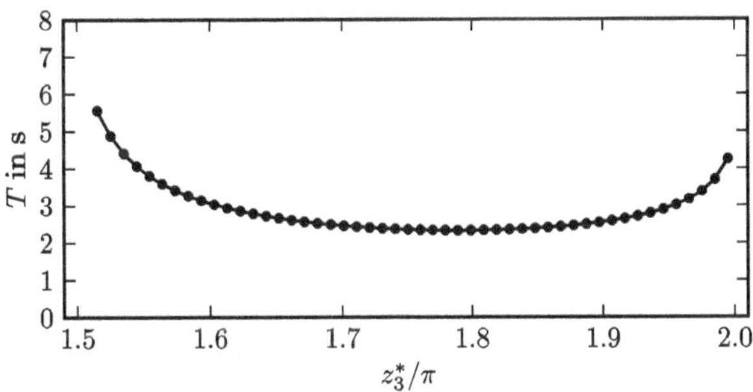

Figure 6. Simulation results for maneuver A2 with $z_3^* \in (\frac{3}{2}\pi, 2\pi)$. Obviously, the maneuver can be executed with acceptable duration everywhere, except near the interval boundary.

system is stabilized onto an suitable sliding surface, see Figures 5 and 6 for an overview and the simulation source code (Knoll, 2014a) for details.

Together the maneuvers A1, A2 and D1, D2 can be used to leave or reach, respectively, almost any equilibrium point. However, the values $z_3 = (k/2)\pi,\ k \in \mathbb{Z}$ must be excluded due to the singular drift conditions in these cases. Additionally, z_3^{eq} values close to these singularities might cause too long maneuver durations, see Figures 4 and 6.

5.3. *Transition of z_4 (Maneuver B)*

When leaving or reaching an equilibrium, due to the drift conditions (see Figure 2) the equilibrium-angle of the second joint determines the sign of z_4 in the associated parking regime. If the desired initial and final states are given such that the z_4^{p}-values, associated to the appropriate maneuvers A and D, do not match, an additional maneuver ("B") has to perform the transfer $z_4^{\mathrm{pA}} = \pm|z_4^{\mathrm{p}}| \rightarrow z_4^{\mathrm{pD}} = \mp|z_4^{\mathrm{p}}|$. As stated above, during parking regime z_4 is equivalent to $\dot{z}_3$, that is, to the angular velocity of the second joint. Thus, the task of maneuver B is the inversion of the rotating direction of that joint.

Although there naturally exists an infinite variety of possible motion patterns for this goal, in the present context it is easily achieved by the combination of maneuvers D1 and A1, described above. With this choice and the assumptions $z_4^{\mathrm{pA}} < 0$ and $z_4^{\mathrm{pD}} > 0$, we have the following situation: The system is in parking regime before maneuver B begins, which makes it possible to wait until $z_3 = z_3^{*\mathrm{B}} \in (\pi, \frac{3}{2}\pi)$. From Figure 2 it is obvious, that the third quadrant can be used for D1-breaking. Furthermore, we can choose a suitable value for $z_3^{*\mathrm{B}}$ such that the equilibrium reached by this (intermediate) breaking maneuver has the same z_3-value. This is expressed in Figure 4(a) by the intersection (marked with a star) of the simulated curve with the bisecting line of the first quadrant (dash-dotted). After reaching the z_2–z_4-origin, the manipulator is immediately re-accelerated by maneuver A1 (as second part of maneuver B) in the first quadrant. Due to time reversal symmetry this motion is the exact inversion of the first part. Consequently, the z_3 value at the beginning and the end are identical as well. Finally, the desired parking regime with z_4^{pD} is reached with $z_3(t^{\mathrm{B}\dagger}) = z_3^{*\mathrm{B}}$.

5.4. *Transition of z_1 (Maneuver C)*

The last missing design step for a transition from almost any desired equilibrium to almost any other is the proper adaption of the actuated joint angle z_1, which is the task of maneuver C. As for the previously treated maneuvers we initially assume a certain situation (here: $z_4^{\mathrm{pD}} > 0$), and later extend the results by applying the symmetry properties.

For $z_4^{\mathrm{pD}} > 0$ Figure 2 implies that the subsequent maneuver D takes place in the first quadrant, thus we have

$$\Delta z_1^{\mathrm{D}} := \int_{t^{\mathrm{D}*}}^{t^{\mathrm{D}\dagger}} z_2(t)\,\mathrm{d}t > 0. \tag{32}$$

That value is known a priory and only depends on $z_3^{\dagger}$ because z_1 is cyclic.

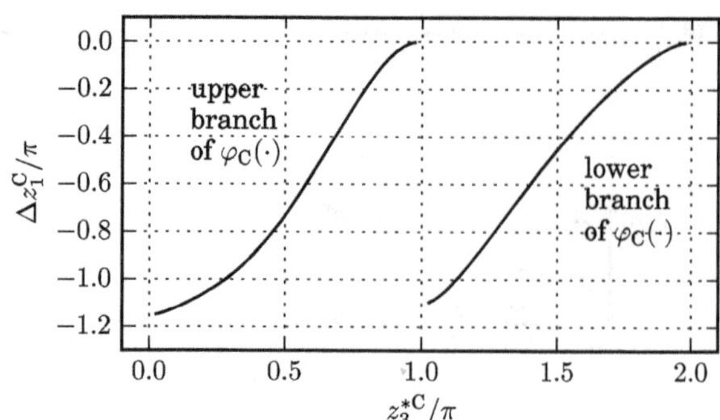

Figure 8. Value of Δz_1^{C} (displacement of active joint) in dependency of the z_3-start value of maneuver C (obtained by numerical simulation). If $z_3^{*\mathrm{C}} < \pi$ according to Figure 2 the drift in the second quadrant points upwards, therefore the upper branch of the sliding surface (cf. (35) and Figure 7(a)) is active. For $z_3^{*\mathrm{C}} > \pi$, the drift points downwards and thus the lower branch is used. For motion planning this mapping can be inverted piecewise to obtain $z_3^{*\mathrm{C}}$ for a given Δz_1^{C}.

To reach the desired final equilibrium $(z_1^{\dagger}, 0, z_3^{\dagger}, 0)$, it has to be ensured that at the beginning of maneuver D

$$z_1^{\mathrm{D}*} = z_1^{\dagger} - \Delta z_1^{\mathrm{D}} \tag{33}$$

holds. During parking regime, in which the system is between maneuvers B and C as well as between C and D, respectively, $z_1(\cdot)$ is constant. Therefore, the necessary displacement Δz_1^{C} of the active joint during maneuver C is easily determined by

$$\Delta z_1^{\mathrm{C}} = z_1^{\mathrm{C}\dagger} - z_1^{\mathrm{C}*} = z_1^{\mathrm{D}*} - z_1^{\mathrm{B}\dagger} \overset{33}{=} z_1^{\dagger} - \Delta z_1^{\mathrm{D}} - z_1^{\mathrm{B}\dagger}. \tag{34}$$

To achieve this displacement the recurrence property can be exploited (see Section 4.3). For a suitable sliding surface (i.e. $\varphi(\cdot)$ is bounded and does not pass through the origin) the resulting motion in z_2–z_4-plane is recurrent: every point on the sliding surface is reached twice. Hence, starting the maneuver in parking regime implies ending it in parking regime. Obviously, the actual shape of the sliding surface for maneuver C is a (function-valued and thus infinite dimensional) degree of freedom. For convenience, we again use a power function similar to Equation (29), namely

$$\varphi_{\mathrm{C}}(z_4) = -(\mu\beta|z_4^{\mathrm{p}} - z_4|)^{1/\beta}. \tag{35}$$

Note that, due to the absolute value $|\cdot|$ the graph of this function has two branches.[9] The negative sign indicates that the curve is located in the left half of the z_2–z_4-plane, see Figure 7. Consequently, during the motion through the left half-plane we have $z_2 < 0$ and thus $\Delta z_1^{C} < 0$. The actual value of Δz_1^{C} depends on the starting conditions of maneuver C, that is, on $z_3^{*\mathrm{C}}$. While this dependency cannot be obtained in closed form, it is straightforward to calculate it numerically, see Figure 8.

For any given value of $\Delta z_1^{C} \in [-\pi, 0)$ the displacement can be achieved with a single instance of maneuver C. In

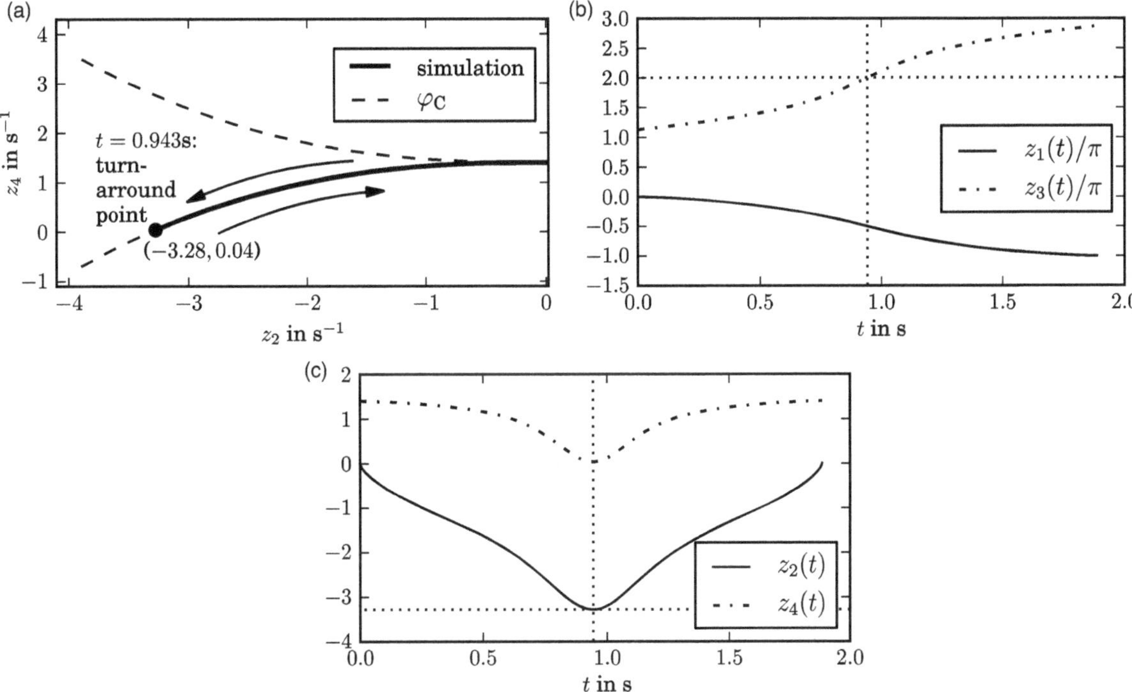

Figure 7. Exemplary simulation result for maneuver C (with $\Delta z_1^C \stackrel{!}{=} \pi$). Note that the sliding surface according to (35) has two branches(dashed lines in subfigure (a)). During one run of maneuver C only one of these branches is used, depending on the sign of the drift at the beginning of the maneuver. In this example the lower branch is used.

fact, there are two possible solutions for z_3^{*C} corresponding to the two branches of the sliding surface (35). From these, the "nearest" one (considering the $\mathbb{S}_1$-Topology) should be chosen, to minimize the waiting time in parking regime. If a bigger displacement is desired, that is, $\Delta z_1^C \in (-2\pi, -\pi)$ is needed, then the simplest approach is to perform two instances of maneuver C, each one for achieving a displacement of $\Delta z_1^C/2$. Finally, note that every value from outside the interval $(-2\pi, 0]$ can be mapped to its interior by

$$\tilde{\Delta} z_1 := (\Delta z_1 \bmod 2\pi) - 2\pi. \qquad (36)$$

Because z_1 is the angle of a revolute joint, this mapping does not affect the actual configuration of the manipulator, cf. Section 3.1.

Due to the quadrant equivalence (Symmetry property 1) the extension of maneuver C to $z_4^P < 0$ with a sliding surface in the right half-plane is straightforward.

5.5. *Maneuver summary*

All four maneuvers (A, B, C, D) enable us to perform almost arbitrary equilibrium transitions. Due to adverse drift conditions equilibria with z_3-values close to integral multiples of $\pi/2$ must be excluded. The actual sizes of the excluded z_3 intervals depend on the maneuver duration which can be tolerated, cf. for example, Figure 4(b). In Figure 9 a complete equilibrium transition with all maneuvers and parking regimes is depicted. See (Knoll, 2014b) for the respective animations.

Remark 4 Similar to the simple ODE $\dot{x} = \sqrt{x}$ there is an issue with the uniqueness of the solution of the reduced dynamics which is related with the "infinite slope" $\mathrm{d}z_2/\mathrm{d}z_4$ of the used sliding surfaces for $z_2 = 0$ (see Figure 3 and Equation (29)). The respective intersection point of the sliding surface with the z_4-axis (either an equilibrium or an parking regime), cannot be left by sliding mode control. However, if once $|z_2| > 0$ then the solution of the reduced dynamics is unique due to $\varphi(\cdot)$ being locally Lipschitz. For the (numerical) realization of the maneuvers this suggests to choose $|z_2(t = 0)| = \varepsilon > 0$. For ε small enough, its actual value becomes irrelevant to the solution of the dynamics and the input trajectories obtained by this "trick" nevertheless lead the unspoilt system to the desired state.

Remark 5 Of course the proposed motion schemes are by far not unique. In fact, for every single part of the motion (i.e. for every maneuver) there exists an infinite degree of freedom for the shape of the sliding surface, which might be used to optimize the motion w.r.t. overall duration or other suitable criteria. However, the aim of the presented solution is to show the principle applicability of the concept.

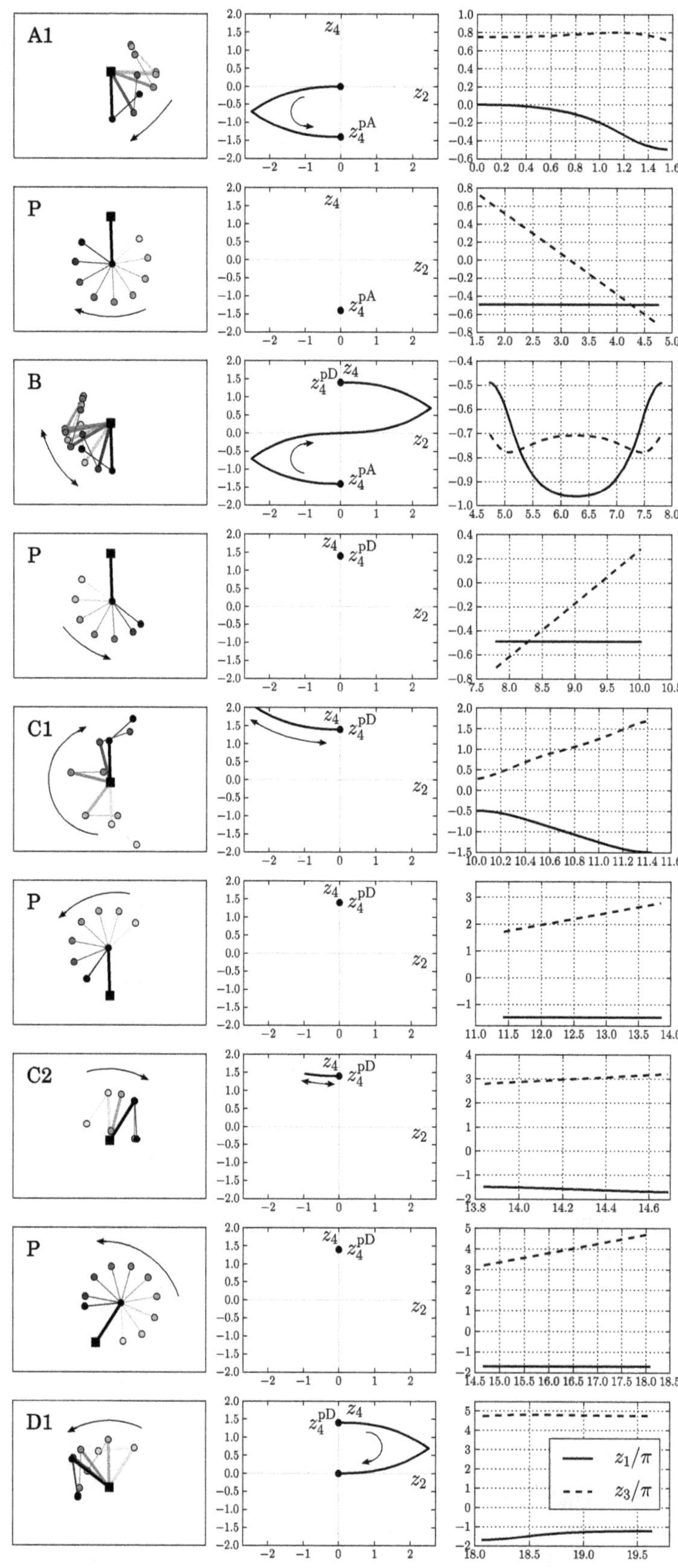

Figure 9. Simulation results for a complete equilibrium transition. Left column: Motion in working space. Middle column: Projection of the trajectory (and hence the switching functions) to the z_2-z_4-subspace. Right column: Time evolution of the joint angles $z_1(t), z_3(t)$. For a video of these results see (Knoll, 2014b).

6. Conclusion and outlook

In this contribution, we defined the LBINF as a special case of the Byrnes–Isidori normal form for generic Lagrangian mechanical systems. Furthermore, we constructively proved its existence for systems given by Equation (1). Then, we analyzed the frictionless underactuated 2-DOF manipulator in that normal form, especially the drift conditions in the z_2–z_4-subspace. On that basis, the controller (20) was defined which restricts the system dynamics to a submanifold (sliding surface) of the state space, determined by $z_2 = \varphi(z_4)$. Additionally, useful symmetries and the parking regime were identified. Finally, the equilibrium transition was decomposed into several subtasks, each of which is complied by a suitable maneuver, that is, by the suitable choice of the sliding function and appropriate initial conditions (obtained during parking regime).

The feasibility of the proposed controller is demonstrated by detailed simulation studies (Knoll, 2014a, 2014b) and first experimental results (Knoll et al., 2011). Currently, the experimental setup is reworked in order to implement the full equilibrium transition.

Besides the experimental realization there are some more questions for further investigation. One research direction concerns the application of the LBINF to other underactuated control systems and its relation to important control properties like linearizability by static or dynamic feedback, differential flatness and configuration flatness (see Knoll & Röbenack, 2014; Rathinam & Murray, 1998 and the references therein) or accessibility.

Another direction deals with the consequences of the Brockett condition, mentioned in Section 3.1. This result precludes the existence of a continuous differentiable feedback of the state of the original system to stabilize an isolated equilibrium. However, it does not preclude a nonsmooth state feedback or a dynamic extension of the system.

Acknowledgments

The authors would like to thank Matthias Franke, Chenzi Huang and Christian Mächler for valuable discussions and the anonymous reviewers for their useful comments and suggestions.

Funding

This work has been supported by Deutsche Forschungsgemeinschaft (DFG) under research [grant number RO 2427/2-1].

Notes

1. Although underactuated systems are sometimes referenced as "second-order nonholonomic", for example, in Oriolo and Nakamura (1991), many of them are holonomic in a physical sense.
2. Note the redefinition of $\mathbf{q}$ along with the elimination of $\boldsymbol{B}(\mathbf{q})$.
3. In such a case (Oriolo & Nakamura, 1991) suggest either nonsmooth feedback or a different control objective. In particular, the authors propose a smooth controller, which stabilizes a 1-dimensional submanifold of all equilibrium points. The case of a dynamic extension of the state, which might be another possibility to cope with the violation of Brocketts condition, is not discussed.
4. Then $\bar{f}_3(\cdot,\cdot)$ and $\bar{f}_4(\cdot,\cdot)$ are globally Lipschitz and hence finite escape time cannot occur.
5. This choice for φ is made for the sake of simplicity. In regions which are not near to the z_4-axis the curve $\varphi(\cdot)$ could have an arbitrary shape as long as it is continuous, (piecewise) differentiable and takes finite positive values.
6. Note that, due to time reversal symmetry, this also could be interpreted as an execution of maneuver A1 with appropriate adaption of the state.
7. As numerical values $\kappa = 0.9$ and $|z_4|^{\mathrm{p}} = 1.4\mathrm{s}^{-1}$ are used for all simulations.
8. For convenience, we define $\mathrm{sign}(0) := 1$.
9. They are called "upper" and "lower" branch because in the z_2–z_4-plane the argument of φ_{C} runs vertically.

References

Arai, H., & Tachi, S. (1991). Position control of a two degree of freedom manipulator with passive joint. *IEEE Transaction on Industrial Electronics*, *38*(1), 15–20.

Brockett, R. W. (1983). Asymptotic stability and feedback stabilization. In R. Brockett, R. S. Millmann, & H. J. Sussmann (Eds.), *Differential geometric control theory* (pp. 181–191). Boston, MA: Birkhäuser.

Cameron, J. M., & Book, W. J. (1997). Modeling mechanisms with nonholonomic joints using the Boltzmann–Hamel equations. *The International Journal of Robotics Research*, *16*(1), 47–59.

Choukchou-Braham, A., Cherki, B., Djemai, M., & Busawon, K. (2014). *Analysis and control of underactuated mechanical systems*. Cham: Springer International Publishing Switzerland.

De Luca, A., Iannitti, S., Mattone, R., & Oriolo, G. (2002). Underactuated manuipulators: Control properties and techniques. *Machine Intelligence and Robotic Control*, *4*, 113–125.

De Luca, A., Mattone, R., & Oriolo, G. (2000). Stabilization of an underactuated planar 2R manipulator. *International Journal of Robust and Nonlinear Control*, *10*, 181–198.

Fantoni, I., & Lozano, R. (2001). *Non-linear control for underactuated mechanical systems*. London: Springer.

Franch, J., Reyes, A., & Agrawal, S. K. (2013). *Differential flatness of a class of n – DOF planar manipulators driven by an arbitrary number of actuators*. Proceedings of the European Control Conference (ECC), 2013, Zurich (pp. 161–166).

Glück, T., Eder, A., & Kugi, A. (2013). Swing-up control of a triple pendulum on a cart with experimental validation. *Automatica*, *49*(3), 801–808.

Graichen, K., Treuer, M., & Zeitz, M. (2007). Swing-up of the double pendulum on a cart by feedforward and feedback control with experimental validation. *Automatica*, *43*(1), 63–71.

Ince, D. C., Hatton, L., & Graham-Cumming, J. (2012). The case for open computer programs. *Nature*, *482*(7386), 485–488.

Isidori, A. (1995). *Nonlinear control systems: An introduction* (3rd ed.). London: Springer.

Knoll, C. (2014a). Python source code for simulation and visualization of the controlled 2-DOF manipulator. Retrieved from http://www.tu-dresden.de/rst/software

Knoll, C. (2014b). Video sequence: Equilibrium transition of the underactuated 2-DOF manipulator. Retrieved from http://www.tu-dresden.de/rst/software

Knoll, C., Leist, B., & Röbenack, K. (2011). Konzeption und prototypische Realisierung eines Versuchsstandes zur Regelung eines unteraktuierten Manipulators. In T. Bertram, B. Corves, & K. Janschek, (Eds.), *Tagungsband Mechatronik 2011* (pp. 241–246), Dresden.

Knoll, C., & Röbenack, K. (2010). Sliding mode control of an underactuated two-link manipulator. *Proc in Applied Mathematics and Mechanics*, *10*(1), 615–616.

Knoll, C., & Röbenack, K. (2011a). *Control of an underactuated manipulator using similarities to the double integrator*. Proceedings of the 18th IFAC World Congress, Milano, Italy (pp. 11501–11507).

Knoll, C., & Röbenack, K. (2011b). Trajectory planning for a non flat mechanical system using time-reversal symmetry. *Proceeding in Applied Mathematics and Mechanics*, *11*(1), 819–820.

Knoll, C., & Röbenack, K. (2013). Stable limit cycles with specified oscillation parameters induced by feedback: Theoretical and experimental results. *Transactions on Systems, Signals and Devices*, *8*(1), 127–147.

Knoll, C., & Röbenack, K. (2014). *On configuration flatness of linear mechanical systems*. Proceeding of the European Control Conference (ECC), 2014, Strasbourg (pp. 1416–1421).

Mahindrakar, A. D., Rao, S., & Banavar, R. N. (2006). Point-to-point control of a 2R planar horizontal underactuated manipulator. *Mechanism and Machine Theory*, *41*, 838–844.

Mareczek, J., Buss, M., & Schmidt, G. (1999). Robust control of a non-holonomic underactuated SCARA robot. In S. G. Tzafestas, & G. Schmidt (Eds.), *Progress in system and robot analysis and control design* (pp. 381–396). Lecture Notes in Control and Information Science, Vol. 243. London: Springer.

Nakamura, Y., Suzuki, T., & Koinuma, M. (1997). Nonlinear behavior and control of a nonholonomic free-joint manipulator. *IEEE Transactions on Robotics and Automation*, *13*(6), 853–862.

Olfati-Saber, R. (2001). *Nonlinear control of underactuated mechanical systems with application to robotics and aerospace vehicles* (Ph.D. thesis). Massachusetts Institute of Technology, Cambridge, MA.

Oriolo, G., & Nakamura, Y. (1991). *Control of mechanical systems with second-order nonholonomic constraints: Underactuated manipulators*. Proceedings of the 30th conference on decision and control, Brighton, England (pp. 2398–2403).

Rathinam, M., & Murray, R. M. (1998). Configuration flatness of Lagrangian systems underactuated by one control. *SIAM J. Control and Optimization*, *36*(1), 164–179.

Riachy, S., Orlov, Y., Floquet, T., Santiesteban, R., & Richard, J. P. (2008). Second-order sliding mode control of underactuated mechanical systems I: Local stabilization with application to an inverted pendulum. *International Journal of Robust and Nonlinear Control*, *18*(4), 529–543.

Sastry, S. (1999). *Nonlinear systems: Analysis, stability, and control*. New York: Springer.

Scherm, N., & Heimann, B. (2000). Dynamics and control of underactuated manipulation systems: A discrete-time approach. *Robotics and Autonomous Systems*, *30*(3), 237–248.

Slotine, J. J. E., & Li, W. (1991). *Applied nonlinear control*. Upper Saddle River, NJ: Prentice-Hall.

Spong, M. W. (1998). Underactuated mechanical systems. In B. Siciliano, & K. P. Valavanis (Eds.), *Control problems in robotics* (pp. 135–150). Lecture Notes in Control and Information Science, Vol. 230. London: Springer.

Spong, M. W. (2000). Some aspects of switching control in robot locomotion. *Automatisierungstechnik*, *48*(4).

Yesiloglu, S. M., & Temeltas, H. (2010). Dynamical modeling of cooperating underactuated manipulators for space manipulation. *Advanced Robotics*, *24*(3), 325–341.

Yoshida, K. (1997). *A general formulation for underactuated manipulators*. Proceedings of the 1997 IEEE/RSJ International Conference on Intelligent Robots and System (IROS '97), Grenoble, France (Vol. 3, pp. 1651–1657).

The paradigm of complex probability and the Brownian motion

Abdo Abou Jaoude*

Department of Mathematics and Statistics, Faculty of Natural and Applied Sciences, Notre Dame University – Louaize, Zouk Mosbeh, Lebanon

Andrey N. Kolmogorov's system of axioms can be extended to encompass the imaginary set of numbers and this by adding to his original five axioms an additional three axioms. Hence, any experiment can thus be executed in what is now the complex probability set $\mathscr{C}$ which is the sum of the real set $\mathscr{R}$ with its corresponding real probability, and the imaginary set $\mathscr{M}$ with its corresponding imaginary probability. The objective here is to evaluate the complex probabilities by considering supplementary new imaginary dimensions to the event occurring in the 'real' laboratory. Whatever the probability distribution of the input random variable in $\mathscr{R}$ is, the corresponding probability in the whole set $\mathscr{C}$ is always one, so the outcome of the random experiment in $\mathscr{C}$ can be predicted totally. The result indicates that chance and luck in $\mathscr{R}$ is replaced now by total determinism in $\mathscr{C}$. This is the consequence of the fact that the probability in $\mathscr{C}$ is got by subtracting the chaotic factor from the degree of our knowledge of the stochastic system. This novel complex probability paradigm will be applied to the classical theory of Brownian motion and to prove as well the law of large numbers in an original way.

Keywords: extended Kolmogorov's axioms; complex set; probability norm; degree of our knowledge; chaotic factor; Gauss–Laplace distribution; diffusion; entropy; resultant complex random vector

Nomenclature

$\mathscr{R}$	real set of events
$\mathscr{M}$	imaginary set of events
$\mathscr{C}$	complex set of events
i	the imaginary number, where $i^2 = -1$
EKA	extended Kolmogorov's axioms
P_{rob}	probability of any event
P_r	probability in the real set $\mathscr{R}$
P_m	probability in the imaginary set $\mathscr{M}$ corresponding to the real probability in $\mathscr{R}$
Pc	probability of an event in $\mathscr{R}$ with its associated event in $\mathscr{M}$ probability in the complex set $\mathscr{C}$
z	complex number = sum of P_r and P_m = complex random vector
DOK	$\lvert z\rvert^2$ = the degree of our knowledge of the random experiment it is the square of the norm of z.
Chf	the chaotic factor of z
MChf	magnitude of the chaotic factor of z
N	number of random variables
Z	the resultant complex random vector
$\text{DOK}_Z = \frac{\lvert Z\rvert^2}{N^2}$	the degree of our knowledge of Z
$\text{Chf}_Z = \frac{\text{Chf}}{N^2}$	the chaotic factor of Z
MChf_Z	magnitude of the chaotic factor of Z
S_R	entropy in $\mathscr{R}$
$\bar{S}_R$	entropy of the complementary real probability set to $\mathscr{R}$
S_M	entropy in $\mathscr{M}$
S_C	entropy in $\mathscr{C}$

1. Introduction

The terms 'pedesis' ($\pi\acute{\eta}\delta\eta\sigma\iota\varsigma$ =pĕ : dε : sis= 'leaping') in Greek or 'Brownian motion' in English, both refer to the random motion of particles suspended in a liquid or a gas. This motion is the result of the particles collision with the rapid atoms or molecules in the fluid. Moreover, the term Brownian motion can also refer to the mathematical model used to describe such random movements. This model is usually called a 'particle theory' (Kuhn, 1970).

This phenomenon, also called a 'transport phenomenon', is named after the botanist Robert Brown. In the nineteenth century and precisely in1827, while Brown was looking through a microscope at particles found in pollen grains in water, he noticed that the particles moved through the water but was not able to determine the mechanisms that caused this motion. It had long been theorized that atoms and molecules are the constituents of matter, and many decades later (precisely in 1905), Albert Einstein published a paper that explained in accurate detail the

*Email: abdoaj@idm.net.lb

Brownian motion. In fact the motion that Brown had perceived was the consequence of the pollen being moved by individual water molecules. Consequently, this explanation of Brownian motion yielded a decisive confirmation of the actual existence of atoms and molecules. Furthermore, in 1908, Jean Perrin experimentally verified the theory. As a result, in 1926, the Nobel Prize in Physics was attributed to Perrin 'for his work on the discontinuous structure of matter', whereas five years earlier, Einstein had received the prize 'for his services to theoretical physics' with specific citation of different research. In fact, the seemingly random nature of the Brownian motion was due to the constantly changing direction of the force of atomic bombardment and therefore the particle is hit more on one side than another at different times (Poincaré, 1968; Stewart, 1996).

Additionally, there are various real-world applications to the Brownian motion mathematical model. As an example, we frequently cite the stock market fluctuations, even though Benoit Mandelbrot rejected to apply it to the movements of stock price in part due to the discontinuous nature of these movements (Stewart, 2012).

We consider Brownian motion among the simplest of the continuous-time stochastic processes. In fact, it is a limit of both simpler and more complicated stochastic processes (random walk and Donsker's theorem). In both cases, Brownian motion is usually more convenient rather than more accurate of the models. Consequently, this motivates the use of the Brownian probabilistic model (Barrow, 1992; Warusfel & Ducrocq, 2004).

Around 60 BC, in his scientific poem 'On the Nature of Things', the Roman Lucretius described remarkably in the verses 113–140 from Book II, the Brownian motion of dust particles. In fact, Lucretius considered this motion as an inevitable proof of the existence of atoms. Even though air currents largely cause the interspersing motion of dust particles, the tumbling motion of small dust particles is certainly and primarily due to the true Brownian dynamics (Greene, 2003; Hawking, 2005).

Moreover, in the eighteenth century, precisely in 1785, Jan Ingenhousz had described the irregular motion of coal dust particles on the surface of alcohol. But the discovery is usually credited to the botanist Robert Brown in 1827. In fact, while studying pollen grains of the plant 'Clarkia pulchella' suspended in water under a microscope, Brown noticed tiny particles which were ejected by the pollen grains and performing a jittery motion. Although the origin of the motion was yet to be explained, Brown repeated the experiment with particles of inorganic matter and concluded that the motion was not life-related (Bogdanov & Bogdanov, 2013; Penrose, 1999).

In 1880, in a paper on the method of least squares, Thorvald N. Thiele was the first person to describe mathematically the Brownian motion. In 1900, Louis Bachelier, and independently, followed Thiele and explained the motion in his Ph.D. thesis entitled 'The theory of speculation' and hence presented a stochastic analysis of the stock market and option markets. In 1905, Albert Einstein's miraculous year, and Marian Smoluchowski in 1906, using Brownian motion, they confirmed indirectly the existence of atoms and molecules. Both scientists' equations that described Brownian motion were verified experimentally in 1908 by Jean Perrin (Bogdanov & Bogdanov, 2012).

Moreover, let us consider the Brownian motion of pollen grain travelling randomly in water and emitting erratic particles. It is well known that a water molecule has the size of about 0.1×0.2 nm; however, Robert Brown observed particles of the order of a few micrometers in size. Notice that they are not to be confused with the actual pollen particle which has a size of about 100 micrometers. The instantaneous imbalance in the combined forces exerted by the collisions of the particle with the much smaller liquid molecules which are in random thermal motion yields the Brownian motion of the surrounded particle in a liquid (Bogdanov & Bogdanov, 2010).

The time evolution of the probability density function (PDF) associated with the position of the particle in Brownian motion is approximated by the diffusion equation, knowing that this approximation is valid on short timescales. In addition, Langevin equation which contains a random force field representing the effect of the thermal fluctuations of the solvent on the particle is the best description of the time evolution of the position of the Brownian particle itself (Bogdanov & Bogdanov, 2009).

Furthermore, by solving the diffusion equation under appropriate boundary conditions and hence by finding its solution, we acquire the displacement of a particle in Brownian motion. Consequently, the solution indicates that the displacement varies as the square root of the time and not linearly. This leads to the explanation why the results concerning the Brownian particles gave nonsensical results. In fact, the assumption of a linear time dependence was incorrect (Davies, 1993).

Additionally, the particle motion is governed by its inertia and its displacement will be linearly dependent on time, and this at very short-time scales. Hence, we obtained $\Delta x = v\Delta t$, and the instantaneous velocity of the Brownian motion is expressed by $v = \Delta x/\Delta t$, when $\Delta t \ll \tau$, where τ is the momentum relaxation time. For a glass microsphere trapped in air with an optical tweezer, we succeeded in 2010 to measure the instantaneous velocity of a Brownian particle. This important experiment, verified the distribution of Maxwell–Boltzmann, and the Brownian particle equipartition theorem (Wikipedia, *Brownian Motion*; Wikipedia, *Entropy*).

Also, a random walk can model the Brownian motion. We must mention that random walks in porous media or fractals are anomalous. In the general case, Brownian motion is considered as a non-Markov random process

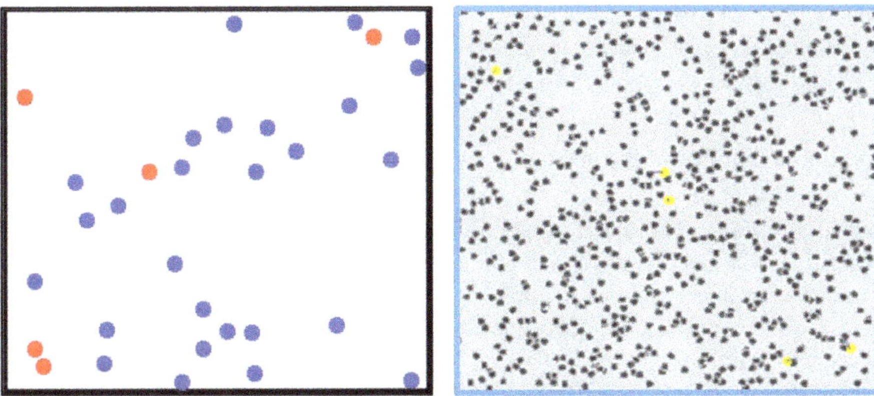

Figure 1. (a) and (b) This is an example of the Brownian motion of five particles (red at left and yellow at right) that collide with a large set of 800 particles (blue at left and black at right) (Wikipedia, *Brownian Motion*).

and can be expressed by stochastic integral equations (Figure 1(a) and 1(b)) (Wikipedia, *Mass Diffusivity*).

2. Albert Einstein's contribution

Einstein's theory is divided into two parts (Aczel, 2000; Balibar, 2002; Hawking, 2002, 2011; Hoffmann, 1975; Pickover, 2008; Reeves, 1988; Ronan, 1988; Wikipedia, *Brownian Motion*; Wikipedia, *Entropy*; Wikipedia, *Mass Diffusivity*):

(1) The first part consists in the formulation of the Brownian particles diffusion equation. The diffusion coefficient is linked to the mean-squared displacement of a Brownian particle.
(2) The second part consists in connecting the diffusion coefficient to measurable physical quantities.

Consequently, Einstein succeeded in determining the size of atoms, the number of atoms in a mole, or the weight of molecules in grams of a gas. As fixed by Avogadro's law, this volume is the same for all ideal gases, and which is 22.414 L at standard temperature and pressure. It is well known that the number of atoms contained in this volume is denoted by Avogadro's number. This leads to determining the mass of an atom which can be computed by dividing the mass of a mole of the gas by Avogadro's number.

In the first part of Einstein's theory, we are able to determine the distance that a Brownian particle travels in a given interval of time. Knowing that due to the enormous number of collisions that a Brownian particle undergoes (approximately of the order of 10^{21} collisions per second), classical mechanics is incapable to determine this distance. Consequently, Einstein considered the Brownian particles collective motion.

Moreover, the type of dynamical equilibrium suggested by Einstein was not new. In fact, in his series of lectures at Yale University in May 1903, J. J. Thomson proposed that

> the dynamic equilibrium between the velocity generated by a concentration gradient given by Fick's law and the velocity due to the variation of the partial pressure due to the ions set in motion gives us a method of determining Avogadro's Constant which is independent of any hypothesis as to the shape or size of molecules, or of the way in which they act upon each other. (Wikipedia, *Brownian Motion*)

Furthermore, in 1888, Walter Nernst found a similar expression to Einstein's formula for the diffusion coefficient. In fact, he formulated that the diffusion coefficient is the ratio of the osmotic pressure to the ratio of the frictional force and the velocity to which it gives rise. The former was equated to the law of Van't Hoff, while the latter was given by Stokes's law. Nernst wrote

$$k' = \frac{p_0}{k},$$

where k' is the diffusion coefficient, p_0 is the osmotic pressure, and k is the ratio of the frictional force to the molecular viscosity which he assumes is given by Stokes's formula for the viscosity.

> By introducing the ideal gas law per unit volume for the osmotic pressure, Nernst formula is consequently identical to Einstein's formula. The use of Stokes's law in Nernst's case, as well as in Einstein and Smoluchowski, is not strictly applicable since it does not apply to the case where the radius of the sphere is small in comparison with the mean free path. (Wikipedia, *Brownian Motion*)

The series of experiments conducted in 1906 and 1907 by Svedberg, refuted seemingly at first Einstein's formula predictions. These experiments gave the displacements value as 4–6 times Einstein's value. In 1908, Henri's experiments yielded displacements three times greater than those predicted by Einstein's formula. Finally, in 1908 and in 1909, Chaudesaigues and Perrin, respectively, confirmed in a series of experiments Einstein's predictions.

> The confirmation of Einstein's theory constituted empirical progress for the kinetic theory of heat. In essence, Einstein showed that the motion can be predicted directly from the kinetic model of thermal equilibrium. The importance of the theory lays in the fact that it confirmed the kinetic

theory's account of the second law of thermodynamics as being an essentially statistical law. (Wikipedia, *Brownian Motion*)

3. The purpose and the advantages of the present work

All our work in classical probability theory is to compute probabilities. The original idea in this paper is to add new dimensions to our random experiment and this will make the work deterministic. In fact, the probability theory is a nondeterministic theory by nature, meaning that the outcome of the events is due to chance and luck. By adding new dimensions to the event in $\mathcal{R}$, we make the work deterministic and hence a random experiment will have a certain outcome in the complex set of probabilities $\mathcal{C}$. It is of great importance that the stochastic system becomes totally predictable since we will be totally knowledgeable to foretell the outcome of chaotic and random events that occur in nature like, for example, in statistical mechanics or in all stochastic processes. Therefore, the work that should be done is to add to the real set of probabilities $\mathcal{R}$, the contributions of $\mathcal{M}$ which is the imaginary set of probabilities, that makes the event in $\mathcal{C} = \mathcal{R} + \mathcal{M}$ deterministic. If this is found to be fruitful, then a new theory in statistical sciences and prognostic is elaborated and this to understand deterministically those phenomena that used to be random phenomena in $\mathcal{R}$. This is what I called 'the complex probability paradigm' that was initiated and elaborated in my five previous papers (Abou Jaoude, 2013a, 2013b, 2014, 2015; Abou Jaoude, El-Tawil, & Kadry, 2010).

Consequently, the purpose and the advantages of the present work are to

(1) Extend classical probability theory to the set of complex numbers, hence to relate probability theory to the field of complex analysis. This task was initiated and elaborated in my five previous papers.
(2) Apply the new probability axioms and paradigm to the Brownian motion.
(3) Prove that all random phenomena can be expressed deterministically in the complex set $\mathcal{C}$.
(4) Quantify both the *d*egree of *o*ur *k*nowledge (DOK for short) and the chaos of the random walk of particles.
(5) Draw and represent graphically the functions and parameters of the novel paradigm associated with the Brownian motion.
(6) Show that the classical concept of entropy is always equal to zero in the complex set; hence, no chaos and disorder exist in $\mathcal{C}$ (complex set) $= \mathcal{R}$ (real set) $+ \mathcal{M}$ (imaginary set).
(7) Prove the very well-known law of large numbers using the newly defined axioms.
(8) Pave the way to apply the original paradigm to other topics in statistical mechanics, in stochastic processes, and to the field of prognostics in engineering. These will be the subjects of my subsequent research papers.

3.1. The original Andrey N. Kolmogorov set of axioms

The simplicity of Kolmogorov's system of axioms may be surprising. Let E be a collection of elements $\{E_1, E_2, \dots\}$ called elementary events and let F be a set of subsets of E called random events. The five axioms for a finite set E are (Benton, 1966a, 1966b; Feller, 1968; Montgomery & Runger, 2003; Walpole, Myers, Myers, & Ye, 2002) as follows:

Axiom 1: F is a field of sets.
Axiom 2: F contains the set E.
Axiom 3: A non-negative real number $P_{\text{rob}}(A)$, called the probability of A, is assigned to each set A in F. We have always $0 \leq P_{\text{rob}}(A) \leq 1$.
Axiom 4: $P_{\text{rob}}(E)$ equals 1.
Axiom 5: If A and B have no elements in common, the number assigned to their union is as follows:

$$P_{\text{rob}}(A \cup B) = P_{\text{rob}}(A) + P_{\text{rob}}(B).$$

Hence, we say that A and B are disjoint; otherwise, we have:

$$P_{\text{rob}}(A \cup B) = P_{\text{rob}}(A) + P_{\text{rob}}(B) - P_{\text{rob}}(A \cap B).$$

And we say also that $P_{\text{rob}}(A \cap B) = P_{\text{rob}}(A) \times P_{\text{rob}}(B|A) = P_{\text{rob}}(B) \times P_{\text{rob}}(A|B)$ which is the conditional probability. If both A and B are independent, then $P_{\text{rob}}(A \cap B) = P_{\text{rob}}(A) \times P_{\text{rob}}(B)$.

In addition, we can generalize and say that, for N disjoint events $A_1, A_2, \dots, A_j, \dots, A_N$ (for $1 \leq j \leq N$), we have

$$P_{\text{rob}}\left(\bigcup_{j=1}^{N} A_j\right) = \sum_{j=1}^{N} P_{\text{rob}}(A_j).$$

3.2. Adding the imaginary part $\mathcal{M}$

Now, we can add to this system of axioms an imaginary part such that

Axiom 6: Let $P_m = \mathrm{i}(1 - P_r)$ be the probability of an associated event in $\mathcal{M}$ (the imaginary part) to the event A in $\mathcal{R}$ (the real part). It follows that $P_r + P_m/\mathrm{i} = 1$, where i is the imaginary number with $\mathrm{i}^2 = -1$.
Axiom 7: We construct the complex number or vector $Z = P_r + P_m = P_r + \mathrm{i}(1 - P_r)$ having a norm $|Z|$ such that $|Z|^2 = P_r^2 + (P_m/\mathrm{i})^2$.
Axiom 8: Let Pc denote the probability of an event in the universe $\mathcal{C}$ where $\mathcal{C} = \mathcal{R} + \mathcal{M}$.

We say that Pc is the probability of an event A in $\mathscr{R}$ with its associated event in $\mathscr{M}$ such that

$$Pc^2 = \left(P_r + \frac{P_m}{\mathrm{i}}\right)^2 = |Z|^2 - 2\mathrm{i}P_rP_m$$

and is always equal to 1.

We can see that the system of axioms defined by Kolmogorov could be hence expanded to take into consideration the set of imaginary probabilities by adding three new axioms (Abou Jaoude, 2013a, 2013b, 2014, 2015; Abou Jaoude et al., 2010).

3.3. *The purpose of extending the axioms*

It is apparent from the set of axioms that the addition of an imaginary part to the real event makes the probability of the event in $\mathscr{C}$ always equal to 1. In fact, if we begin to see the set of probabilities as divided into two parts, one real and the other imaginary, understanding will directly follow. The random event that occurs in $\mathscr{R}$ (like tossing a coin and getting a head) has a corresponding probability P_r. Now, let $\mathscr{M}$ be the set of imaginary probabilities and let $|Z|^2$ be the DOK of this phenomenon. P_r is always, and according to Kolmogorov's axioms, the probability of an event.

A total ignorance of the set $\mathscr{M}$ makes

$P_r = 0.5$ and $|Z|^2$ in this case is equal to $1 - 2P_r(1 - P_r) = 1 - (2 \times 0.5) \times (1 - 0.5) = 0.5$.

Conversely, a total knowledge of the set in $\mathscr{R}$ makes

$P_{\text{rob}}(\text{event}) = P_r = 1$ and $P_m = P_{\text{rob}}(\text{imaginary part}) = 0$. Here we have $|Z|^2 = 1 - (2 \times 1) \times (1 - 1) = 1$ because the phenomenon is totally known, that is, its laws and variables are completely determined, hence; our DOK of the system is 1 or 100%.

Now, if we can tell for sure that an event will never occur, that is, like 'getting nothing' (the empty set), P_r is accordingly $= 0$, that is the event will never occur in $\mathscr{R}$. P_m will be equal to

$\mathrm{i}(1 - P_r) = \mathrm{i}(1 - 0) = \mathrm{i}$, and $|Z|^2 = 1 - (2 \times 0) \times (1 - 0) = 1$, because we can tell that the event of getting nothing surely will never occur; thus, the DOK of the system is 1 or 100% (Abou Jaoude et al., 2010).

We can infer that we have always

$$0.5 \le |Z|^2 \le 1 \quad \forall P_r: \quad 0 \le P_r \le 1$$

and

$$|Z|^2 = \text{DOK} = P_r^2 + \left(\frac{P_m}{\mathrm{i}}\right)^2, \tag{1}$$

where $0 \le P_r, P_m/i \le 1$.

And what is important is that in all cases we have

$$Pc^2 = \left(P_r + \frac{P_m}{\mathrm{i}}\right)^2 = |Z|^2 - 2\mathrm{i}P_rP_m = [P_r + (1 - P_r)]^2 = 1^2 = 1. \tag{2}$$

In fact, according to an experimenter in $\mathscr{R}$, the game is a game of luck: the experimenter does not know the output of the event. He will assign to each outcome a probability P_r and he will say that the output is nondeterministic. But in the universe $\mathscr{C} = \mathscr{R} + \mathscr{M}$, an observer will be able to predict the outcome of the game of chance since he takes into consideration the contribution of $\mathscr{M}$, so we write

$$Pc^2 = \left(P_r + \frac{P_m}{\mathrm{i}}\right)^2.$$

Hence Pc is always equal to 1. In fact, the addition of the imaginary set to our random experiment resulted to the abolition of ignorance and indeterminism. Consequently, the study of this class of phenomena in $\mathscr{C}$ is of great usefulness since we will be able to predict with certainty the outcome of experiments conducted. In fact, the study in $\mathscr{R}$ leads to unpredictability and uncertainty. So instead of placing ourselves in $\mathscr{R}$, we place ourselves in $\mathscr{C}$ then study the phenomena, because in $\mathscr{C}$ the contributions of $\mathscr{M}$ are taken into consideration and therefore a deterministic study of the phenomena becomes possible. Conversely, by taking into consideration the contribution of the set $\mathscr{M}$, we place ourselves in $\mathscr{C}$ and by ignoring $\mathscr{M}$, we restrict our study to nondeterministic phenomena in $\mathscr{R}$ (Bell, 1992; Boursin, 1986; Dacunha-Castelle, 1996; Dalmedico-Dahan, Chabert, & Chemla, 1992; Srinivasan & Mehata, 1988; Stewart, 2002; Van Kampen, 2006).

Moreover, it follows from the above definitions and axioms that (Abou Jaoude et al., 2010):

$$\begin{aligned} 2\mathrm{i}P_rP_m &= 2\mathrm{i} \times P_r \times \mathrm{i} \times (1 - P_r) \\ &= 2\mathrm{i}^2 \times P_r \times (1 - P_r) = -2P_r(1 - P_r) \\ &\Rightarrow 2\mathrm{i}P_rP_m = \text{Chf}. \end{aligned} \tag{3}$$

$2\mathrm{i}P_rP_m$ will be called the *Chaotic factor* in our experiment and will be denoted accordingly by 'Chf'. We will see why we have called this term the chaotic factor; in fact:

In case $P_r = 1$, that is the case of a certain event, then the chaotic factor of the event is equal to 0.

In case $P_r = 0$, that is the case of an impossible event, then $\text{Chf} = 0$. Hence, in both two last cases, there is no chaos since the outcome is certain and is known in advance.

In case $P_r = 0.5$, $\text{Chf} = \quad 0.5$ (Figure 2).

We notice that $-0.5 \le \text{Chf} \le 0 \quad \forall P_r: \ 0 \le P_r \le 1$.

What is interesting here is thus we have quantified both the DOK and the chaotic factor of any random event and hence we write now

$$Pc^2 = |Z|^2 - 2\mathrm{i}P_rP_m = \text{DOK} - \text{Chf}. \tag{4}$$

Then, we can conclude that

$Pc^2 =$ DOK of the system $-$ chaotic factor $= 1$, therefore $Pc = 1$.

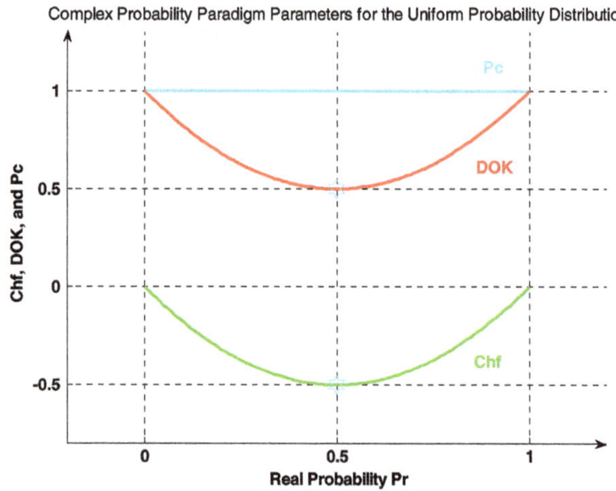

Figure 2. Chf, DOK, and Pc for a uniform probability distribution.

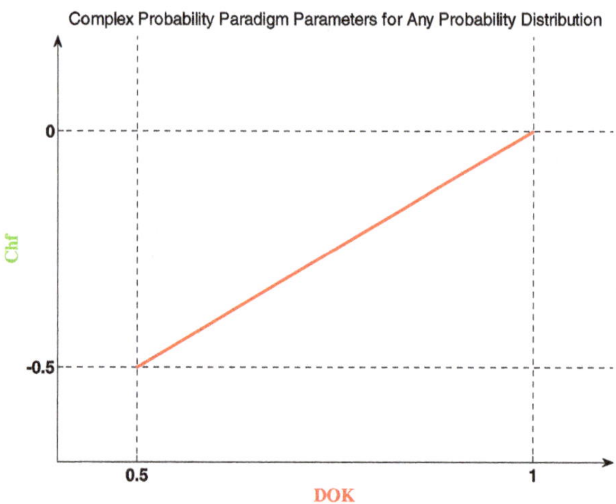

Figure 3. Graph of DOK − Chf = 1.

This directly means that if we succeed to subtract and eliminate the chaotic factor in any random experiment, then the output will always be with the probability equal to 1 (Dalmedico-Dahan & Peiffer, 1986; Ekeland, 1991; Gleick, 1997; Gullberg, 1997; Science et vie, 1999).

The graph in Figure 3 shows the linear relation between both DOK and Chf.

Furthermore, we need in our current study the absolute value of the chaotic factor that will give us the magnitude of the chaotic and random effects on the studied system materialized by the PDF, and which leads to an increasing system chaos in $\mathscr{R}$. This new term will be denoted accordingly MChf or *M*agnitude of the *Ch*aotic *f*actor (Figures 4–6). Hence, we can deduce the following:

$$\begin{aligned} \text{MChf} &= |\text{Chf}| = |2iP_rP_m| = -2iP_rP_m \\ &= 2P_r(1-P_r) \geq 0 \quad \forall P_r:\ 0 \leq P_r \leq 1, \end{aligned} \tag{5}$$

and

$$\begin{aligned} Pc^2 &= \text{DOK} - \text{Chf} \\ &= \text{DOK} + |\text{Chf}|, \text{since} -0.5 \leq \text{Chf} \leq 0 \end{aligned}$$

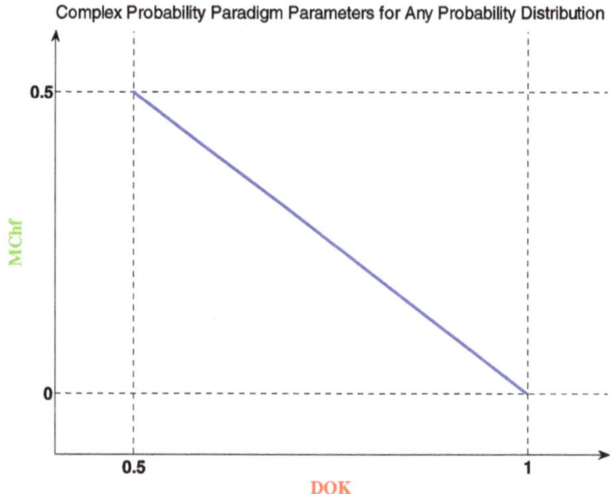

Figure 4. Graph of DOK + MChf = 1.

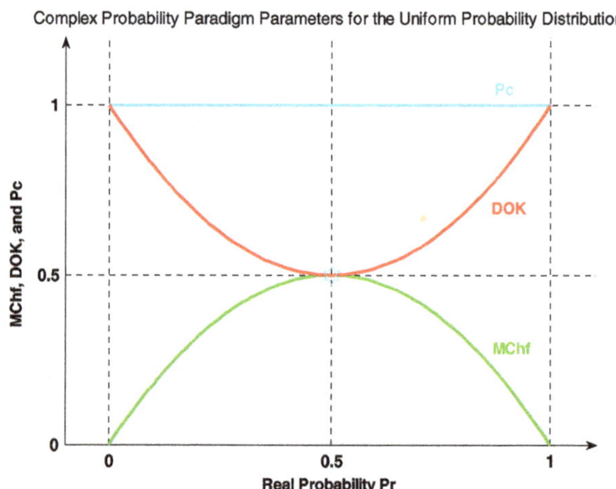

Figure 5. MChf, DOK, and Pc for a uniform probability distribution.

$$\begin{aligned} &= \text{DOK} + \text{MChf} = 1, \\ &\Leftrightarrow 0 \leq \text{MChf} \leq 0.5, \end{aligned}$$

where $0.5 \leq \text{DOK} \leq 1$

The graph in Figure 4 shows the linear relation between both DOK and MChf. Moreover, the graphs in Figures 5 and 6 show the Chf, MChf, DOK, and Pc as functions of the real probability P_r for a uniform probability distribution.

To summarize and to conclude, as the *D*egree of *O*ur certain *K*nowledge or DOK in the real universe $\mathscr{R}$ is unfortunately incomplete, the extension to the complex set $\mathscr{C}$ includes the contributions of both the real set of probabilities $\mathscr{R}$ and the imaginary set of probabilities $\mathscr{M}$. Consequently, this will result in a complete and perfect DOK in $\mathscr{C} = \mathscr{R} + \mathscr{M}$ ($Pc = 1$). In fact, in order to have a certain prediction of any random event, it is necessary to work in the complex set $\mathscr{C}$ in which the chaotic factor is quantified and subtracted from the computed DOK to lead to a probability in $\mathscr{C}$

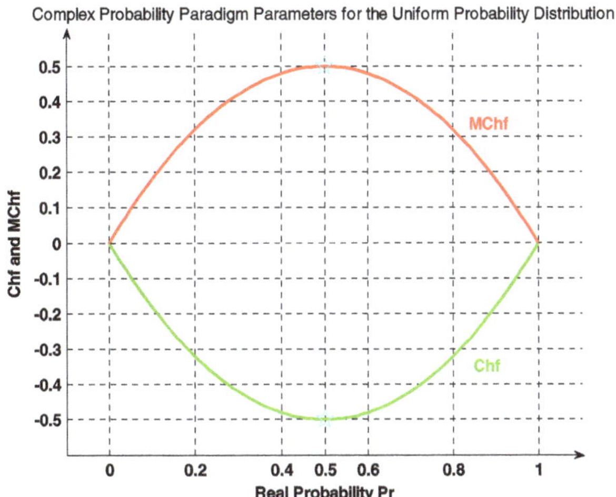

Figure 6. Chf and MChf for a uniform probability distribution.

equal to one ($Pc^2 = \text{DOK} - \text{Chf} = \text{DOK} + \text{MChf} = 1$). This hypothesis is verified in my five previous research papers by the mean of many examples encompassing both discrete and continuous distributions (Abou Jaoude et al., 2010, 2013a, 2013b, 2014, 2015). The *e*xtended *K*olmogorov *a*xioms (EKA for short) or the complex probability paradigm can be illustrated in Figure 7:

4. The new paradigm and the diffusion equation

The continuous law that we will illustrate here is the normal diffusion distribution of one particle starting from the origin at the initial time $t = 0$ (Wikipedia, *Brownian Motion*):

$$dF = \rho(x,t)dx = \frac{1}{\sqrt{4\pi Dt}} \exp\left(\frac{-x^2}{4Dt}\right) dx \quad \text{for} \quad -\infty < x < +\infty, \tag{6}$$

where $\rho(x,t)$ = the PDF of diffusion, $F(x,t)$ = the probability cumulative distribution function (CDF) of diffusion, D = the diffusion factor, x = the displacement of the particle, t = the time of displacement, $\bar{x} = 0$ = the mean value of x, $\sigma = \sqrt{2Dt}$ = the standard deviation of x (Abou Jaoude, 2004, 2005, 2007, 2013a, 2013b, 2014, 2015; Abou Jaoude et al., 2010; Bidabad, 1992; Chan Man Fong, De Kee, & Kaloni, 1997; Cox, 1955; Fagin, Halpern, & Megiddo, 1990; Ognjanović, Marković, Rašković, Doder, & Perović, 2012; Stepić, & Ognjanović, 2014; Weingarten, 2002; Wikipedia, *Brownian Motion*; Wikipedia, *Entropy*; Wikipedia, *Mass Diffusivity*; Youssef, 1994).

This bell-shaped density function has been taken from thermodynamics and statistical mechanics and it is the normal law of *Karl Friedrich Gauss* (the Prince of Mathematicians) and the *Marquis Pierre-Simon de Laplace*, or for short, the Gauss–Laplace distribution.

To illustrate the novel probability paradigm, I will consider in the simulations of the Brownian motion throughout the whole paper the diffusion of oxygen gas in air gas; hence, $D = 0.176\ \text{cm}^2/\text{s}$ at a temperature $T = 25\ °\text{C}$ (Wikipedia, *Mass Diffusivity*).

Figure 8 shows the characteristic bell-shaped curves of the diffusion of Brownian particles. The distribution begins as a Dirac delta function, indicating that all the particles are located at the origin at time $t = 0$ s, and for increasing

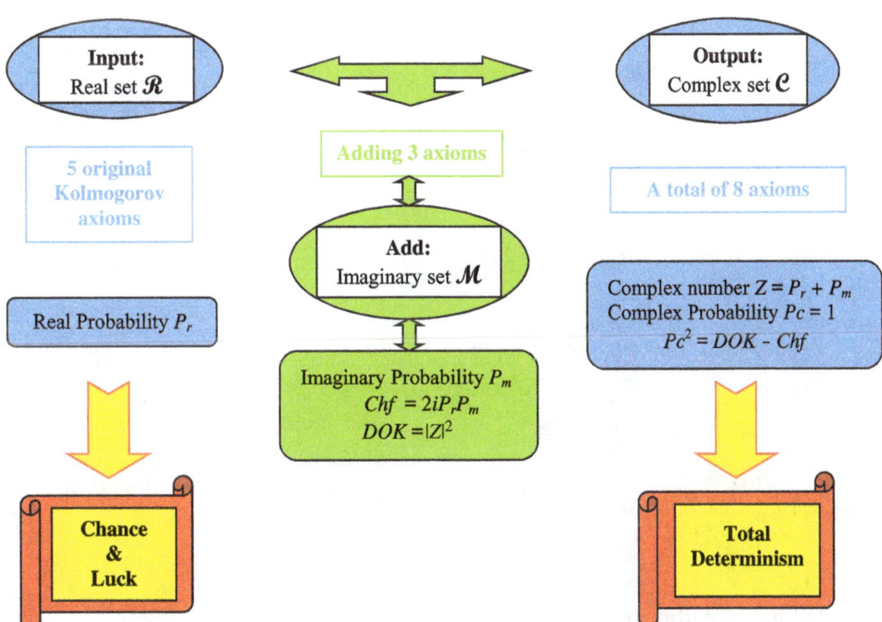

Figure 7. The EKA or the complex probability paradigm diagram.

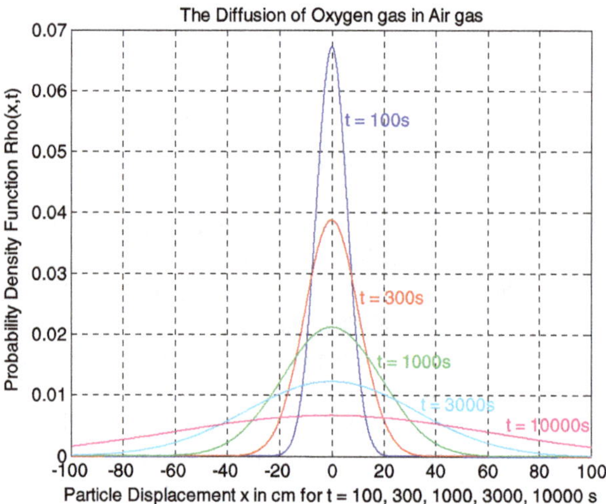

Figure 8. The PDF $\rho(x, t)$ of the diffusion of oxygen gas in air gas for different times t.

times they become flatter and flatter until the distribution becomes uniform in the asymptotic time limit.

To find the probability of an event following the normal distribution, we use the density function. We should have in the discrete case $\sum_j p_j = 1$, or in the continuous case we should have the integration over the whole domain equals to 1 like in this way:

$$\int_{-\infty}^{+\infty} \mathrm{d}F = \int_{-\infty}^{+\infty} \rho(x,t)\mathrm{d}x = \int_{-\infty}^{+\infty} \frac{1}{\sqrt{4\pi Dt}} \exp\left(\frac{-x^2}{4Dt}\right)\mathrm{d}x$$
$$= \frac{1}{\sqrt{4\pi Dt}} \int_{-\infty}^{+\infty} \exp\left(\frac{-x^2}{4Dt}\right)\mathrm{d}x = \frac{1}{\sqrt{4\pi Dt}} \times \sqrt{4\pi Dt} = 1.$$

Now, the real probability P_{rL} in $\mathscr{R}$ of finding the particle in the interval between $-\infty$ and L is as follows:

$$\int_{-\infty}^{L} \mathrm{d}F = \int_{-\infty}^{L} \rho(x,t)\mathrm{d}x = \int_{-\infty}^{L} \frac{1}{\sqrt{4\pi Dt}} \exp\left(\frac{-x^2}{4Dt}\right)\mathrm{d}x$$
$$= P_{\text{rob}}[x \le L] = F(x = L) = P_{rL} = p_L. \quad (7)$$

We note that P_{rL} is a non-decreasing function (Figure 9).

The associated imaginary probability P_{mL} in $\mathscr{M}$ is as follows:

$$P_{mL} = \mathrm{i} \times (1 - P_{rL}) = \mathrm{i} \times P_{\text{rob}}[x > L] = \mathrm{i}\int_{L}^{+\infty} \mathrm{d}F$$
$$= \mathrm{i}\int_{L}^{+\infty} \rho(x,t)\,\mathrm{d}x = \mathrm{i}\int_{L}^{+\infty} \frac{1}{\sqrt{4\pi Dt}} \exp\left(\frac{-x^2}{4Dt}\right)\mathrm{d}x$$
$$= \mathrm{i} \times [1 - F(x = L)]$$
$$\Rightarrow P_{mL} = \mathrm{i} \times P_{\text{rob}}[\text{to find the particle in the interval between } L \text{ and } +\infty]. \quad (8)$$

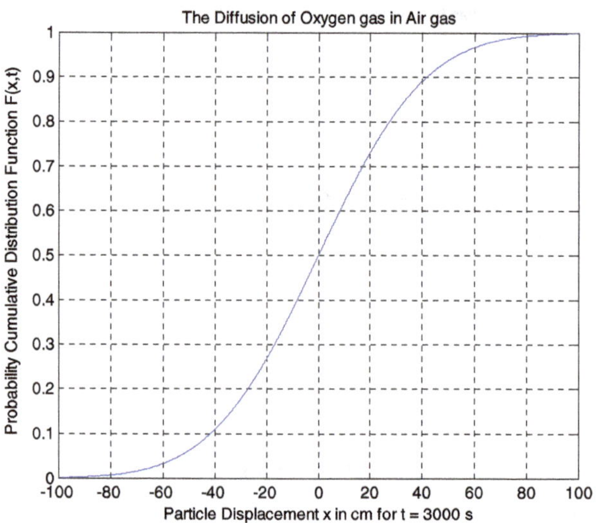

Figure 9. The CDF $F(x,t)$ of the diffusion of oxygen gas in air gas for $t = 3000$ s.

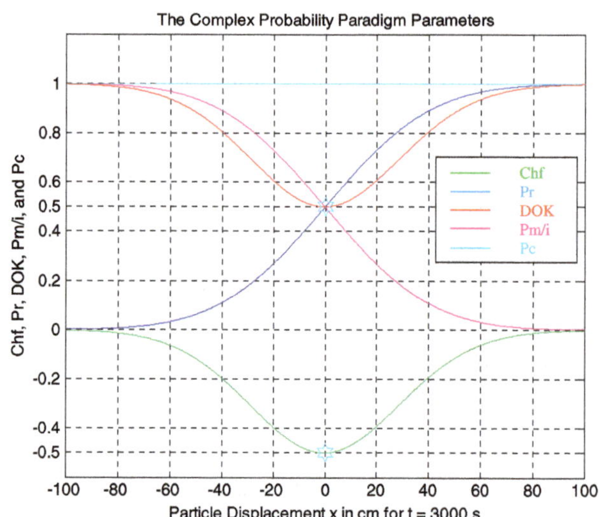

Figure 10. The complex probability parameters of the diffusion of oxygen gas in air gas for $t = 3000$ s.

The complementary probability P_{mL}/i is as follows:

$$\frac{P_{mL}}{\mathrm{i}} = 1 - P_{rL} = 1 - F(x = L)$$
$$= \int_{L}^{+\infty} \frac{1}{\sqrt{4\pi Dt}} \exp\left(\frac{-x^2}{4Dt}\right)\mathrm{d}x. \quad (9)$$

We note that P_{mL}/i is a non-increasing function (Figure 10).

Now, if we compute the norm of the complex number $Z_L = P_{rL} + P_{mL}$, we obtain

$$|Z_L|^2 = P_{rL}^2 + \left(\frac{P_{mL}}{\mathrm{i}}\right)^2 = p_L^2 + (1 - p_L)^2$$
$$= 1 + 2p_L(p_L - 1) = 1 - 2p_L(1 - p_L). \quad (10)$$

This implies that

$$1 = |Z_L|^2 + 2p_L(1-p_L) = |Z_L|^2 - 2i^2 p_L(1-p_L) = |Z_L|^2 - 2ip_L i(1-p_L) = |Z_L|^2 - 2iP_{rL}P_{mL} = P_{rL}^2 + \left(\frac{P_{mL}}{i}\right)^2 - 2iP_{rL}P_{mL} = P_{rL}^2 + \left(\frac{P_{mL}}{i}\right)^2 + 2P_{rL}\frac{P_{mL}}{i} = \left(P_{rL} + \frac{P_{mL}}{i}\right)^2 = Pc_L^2 \Rightarrow Pc_L = 1, \quad (11)$$

where $Z_L = P_{rL} + P_{mL} = \int_{-\infty}^{L} \rho(x,t)dx + i\int_{L}^{+\infty} \rho(x,t)$ dx, written for short $Z_L = \int_{-\infty}^{L} + i\int_{L}^{+\infty}$; and with the imaginary number i having $i^2 = -1 \Rightarrow (1/i) = -i$.

We can deduce from the above that

$$Pc_L^2 = \left(P_{rL} + \frac{P_{mL}}{i}\right)^2 = \left(\int_{-\infty}^{L} + \frac{i\int_{L}^{+\infty}}{i}\right)^2 = \left(\int_{-\infty}^{L} + \int_{L}^{+\infty}\right)^2 = \left(\int_{-\infty}^{+\infty}\right)^2 = 1^2 = 1. \quad (12)$$

This is also coherent with the three added axioms previously defined.

The DOK is as follows:

$$\text{DOK}_L = |Z_L|^2 = 1 - 2p_L(1-p_L) = 1 - 2 \times \int_{-\infty}^{L} \times \left(1 - \int_{-\infty}^{L}\right); \quad (13)$$

which is a curve concave upward having a minimum at $p_L = 0.5 \Leftrightarrow \text{At}(L = \bar{x} = 0, 0.5)$ since the diffusion equation considered is a normal distribution symmetric about the mean which is $\bar{x} = 0$ cm (Figure 10).The chaotic factor is as follows:

$$\text{Chf}_L = 2iP_{rL}P_{mL} = 2i \times \int_{-\infty}^{L} \times i \times \int_{L}^{+\infty} = -2 \times \int_{-\infty}^{L} \times \int_{L}^{+\infty} = -2 \times \int_{-\infty}^{L} \times \left(1 - \int_{-\infty}^{L}\right), \quad (14)$$

which is a curve concave upward having a minimum at $p_L = 0.5 \Leftrightarrow \text{At}(L = \bar{x} = 0, -0.5)$ since the diffusion equation considered is a normal distribution symmetric about the mean which is $\bar{x} = 0$ cm (Figure 10).

Moreover, the MChf is as follows:

$$\text{MChf}_L = -2iP_{rL}P_{mL} = -2i \times \int_{-\infty}^{L} \times i \times \int_{L}^{+\infty} = 2 \times \int_{-\infty}^{L} \times \int_{L}^{+\infty} = 2 \times \int_{-\infty}^{L} \times \left(1 - \int_{-\infty}^{L}\right); \quad (15)$$

which is a curve concave downward having a maximum at

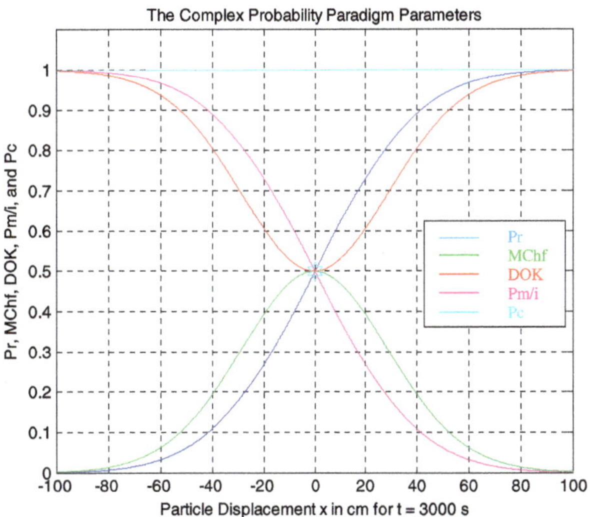

Figure 11. The complex probability parameters of the diffusion of oxygen gas in air gas with MChf for $t = 3000$ s.

$p_L = 0.5 \Leftrightarrow \text{At}(L = \bar{x} = 0, 0.5)$ since the diffusion equation considered is a normal distribution symmetric about the mean which is $\bar{x} = 0$cm (Figure 11).

One can directly see that $\text{DOK}_L = 1$ and $\text{Chf}_L = \text{MChf}_L = 0$ if $L \to -\infty$ or $L \to +\infty$, that means that we will not find the particle anywhere ($P_{rL} = 0$; impossible event) or we will find always the particle somewhere ($P_{rL} = 1$; certain event), respectively (Figures 10 and 11).

Furthermore, the intersection point of the complex probability model functions can be computed as follows:

$$P_{rL} = \frac{P_{mL}}{i} \Leftrightarrow \int_{-\infty}^{L} \frac{1}{\sqrt{4\pi Dt}} \exp\left(\frac{-x^2}{4Dt}\right) dx = 1 - \int_{-\infty}^{L} \frac{1}{\sqrt{4\pi Dt}} \exp\left(\frac{-x^2}{4Dt}\right) dx \Leftrightarrow 2\int_{-\infty}^{L} \frac{1}{\sqrt{4\pi Dt}} \exp\left(\frac{-x^2}{4Dt}\right) dx = 1 \Leftrightarrow \int_{-\infty}^{L} \frac{1}{\sqrt{4\pi Dt}} \exp\left(\frac{-x^2}{4Dt}\right) dx = \frac{1}{2} = 0.5 \Leftrightarrow L = \bar{x} = 0 \text{ cm}.$$

Notice that $P_{rL}(L = \bar{x} = 0) = 0.5$ and $P_{mL}(L = \bar{x} = 0)/i = 1 - 0.5 = 0.5$, so P_{rL} and P_{mL}/i intersect at (0, 0.5).

Moreover, we can deduce mathematically that the minimum of DOK and the maximum of MChf occur at (0, 0.5).

So we conclude that P_{rL}, P_{mL}/i, DOK, and MChf all intersect at (0, 0.5) (Figure 11).

Finally, we state that

$$Pc_L^2 = |Z_L|^2 - 2iP_{rL}P_{mL} = \text{degree of our knowledge} - \text{chaotic factor}$$

$$= \text{degree of our knowledge}$$
$$+ \text{magnitude of the chaotic factor}$$
$$\Rightarrow Pc_L^2 = 1 \quad \text{for} \quad -\infty < L < +\infty, \tag{16}$$

where $|Z_L|$ is the norm of the complex number Z_L and it combines here both the contributions of $\mathcal{R}$ and $\mathcal{M}$. Note that, it is the chaotic factor that makes the study of an event in $\mathcal{R}$ a random process. Hence, any event in $\mathcal{C}$ is deterministic since $Pc_L = 1$. Consequently, we deduce that in the set $\mathcal{C}$, we have a complete knowledge of the random variable and therefore this is the advantage of working in $\mathcal{C} = \mathcal{R} + \mathcal{M}$.

5. The evolution of DOK, Chf, and MChf

Since $Pc^2(x,t) = \text{DOK}(x,t) - \text{Chf}(x,t) = 1$ then $(\partial\text{DOK}/\partial x) - (\partial\text{Chf}/\partial x) = (\partial(1)/\partial x) = 0$

$$\Rightarrow \frac{\partial\text{DOK}}{\partial x} = \frac{\partial\text{Chf}}{\partial x}. \tag{17}$$

Similarly,

$$\frac{\partial\text{DOK}}{\partial t} = \frac{\partial\text{Chf}}{\partial t}. \tag{18}$$

That means if DOK increases (or decreases) with displacement and time evolution, then Chf increases (or decreases).

Moreover, since $Pc^2(x,t) = \text{DOK}(x,t) + \text{MChf}(x,t) = 1$ then $(\partial\text{DOK}/\partial x) + (\partial\text{MChf}/\partial x) = (\partial(1)/\partial x) = 0$

$$\Rightarrow \frac{\partial\text{DOK}}{\partial x} = -\frac{\partial\text{MChf}}{\partial x}. \tag{19}$$

Similarly,

$$\frac{\partial\text{DOK}}{\partial t} = -\frac{\partial\text{MChf}}{\partial t}. \tag{20}$$

That means if DOK increases (or decreases) with displacement and time evolution, then MChf decreases (or increases).

In addition, since $\text{Chf}(x,t) + \text{MChf}(x,t) = 0$, therefore:

$$\frac{\partial\text{Chf}}{\partial x} = -\frac{\partial\text{MChf}}{\partial x} \tag{21}$$

and

$$\frac{\partial\text{Chf}}{\partial t} = -\frac{\partial\text{MChf}}{\partial t}. \tag{22}$$

That means if Chf increases (or decreases) with displacement and time evolution, then MChf decreases (or increases).

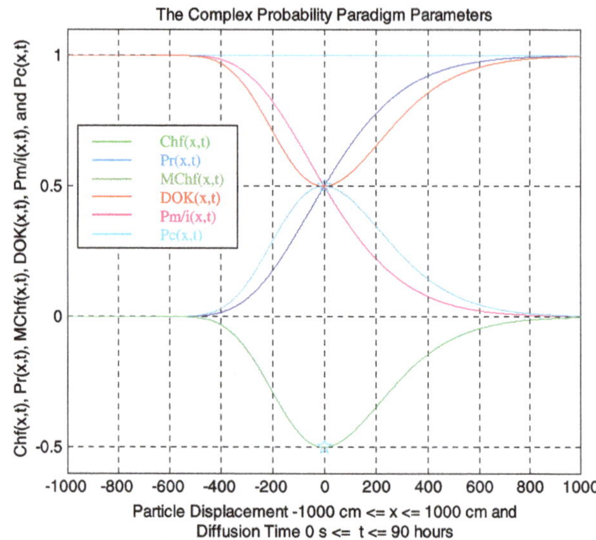

Figure 12. The complex probability parameters of the diffusion of oxygen gas in air gas whenever both x and t vary simultaneously.

Moreover, the mixed partial derivatives are the following:

$$\frac{\partial^2\text{DOK}}{\partial x\partial t} = \frac{\partial^2\text{Chf}}{\partial x\partial t} \Rightarrow \text{both DOK and Chf are concave upward,} \tag{23}$$

$$\frac{\partial^2\text{DOK}}{\partial x\partial t} = -\frac{\partial^2\text{MChf}}{\partial x\partial t} \Rightarrow \text{DOK and MChf are of opposite concavity,} \tag{24}$$

$$\frac{\partial^2\text{Chf}}{\partial x\partial t} = -\frac{\partial^2\text{MChf}}{\partial x\partial t} \Rightarrow \text{Chf and MChf are of opposite concavity.} \tag{25}$$

Figure 12 shows that all the complex probability paradigm functions conserve their characteristics and behaviour whenever both x and t vary simultaneously:

$$-1000\ \text{cm} \le x \le 1000\ \text{cm} \quad \text{and} \quad 0\,\text{s} \le t \le 90\,\text{h}.$$

But as we can notice, all the graphs are skewed to the right. Furthermore, we can write

$$\int_{t=0\,\text{s}}^{t=90\,\text{h}} \int_{x=-1000\,\text{cm}}^{x=g(t)} f(x,t)\,\mathrm{d}x\,\mathrm{d}t = F(x = 1000\,\text{cm}, t = 90\,\text{h}) = 1,$$

where

$$x = g(t) = t/162 - 1000$$
$$\Rightarrow \begin{cases} x = g(t = 0\,\text{s}) = 0/162 - 1000 = -1000\,\text{cm}, \\ x = g(t = 90\,\text{h} = 324,000\,\text{s}) = 324,000/ \\ \quad 162 - 1000 = 1000\ \text{cm}. \end{cases}$$

6. Numerical example

Consider for this numerical example always the diffusion of oxygen gas in air gas (Wikipedia, *Mass Diffusivity*), hence: $\bar{x}$ = mean value of $x = 0$ cm.

The diffusion factor $D = 0.176\ \text{cm}^2/\text{s}$ at a temperature $T = 25\,°\text{C}$.

The diffusion time: $t = 3000$ s.

This implies that the standard deviation of the particles Brownian motion is as follows:

$$\sigma = \sqrt{2Dt} = \sqrt{2 \times 0.176 \times 3000} = 32.4962\ \text{cm}.$$

We can compute from the CDF the following:

$P_{\text{rob}}[7\,\text{cm} \le x \le 21\,\text{cm}] = 0.1557$, $P_{\text{rob}}[-\infty < x \le 21\,\text{cm}] = 0.7409$, and $P_{\text{rob}}[21\,\text{cm} \le x < +\infty] = 0.2591$.

We can see that $P_{\text{rob}}[L_a \le x \le L_b] = F(u_b) - F(u_a)$, where $u_b = \frac{L_b - \bar{x}}{\sigma}$ and $u_a = \frac{L_a - \bar{x}}{\sigma}$.

If $L_a = 7$ cm and $L_b = 21$ cm, then the probability $= F\left(\frac{21-0}{32.4962}\right) - F\left(\frac{7-0}{32.4962}\right) = 0.1557$.

Note that

$$F(u_L) = \int_{-\infty}^{u_L} \frac{1}{\sqrt{2\pi}} \exp\left(\frac{-u^2}{2}\right) du = P_{\text{rob}}[u \le u_L],$$

where $u = \frac{x - \bar{x}}{\sigma}$.

In the real domain $\mathscr{R}$, we have $dF = f(u)\,du = (1/\sqrt{2\pi}) \exp(-u^2/2)\,du$, and which is the standard normal distribution of diffusion having $\bar{u} = 0$ and $\sigma_u = 1$.

And we know that

$$\int_{-\infty}^{+\infty} dF = \int_{-\infty}^{+\infty} \frac{1}{\sqrt{2\pi}} \exp\left(\frac{-u^2}{2}\right) du$$
$$= \int_{-\infty}^{+\infty} \frac{1}{\sqrt{2\pi}\sigma} \exp\left[-\frac{1}{2}\left(\frac{x - \bar{x}}{\sigma}\right)^2\right] dx = 1.$$

Now, as a numerical example, take $L = 21$ cm, then in the real domain $\mathscr{R}$ we have

$$P_{\text{rob}}[-\infty < x \le 21] = P_{rL}$$
$$= \int_{-\infty}^{L=21} \frac{1}{\sqrt{2\pi} \times 32.4962} \times \exp\left[-\frac{1}{2}\left(\frac{x - 0}{32.4962}\right)^2\right] dx$$
$$= 0.7409.$$

The correspondent probability in the imaginary domain $\mathscr{M}$ is as follows:

$$P_{mL} = i(1 - P_{rL}) = iP_{\text{rob}}[x > 21]$$
$$= i \int_{21}^{+\infty} \frac{1}{\sqrt{2\pi} \times 32.4962} \exp\left[-\frac{1}{2}\left(\frac{x - 0}{32.4962}\right)^2\right] dx$$
$$= i \times 0.2591.$$

And the complementary probability is $P_{mL}/i = 0.2591$. We note that

$$Z_L = P_{rL} + P_{mL} = \int_{-\infty}^{L=21} f(u)\,du + i \int_{L=21}^{+\infty} f(u)\,du$$
$$= 0.7409 + i \times 0.2591.$$

We also have

$$Pc_L^2 = \left(P_{rL} + \frac{P_{mL}}{i}\right)^2 = \left(\int_{-\infty}^{L=21} + \int_{L=21}^{+\infty}\right)^2$$
$$= \left(\int_{-\infty}^{+\infty}\right)^2 = 1^2 = 1.$$

And the DOK is as follows:

$$\text{DOK}_L = |Z_L|^2 = 1 - 2P_{rL}(1 - P_{rL}) = 1 - 2 \times \int_{-\infty}^{L=21}$$
$$\times \left(1 - \int_{-\infty}^{L=21}\right),$$

where

$$\text{DOK}_L = 1 \quad \text{if} \begin{cases} L \to -\infty & \text{hence } P_{rL} = 0, \\ L \to +\infty & \text{hence } P_{rL} = 1. \end{cases}$$

Furthermore, the chaotic factor is as follows:

$$\text{Chf}_L = 2i \times P_{rL} \times P_{mL} = 2i \times \int_{-\infty}^{L=21} \times i \times \int_{L=21}^{+\infty}$$
$$= -2 \times \int_{-\infty}^{21} \times \left(1 - \int_{-\infty}^{21}\right),$$

where

$$\text{Chf}_L = 0 \text{ if} \begin{cases} L \to -\infty & \text{hence } P_{rL} = 0, \\ L \to +\infty & \text{hence } P_{rL} = 1. \end{cases}$$

Moreover, the MChf is as follows:

$$\text{MChf}_L = -2i \times P_{rL} \times P_{rL} = -2i \times \int_{-\infty}^{L=21} \times i \times \int_{L=21}^{+\infty}$$
$$= 2 \times \int_{-\infty}^{21} \times \left(1 - \int_{-\infty}^{21}\right),$$

where

$$\text{MChf}_L = 0 \text{ if } \begin{cases} L \to -\infty & \text{hence } P_{rL} = 0, \\ L \to +\infty & \text{hence } P_{rL} = 1. \end{cases}$$

Numerically, we write

$$\text{DOK}_L = |Z_L|^2 = (0.7409)^2 + (0.2591)^2 = 0.5489 + 0.0671 = 0.6160 \Rightarrow |Z_L| = 0.7849$$

$$\Rightarrow \text{Chf}_L \neq 0 \quad \text{notice that } \frac{1}{2} \leq \text{DOK}_L \leq 1.$$

Hence,

$$\text{Chf}_L = -2 \times 0.7409 \times 0.2591 = -0.3840 \quad \text{notice that } -\frac{1}{2} \leq \text{Chf}_L \leq 0.$$

And

$$\text{MChf}_L = 2 \times 0.7409 \times 0.2591 = 0.3840 \quad \text{notice that } 0 \leq \text{MChf}_L \leq \frac{1}{2}.$$

What is interesting here and throughout the whole original paradigm simulation, is that we have quantified both the DOK and the chaotic factor of the random event in $\mathscr{R}$ which is the stochastic Brownian motion of the oxygen gas particles in air gas.

Moreover, notice that

$$\text{the DOK} - \text{the chaotic factor} = 0.6160 - (-0.3840) = 0.6160 + 0.3840 = 1 = Pc_L.$$

And

$$\text{the DOK} + \text{the MChf} = 0.6160 + 0.3840 = 1 = Pc_L.$$

Therefore, we state that we have always

$$\begin{aligned} Pc_L^2 &= |Z_L|^2 - 2iP_{rL}P_{mL} \\ &= \text{degree of our knowledge} - \text{chaotic factor} \\ &= \text{degree of our knowledge} \\ &\quad + \text{magnitude of the chaotic factor} \\ &= 1. \end{aligned}$$

And if $\text{Chf}_L = \text{MChf}_L = 0 \Rightarrow |Z_L|^2 = 1$, in other words, if the chaotic factor or the MChf are zero, then the DOK in $\mathscr{R}$ is 1 or 100%.

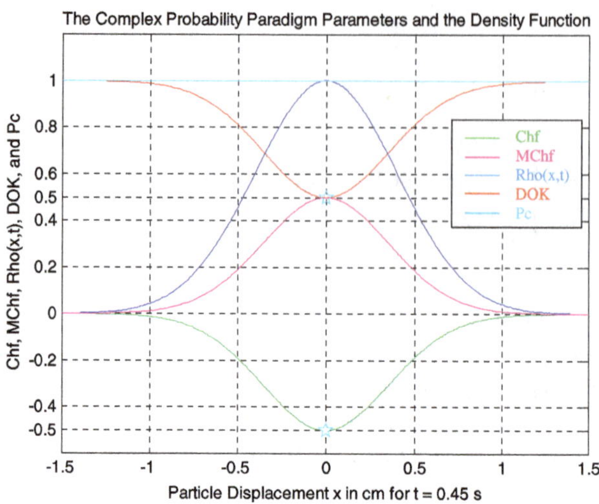

Figure 13. The complex probability parameters of the diffusion of oxygen gas in air gas with $\rho(x,t)$ at $t = 0.45$ s.

Conversely, if we assume that

$$\text{Chf}_L = 0 \Rightarrow |Z_L|^2 = 1 \Rightarrow P_{rL}^2 + \left(\frac{P_{mL}}{i}\right)^2 = 1$$

$$\Rightarrow 2P_{rL}(1 - P_{rL}) = 0 \Rightarrow \begin{cases} P_{rL} = 0 \\ \text{or} \\ P_{rL} = 1 \end{cases}$$

$$\Rightarrow \begin{cases} L \to -\infty \\ \text{or} \\ L \to +\infty \end{cases}$$

And if

$$\text{Chf}_L = -\frac{1}{2} \quad \text{or} \quad \text{MChf}_L = \frac{1}{2}$$

$$\Rightarrow L = \bar{x} = 0 \quad \text{and} \quad \text{DOK}_L = \frac{1}{2}.$$

Now, if L increases from 21 cm to become equal to 50 cm for example, then: DOK_L and Chf_L both increase, and MChf_L decreases.

Therefore, we can deduce that

$$\lim_{L \to \pm\infty} \text{Chf}_L = \lim_{L \to \pm\infty} \text{MChf}_L = 0 \text{ and } \lim_{L \to \pm\infty} \text{DOK}_L = 1,$$

where

$$Pc_L^2 = \text{DOK}_L - \text{Chf}_L = \text{DOK}_L + \text{MChf}_L = 1$$

for every value of L in the real set $\mathscr{R}$ and for any diffusion time $0 \leq t < +\infty$, specifically for $t = 0.45$ s as in Figure 13, and as we will see in Section 8 for other values of t.

7. Flowchart of the complex probability paradigm

The following flowchart summarizes all the procedures of the proposed complex probability prognostic model:

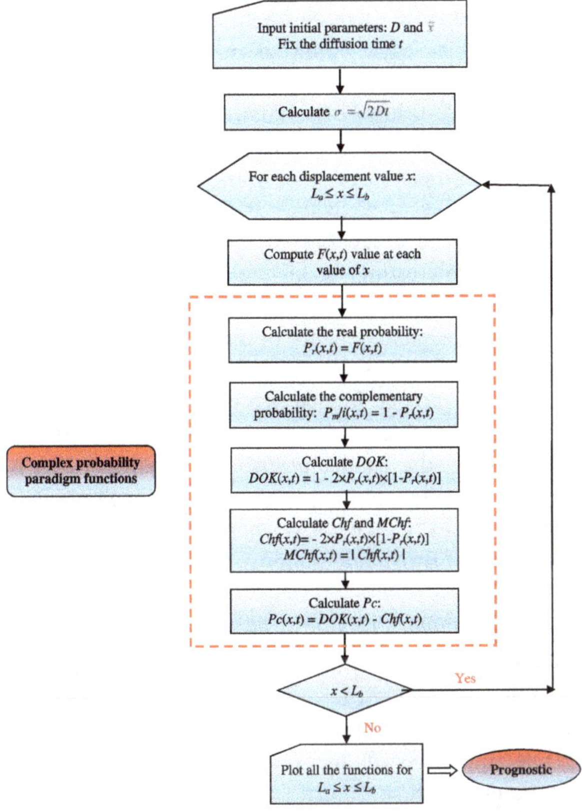

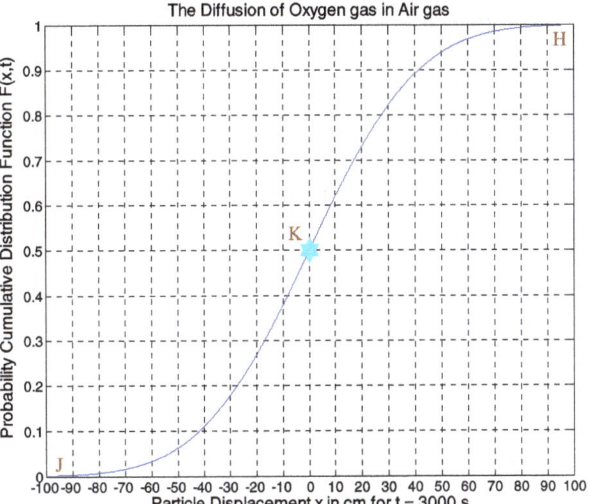

Figure 14. The CDF of the diffusion of oxygen gas in air gas for $t = 3000$ s.

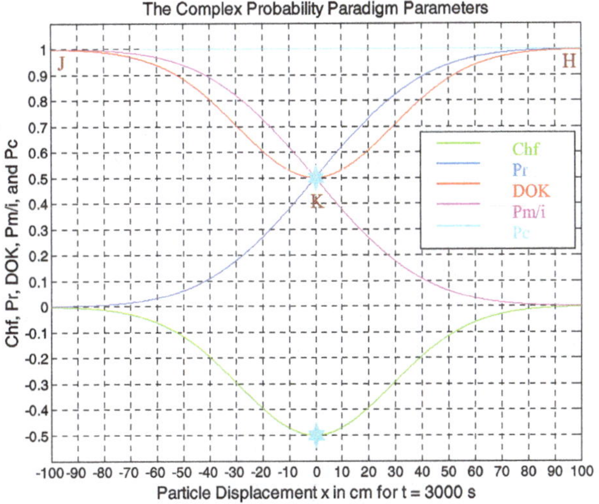

Figure 15. The complex probability parameters for $t = 3000$ s.

8. Simulation of the new paradigm

Note that all the numerical values found in the paradigm functions analysis for $t = 3000$ s or $t = 1000$ s or $t = 100$ s were computed using the MATLAB version 2015 software.

8.1. *The paradigm functions analysis for t = 3000 s*

We notice from Figures 14–16 that the DOK is maximum (DOK $= 1$) when MChf is minimum (MChf $= 0$) (points J & H) and that means when the MChf decreases our certain knowledge in $\mathcal{R}$ increases.

At the beginning $P_r(x < -100\,\text{cm}) \approx 0$, the system is nearly intact and has nearly zero chaotic factor (Chf $=$ MChf ≈ 0) before any diffusion, hence at this instant DOK ≈ 1. Here $P_m/\text{i} \approx 1$ with $Pc = P_r + P_m/\text{i} \approx 0 + 1 = 1$ and $Pc = \text{DOK} + \text{MChf} \approx 1 + 0 = 1$.

Note that if $x \to -\infty$ cm, then DOK $\to 1$, Chf $\to 0$, MChf $\to 0$, $P_r \to 0$, $P_m/\text{i} \to 1$, with $Pc = P_r + P_m/\text{i} \to 0 + 1 = 1$ and $Pc = \text{DOK} + \text{MChf} \to 1 + 0 = 1$.

Afterward, with diffusion, x starts to increase with $-\infty < x < +\infty$, $P_r(L) = \int_{-\infty}^{L} \rho(x,t)\,dx \neq 0$ and we have always $Pc(L) = P_r(L) + P_m/\text{i}(L) = \text{DOK}(L) + \text{MChf}(L) = 1$, thus MChf starts to increase also during the diffusion due to the environment and intrinsic conditions thus leading to a decrease in DOK.

If $x = -100$ cm (point J), then DOK $= 0.997913$, Chf $= -0.002087$, MChf $= 0.002087$, $P_r = 0.001044$, $P_m/\text{i} = 0.998956$, with $Pc = P_r + P_m/\text{i} = 0.001044 + 0.998956 = 1$ and $Pc = \text{DOK} + \text{MChf} = 0.997913 + 0.002087 = 1$.

If $x = -60$ cm, then DOK $= 0.93726$, Chf $= -0.06274$, MChf $= 0.06274$, $P_r = 0.03242$, $P_m/\text{i} = 0.96758$, with $Pc = P_r + P_m/\text{i} = 0.03242 + 0.96758 = 1$ and $Pc = \text{DOK} + \text{MChf} = 0.93726 + 0.06274 = 1$. We can see that with the increase of x, the real probability P_r also increases.

If $x = -20$ cm, then DOK $= 0.6066$, Chf $= -0.3934$, MChf $= 0.3934$, $P_r = 0.2691$, $P_m/\text{i} = 0.7309$, with $Pc = P_r + P_m/\text{i} = 0.2691 + 0.7309 = 1$ and $Pc = \text{DOK} + \text{MChf} = 0.6066 + 0.3934 = 1$. We can see here

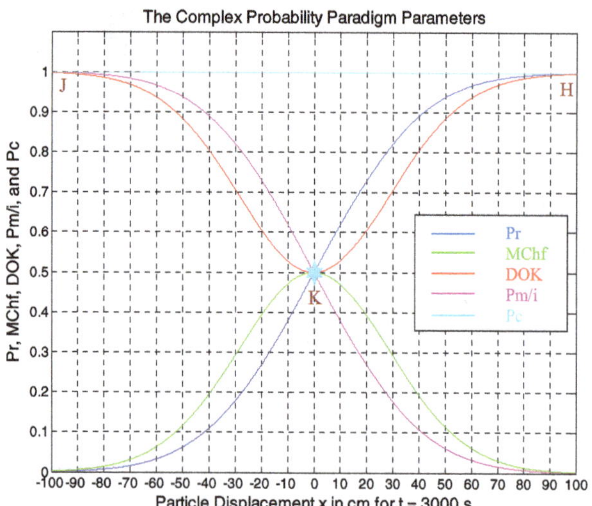

Figure 16. The complex probability parameters with MChf for $t = 3000$ s.

that with the increase of x, the MChf also increases, whereas DOK decreases.

If $x = 0$ cm (point K), both DOK (minimum) and MChf (maximum) reach 0.5, $P_r = 0.5$, $P_m/\text{i} = 0.5$, and Chf $= -0.5$ with $Pc = P_r + P_m/\text{i} = 0.5 + 0.5 = 1$ and $Pc = \text{DOK} + \text{MChf} = 0.5 + 0.5 = 1$, as always. Hence, all the EKA parameters will intersect at the point K. We have here maximum chaos and the minimum of the system knowledge in $\mathscr{R}$; therefore, the real probability is $P_r = 1/2$ = probability half way to full diffusion. Furthermore, notice in Figure 16 the complete symmetry at the vertical axis $x = 0$ cm = half way to complete dispersion.

If $x = 30$ cm, then DOK $= 0.7074$, Chf $= -0.2926$, MChf $= 0.2926$, $P_r = 0.822$, $P_m/\text{i} = 0.178$, with $Pc = P_r + P_m/\text{i} = 0.822 + 0.178 = 1$ and $Pc = \text{DOK} + \text{MChf} = 0.7074 + 0.2926 = 1$. We can see that with the increase of x and consequently the decrease of MChf, the real probability P_r also increases.

If $x = 70$ cm, then DOK $= 0.96926$, Chf $= -0.03074$, MChf $= 0.03074$, $P_r = 0.98438$, $P_m/\text{i} = 0.01562$, with $Pc = P_r + P_m/\text{i} = 0.98438 + 0.01562 = 1$ and $Pc = \text{DOK} + \text{MChf} = 0.96926 + 0.03074 = 1$.

If $x = 100$ cm (point H), then DOK $= 0.997913$, Chf $= -0.002087$, MChf $= 0.002087$, $P_r = 0.998956$, $P_m/\text{i} = 0.001044$, with $Pc = P_r + P_m/\text{i} = 0.998956 + 0.001044 = 1$ and $Pc = \text{DOK} + \text{MChf} = 0.997913 + 0.002087 = 1$. Here, since $x = 100$ cm which is relatively very big in this case, then the real probability P_r is very near to 1 since we will reach total and full dispersion very soon.

With the increase of displacement beyond 100 cm, MChf and Chf are nearly zero, DOK returns to 1; hence we have the total dispersion of the system of the oxygen gas particles in air gas. At this last point, $P_r \approx 1$, $P_m/\text{i} \approx 0$, with $Pc = P_r + P_m/\text{i} \approx 1 + 0 = 1$ and $Pc = \text{DOK} + \text{MChf} \approx 1 + 0 = 1$; thus, the logical explanation of the value of DOK ≈ 1 follows.

Note that if $x \to +\infty$ cm, then DOK $\to 1$, Chf $\to 0$, MChf $\to 0$, $P_r \to 1$, $P_m/\text{i} \to 0$, with $Pc = P_r + P_m/\text{i} \to 1 + 0 = 1$ and $Pc = \text{DOK} + \text{MChf} \to 1 + 0 = 1$.

Moreover, at each value of x, the probability to find the particle $Pc(x,t)$ is certainly predicted in the complex set $\mathscr{C}$ with Pc maintained as equal to one through a continuous compensation between DOK and Chf. This compensation is from $x = -\infty$, where $P_r = 0$ until $x = +\infty$ where $P_r = 1$, keeping always $Pc = \text{DOK} - \text{Chf} = \text{DOK} + \text{MChf} = 1$. Furthermore, what is truly interesting for all the particle displacement simulations is that we have quantified and visualized both the DOK and the chaotic factor of the Brownian motion.

It becomes clear from the simulations that DOK is the measure of our certain knowledge in $\mathscr{R}$ (100% probability) about the expected event and it does not include any uncertain knowledge (with a probability less than 100%).

We note that the same logic and analysis concerning the diffusion as well as all the EKA parameters apply for all the three instants of Brownian motion.

8.1.1. The complex probability cubes

In Figure 17, we can see the simulation of DOK and Chf as functions of x and of each other for $t = 3000$ s. The line in cyan is $Pc^2(x,t) = \text{DOK}(x,t) - \text{Chf}(x,t) = 1 = Pc(x,t)$. This line, projected on the $x = -100$ cm plane, starts at the point (DOK $= 1$, Chf $= 0$) when $x = -100$ cm (point J in Figures 14–16), reaches the point (DOK $= 0.5$, Chf $= -0.5$) when $x = 0$ cm (point K in Figures 14–16), and returns at the end to the point (DOK $= 1$, Chf $= 0$) when $x = 100$ cm (point H in Figures 14–16). The other curves are the graphs of DOK(x,t) and Chf(x,t) in different planes. Notice that they all have a minimum at $x = 0$ cm (point K in Figures 14–16), as explained previously.

In Figure 18, we can notice the simulation of the real probability $P_r(x,t)$ and its complementary probability $P_m/\text{i}(x,t)$ as functions of x for $t = 3000$ s. The line in cyan is $Pc^2(x,t) = P_r(x,t) + P_m/\text{i}(x,t) = 1 = Pc(x,t)$. This line, projected on the $x = -100$ plane, starts at the point ($P_r = 0$, $P_m/\text{i} = 1$) (point J in Figures 14–16) and ends at the point ($P_r = 1$, $P_m/\text{i} = 0$) (point H in Figures 14–16). The blue curve represents $P_r(x,t)$ in the plane $P_r = P_m/\text{i}$ and the red curve represents $P_m/\text{i}(x,t)$ in the plane $P_r + P_m/\text{i} = 1$. Notice the importance of the point ($P_r = 0.5$, $P_m/\text{i} = 0.5$) corresponding to $x = 0$ cm (point K in Figures 14–16).

Note that similar cubes can be drawn for the instants $t = 1000$ s and for $t = 100$ s with their corresponding points J, K, and H.

8.2. The paradigm functions analysis for t = 1000 s

We notice from Figures 19–21 that the DOK is maximum (DOK $= 1$) when MChf is minimum (MChf $= 0$) (points

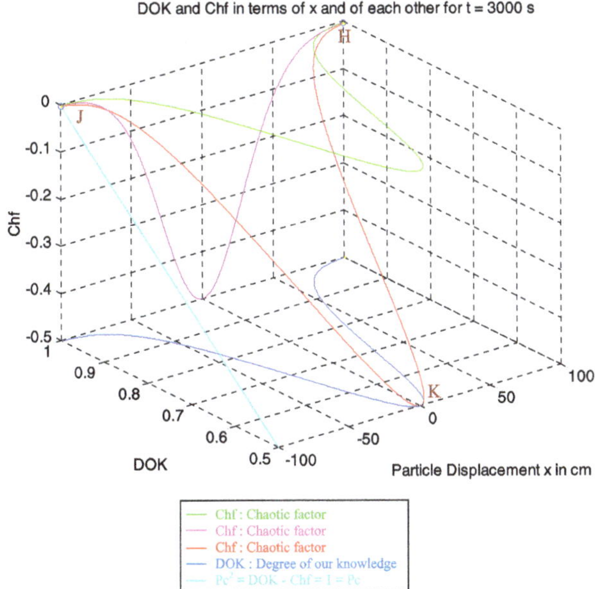

Figure 17. DOK Chf and Pc, in terms of x and of each other for the diffusion time $t = 3000$ s.

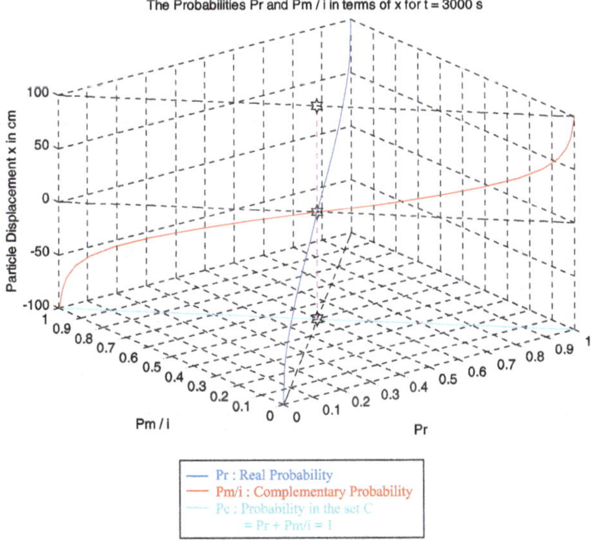

Figure 18. P_r, P_m/i, and Pc, in terms of x for the diffusion time $t = 3000$ s.

J & H) and that means when the MChf decreases our certain knowledge in $\mathscr{R}$ increases.

At the beginning $P_r(x < -60\,\text{cm}) \approx 0$, the system is nearly intact and has nearly zero chaotic factor (Chf = MChf ≈ 0) before any diffusion, hence at this instant DOK ≈ 1. Here $P_m/\text{i} \approx 1$ with $Pc = P_r + P_m/\text{i} \approx 0 + 1 = 1$ and $Pc = \text{DOK} + \text{MChf} \approx 1 + 0 = 1$.

Note that if $x \to -\infty$ cm, then DOK $\to 1$, Chf $\to 0$, MChf $\to 0$, $P_r \to 0$, $P_m/\text{i} \to 1$, with $Pc = P_r + P_m/\text{i} \to 0 + 1 = 1$ and $Pc = \text{DOK} + \text{MChf} \to 1 + 0 = 1$.

Afterward, with diffusion, x starts to increase with $-\infty < x < +\infty, P_r(L) = \int_{-\infty}^{L} \rho(x,t)\,\mathrm{d}x \neq 0$ and we have always $Pc(L) = P_r(L) + P_m/\text{i}(L) = \text{DOK}(L) + \text{MChf}(L)$

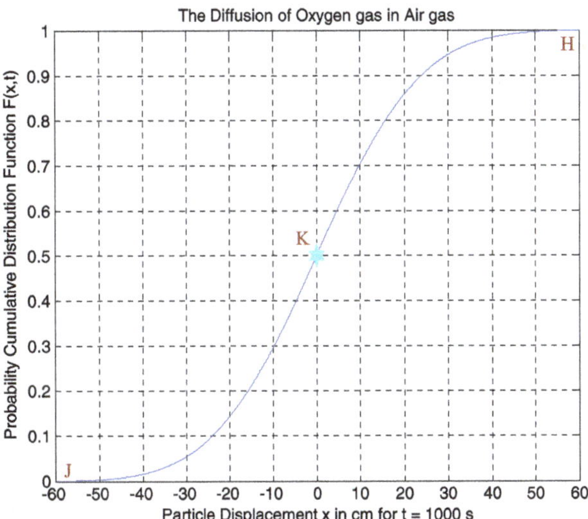

Figure 19. The CDF of the diffusion of oxygen gas in air gas for $t = 1000$ s.

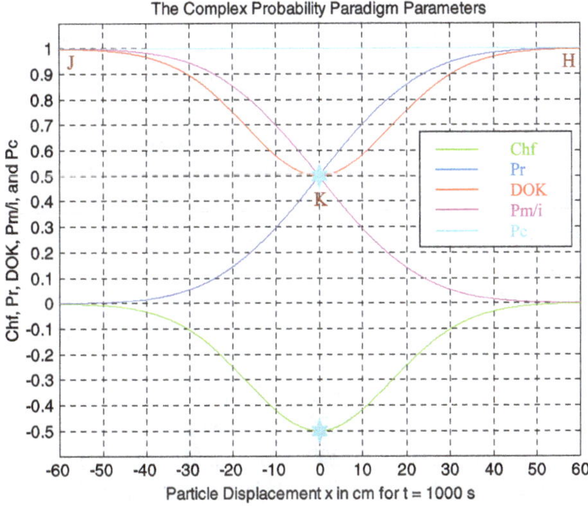

Figure 20. The complex probability parameters for $t = 1000$ s.

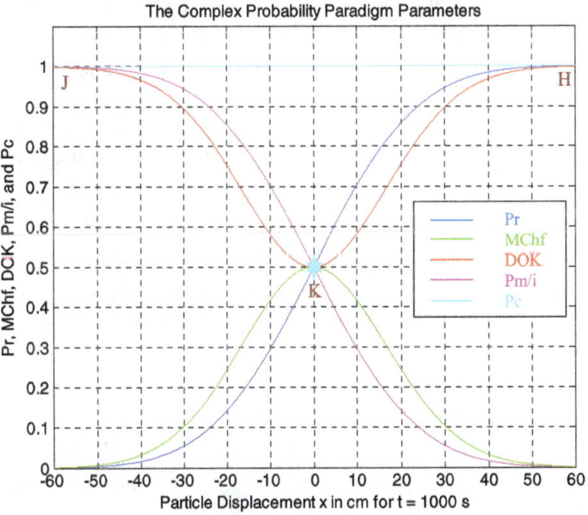

Figure 21. The complex probability parameters with MChf for $t = 1000$ s.

= 1, thus MChf starts to increase also during the diffusion due to the environment and intrinsic conditions thus leading to a decrease in DOK.

If $x = -60$ cm (point J), then DOK = 0.998617, Chf = −0.001383, MChf = 0.001383, $P_r = 0.0006919$, $P_m/i = 0.9993081$, with $Pc = P_r + P_m/i = 0.0006919 + 0.9993081 = 1$ and Pc = DOK + MChf = 0.998617 + 0.001383 = 1.

If $x = -40$ cm, then DOK = 0.96754, Chf = −0.03246, MChf = 0.03246, $P_r = 0.0165$, $P_m/i = 0.9835$, with $Pc = P_r + P_m/i = 0.0165 + 0.9835 = 1$ and Pc = DOK + MChf = 0.96754 + 0.03246 = 1. We can see that with the increase of x, the real probability P_r also increases.

If $x = -20$ cm, then DOK = 0.7546, Chf = −0.2454, MChf = 0.2454, $P_r = 0.1432$, $P_m/i = 0.8568$, with $Pc = P_r + P_m/i = 0.1432 + 0.8568 = 1$ and Pc = DOK + MChf = 0.7546 + 0.2454 = 1. We can see here that with the increase of x, the MChf also increases, whereas DOK decreases.

If $x = 0$ cm (point K), both DOK (minimum) and MChf (maximum) reach 0.5, $P_r = 0.5$, $P_m/i = 0.5$, and Chf = −0.5 with $Pc = P_r + P_m/i = 0.5 + 0.5 = 1$ and Pc = DOK + MChf = 0.5 + 0.5 = 1, as always. Hence, all the EKA parameters will intersect at the point K. We have here maximum chaos and the minimum of the system knowledge in $\mathscr{R}$; therefore, the real probability is $P_r = 1/2$ = probability half way to full diffusion. Furthermore, notice in the last figure (Figure 21) the complete symmetry at the vertical axis $x = 0$ cm = half way to complete dispersion.

If $x = 10$ cm, then DOK = 0.5824, Chf = −0.4176, MChf = 0.4176, $P_r = 0.703$, $P_m/i = 0.297$, with $Pc = P_r + P_m/i = 0.703 + 0.297 = 1$ and Pc = DOK + MChf = 0.5824 + 0.4176 = 1. We can see here that with the increase of x and consequently the decrease of MChf, the real probability P_r also increases.

If $x = 30$ cm, then DOK = 0.8962, Chf = −0.1038, MChf = 0.1038, $P_r = 0.94509$, $P_m/i = 0.05491$, with $Pc = P_r + P_m/i = 0.94509 + 0.05491 = 1$ and Pc = DOK + MChf = 0.8962 + 0.1038 = 1.

If $x = 60$ cm (point H), then DOK = 0.998617, Chf = −0.001383, MChf = 0.001383, $P_r = 0.9993081$, $P_m/i = 0.0006919$, with $Pc = P_r + P_m/i = 0.9993081 + 0.0006919 = 1$ and Pc = DOK + MChf = 0.998617 + 0.001383 = 1. Here, since $x = 60$ cm which is relatively very big in this case, then the real probability P_r is very near to 1 since we will reach total and full dispersion very soon.

With the increase of displacement beyond 60 cm, MChf and Chf are nearly zero, DOK returns to 1, hence we have the total dispersion of the system of the oxygen gas particles in air gas. At this last point, $P_r \approx 1$, $P_m/i \approx 0$, with $Pc = P_r + P_m/i \approx 1 + 0 = 1$ and Pc = DOK + MChf ≈ 1 + 0 = 1; thus, the logical explanation of the value of DOK ≈ 1 follows.

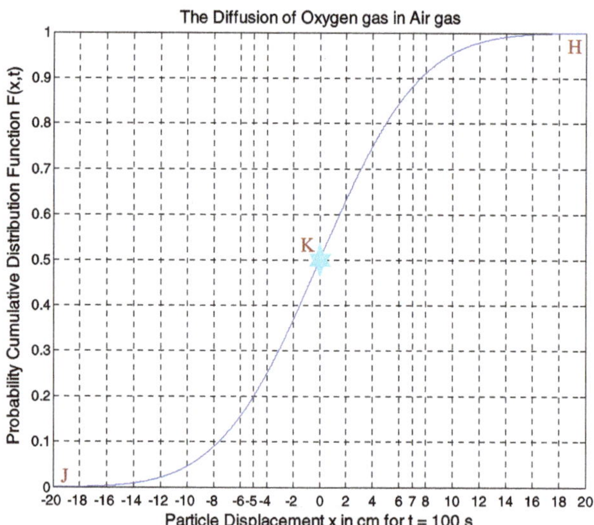

Figure 22. The CDF of the diffusion of oxygen gas in air gas for $t = 100$ s.

Note that, if $x \to +\infty$ cm, then DOK → 1, Chf → 0, MChf → 0, $P_r \to 1$, $P_m/i \to 0$, with $Pc = P_r + P_m/i \to 1 + 0 = 1$ and Pc = DOK + MChf → 1 + 0 = 1.

Moreover, at each value of x, the probability to find the particle $Pc(x,t)$ is certainly predicted in the complex set $\mathscr{C}$ with Pc maintained as equal to one through a continuous compensation between DOK and Chf. This compensation is from $x = -\infty$ where $P_r = 0$ until $x = +\infty$ where $P_r = 1$, keeping always Pc = DOK − Chf = DOK + MChf = 1. We can understand now that DOK is the measure of our certain knowledge in $\mathscr{R}$ (100% probability) about the expected event, it does not include any uncertain knowledge (with a probability less than 100%). Furthermore, what is truly interesting in the particle displacement simulations is that we have quantified and visualized both the DOK and the chaotic factor of the Brownian motion.

We note that the same logic and analysis for the first instant of Brownian motion were applied to the second instant concerning the diffusion as well as all the EKA parameters.

8.3. The paradigm functions analysis for t = 100 s

We notice from Figures 22–24 that the DOK is maximum (DOK = 1) when MChf is minimum (MChf = 0) (points J & H) and that means when the MChf decreases our certain knowledge in $\mathscr{R}$ increases.

At the beginning $P_r(x < -20\text{ cm}) \approx 0$, the system is nearly intact and has nearly zero chaotic factor (Chf = MChf ≈ 0) before any diffusion, hence at this instant DOK ≈ 1. Here $P_m/i \approx 1$ with $Pc = P_r + P_m/i \approx 0 + 1 = 1$ and Pc = DOK + MChf ≈ 1 + 0 = 1.

Note that, if $x \to -\infty$ cm, then DOK → 1, Chf → 0, MChf → 0, $P_r \to 0$, $P_m/i \to 1$, with $Pc = P_r + P_m/i \to 0 + 1 = 1$ and Pc = DOK + MChf → 1 + 0 = 1.

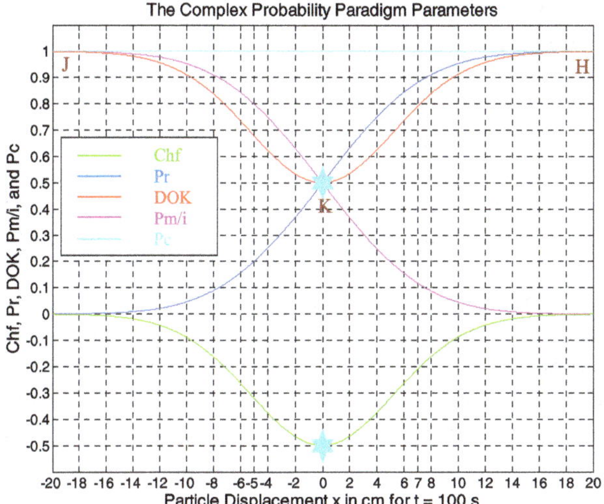

Figure 23. The complex probability parameters for $t = 100$ s.

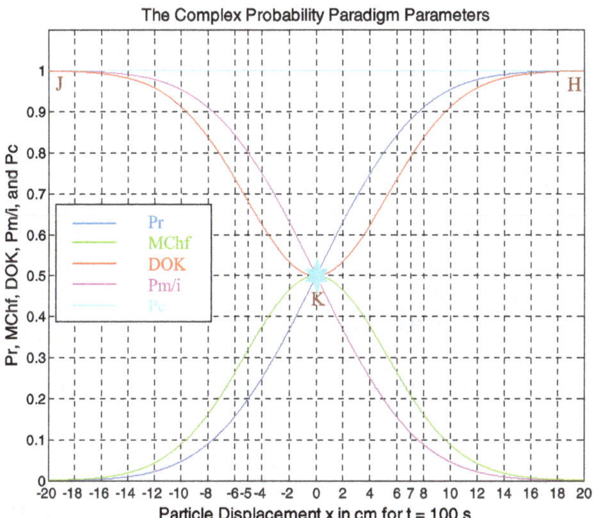

Figure 24. The complex probability parameters with MChf for $t = 100$ s.

Afterward, with diffusion, x starts to increase with $-\infty < x < +\infty, P_r(L) = \int_{-\infty}^{L} \rho(x,t)\,dx \neq 0$ and we have always $Pc(L) = P_r(L) + P_m/\mathrm{i}(L) = \mathrm{DOK}(L) + \mathrm{MChf}(L) = 1$, thus MChf also starts to increase during the diffusion due to the environment and intrinsic conditions thus leading to a decrease in DOK.

If $x = -20$ cm (point J), then DOK $= 0.9992513$, Chf $= -0.0007487$, MChf $= 0.0007487$, $P_r = 0.0003745$, $P_m/\mathrm{i} = 0.9996255$, with $Pc = P_r + P_m/\mathrm{i} = 0.0003745 + 0.9996255 = 1$ and $Pc =$ DOK + MChf $= 0.9992513 + 0.0007487 = 1$.

If $x = -10$ cm, then DOK $= 0.91233$, Chf $= -0.08767$, MChf $= 0.08767$, $P_r = 0.04595$, $P_m/\mathrm{i} = 0.95405$, with $Pc = P_r + P_m/\mathrm{i} = 0.04595 + 0.95405 = 1$ and $Pc =$ DOK + MChf $= 0.91233 + 0.08767 = 1$. We can see that with the increase of x, the real probability P_r also increases.

If $x = -5$ cm, then DOK $= 0.6804$, Chf $= -0.3196$, MChf $= 0.3196$, $P_r = 0.1997$, $P_m/\mathrm{i} = 0.8003$, with $Pc = P_r + P_m/\mathrm{i} = 0.1997 + 0.8003 = 1$ and $Pc =$ DOK + MChf $= 0.6804 + 0.3196 = 1$. We can see here that with the increase of x, the MChf also increases, whereas DOK decreases.

If $x = 0$ cm (point K), both DOK (minimum) and MChf (maximum) reach 0.5, $P_r = 0.5$, $P_m/\mathrm{i} = 0.5$, and Chf $= -0.5$ with $Pc = P_r + P_m/\mathrm{i} = 0.5 + 0.5 = 1$ and $Pc =$ DOK + MChf $= 0.5 + 0.5 = 1$, as always. Hence, all the EKA parameters will intersect at the point K. We have here maximum chaos and the minimum of the system knowledge in $\mathcal{R}$; therefore, the real probability is $P_r = 1/2 =$ probability half way to full diffusion. Furthermore, notice in Figure 24 the complete symmetry at the vertical axis $x = 0$ cm = half way to complete dispersion.

If $x = 7$ cm, then DOK $= 0.7903$, Chf $= -0.2097$, MChf $= 0.2097$, $P_r = 0.881$, $P_m/\mathrm{i} = 0.119$, with $Pc = P_r + P_m/\mathrm{i} = 0.881 + 0.119 = 1$ and $Pc =$ DOK + MChf $= 0.7903 + 0.2097 = 1$. We can see here that with the increase of x and consequently the decrease of MChf, the real probability P_r also increases.

If $x = 12$ cm, then DOK $= 0.95781$, Chf $= -0.04219$, MChf $= 0.04219$, $P_\mathrm{r} = 0.97844$, $P_\mathrm{m}/\mathrm{i} = 0.02156$, with $Pc = P_\mathrm{r} + P_\mathrm{m}/\mathrm{i} = 0.97844 + 0.02156 = 1$ and $Pc =$ DOK + MChf $= 0.95781 + 0.04219 = 1$.

If $x = 20$ cm (point H), then DOK $= 0.9992513$, Chf $= -0.0007487$, MChf $= 0.0007487$, $P_r = 0.9996255$, $P_m/\mathrm{i} = 0.0003745$, with $Pc = P_r + P_m/\mathrm{i} = 0.9996255 + 0.0003745 = 1$ and $Pc =$ DOK + MChf $= 0.9992513 + 0.0007487 = 1$. Here, since $x = 20$ cm which is relatively very big in this case, then the real probability P_r is very near to 1 since we will reach total and full dispersion very soon.

With the increase of displacement beyond 20 cm, MChf and Chf are nearly zero, DOK returns to 1; hence, we have the total dispersion of the system of the oxygen gas particles in air gas. At this last point, $P_r \approx 1$, $P_m/\mathrm{i} \approx 0$, with $Pc = P_r + P_m/\mathrm{i} \approx 1 + 0 = 1$ and $Pc =$ DOK + MChf $\approx 1 + 0 = 1$; thus, the logical explanation of the value of DOK ≈ 1 follows.

Note that if $x \to +\infty$ cm, then DOK $\to 1$, Chf $\to 0$, MChf $\to 0$, $P_r \to 1$, $P_m/\mathrm{i} \to 0$, with $Pc = P_r + P_m/\mathrm{i} \to 1 + 0 = 1$ and $Pc =$ DOK + MChf $\to 1 + 0 = 1$.

Moreover, at each value of x, the probability to find the particle $Pc(x,t)$ is certainly predicted in the complex set $\mathcal{C}$ with Pc maintained as equal to one through a continuous compensation between DOK and Chf. This compensation is from $x = -\infty$, where $P_r = 0$ until $x = +\infty$, where $P_r = 1$, keeping always throughout the whole process $Pc =$ DOK $-$ Chf $=$ DOK + MChf $= 1$.

It is clear from the particle displacements simulations that DOK is the measure of our certain knowledge in $\mathcal{R}$ (100% probability) about the expected event and it does not include any uncertain knowledge (with a probability less than 100%).

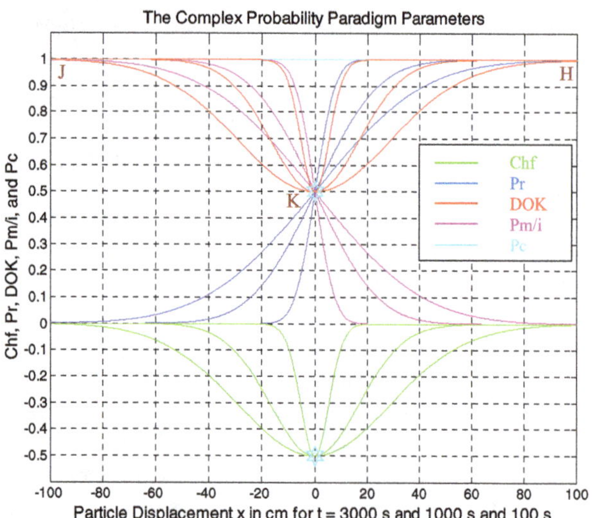

Figure 25. The complex probability parameters for $t = 3000$ s, $t = 1000$ s, and $t = 100$ s.

Furthermore, what is truly interesting in three instants simulations is that we have quantified and visualized both the DOK and the chaotic factor of the Brownian motion.

We note that the same methodology and analysis for the first and second instants were applied to the third instant concerning the diffusion, as well as all the EKA parameters. Thus, we can consequently conclude that whatever the instant is, both the logic and the method implemented are similar. This proves the validity of the new axioms developed and of the novel prognostic model adopted.

Figure 25 summarizes what has been previously explained.

9. The new paradigm and entropy

In the nineteenth century, specifically in the 1870s, Ludwig Boltzmann developed the statistical definition of entropy as a consequence of analysing the statistical behaviour of the microscopic components of a system (Boltzmann, 1995; Cercignani, 2010; Planck, 1969; Wikipedia, *Brownian Motion*; Wikipedia, *Entropy*; Wikipedia, *Mass Diffusivity*). The definition of entropy proposed by Boltzmann proved that entropy was equivalent to the thermodynamic entropy to within a constant number which is designated by Boltzmann's constant. This result was ascertained by Boltzmann himself. To summarize, the experimental definition of entropy was provided by the definition of the thermodynamic definition, whereas the definition of the statistical entropy extends the concept and yields a deeper explanation and a profound understanding of its nature.

Moreover, in statistical mechanics, the interpretation of entropy is the measure of uncertainty. Gibbs used the phrase *mixedupness* to designate entropy. When we determine the system set of macroscopic variables, the entropy measures the degree to which the probability of the system is spread out over different microstates. A macrostate characterizes simply observable average quantities, whereas a microstate expresses all the details of molecules in a system and this includes the velocity and the position of every molecule, knowing that the greater the available system states probability, the bigger the system entropy. In addition, we state that in statistical mechanics, entropy is a measure of the number of ways in which a system may be arranged, usually it is considered as a measure of 'disorder', that means that the higher the entropy, the greater the disorder. This definition describes the entropy as being proportional to the natural logarithm of the number of possible microscopic configurations of the individual atoms and molecules of the system (microstates) which could give rise to the observed macroscopic state (macro-state) of the system. The constant of proportionality is the Boltzmann constant.

More specifically, entropy is a logarithmic measure of the number of states with significant probability of being occupied; so mathematically we write

$$S = -k_B \sum_j p_j \operatorname{Ln}(p_j), \tag{26}$$

where k_B is the Boltzmann constant, equal to 1.38065×10^{-23} J/K or 8.6173324×10^{-5} eV/K. The summation is over all the possible microstates of the system, and p_j is the probability that the system is in the jth microstate. This definition assumes that the basis set of states has been picked so that there is no information on their relative phases.

In a different basis set, the more general expression is as follows:

$$S = -k_B \operatorname{Tr}(\hat{\rho} \operatorname{Ln}(\hat{\rho})), \tag{27}$$

where $\hat{\rho}$ is the density matrix and Ln is the matrix logarithm.

> This density matrix formulation is not needed in cases of thermal equilibrium so long as the basis states are chosen to be energy eigen-states. For most practical purposes, this can be taken as the fundamental definition of entropy since all other formulas for S can be mathematically derived from it, but not vice versa. (Wikipedia, *Entropy*)

In addition, the *fundamental assumption of statistical thermodynamics* or *the fundamental postulate in statistical mechanics* states that the occupation of any microstate is assumed to be equally probable, that is: $p_j = 1/\Omega$, where Ω is the number of microstates. It is important to mention that this assumption is usually justified for an isolated system in equilibrium. Then, Equation (26) becomes equal to

$$S = k_B \operatorname{Ln} \Omega. \tag{28}$$

Furthermore,

> the most general interpretation of entropy is as a measure of our uncertainty about a system. The equilibrium state

> of a system maximizes the entropy because we have lost all information about the initial conditions except for the conserved variables; maximizing the entropy maximizes our ignorance about the details of the system. This uncertainty is not of the everyday subjective kind, but rather the uncertainty inherent to the experimental method and interpretative model. In addition, entropy can be defined for any Markov processes with reversible dynamics and the detailed balance property. (Wikipedia, *Entropy*)

In 1896, in his book entitled 'Lectures on Gas Theory', Ludwig Boltzmann showed that this expression yields a measure of the entropy for systems of atoms and molecules in the gas phase, and consequently it provided a measure for the classical thermodynamics entropy.

Also, the very well-known second law of thermodynamics states that in general any system total entropy will not decrease other than by increasing the entropy of some other system. Therefore, in a system isolated from its environment, the entropy of that system will tend not to decrease. So mathematically we write

$$dS \geq 0. \tag{29}$$

It follows that

> heat will not flow from a colder body to a hotter body without the application of work (the imposition of order) to the colder body. Secondly, it is impossible for any device operating on a cycle to produce network from a single temperature reservoir; the production of network requires flow of heat from a hotter reservoir to a colder reservoir, or a single expanding reservoir undergoing adiabatic cooling, which performs adiabatic work. As a result, there is no possibility of a perpetual motion system. It follows that a reduction in the increase of entropy in a specified process, such as a chemical reaction, means that it is energetically more efficient. (Wikipedia, *Entropy*)

Additionally, statistical mechanics shows that entropy is governed by probability. In fact, it allows a disorder decrease even in an isolated system. Although this is mathematically possible, such an event has a small probability of occurring, making it unlikely.

Hence, in the novel complex probability paradigm, we can deduce the following consequences.

In the set $\mathscr{R}$, we denote the corresponding real entropy by S_R and $\Omega =$ the number of microstates. We have for an isolated system in equilibrium the real probability equals to

$$\begin{aligned} p_j &= P_r = \frac{1}{\Omega} \Rightarrow S_R = -k_B \sum_{j=1}^{\Omega} \frac{1}{\Omega} \mathrm{Ln}\left(\frac{1}{\Omega}\right) \\ &= -k_B\left(\Omega \times \frac{1}{\Omega}\right)(-\mathrm{Ln}\,\Omega) \\ &\Rightarrow S_R = k_B \mathrm{Ln}\,\Omega \end{aligned} \tag{30}$$

and is a divergent non-decreasing series

$$\Rightarrow dS_R \geq 0 \quad \text{and} \quad \lim_{\Omega \to +\infty} dS_R = +\infty \tag{31}$$

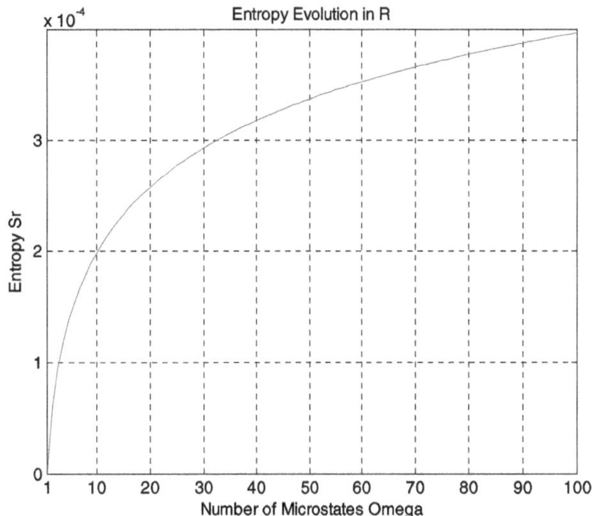

Figure 26. The real entropy S_R in $\mathscr{R}$ as function of the number of microstates Ω.

that means in $\mathscr{R}$, chaos and disorder are increasing with time (Figure 26).

In the set $\mathscr{M}$, we denote the corresponding imaginary entropy by S_M. We have for an isolated system in equilibrium the imaginary probability equals to

$$\begin{aligned} p_j &= P_m = \mathrm{i}(1 - P_r) = \mathrm{i}\left(1 - \frac{1}{\Omega}\right) \\ &\Rightarrow S_M = -k_B \sum_{j=1}^{\Omega} \mathrm{i}\left(1 - \frac{1}{\Omega}\right) \mathrm{Ln}\left[\mathrm{i}\left(1 - \frac{1}{\Omega}\right)\right]. \end{aligned} \tag{32}$$

In the complementary real probability set to $\mathscr{R}$, we denote the corresponding real entropy by $\bar{S}_R$.

The meaning of $\bar{S}_R$ is the following: it is the real entropy in the real set $\mathscr{R}$ and which is related to the complementary real probability $p_j = P_m/\mathrm{i} = 1 - P_r$. We have for an isolated system in equilibrium the complementary probability equals to

$$\begin{aligned} p_j &= \frac{P_m}{\mathrm{i}} = 1 - P_r = 1 - \frac{1}{\Omega} \Rightarrow \\ \bar{S}_R &= -k_B \sum_{j=1}^{\Omega}\left(1 - \frac{1}{\Omega}\right) \mathrm{Ln}\left(1 - \frac{1}{\Omega}\right) = -k_B \times \Omega \\ &\quad \times \left(1 - \frac{1}{\Omega}\right) \times \left[\mathrm{Ln}\left(\frac{\Omega - 1}{\Omega}\right)\right] \\ &= -k_B \times (\Omega - 1) \times \left[\mathrm{Ln}\left(\frac{\Omega - 1}{\Omega}\right)\right] \\ &\Rightarrow \bar{S}_R = k_B \times (\Omega - 1) \times \left[\mathrm{Ln}\left(\frac{\Omega}{\Omega - 1}\right)\right] \end{aligned} \tag{33}$$

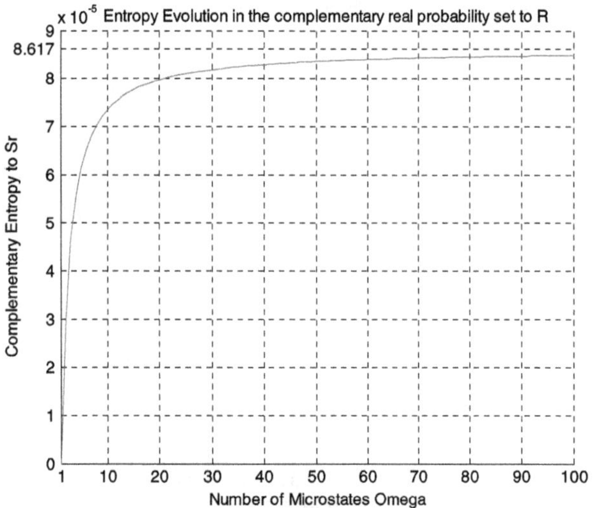

Figure 27. The complementary entropy $\bar{S}_R$ to S_R in $\mathscr{R}$ as function of the number of microstates Ω.

and is a convergent non-decreasing series. In fact,

$$\lim_{\Omega\to+\infty}\sum_{j=1}^{\Omega}\left(1-\frac{1}{\Omega}\right)\mathrm{Ln}\left(1-\frac{1}{\Omega}\right)=-1\Rightarrow\bar{S}_R$$

$$\to[-k_B\times(-1)]=k_B$$

$$\Rightarrow d\bar{S}_R\geq 0 \quad\text{and}\quad \lim_{\Omega\to+\infty}d\bar{S}_R=0, \tag{34}$$

that means in the complementary real probability set to $\mathscr{R}$, chaos is increasing with time, and its corresponding entropy is converging to the Boltzmann constant $k_B = 8.6173324\times10^{-5}$ eV/K (Figure 27).

In the set $\mathscr{C}$, we denote the corresponding real entropy by S_C. We have for an isolated system in equilibrium the real probability equals to

$$p_j = Pc = 1 \Rightarrow S_C = -k_B\sum_j p_j\mathrm{Ln}(p_j)$$

$$= -k_B\sum_j 1\times \mathrm{Ln}1 \Rightarrow S_C = 0 \tag{35}$$

and is a constant series $\Rightarrow dS_C = 0 \quad \forall\Omega\in[1,+\infty)$ (36)

that means, in $\mathscr{C}=\mathscr{R}+\mathscr{M}$, we have complete order, no chaos, and no ignorance since all measurements are completely deterministic (Figure 28).

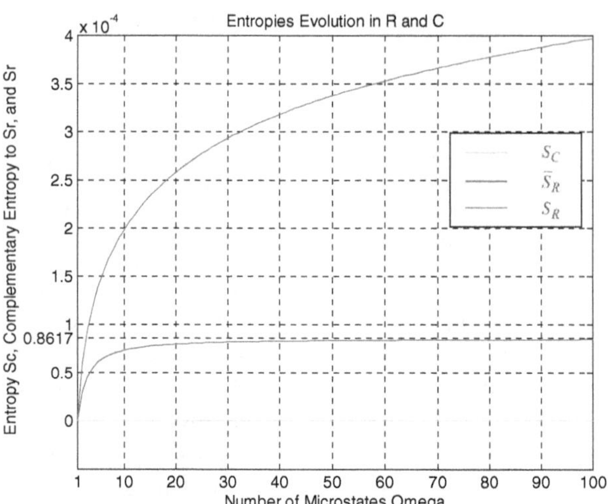

Figure 28. The entropies S_C, $\bar{S}_R$, and S_R as functions of the number of microstates Ω.

10. The resultant complex random vector *Z*

I will describe in this section a powerful tool, developed in a personal previous paper, based on the concept of a complex random vector which is a vector representing the real and the imaginary probabilities of an outcome, defined in the added axioms by the term $z = P_r + P_m$ (Abou Jaoude, 2013a, 2013b, 2014, 2015; Abou Jaoude et al., 2010). Then express the resultant complex random vector Z as the vector which is the sum of all the complex random vectors in the complex probability space $\mathscr{C}$. I will illustrate this methodology by considering a general Bernoulli distribution first, then a discrete distribution with N random variables as a general case. Afterward, I will prove the very well-known law of large numbers using this new powerful concept. In fact, if z represents one particle in Brownian motion, then Z represents the whole system of particles in a gas or liquid that means the whole random distribution in the complex probability space $\mathscr{C}$. So it follows directly that a Bernoulli distribution can be understood as a simplified system with two random particles (paragraph 10.1), whereas the general case is a random system with N particles (paragraph 10.2).

10.1. The resultant complex random vector Z of a general Bernoulli distribution

First, let us define the complex random vectors and their resultant by considering the following general Bernoulli distribution (Table 1):

We have

$$\sum_{j=1}^{2}P_{rj}=P_{r1}+P_{r2}=p+q=1$$

and

$$\sum_{j=1}^{2}P_{mj}=P_{m1}+P_{m2}=iq+ip=i(1-p)+ip$$

$$=i-ip+ip=i=i(2-1)=i(N-1),$$

where N is the number of random variables which is equal to 2 for a Bernoulli distribution.

Table 1. A general Bernoulli distribution.

Outcome	x_j	x_1	x_2
In $\mathcal{R}$	P_{rj}	$P_{r1} = p$	$P_{r2} = q$
In $\mathcal{M}$	P_{mj}	$P_{m1} = \mathrm{i}(1-p) = \mathrm{i}q$	$P_{m2} = \mathrm{i}(1-q) = \mathrm{i}p$
In $\mathcal{C} = \mathcal{R} + \mathcal{M}$	z_j	$z_1 = P_{r1} + P_{m1}$	$z_2 = P_{r2} + P_{m2}$

Notes: x_1 and x_2 are the outcomes of the first and second random variables, respectively. P_{r1} and P_{r2} are the real probabilities of x_1 and x_2, respectively. P_{m1} and P_{m2} are the imaginary probabilities of x_1 and x_2, respectively.

The complex random vector corresponding to the random outcome x_1 is as follows:

$$z_1 = P_{r1} + P_{m1} = p + \mathrm{i}(1-p) = p + \mathrm{i}q.$$

The complex random vector corresponding to the random outcome x_2 is as follows:

$$z_2 = P_{r2} + P_{m2} = q + \mathrm{i}(1-q) = q + \mathrm{i}p.$$

The resultant complex random vector is defined as follows:

$$\begin{aligned} Z = z_1 + z_2 &= \sum_{j=1}^{2} P_{rj} + \sum_{j=1}^{2} P_{mj} \\ &= (p + \mathrm{i}q) + (q + \mathrm{i}p) = (p+q) + \mathrm{i}(p+q) \\ &= 1 + \mathrm{i} = 1 + \mathrm{i}(2-1) \\ &\Rightarrow Z = 1 + \mathrm{i}(N-1). \end{aligned} \tag{37}$$

The probability Pc_1 in the complex space $\mathcal{C} = \mathcal{R} + \mathcal{M}$ which corresponds to the complex random vector z_1 is computed as follows:

$$|z_1|^2 = P_{r1}^2 + \left(\frac{P_{m1}}{\mathrm{i}}\right)^2 = p^2 + q^2,$$

$$\begin{aligned} \mathrm{Chf}_1 &= -2P_{r1}\frac{P_{m1}}{\mathrm{i}} = -2pq, \Rightarrow Pc_1^2 = |z_1|^2 - \mathrm{Chf}_1 \\ &= p^2 + q^2 + 2pq = (p+q)^2 = 1^2 = 1 \Rightarrow Pc_1 = 1. \end{aligned}$$

This is coherent with the three new complementary axioms defined for the extended Kolmogorov's system.

Similarly, Pc_2 corresponding to z_2 is as follows:

$$|z_2|^2 = P_{r2}^2 + \left(\frac{P_{m2}}{\mathrm{i}}\right)^2 = q^2 + p^2,$$

$$\begin{aligned} \mathrm{Chf}_2 &= -2P_{r2}\frac{P_{m2}}{\mathrm{i}} = -2qp \Rightarrow Pc_2^2 = |z_2|^2 - \mathrm{Chf}_2 \\ &= q^2 + p^2 + 2qp = (q+p)^2 = 1^2 = 1 \Rightarrow Pc_2 = 1. \end{aligned}$$

The probability Pc in the complex space $\mathcal{C}$ which corresponds to the resultant complex random vector $Z = 1 + \mathrm{i}$ is computed as follows:

$$|Z|^2 = \left(\sum_{j=1}^{2} P_{rj}\right)^2 + \left(\sum_{j=1}^{2} \frac{P_{mj}}{\mathrm{i}}\right)^2 = 1^2 + 1^2 = 2,$$

$$\mathrm{Chf} = -2\sum_{j=1}^{2} P_{rj} \sum_{j=1}^{2} \frac{P_{mj}}{\mathrm{i}} = -2(1)(1) = -2.$$

Let $s^2 = |Z|^2 - \mathrm{Chf} = 2 + 2 = 4 \Rightarrow s = 2 \Rightarrow Pc^2 = \frac{s^2}{N}$

$$= \frac{|Z|^2 - \mathrm{Chf}}{N^2} = \frac{|Z|^2}{N^2} - \frac{\mathrm{Chf}}{N^2} \Rightarrow Pc = \frac{s}{N} = \frac{2}{2} = 1,$$

where s is an intermediary quantity used in our computation of Pc.

Pc is the probability corresponding to the resultant complex random vector Z in the universe $\mathcal{C} = \mathcal{R} + \mathcal{M}$ and is also equal to 1. In fact, Z represents both z_1 and z_2 that means the whole distribution of random variables in the complex space $\mathcal{C}$ and its probability Pc is computed in the same way as Pc_1 and Pc_2.

By analogy, for the case of one random variable z_j we have

$$Pc_j^2 = |z_j|^2 - \mathrm{Chf}_j \quad \text{with } (N = 1).$$

In general, for the vector Z we have

$$Pc^2 = \frac{|Z|^2}{N^2} - \frac{\mathrm{Chf}}{N^2}; \quad (N \geq 1), \tag{38}$$

where the DOK of the whole distribution is equal to $(|Z|^2/N^2)$ and its relative chaotic factor is (Chf/N^2).

Notice, if $N = 1$ in Equation (38), then:

$$\begin{aligned} Pc^2 &= \frac{|Z|^2}{N^2} - \frac{\mathrm{Chf}}{N^2} = \frac{|Z|^2}{1^2} - \frac{\mathrm{Chf}}{1^2} = |Z|^2 - \mathrm{Chf} = |z_j|^2 \\ &\quad - \mathrm{Chf}_j = Pc_j^2, \end{aligned}$$

which is coherent with the calculations already done.

To illustrate the concept of the resultant complex random vector Z, I will use Figure 29.

10.2. *The general case: a discrete distribution with N random variables*

As a general case, let us consider then this discrete probability distribution with N equiprobable random variables (Table 2).

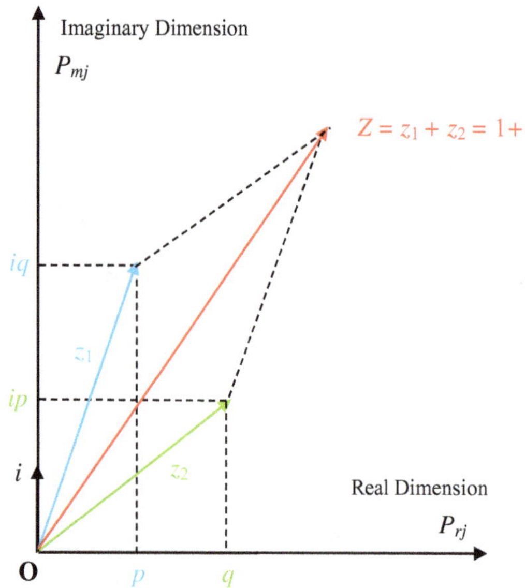

Figure 29. The resultant complex random vector $Z = z_1 + z_2$ for a general Bernoulli distribution in the complex probability space $\mathscr{C}$.

We have here in $\mathscr{C} = \mathscr{R} + \mathscr{M}$:

$$z_j = P_{rj} + P_{mj} \quad \forall j: \quad 1 \leq j \leq N$$

and

$$z_1 = z_2 = \cdots = z_N = \frac{1}{N} + \frac{\mathrm{i}(N-1)}{N} \Rightarrow Z = z_1 + z_2 + \cdots + z_N = N z_j = N\left(\frac{1}{N} + \frac{\mathrm{i}(N-1)}{N}\right) = 1 + \mathrm{i}(N-1).$$

Moreover, we can notice that $|z_1| = |z_2| = \cdots = |z_N|$, hence,

$$|Z| = |z_1 + z_2 + \cdots + z_N| = N|z_1| = N|z_2| = \cdots = N|z_N| \Rightarrow |Z|^2 = N^2|z_j|^2 = N^2\left(\frac{1}{N^2} + \frac{(N-1)^2}{N^2}\right) = 1 + (N-1)^2 \quad \text{where } 1 \leq j \leq N$$

and

$$Chf = N^2 \times Chf_j = -2 \times P_{r_j} \times \frac{P_{m_j}}{i} \times N^2 = -2N^2 \times \left(\frac{1}{N}\right)\left(\frac{N-1}{N}\right) = -2(1)(N-1) = -2(N-1)$$

$$= -2(N-1) \Rightarrow s^2 = |Z|^2 - \text{Chf} = 1 + (N-1)^2 + 2(N-1) = N^2 \Rightarrow Pc^2 = \frac{s^2}{N^2} = \frac{N^2}{N^2} = 1 \Rightarrow Pc = 1,$$

where s is an intermediary quantity used in our computation of Pc.

Therefore, the DOK corresponding to the resultant complex vector Z representing the whole distribution is as follows:

$$\text{DOK}_Z = \frac{|Z|^2}{N^2} = \frac{1 + (N-1)^2}{N^2} \tag{39}$$

and its relative chaotic factor is as follows:

$$\text{Chf}_Z = \frac{\text{Chf}}{N^2} = -\frac{2(N-1)}{N^2}. \tag{40}$$

Similarly, its relative MChf is as follows:

$$\text{MChf}_Z = |\text{Chf}_Z| = \left|\frac{\text{Chf}}{N^2}\right| = \left|-\frac{2(N-1)}{N^2}\right| = \frac{2(N-1)}{N^2}. \tag{41}$$

Thus, we can verify that we always have

$$Pc^2 = \frac{|Z|^2}{N^2} - \frac{\text{Chf}}{N^2} = \text{DOK}_Z - \text{Chf}_Z = \text{DOK}_Z + \text{MChf}_Z = 1. \tag{42}$$

What is important here is that we can notice the following: Take for example:

$$N = 2 \Rightarrow \frac{|Z|^2}{N^2} = \frac{1 + (2-1)^2}{2^2} = 0.5 \quad \text{and} \quad \frac{\text{Chf}}{N^2} = \frac{-2(2-1)}{2^2} = -0.5,$$

$$N = 4 \Rightarrow \frac{|Z|^2}{N^2} = \frac{1 + (4-1)^2}{4^2} = 0.625 \geq 0.5 \quad \text{and} \quad \frac{\text{Chf}}{N^2} = \frac{-2(4-1)}{4^2} = -0.375 \geq -0.5,$$

$$N = 5 \Rightarrow \frac{|Z|^2}{N^2} = \frac{1 + (5-1)^2}{5^2} = 0.68 \geq 0.625 \quad \text{and} \quad \frac{\text{Chf}}{N^2} = \frac{-2(5-1)}{5^2} = -0.32 \geq -0.375,$$

Table 2. A discrete distribution with N equiprobable random variables.

Outcome	x_j	x_1	x_2	...	x_N
In $\mathscr{R}$	P_{rj}	$P_{r1} = \frac{1}{N}$	$P_{r2} = \frac{1}{N}$	...	$P_{rN} = \frac{1}{N}$
In $\mathscr{M}$	P_{mj}	$P_{m1} = \mathrm{i}\left(1 - \frac{1}{N}\right)$	$P_{m2} = \mathrm{i}\left(1 - \frac{1}{N}\right)$	...	$P_{mN} = \mathrm{i}\left(1 - \frac{1}{N}\right)$

$$N = 10 \Rightarrow \frac{|Z|^2}{N^2} = \frac{1 + (10 - 1)^2}{10^2} = 0.82 \geq 0.68 \quad \text{and}$$

$$\frac{\text{Chf}}{N^2} = \frac{-2(10 - 1)}{10^2} = -0.18 \geq -0.32,$$

$$N = 100 \Rightarrow \frac{|Z|^2}{N^2} = \frac{1 + (100 - 1)^2}{100^2}$$

$$= 0.9802 \geq 0.82 \quad \text{and} \quad \frac{\text{Chf}}{N^2} = \frac{-2(100 - 1)}{100^2}$$

$$= -0.0198 \geq -0.18,$$

$$N = 1000 \Rightarrow \frac{|Z|^2}{N^2} = \frac{1 + (1000 - 1)^2}{1000^2}$$

$$= 0.998002 \geq 0.9802, \quad \text{and}$$

$$\frac{\text{Chf}}{N^2} = \frac{-2(1000 - 1)}{1000^2} = -0.001998 \geq -0.0198.$$

We can deduce mathematically that

$$\lim_{N \to +\infty} \frac{|Z|^2}{N^2} = \lim_{N \to +\infty} \frac{1 + (N - 1)^2}{N^2} = 1 \qquad (43)$$

and

$$\lim_{N \to +\infty} \frac{\text{Chf}}{N^2} = \lim_{N \to +\infty} -\frac{2(N - 1)}{N^2} = 0. \qquad (44)$$

From the above, we can also deduce this conclusion.

As much as N increases, as much as the DOK in $\mathscr{R}$ corresponding to the resultant complex vector is perfect, that is, it is equal to 1, and as much as the chaotic factor that forbids us from predicting exactly the result of the random experiment in $\mathscr{R}$ approaches 0. Mathematically, we say: if N tends to infinity then the DOK in $\mathscr{R}$ tends to 1 and the chaotic factor tends to 0.

Moreover,

$$\text{For } N = 1 \Rightarrow \frac{|Z|^2}{N^2} = \frac{1 + (1 - 1)^2}{1^2} = 1 \quad \text{and}$$

$$\frac{\text{Chf}}{N^2} = \frac{-2(1 - 1)}{1^2} = 0.$$

This means that we have a random experiment with only one outcome, hence, either $P_r = 1$ or $P_r = 0$, that means we have, respectively, either a sure event or an impossible event in $\mathscr{R}$. For this, we have surely the DOK is 1 and the chaotic factor is 0 since the experiment is either certain or impossible, which is absolutely logical.

10.3. The resultant complex random vector Z and the law of large numbers

The law of large numbers states that:

'As N increases, then the probability that the value of sample mean to be close to population mean approaches 1.'

We can deduce now the following conclusion related to the law of large numbers.

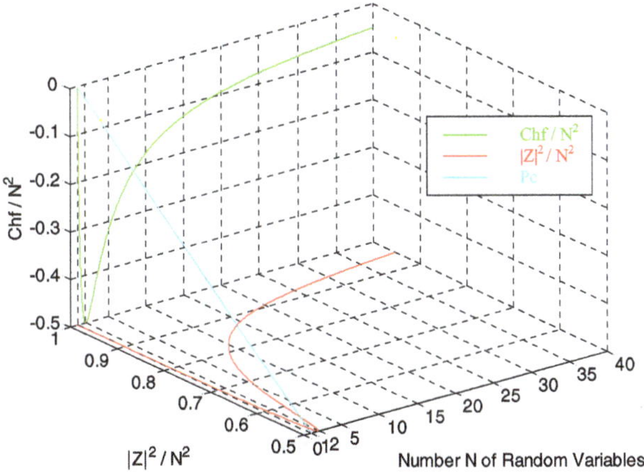

Figure 30. $\text{Chf}_Z = (\text{Chf}/N^2)$, $\text{DOK}_Z = (|Z|^2/N^2)$, and Pc, as functions of N in 3D.

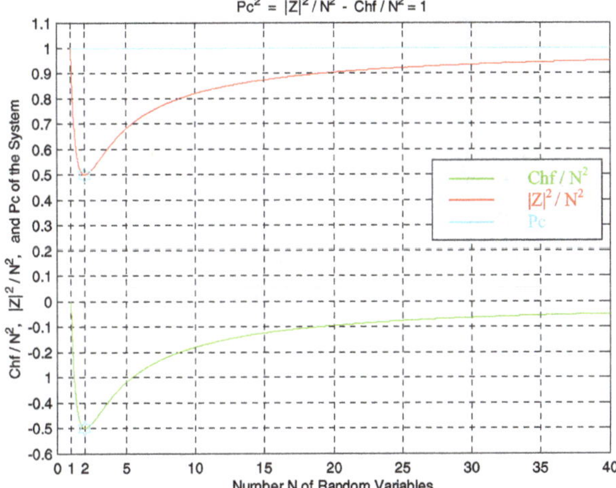

Figure 31. $\text{Chf}_Z = (\text{Chf}/N^2)$, $\text{DOK}_Z = (|Z|^2/N^2)$, and Pc, as functions of N in 2D.

We can see, as we have proved, that as much as N increases, as much as the DOK of the resultant complex vector $\text{DOK}_Z = (|Z|^2/N^2)$ tends to 1 and its relative chaotic factor $\text{Chf}_Z = (\text{Chf}/N^2)$ tends to 0. Assume now that the random variables x_j's correspond to the particles or molecules moving randomly in a gas or a liquid. So if we study a gas or a liquid with billions of such particles, N is big enough (e.g. Avogadro number $\approx 6.02214 \times 10^{23}$/mole in the International System of Units) to allow that its corresponding temperature, pressure, energy, etc. tend to the mean of these quantities corresponding to the whole system. This is because the chaotic factor of the whole gas or liquid, that is, of the resultant complex random vector Z representing all the random particles or vectors, tends to 0; thus, the behaviour and characteristics of the whole system in $\mathscr{R}$ is predictable with great precision since the DOK of the whole gas or liquid tends to 1. Figures 30 and 31 illustrate this result.

Hence, we have joined here two different key concepts which are as follows: the law of large numbers and the

Table 3. Computation of Pc for different of values of z_1 and z_2 which are the complex random vectors of a Bernoulli distribution and which are chosen at random.

The complex random vectors and their probabilities	z_1	Pc_1	z_2	Pc_2	Z	Pc
Simulation #1	0.8106 + i(0.1894)	1	0.1894 + i(0.8106)	1	1 + i	1
Simulation #2	0.0084 + i(0.9916)	1	0.9916 + i(0.0084)	1	1 + i	1
Simulation #3	0.4558 + i(0.5442)	1	0.5442 + i(0.4558)	1	1 + i	1
Simulation #4	0.5225 + i(0.4775)	1	0.4775 + i(0.5225)	1	1 + i	1
Simulation #5	0.3723 + i(0.6277)	1	0.6277 + i(0.3723)	1	1 + i	1
Simulation #6	0.1908 + i(0.8092)	1	0.8092 + i(0.1908)	1	1 + i	1
Simulation #7	0.208 + i(0.792)	1	0.792 + i(0.208)	1	1 + i	1

Notes: In this case, the resultant complex random vector is $Z = z_1 + z_2$ and is always equal to $1 + \mathrm{i}$. The corresponding probability of Z in $\mathcal{C}$ is always 1, just as postulated and expected.

Table 4. Computation of Pc for different of values of z_1, z_2, z_3 which are the complex random vectors of the distribution and which are chosen at random.

The complex random vectors and their probabilities	z_1	Pc_1	z_2	Pc_2	z_3	Pc_3	Z	Pc
Simulation #1	0.636 + i(0.364)	1	0.136 + i(0.864)	1	0.228 + i(0.772)	1	1 + i(2)	1
Simulation #2	0.8393 + i(0.1607)	1	0.0402 + i(0.9598)	1	0.1205 + i(0.8795)	1	1 + i(2)	1
Simulation #3	0.7802 + i(0.2198)	1	0.0220 + i(0.978)	1	0.1978 + i(0.8022)	1	1 + i(2)	1
Simulation #4	0.3619 + i(0.6381)	1	0.1381 + i(0.8619)	1	0.5 + i(0.5)	1	1 + i(2)	1
Simulation #5	0.9909 + i(0.0091)	1	0.0015 + i(0.9985)	1	0.0076 + i(0.9924)	1	1 + i(2)	1
Simulation #6	0.5205 + i(0.4795)	1	0.0533 + i(0.9467)	1	0.4262 + i(0.5738)	1	1 + i(2)	1
Simulation #7	0.0651 + i(0.9349)	1	0.2349 + i(0.7651)	1	0.7 + i(0.3)	1	1 + i(2)	1

Notes: In this case, the resultant complex random vector is $Z = z_1 + z_2 + z_3$ and is always equal to $1 + 2\mathrm{i}$. The corresponding probability of Z in $\mathcal{C}$ is always 1, just as postulated and expected.

Table 5. The resultant complex random vector $Z = z_1 + z_2 + \cdots + z_j + \cdots + z_N = Nz_j = 1 + i(N-1)$ with $1 \le j \le N$ and the verification of the law of large numbers.

Complex random vectors and their characteristics	N	z_j	Z	$\mathrm{DOK}_Z = \frac{\lvert Z\rvert^2}{N^2}$	$\mathrm{Chf}_Z = \frac{\mathrm{Chf}}{N^2}$	Pc
Simulation #1	1	1 + i(0)	1 + i(0)	1	0	1
Simulation #2	2	0.5 + i(0.5)	1 + i(1)	0.5	− 0.5	1
Simulation #3	3	0.3333 + i(0.6667)	1 + i(2)	0.5556	− 0.4444	1
Simulation #4	5	0.2 + i(0.8)	1 + i(4)	0.68	− 0.32	1
Simulation #5	10	0.1 + i(0.9)	1 + i(9)	0.82	− 0.18	1
Simulation #6	100	0.01 + i(0.99)	1 + i(99)	0.9802	− 0.0198	1
Simulation #7	1000	0.001 + i(0.999)	1 + i(999)	0.998002	− 0.001998	1
Simulation #8	10000	0.0001 + i(0.9999)	1 + i(9999)	0.99980002	− 0.00019998	1
Simulation #9	100000	1e − 005 + i(0.99999)	1 + i(99999)	0.9999800002	− 1.99998e − 005	1
Simulation #10	1000000	1e − 006 + i(0.999999)	1 + i(999999)	0.999998	− 1.999998e − 006	1
Simulation #11	1000000000	1e − 009 + i(0.999999999)	1 + i(999999999)	0.999999998	− 1.999999998e − 009	1
Simulation #12	10^{12} = 1e + 12	≈ 1e − 012 + i(1)	≈ 1 + i(1e + 012)	≈ 1	≈ − 2e − 012 ≈ 0	1

resultant complex random vector. The first concept comes from ordinary statistics, classical probability theory, and statistical mechanics, whereas the second concept comes from the new theory of complex probability and statistics. This looks very interesting and fruitful and shows the validity and the benefits of extending Kolmogorov's axioms to the complex set $\mathcal{C}$.

10.4. Numerical simulations

Numerical simulations verify what has been found earlier (Gentle, 2003). We will use Monte Carlo simulation method with the help of the programming language C++ with its predefined pseudorandom function rand() that generates random numbers with a uniform distribution. Table 3 is a simulation of a Bernoulli distribution where the complex random vectors are chosen randomly by C++. Table 4 is a simulation of a uniform distribution with three random variables having their complex random vectors also chosen randomly by C++. Table 5 is a simulation that confirms the direct relation between the resultant complex vector Z and the law of large numbers.

11. Conclusion and perspectives

In the current paper, I applied the theory of EKA to the classical theory of Brownian motion. Hence, I established a tight link between the new paradigm and diffusion. Thus, I developed the theory of 'Complex Probability' beyond the scope of my previous five papers on this topic.

As it was proved and illustrated, before any diffusion and after the full particles dispersion, the DOK is one and the chaotic factor (Chf and MChf) is 0 since the state of the system is totally known. During the process of gas diffusion $(-\infty < x < +\infty, 0 < t < +\infty,)$ we have $0.5 \leq \text{DOK} < 1$, $-0.5 \leq \text{Chf} < 0$, and $0 < \text{MChf} \leq 0.5$. Notice that during the whole process of the Brownian motion, we have $Pc = \text{DOK} - \text{Chf} = \text{DOK} + \text{MChf} = 1$, that means that the phenomenon which seems to be random and stochastic in $\mathscr{R}$ is now deterministic and certain in $\mathscr{C} = \mathscr{R} + \mathscr{M}$, and this after adding to $\mathscr{R}$ the contributions of $\mathscr{M}$ and hence after subtracting the chaotic factor from the DOK. Moreover, for each value of the diffusion instant t, I have determined its corresponding probability of finding the gas particle in the interval $(-\infty, L]$. Therefore, at each instant, this probability is certainly predicted in the complex set $\mathscr{C}$ with Pc maintained as equal to one through a continuous compensation between DOK and Chf. This compensation is from the instant $t = 0$ (before any diffusion) until the instant of full gas dispersion when $t \to +\infty$. Furthermore, what is important, is that using all these graphs and simulations illustrated throughout the whole paper, we can visualize and quantify both the system chaos (Chf and MChf) and the system certain knowledge (DOK and Pc). I applied this novel methodology in the current research paper to the diffusion of oxygen gas in air gas, knowing that this novel paradigm can be applied to any diffusion phenomenon not discussed here.

Additionally, I have used in this paper a new powerful tool, already defined in a personal previous paper, which is the concept of the complex random vector that is a vector representing the real and the imaginary probabilities of an outcome, identified in the added axioms as being the term $z = P_r + P_m$. Then I have defined and expressed the resultant complex random vector as the vector which is the sum of all the complex random vectors and representing the whole distribution and system in the complex space $\mathscr{C}$. I have illustrated this methodology by considering a general Bernoulli distribution, then a discrete distribution with N random variables as a general case. Afterward, I have proven that there is a direct correlation between the concept of the resultant complex random vector and the very well-known law of large numbers. In fact, using this original concept and tool, I have succeeded to demonstrate the law of large numbers in a new way.

All the results found are certainly very interesting and fruitful and show once again the benefits of extending Kolmogorov's axioms and thus the originality and usefulness of this new field in applied mathematics and prognostic that can be called verily: 'The Complex Probability Paradigm.'

Additional development of this new paradigm will be done in subsequent works. And as prospective and future works, it is planned to more elaborate the novel proposed prognostic methodology and to apply it to a wide set of random dynamic systems and stochastic processes.

Disclosure statement

No potential conflict of interest was reported by the author.

References

Abou Jaoude, A. (2004). *Numerical methods and algorithms for applied mathematicians* (Ph.D. thesis in applied mathematics). Bircham International University. Retrieved August 1, 2004, from http://www.bircham.edu

Abou Jaoude, A. (2005). *Computer simulation of Montle Carlo methods and random phenomena* (Ph.D. thesis in computer science). Bircham International University. Retrieved October 14, 2005, from http://www.bircham.edu

Abou Jaoude, A. (2007). *Analysis and algorithms for the statistical and stochastic paradigm* (Ph.D. thesis in applied statistics and probability). Bircham International University. Retrieved April 27, 2007, from http://www.bircham.edu

Abou Jaoude, A. (2013a). The complex statistics paradigm and the law of large numbers. *Journal of Mathematics and Statistics*, *9*(4), 289–304.

Abou Jaoude, A. (2013b). The theory of complex probability and the first order reliability method. *Journal of Mathematics and Statistics*, *9*(4), 310–324.

Abou Jaoude, A. (2014). Complex probability theory and prognostic. *Journal of Mathematics and Statistics*, *10*(1), 1–24.

Abou Jaoude, A. (2015). The complex probability paradigm and analytic linear prognostic for vehicle suspension systems. *American Journal of Engineering and Applied Sciences*, *8*(1), 147–175.

Abou Jaoude, A., El-Tawil, K., & Kadry, S. (2010). Prediction in complex dimension using Kolmogorov's set of axioms. *Journal of Mathematics and Statistics*, *6*(2), 116–124.

Aczel, A. (2000). *God's equation*. New York: Dell.

Balibar, F. (2002). *Albert Einstein: Physique, Philosophie, Politique* (1st ed.). Paris: Le Seuil.

Barrow, J. (1992). *Pi in the sky*. London: Oxford University Press.

Bell, E. T. (1992). *The development of mathematics*. New York: Dover.

Benton, W. (1966a). *Probability, Encyclopedia Britannica* (Vol. 18, pp. 570–574). Chicago, IL: Encyclopedia Britannica.

Benton, W. (1966b). *Mathematical probability, Encyclopedia Britannica* (Vol. 18, pp. 574–579). Chicago, IL: Encyclopedia Britannica.

Bidabad, B. (1992). *Complex probability and Markov stochastic processes*. Proceedings first Iranian statistics conference, Isfahan University of Technology, Tehran.

Bogdanov, I., & Bogdanov, G. (2009). *Au Commencement du Temps*. Paris: Flammarion.

Bogdanov, I., & Bogdanov, G. (2010). *Le Visage de Dieu*. Paris: Editions Grasset et Fasquelle.

Bogdanov, I., & Bogdanov, G. (2012). *La Pensée de Dieu*. Paris: Editions Grasset et Fasquelle.

Bogdanov, I., & Bogdanov, G. (2013). *La Fin du Hasard*. Paris: Editions Grasset et Fasquelle.

Boltzmann, L. (1995). *Lectures on gas theory*. New York, NY: Dover.

Boursin, J.-L. (1986). *Les structures du Hasard*. Paris: Editions du Seuil.

Cercignani, C. (2010). *Ludwig Boltzmann, the man who trusted atoms*. Oxford: Oxford University Press.

Chan Man Fong, C. F., De Kee, D., & Kaloni, P. N. (1997). *Advanced mathematics for applied and pure sciences*. Amsterdam: Gordon and Breach Sciences.

Cox, D. R. (1955). A use of complex probabilities in the theory of stochastic processes. *Mathematical Proceedings of the Cambridge Philosophical Society, 51*, 313–319.

Dacunha-Castelle, D. (1996). *Chemins de l'Aléatoire*. Paris: Flammarion.

Dalmedico-Dahan, A., Chabert, J-L., & Chemla, K. (1992). *Chaos Et Déterminisme*. Paris: Edition du Seuil.

Dalmedico-Dahan, A., & Peiffer, J. (1986). *Une Histoire des Mathématiques*. Paris: Edition du Seuil.

Davies, P. (1993). *The mind of god*. London: Penguin Books.

Ekeland, I. (1991). *Au Hasard. La Chance, la Science et le Monde*. Paris: Editions du Seuil.

Fagin, R., Halpern, J., & Megiddo, N. (1990). A logic for reasoning about probabilities. *Information and Computation, 87*, 78–128.

Feller, W. (1968). *An introduction to probability theory and its applications* (3rd ed.). New York, NY: Wiley.

Gentle, J. (2003). *Random number generation and Monte Carlo methods* (2nd ed.). Sydney: Springer.

Gleick, J. (1997). *Chaos, making a new science*. New York: Penguin Books.

Greene, B. (2003). *The elegant universe*. New York: Vintage.

Gullberg, J. (1997). *Mathematics from the birth of numbers*. New York: W.W. Norton.

Hawking, S. (2002). *On the shoulders of giants*. London: Running Press.

Hawking, S. (2005). *God created the integers*. London: Penguin Books.

Hawking, S. (2011). *The dreams that stuff is made of*. London: Running Press.

Hoffmann, B. (1975). *In collaboration with Helen Dukas, Albert Einstein, Créateur et Rebelle* (1st ed.). Paris: Editions du Seuil.

Kuhn, T. (1970). *The structure of scientific revolutions* (2nd ed.). Chicago: Chicago Press.

Montgomery, D. C., & Runger, G. C. (2003). *Applied statistics and probability for engineers* (3rd ed.). New York: John Wiley & Sons.

Ognjanović, Z., Marković, Z., Rašković, M., Doder, D., & Perović, A. (2012). A connection between the Cantor–Bendixson derivative and the well-founded semantics of finite logic programs. *Annals of Mathematics and Artificial Intelligence, 65*, 1–24.

Penrose, R. (1999). *traduction Française: Les Deux Infinis et L'Esprit Humain*. Paris: Roland Omnès, Flammarion.

Pickover, C. (2008). *Archimedes to Hawking*. Oxford: Oxford University Press.

Planck, M. (1969). *Treatise on thermodynamics*. New York, NY: Dover.

Poincaré, H. (1968). *La Science et l'Hypothèse* (1st ed.). Paris: Flammarion.

Reeves, H. (1988). *Patience dans L'Azur, L'Evolution Cosmique*. Paris: Le Seuil.

Ronan, C. (1988). *traduction Française: Histoire Mondiale des Sciences*. Paris: Claude Bonnafont, Le Seuil.

Science et vie. (1999). *Le Mystère des Mathématiques. Numéro, 984*, 1–1.

Srinivasan, S. K., & Mehata, K. M. (1988). *Stochastic processes* (2nd ed.). New Delhi: McGraw-Hill.

Stepić, A. I., & Ognjanović, Z. (2014). Complex valued probability logics. *Publications De L'institut Mathématique, Nouvelle Série, tome, 95*(109), 73–86. doi:10.2298/PIM1409073I

Stewart, I. (1996). *From here to infinity* (2nd ed.). Oxford: Oxford University Press.

Stewart, I. (2002). *Does god play dice?* (2nd ed.). Oxford: Blackwell.

Stewart, I. (2012). *In pursuit of the unknown*. Oxford: Basic Books.

Van Kampen, N. G. (2006). *Stochastic processes in physics and chemistry* (Revised and Enlarged ed.). Sydney: Elsevier.

Walpole, R., Myers, R., Myers, S., & Ye, K. (2002). *Probability and statistics for engineers and scientists* (7th ed.). Upper Saddle River, NJ: Prentice Hall.

Warusfel, A., & Ducrocq, A. (2004). *Les Mathématiques, Plaisir et Nécessité* (1st ed.). Paris: Edition Vuibert.

Weingarten, D. (2002). Complex probabilities on RN as real probabilities on CN and an application to path integrals. *Physical Review Letters, 89*, 1–1.

Wikipedia, the free encyclopedia, *Brownian Motion*. Retrieved from https://en.wikipedia.org/

Wikipedia, the free encyclopedia, *Entropy*. Retrieved from https://en.wikipedia.org/

Wikipedia, the free encyclopedia, *Mass Diffusivity*. Retrieved from https://en.wikipedia.org/

Youssef, S. (1994). Quantum mechanics as complex probability theory. *Modern Physics Letters A, 9*, 2571–2586.

Permissions

All chapters in this book were first published in SSCE, by Taylor & Francis; hereby published with permission under the Creative Commons Attribution License or equivalent. Every chapter published in this book has been scrutinized by our experts. Their significance has been extensively debated. The topics covered herein carry significant findings which will fuel the growth of the discipline. They may even be implemented as practical applications or may be referred to as a beginning point for another development.

The contributors of this book come from diverse backgrounds, making this book a truly international effort. This book will bring forth new frontiers with its revolutionizing research information and detailed analysis of the nascent developments around the world.

We would like to thank all the contributing authors for lending their expertise to make the book truly unique. They have played a crucial role in the development of this book. Without their invaluable contributions this book wouldn't have been possible. They have made vital efforts to compile up to date information on the varied aspects of this subject to make this book a valuable addition to the collection of many professionals and students.

This book was conceptualized with the vision of imparting up-to-date information and advanced data in this field. To ensure the same, a matchless editorial board was set up. Every individual on the board went through rigorous rounds of assessment to prove their worth. After which they invested a large part of their time researching and compiling the most relevant data for our readers.

The editorial board has been involved in producing this book since its inception. They have spent rigorous hours researching and exploring the diverse topics which have resulted in the successful publishing of this book. They have passed on their knowledge of decades through this book. To expedite this challenging task, the publisher supported the team at every step. A small team of assistant editors was also appointed to further simplify the editing procedure and attain best results for the readers.

Apart from the editorial board, the designing team has also invested a significant amount of their time in understanding the subject and creating the most relevant covers. They scrutinized every image to scout for the most suitable representation of the subject and create an appropriate cover for the book.

The publishing team has been an ardent support to the editorial, designing and production team. Their endless efforts to recruit the best for this project, has resulted in the accomplishment of this book. They are a veteran in the field of academics and their pool of knowledge is as vast as their experience in printing. Their expertise and guidance has proved useful at every step. Their uncompromising quality standards have made this book an exceptional effort. Their encouragement from time to time has been an inspiration for everyone.

The publisher and the editorial board hope that this book will prove to be a valuable piece of knowledge for researchers, students, practitioners and scholars across the globe.

List of Contributors

Haidong Lv, Guoliang Wei, Zhugang Ding and Xueming Ding
Department of Control Science and Engineering, University of Shanghai for Science and Technology, Shanghai 200093, People's Republic of China

Thomas Stockley, Kary Thanapalan, Mark Bowkett and Jonathan Williams
Faculty of Computer, Engineering and Sciences, University of South Wales, Pontypridd, Wales

Jinya Su and Wen-Hua Chen
Department of Aeronautical and Automotive Engineering, Loughborough University, Loughborough LE11 3TU, UK

Baibing Li
School of Business and Economics, Loughborough University, Loughborough LE11 3TU, UK

Esam H. Abdelhameed
Faculty of Energy Engineering, Aswan University, Aswan, Egypt;

Noritaka Sato and Yoshifumi Morita
Department of Electrical and Computer Engineering, Nagoya Institute of Technology, Nagoya, Japan

Keming Tang
College of Information Science and Technology, Yancheng Teachers University, Jiangsu 224002, People's Republic of China;

Zhudeng Wang
School of Mathematical Sciences, Yancheng Teachers University, Yancheng 224002, People's Republic of China

Jigui Jian and Zhihua Zhao
College of Science, China Three Gorges University, Yichang, Hubei 443002, People's Republic of China

Afef Fekih and Shankar Seelem
Department of Electrical and Computer Engineering, University of Louisiana at Lafayette, Lafayette, LA, USA

Rui Yan, Tao Liu, Fengwei Chen and Shijian Dong
Institute of Advanced Control Technology, Dalian University of Technology, Dalian 116024, People's Republic of China

Bongani Malinga and Gregory D. Buckner
Department of Mechanical and Aerospace Engineering, North Carolina State University, Campus Box 7910, Raleigh, NC 27695, USA

G.-L. Osorio-Gordillo
Tecnológico Nacional de México, Centro Nacional de Investigación y Desarrollo Tecnológico, CENIDET, Interior Internado Palmira S/N, Col. Palmira, 62490 Cuernavaca, Mor. Mexico
CRAN-CNRS (UMR 7039), Université de Lorraine, IUT de Longwy, 186, Rue de Lorraine, 54400 Cosnes et Romain, France

M. Darouach and L. Boutat-Baddas
CRAN-CNRS (UMR 7039), Université de Lorraine, IUT de Longwy, 186, Rue de Lorraine, 54400 Cosnes et Romain, France

C.-M. Astorga-Zaragoza
Tecnológico Nacional de México, Centro Nacional de Investigación y Desarrollo Tecnológico, CENIDET, Interior Internado Palmira S/N, Col. Palmira, 62490 Cuernavaca, Mor. Mexico

Na Tao
School of Management, Xi'an Jiaotong University, Xi'an 710049, People's Republic of China

Sheng Zhang
School of Public Policy and Administration, Xi'an Jiaotong University, Xi'an, 710049, People's Republic of China

Rajeev Kumar Dohare and Kailash Singh
Department of Chemical Engineering, Malaviya National Institute of Technology, Jaipur 302017, Rajasthan, India

Rajesh Kumar
Department of Electrical Engineering, Malaviya National Institute of Technology, Jaipur 302017, Rajasthan, India

Abdelkarim M. Ertiame, Dingli Yu and J.B. Gomm
Control System Research Group, School of Engineering, Liverpool John Moores University, Byrom Street, Liverpool L3 3AF, UK

Feng Yu
School of Electronic Information, Changchun Architecture &Civil Engineering College, Changchun, Jilin Province, People's Republic of China

P. Muniappan and S. Ganesh
Department of Mathematics, Sathyabama University, Chennai - 600 119, Tamil Nadu, India

R. Uthayakumar
Department of Mathematics, Gandhigram Rural Institute - Deemed University, Dindigul 624 302, Tamil Nadu, India

Mark Dooner and Jihong Wang
School of Engineering, The University of Warwick, Coventry, UK

Alexandros Mouzakitisb
Jaguar Land Rover Product Development Centre, Gaydon, Warwickshire, UK

Mohammed Jamal Alden and Xin Wang
Department of Electrical and Computer Engineering, Southern Illinois University Edwardsville, Edwardsville, IL 62026, USA

Yue Wang, Jihong Wang, Xing Luo and Shen Guo
School of Engineering, The University of Warwick, Coventry CV4 7AL, UK;

Junfu Lv and Qirui Gao
Department of Thermal Engineering, Tsinghua University, Beijing 100084, People's Republic of China

Y.V. Venkatesh
(formerly) Electrical Sciences Division, Indian Institute of Science, Bangalore, India
Department of ECE, National University of Singapore, Singapore

Jalal Javadi Moghaddam and Ahmad Bagheri
Department of Mechanical Engineering, Faculty of Engineering, University of Guilan, PO Box 3756, Rasht, Iran

S. Samiee
Faculty of Mechanical Engineering, K.N. Toosi University of Technology, Tehran, Iran
Institute of Automotive Engineering, Graz University of Technology, Graz, Austria

A. Nahvi, S. Azadi, R. Kazemi, A.R. Hatamian Haghighi and M.R. Ashouri
Faculty of Mechanical Engineering, K.N. Toosi University of Technology, Tehran, Iran

Yoni Mandel
Core Photonics, Tel Aviv 6971035, Israel

George Weiss
School of EE, Tel Aviv University, Tel Aviv 6997801, Israel

Carsten Knoll and Klaus Röbenack
Faculty of Electrical and Computer Engineering, Institute of Control Theory, Technische Universität Dresden, Dresden, Germany

Abdo Abou Jaoude
Department of Mathematics and Statistics, Faculty of Natural and Applied Sciences, Notre Dame University - Louaize, Zouk Mosbeh, Lebanon

www.ingramcontent.com/pod-product-compliance
Lightning Source LLC
Chambersburg PA
CBHW080459200326
41458CB00012B/4023

* 9 7 8 1 6 3 2 3 8 5 0 2 4 *